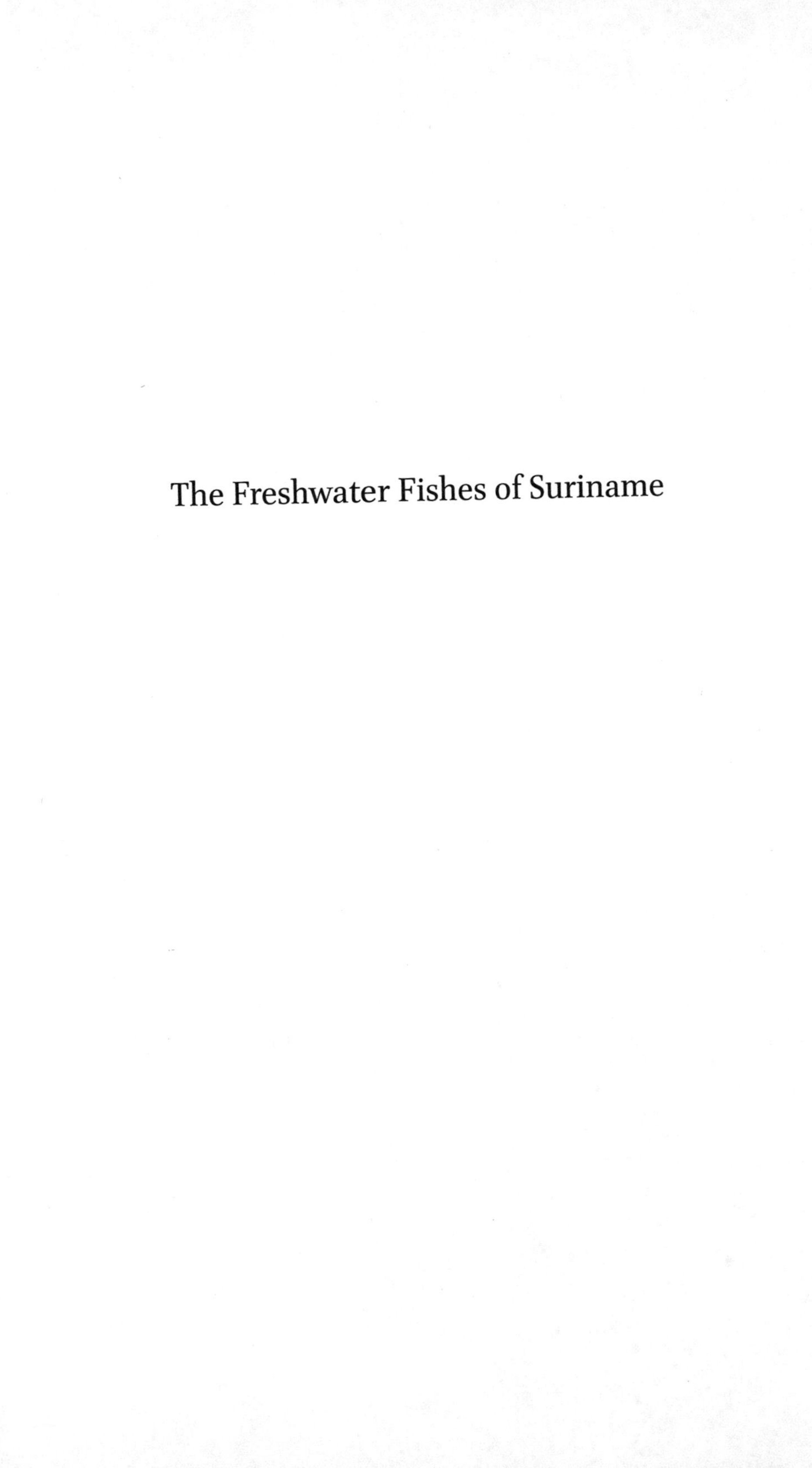

The Freshwater Fishes of Suriname

The Freshwater Fishes of Suriname

By

Jan H.A. Mol

BRILL

LEIDEN · BOSTON

2012

Cover illustration: A black-water stream in Suriname with characteristic fishes. At the surface a splashing tetra *Copella arnoldi* and the small curimatid *Curimatopsis crypticus*, mid-water the leaf fish *Polycentrus schomburgkii* and *Mylopus rubripinnis*, and at the bottom the doradid catfish merkikwikwi *Acanthodoras cataphractus* and *Pristella maxillaris*. The shoreline tree is watrabebe or bloodwood (*Pterocarpus officinalis*). © M. Sabaj Pérez.

Library of Congress Cataloging-in-Publication Data

Mol, Jan H., 1958-
The freshwater fishes of Suriname / by Jan H.A. Mol.
p. cm. — (Fauna of suriname ; v. 2)
Includes bibliographical references and index.
ISBN 978-90-04-20766-0 (hardback : alk. paper) — ISBN 978-90-04-21074-5 (pbk. : alk. paper)
1. Freshwater fishes—Suriname. I. Title.

QL632.S75M65 2012
597.176—dc23

2012009460

NCB naturalis

This publication was subsidized by the Uitvoeringsorganisatie Twinningfaciliteit Suriname – Nederland (UTSN).

This publication has been typeset in the multilingual "Brill" typeface. With over 5,100 characters covering Latin, IPA, Greek, and Cyrillic, this typeface is especially suitable for use in the humanities. For more information, please see www.brill.nl/brill-typeface.

This paperback is also published in hardback under ISBN 978 90 04 20766 0 as volume 2 in the series *Fauna of Suriname* (FOS)
ISBN 978 90 04 21074 5 (paperback)
ISBN 978 90 04 20765 3 (e-book)

This book is printed on acid-free paper.

CONTENTS

ACKNOWLEDGEMENTS

Many dedicated freshwater fish enthusiasts, including professional ichthyologists, fishermen, indigenous people, maroons and aquarium hobbyists, contributed information and this book would not have been possible without their help. The book also owes a considerable debt to Han Nijssen, Gerlof Mees and the late Marinus Boeseman, the first ichthyologists that visited Suriname to explore its freshwater fish fauna. I have profited enormously from their subsequent publications as well as from the publications of C.H. Eigenmann on the freshwater fishes of Guyana and P. Planquette, P. Keith and P.-Y. Le Bail on the freshwater fishes of French Guiana. The publications of Ro Lowe-McConnell and M. Goulding have helped me to understand the complex relations between Neotropical freshwater fishes and their environment.

I have received considerable cooperation from various museums in Europe, the United States of America and Suriname during the work to document the Surinamese freshwater fish fauna. These include the Academy of Natural Sciences Philadelphia (ANSP, Philadelphia), Field Museum of Natural History (FMNH, Chicago), Muséum d'histoire naturelle de la Ville de Genève (MHNG, Geneva), Muséum National d'Histoire Naturelle (MNHN, Paris), Netherlands Centre for Biodiversity Naturalis (NCB Naturalis, Leiden; a fusion of the Zoological Museum Amsterdam [ZMA] and the Rijks Museum voor Natuurlijke Historie [RMNH]), National Zoological Collection Suriname (NZCS, Paramaribo, Suriname), and National Museum of Natural History, Smithsonian Institution (USNM, Washington, D.C.). Staff members who were directly involved include John Lundberg and Mark Sabaj Pérez (ANSP), Phil Willink (FMNH), Sonia Fisch-Muller, Raphael Covain and Claude Weber (MHNG), Martien van Oijen, Han Nijssen and Ronald Vonk (Naturalis), Paul Ouboter (NZCS), and Richard Vari (USNM).

Numerous individuals helped with logistics or collections during field expeditions in Suriname. These include Felix Breden, Barry Chernoff, Raphael Covain, Will Crampton, Adrian Flynn, Jan Jaap de Greef, Jozef Joghi, Bennie Jon, John Lundberg, Bernard de Merona, Juan Montoya-Burgos, Paul Ouboter, Dominique Ponton, Stanley Ramanand, Joyce Ramlal, Mark Sabaj Pérez, Andy Sheldon, Ross Smith, Frank van der Lugt, Kenneth Wan Tong You, Claude Weber and Philip Willink.

Special thanks are reserved for contributing photographers Felix Breden, Adrian Flynn, Erling Holm, Willem Kolvoort, Eelco Kruidenier, Trond Larsen, Anna Lindholm, Mike Littmann, Mark Sabaj Pérez, Ross Smith, Logan Volkmann, Philip Willink, and Kenneth Wan Tong You, and for permission to use fish photographs published in the 'Atlas des poissons d'eau douce de Guyane' (Institut National de la Recherche Agronomique / Museum National d'Histoire Naturelle).

Funding for a 6-month visit to the Naturalis Museum, Leiden, to study the large collection of Surinamese fishes in the museum and simultaneously work on this book were provided by the Uitvoeringsorganisatie Twinningfaciliteit Suriname Nederland (UTSN), Ministry of Foreign Affairs, the Netherlands. Sancia van der Meij and René Dekker were responsible for arrangements in Naturalis that allowed me to work full-time and efficiently in the museum and to meet with visiting colleagues in order to prepare a checklist of the freshwater fishes of Suriname.

The backbone of this book is the checklist of the freshwater fishes of Suriname which was completed in 2012 in collaboration with coauthors and distinguished colleagues Raphaël Covain, Sonia Fisch-Muller, Richard Vari and Philip Willink. The checklist is indebted to numerous specialists whose publications and advice were utilized in finalizing the list, including D. Bloom (Engraulidae), A. Cardoso (Aspredinidae), B. Collette (Belonidae), W. Crampton (Gymnotiformes), J. Huber (Rivulidae), M. Jégu (Serrasalminae), S. Kullander (Cichlidae), F. Lima (Characidae), J. Lundberg (Pimelodidae), M. Marinho (*Copella*), T. Munroe (Pleuronectiformes), F. Ribeiro (*Ageneiosus*), T. Roberts (Synbranchidae), Mark Sabaj Pérez (Doradidae), S. Schaefer (gen. nov. aff. *Parotocinclus*), G. Short (Syngnathidae), B. Sidlauskas (Anostomidae) and C. Weber (Hypostominae).

CHAPTER ONE

INTRODUCTION

Suriname is a country in northern South America situated on the Guiana Shield between French Guiana to the east, Guyana to the west, Brazil to the south, and the Atlantic Ocean to the north. At just under 165,000 km² Suriname is the smallest sovereign state in South America. It has an estimated population of approximately 490,000, most of whom live on the country's north coast, where the capital Paramaribo is located. Lying 2 to 5 degrees north of the equator, Suriname has a humid-tropical climate (Af according to the system of Köppen), and the temperatures do not vary much throughout the year (range 22-31 °C). The year has two wet seasons, from late April to middle of August and from December to February, and two dry seasons, from middle of August to early December and February to late April. With some 481 fresh- and brackish-water fish species (Mol *et al.*, 2012), Suriname has a rich inland fish fauna that is readily recognized as related to the 'hottest', most diverse freshwater fish fauna on planet Earth, i.e. that of the Amazon River.

The Interior of Suriname (Fig. 1.1) consists of Precambrian crystalline rocks of the Guiana Shield, forming a hilly landscape covered by tropical rain forest. Basement rocks dip to the north, and are covered by unconsolidated Coastal Plain deposits (Fig. 1.1). The clays and fine sands of the Coastal Plain are derived from the Andes Mountains and transported to Suriname by the Amazon River and then, from the mouth of the Amazon River to Suriname, by the North Brazil and Guiana ocean currents, and are brought shoreward by bottom waters flowing up the slope and across the shelf (Gibbs, 1976; Eisma *et al.*, 1991). The lowland Coastal Plain with extensive wetlands and cultivated areas has traditionally been divided into the Young Coastal Plain (northern part) with Holocene sediments and the Old Coastal Plain (southern part) with predominantly Pleistocene sediments. In the south the Old Coastal Plain is bounded by the Savanna Belt (Fig. 1.1) which comprises mainly Pliocene terrestrial sediments and minor residual weathering products of the Precambrian basement.

Where there is more than a few buckets of water in Suriname, there are fishes, but the aquatic habitat is very dynamic, expanding and contracting with the seasons. When in the wet season the spring tide prevents adequate draining of excess rainwater (and rivers overflow their banks), many fishes such as poeciliids (*Poecilia vivipara, P. reticulata, Micropoecilia* spp), leaf fish (*Polycentrus schomburgkii*), young snook (*Centropomus* sp) and young mullet (*Mugil* sp) swim in the inundated streets of northern Paramaribo. In the dry season, large swamp fishes like pataka (*Hoplias malabaricus*), walapa (*Hoplerythrinus unitaeniatus*), datrafisi (*Crenicichla saxatilis*) and krobia (*Cichlasoma bimaculatum*) are trapped in large numbers in shallow pools and ditches, the only water bodies that remain after the water has retreated; these fishes can then be caught by hand when 'groping' for them in the muddy water.

However, most freshwater fish species of Suriname live in the flowing waters of rivers and streams of the Interior and many species are endemic to a single river system (Mol *et al.*, 2012; Table 2). The wetlands of the Coastal Plain and the estuarine habitats have fewer fish species and these often have a wide distribution in northern South America. Suriname is drained toward the Atlantic Ocean by seven main river basins; in the south the border with Brazil forms the watershed divide between tributaries of the Amazon River System draining to the south and the Surinamese rivers draining to the north. The seven main river systems of Suriname are from east to west: Marowijne (Maroni), Commewijne, Suriname, Saramacca, Coppename, Nickerie and Corantijn (Corentyne) (Fig. 1.1; Table 3; Amatali, 1993). The large Corantijn and Marowijne rivers have drainage areas of 67,600 and 68,700 km², respectively, and mean discharges of 1,570 and 1,780 m³/s, respectively. The two medium-sized rivers, the Coppename and Suriname, have drainage areas of 21,700 and 16,500 km², respectively, and mean discharges of 500

Fig. 1.1. Map of Suriname showing the three major geographical areas, Interior, Savanna Belt and Coastal Plain, and the seven main river basins from east to west: Marowijne (Maroni), Commewijne, Suriname, Saramacca, Coppename, Nickerie and Corantijn (Corentyne) rivers.

and 430 m^3/s, respectively. Finally, the three small rivers, the Nickerie, Saramacca and Commewijne, all have their lower courses deflected to the west as a consequence of deposition of sediments by the Guiana Current and have drainage areas of 10,100, 9,000 and 6,600 km^2, respectively, and mean discharges of 160, 240 and 120 m^3/s, respectively (Amatali, 1993).

Freshwater fishes can be defined as those fishes living in continental lakes, rivers, streams and swamps. Myers (1949) distinguished between primary, secondary and vicarious (peripheral) freshwater fishes based on their tolerance for salt water. Primary freshwater fishes have little salt tolerance and are thus confined to fresh waters. Salt water is a major barrier for them and their distribution has not depended on passage through the sea. This division includes some old groups (e.g. lungfishes and Osteoglossiformes or bony-tongues) and the otophysan orders Characiformes (tetras, piranhas and allies), Siluriformes (catfishes), Gymnotiformes (Neotropical electric knifefishes) and Cypriniformes (this order is not present in South America). Secondary freshwater fishes are usually confined to fresh waters, but they have some salt tolerance and their distribution may reflect dispersal through coastal waters or across short distances of salt water (e.g. Cyprinodontiformes and cichlids). Vicarious fresh-water fishes are derived from rather recent marine ancestors (who used the oceans as dispersal routes). Examples from the Surinamese freshwater fish fauna are freshwater stingrays (Potamotrygonidae) and some drums (Sciaenidae), anchovies (Engraulidae) and needlefish (Belonidae) (Lovejoy *et al.*, 2006).

Fish species included in this book may spend their entire lives in fresh water (indicated by 'F' in the habitat column of Table 2), at least extend into fresh waters with some regularity and occur in brackish but not marine waters (indicated by 'F(B)'), live in the brackish water of estuaries, canals and coastal lagoons (indicated by 'B'; e.g. the Bigi Pan Lagoon in Nickerie district, northwestern Suriname), or migrate seasonally from the ocean into estuaries and the lower freshwater reaches of rivers (indicated by 'M'). In the absence of information on salinity tolerance and life history, it is difficult to accurately designate many species that live in the 'fuzzy' area between the ocean and the river to one of the above categories. In addition, the near-shore coastal waters off Suriname have an estuarine character (low surface salinity, high suspended sediments) due to the massive discharge of Amazon fresh water carried northwestward by the North Brazil and Guiana currents (the Amazon River freshwater plume; Hu *et al.*, 2004) and fishes caught at sea some kilometers off the Surinamese coast may in fact live in brackish water. This book basically includes species that were documented to occur in the freshwaters of Suriname with the exception of rare marine intruders. The categories F and F(B) are counted as freshwater species.

Interest in the freshwater fishes of Suriname by naturalists and scientists extends back over more than two centuries. Suriname is undoubtedly the site of origin of the oldest extant preserved specimens of South American fishes which were collected in the first half of the 18th century (Kullander and Nijssen, 1989), albeit in some instances with inexact locality information. The earliest contribu-

Fig. 1.2. Dried skin of the platkop kwikwi *Callichthys callichthys* pressed onto paper from the Gronovius collection, now in the collection of the Natural History Museum, London (BMNH1853.11.12.194). This specimen was probably collected in Suriname in the first half of the 18th century and used by Linnaeus to describe the species (© Mark Allen, All Catfish Species Inventory).

tions to Surinamese ichthyology are incorporated into more general treatises of special natural history collections such as that of Gronovius (1754, 1756). Linnaeus, in turn, acquired some of the information for his '*Systema Naturae*' (1758, 1766), the official beginning of modern taxonomic nomenclature, from his students who traveled throughout the world. Although one of these, Daniel Rolander (1725-1793), visited Suriname in 1755 (Holthuis, 1959), it is uncertain whether Linnaeus examined the Surinamese fish specimens collected by Rolander. Linnaeus did, however, examine the collections and read the publications of Peter Artedi (1738) who himself examined Albertus Seba's collection, King Adolf Fredrik, and Laurens Theodorus Gronovius (1754, 1756) all of which included fishes that originated in rivers flowing through what is now Suriname. It is sometimes impossible to determine how or where these specimens were collected, but documentation indicates that Daniel Luyx Massis, Director of the famous West India Company, acquired fishes from Suriname for the Gronovius collection (Wheeler, 1958). Some of these Surinamese specimens are still extant as alcohol preserved samples or dried skins pressed onto paper in the collections of The Natural History Museum, London, the Zoological Museum, Copenhagen, and the Swedish Museum of Natural History, Stockholm (Wheeler, 1958, 1989; Fernholm and Wheeler, 1983; Fig. 1.2). Surinamese fishes described and figured by Linnaeus (1758, 1766), with the original genus in parentheses, are: *Achirus* (*Pleuronectes*) *achirus* (Linnaeus, 1758), *Ageneiosus* (*Silurus*) *inermis* (Linnaeus, 1766), *Apteronotus* (*Gymnotus*) *albifrons* (Linnaeus, 1766), *Astyanax* (*Salmo*) *bimaculatus* (Linnaeus, 1758), *Callichthys* (*Silurus*) *callichthys* (Linnaeus, 1758), *Charax* (*Salmo*) *gibbosus* (Linnaeus, 1758), *Cichlasoma* (*Labrus*) *bimaculatum* (Linnaeus, 1758), *Crenicichla* (*Sparus*) *saxatilis*

Fig. 1.3. Dried specimens of spigrikati *Pseudoplatystoma fasciatum* and soké *Platydoras costatus* collected in Suriname by H.H. Dieperink in the first half of the 19th century. Collection Naturalis Museum, Leiden, the Netherlands (© Naturalis).

(Linnaeus, 1758), *Doras* (*Silurus*) *carinatus* (Linnaeus, 1766), *Electrophorus* (*Gymnotus*) *electricus* (Linnaeus, 1766), *Gasteropelecus* (*Clupea*) *sternicla* (Linnaeus, 1758), *Gymnotus carapo* Linnaeus, 1758, *Hypostomus* (*Acipenser*) *plecostomus* (Linnaeus, 1758), *Loricaria cataphracta* Linnaeus, 1758, *Polycentrus schomburgkii* Müller & Troschel, 1849 (due to circumstances, *Labrus punctatus* Linnaeus, 1758 is a synonym), *Pseudoplatystoma* (*Silurus*) *fasciatum* (Linnaeus, 1766), *Pterengraulis* (*Clupea*) *atherinoides* (Linnaeus, 1766), *Salmo notatus* Linnaeus, 1766 (identity of this species unclear although presumably a characiform) and *Serrasalmus* (*Salmo*) *rhombeus* (Linnaeus, 1766).

Many early scientific papers dealing with Surinamese fishes are based on material collected by non-ichthyologists to whom that discipline is greatly indebted for their collecting efforts. For example, the great Dutch ichthyologist P. Bleeker described some Surinamese fish species (e.g. *Anchovia surinamensis*, *Chasmocranus surinamensis*, *Plagioscion surinamensis*; Bleeker, 1862, 1864, 1873) based on specimens collected by H.H. Dieperink, apothecary in Paramaribo, in the first half of the 19th century (Fig. 1.3). Holthuis (1959) and Hoogmoed (1973) extensively documented the collectors of Crustacea and Amphibia/Reptilia, respectively, from Suriname. Most of these collectors also collected fishes.

The people of Suriname also show a keen interest in the fishes of the country, e.g. the professional fishermen who mainly fish in the estuaries and at sea (Emanuels, 1977; Charlier, 1988), sport fishermen and aquarium hobbyists living in the Coastal Plain, and the Maroons (descendants of escaped slaves) and Amerindians in the Interior that are dependent of the fishes for most of their protein intake. However, ca. 95% of the population of Suriname lives in the northern Coastal Plain and, sadly, it is probably fair to say that most Surinamese have little or no knowledge of the majority of native freshwater fishes that live in the rivers and streams of the Interior of the country. When in the 1970s a large aquarium with native Surinamese freshwater fishes was installed in a large hotel in Paramaribo, a Surinamese man, admiring the new aquarium, was overheard saying to his companion ‘Clearly these beautiful fishes are exotic species; in Suriname we don’t have such colorful fishes’ (J. Joghi, pers. communication). These men, apparently living in Paramaribo or somewhere else in the Coastal Plain, were clearly better acquainted with the often rather dull food fishes of the coastal wetlands than with the many colorful fishes of the Interior of the country. Indeed, most people of the Coastal Plain have more knowledge and experience with the myriad exotic fish available in the aquarium trade. Even people living in the Interior often know only the fishes of the river basin in which their village is located (especially the large-sized food fishes), and fish faunas of the seven main river basins show considerable differences, with many species restricted to one or a few rivers (Table 2). Increasing numbers of tourists now visit the Interior of Suriname and are stunned by the fishes they observe in the clear-water streams or by the larger food fishes that are caught by the Maroons and Amerindians. With the exception of an interesting booklet by H. Heyde (1986), most information on the freshwater fishes of Suriname is found in scientific papers that are often not easily accessible in Suriname (although the internet is catching up fast and increasingly is becoming a big help in literature search; e.g. http://www.fishbase.org). The present book hopes to fill this information gap and thus make both the people of Suriname and visitors more aware of the beautiful fishes of Suriname and the aquatic ecosystems in which they live. An increased awareness of the Surinamese freshwater fishes and their habitats would not only allow one to enjoy even more a visit to the Interior of the country, but presumably also help decision makers to better protect the threatened habitats of the fishes, thus giving future generations the opportunity to enjoy the splendor of the Surinamese fishes like we do today.

CHAPTER TWO

ORIGINS OF THE SURINAMESE FRESHWATER FISH FAUNA

By the standards of biogeography in a global context, the margins of the Neotropical realm are remarkably clear, certainly with respect to freshwater fishes (Myers, 1966). Only a few Neotropical lineages, including several characins, catfishes and cichlids, have dispersed as far north as northern Mexico and southernmost Texas, and there are only a dozen or so Neotropical freshwater fishes known from the northern pampas of Argentina. Similarly, very few fish taxa from other regions of the world are present in the Neotropics (Lundberg *et al.*, 2007). The freshwater fish fauna of Suriname is readily recognized as part of the Guiana Shield fish fauna, e.g. as part of the Guianas ecoregion (Abell *et al.*, 2008; Fig. 2.1) or the Atlantic coastal rivers (excluding Corantijn River; Lujan & Armbruster, 2011). The Guiana Shield is recognized as one of the 11 areas with high freshwater fish endemism within the Neotropical ichthyofauna (Hubert & Renno, 2006; Lujan & Armbruster, 2011) and this was partly explained by Hubert and Renno

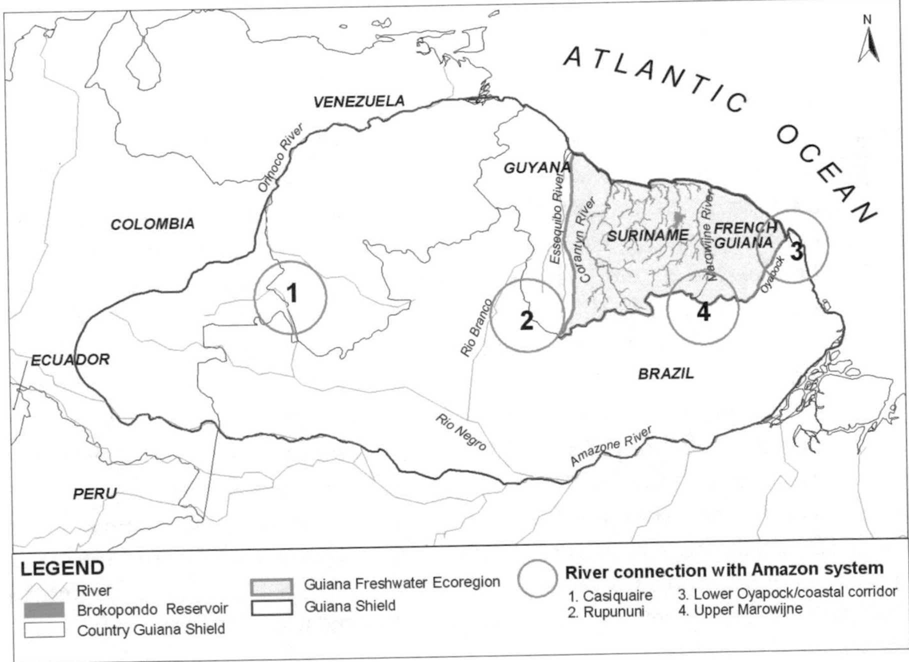

Fig. 2.1. Suriname in Guiana Shield perspective, showing four (possible) connections with the Amazon River system.

(2006) by the shield functioning as a highland freshwater refuge during Miocene marine incursions (see below).

The origins of Neotropical fishes vastly pre-date the arrival of humans in the New World 16,500-13,000 years Before Present [BP] (Bonatta & Salzano 1997; Waters *et al.*, 2011) and in Suriname (earliest charcoal remains of Werehpai caves dated 5,000 BP; http://home.wxs.nl/~vrstg/guianas/werehpai/werehpai-eng.pdf). Until recently, the Pleistocene refuge-allopatric divergence model was the most widely known and invoked mechanism for explaining biological diversity in South America, but it is difficult to apply to freshwater fishes (Weitzman & Weitzman, 1982) and, given the deep temporal framework for fish diversification in the Neotropics (with many modern lineages represented by Miocene fossils; Lundberg *et al.*, 2010), the model could only apply to the most terminal cladogenetic events within the majority of Neotropical fish clades (Lundberg *et al.*, 1998; Rull, 2008).

Data pertaining to the origins (age estimates) of Neotropical freshwater fishes are available from (1) molecular studies such as analyses of DNA sequences (Lovejoy *et al.*, 2010), (2) geographic distributions (e.g. taxa on either side of geographic barriers with known ages), and (3) fossils. Molecular estimates can be problematic for several reasons (e.g. calibration of the molecular clock) and are bounded by wide confidence intervals, biogeographic and paleontological analyses provide only minimum ages, and fossil estimates depend on the vagaries of fossil preservation.

Although fossils generally underestimate the dates of evolutionary origin of taxa, they are the only direct source of evidence on phylogenetic ages and thus are sometimes used as minimum calibration points for molecular rate estimates (Near *et al.*, 2005): fossils of known age are placed into their phylogenetic context in order to infer the ages of close (nested) taxa by using the principle of equal ages of origin for sister group taxa. Unfortunately, the record of Neotropical fossil fishes is relatively sparse, especially considering the very high diversity of this region (Lundberg *et al.*, 2010). This is due to unfavorable conditions for the preservation and recovery of fossils in fluvial systems. Low-energy lacustrine depositional environments, from which most freshwater fossils are known, are rare in the present-day Neotropical hydrological setting. The high current flow and low pH of many tropical rivers combined with high rates of biogenic decomposition also reduce the probability of fossil formation. Further the discovery of sedimentary outcrops is hindered by thickly vegetated landscapes and low topographic relief. To date, fossil fishes from within the watershed of the modern Amazon and Guiana Shield drainage basins are restricted to the Neogene of western Amazonia (Lundberg *et al.*, 2010). However, fossil Amazonian fish faunas are also known from areas currently outside Amazonia, such as Argentina and Andean basins to the west and the north.

Early in the Cretaceous of South America there is an archaic fish fauna known from well-preserved and abundant basal actinopterygian fishes including early teleosts (Maisey, 2000). The Late Cretaceous fish faunas of the Maastrichtian

(c. 71-66 million year ago [Ma]) Maiz Gordo (northern Argentina) and El Molino (eastern Bolivia) formations are still dominated by non-teleost groups (e.g. dipnoans, pycnodonts, polypertiforms, lepisosteids) characteristic of the Cretaceous and some archaic teleosts (Gayet & Meunier, 1998). By contrast, the Paleocene (c. 60-58 Ma) Santa Lucia Formation, overlying the Molino Formation, is dominated by teleosts, especially characiform and siluriform taxa, that characterize modern Neotropical faunas (Malabarba *et al.*, 2006). Many Paleogene fossils of pimelodid and callichthyid catfishes and cheirodontidine and curimatid characoids are readily assigned to modern genera (e.g. the 59 Ma old *Corydoras revelatus* fossil; Cockerell, 1925; Reis, 1998), suggesting diversification within their clades. These faunas may have been some of the earliest known sediments of a large north-flowing 'paleo-Amazonas-Orinoco' river basin and the mega Pebas wetland (Lundberg *et al.*, 1998; Hoorn & Wesselingh, 2010) that was influenced by marine incursions of varying extent.

More recent fossil formations provide some useful materials for estimating minimum ages of certain Neotropical freshwater fish clades. Perhaps the best known is the Middle Miocene (c. 12 Ma) La Venta fauna in what is now the Magdalena valley of Colombia. Fishes of the trans-Andean (west of the eastern Andean cordilleras) La Venta fauna include many living forms that are now known only in the cis-Andean (east of the Andean cordilleras) Amazon and Orinoco basins. Many of these species are indistinguishable from living species (e.g. *Arapaima gigas*, *Colossoma macropomum* and *Lepidosiren paradoxa*) or are closely related to Amazonian species (e.g. the catfish genera *Brachyplatystoma* and *Hoplosternum*) (Lundberg *et al.*, 2010). The geological isolation of the Magdalena from the Amazon drainage basin began with the rise of the Eastern Cordillera of Colombia about 12 Ma (Gregory-Wodzicki, 2000), suggesting a minimum age for the divergence of lineages that occur in both in cis- and trans-Andean basins. Albert *et al.* (2006) list 123 genera of Neotropical freshwater fishes with a *cis-trans*-Andean distribution, 53 of which also occur in Suriname.

Today, South America is situated on the westerly side of the tectonic continental South American Plate (which includes a sizeable region of the Atlantic Ocean seabed extending eastward to the Mid-Atlantic Ridge) and 93% of its freshwater drainage runs into the Atlantic. South America's drainage pattern was shaped by highland areas of the persistent continental shields and the Andes Mountains, the fluctuating foreland basin east of the Andes, and several structural arches (Lundberg *et al.*, 1998). In the Proterozoic (2500-540 Ma; Table 1), South America and Africa were contiguous within the West Gondwana supercontinent (Fig. 2.2; Cordani & Sato, 1999). It was only after the final opening of the Atlantic Ocean in the Early Cretaceous c. 112 Ma that South America was separated from Africa (Maisey, 2000). The occurrence of freshwater catfishes (Siluriformes), characins (Characiformes), cyprinodonts and cichlids (and the much older lungfishes and osteoglossiforms) in both Africa and South America gives a minimum age estimate of these higher-level groups of 112 Ma. On the other hand, the order of Neotropical electric knifefishes (Gymnotiformes) and most South American fami-

Table 1. Geological time table pertaining to fish evolution in the Neotropics and Suriname in particular

Period / Epoch	Age (million years before present Ma)	Formation in Suriname	Environment	Climate	Events
Quaternary / Holocene	0,01-present	Demerara (Coronie / Mara)	chenier plain, Mara mudflats	humid	Quaternary fluvial cycle; oldest remains of humans in Suriname (5,000 BP)
Quaternary / Pleistocene	2,6 –0,01	Coropina	barrier islands, mudflats, ?braided rivers in the Interior	alternating periods of dry and humid climate	migration of first Amerindians (pre-Clovis people?) from east Asia to North America across the Bering Land Bridge (16,500-13,000 BP)
Tertiary (Neogene) / Pliocene	5,3-2,6	Zanderij	braided rivers and alluvial fans	humid	Late Tertiary II peneplain; emergence of the Isthmus of Panama (3 Ma); break up of proto-Berbice River, separation of Corantijn River
Tertiary (Neogene) / Miocene	23,0-5,3	Coesewijne	beach ridges and coastal swamps	relatively dry	uplift Eastern Andean Cordillera (La Venta fauna, 10-12 Ma), Upper Amazon Pebas wetland, marine incursions
Tertiary (Paleogene) / Oligocene	33,9-23,0	A-sands, bauxite hiatus	fluvial deposits; bauxite hiatus	humid	erosion Early Tertiary peneplain, origin of the Brownsberg, Lely and Nassau Mountain Plateaus; Late Tertiary I peneplain
Tertiary (Paleogene) / Eocene	55,8-33,9	Onverdacht	alluvial fans, braided rivers, point bars and back swamps	relatively dry	

Table 1. Continued.

Period / Epoch	Age (million years before present Ma)	Formation in Suriname	Environment	Climate	Events
Tertiary (Paleogene) / Paleocene	65,5-55,8	Onverdacht			Early Tertiary peneplain; proto-Berbice River; Santa Lucia & Maiz Gordo formations; Neotropical fish radiations
Cretaceous / Late Cretaceous	99,6-65,5 (Maastrichtian 70,6-65,5)	Nickerie			El Molino formation with mainly non-teleost groups (dipnoans, polypertiforms, lepisosteids) and some archaic teleosts
Cretaceous / Early Cretaceous	145,5-99,6				isolation of South America after opening of equatorial seaway linking North and South Atlantic (112 Ma); birth of eastern proto-Amazon River (120 Ma)
Jurassic	201,6-145,5				endorheic Lake Maracanata in Takutu Graben, south Guyana
Triassic	251-201,6				first teleost fishes
Silurian	444-416				first jawed fishes
Precambrian, Proterozoic	2500-542				West Gondwana supercontinent (1000-540 Ma), evolution of the Amazon Graben, deposition of Roraima fluviolacustrine sediments, uplift of Guiana Shield (600 Ma) and subsequernt erosion of overlying Roraima sediments, origin of Tafelberg Mountain

Table 1. Continued.

Period / Epoch	Age (million years before present Ma)	Formation in Suriname	Environment	Climate	Events
Precambrian, Archean	3850-2500				South American Platform (3300-1700 Ma; main period of crust formation 2200-2000 Ma)

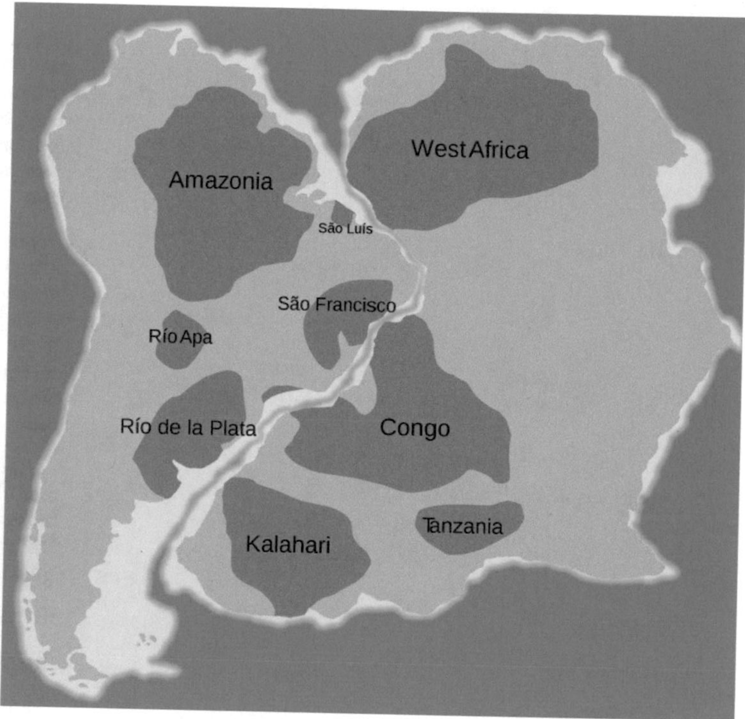

Fig. 2.2. South America and Africa contiguous within the Proterozoic (2500-540 Ma) West-Gondwana supercontinent; Precambrian cratons, including the Amazonian Craton (which includes the Guiana and Brazilian Shields) are shown (© Wikipedia Commons).

lies in the catfish, characin and cyprinodont orders are endemic (restricted in their distribution) to the South American continent and thus must have evolved in situ after the isolation of South America. Lundberg (1998) made it clear that most major Neotropical fish lineages (i.e. families and even genera) were already extant long before the Miocene surge in Andean uplift c. 10-12 Ma (Gregory-Wodzicki, 2000), and the search for geologic events relevant to their evolution should extend deeper in time (i.e. the Late Cretaceous and Paleogene 100-23 Ma).

The largest geological features of the Neotropics are (1) the eastern South American Platform, an ancient Precambrian block of continental crust underlying all of Amazonia and adjacent regions (including the Guiana and Brazilian shields) (Fig. 2.2), (2) the Andean orogenic belt and (3) the foreland basins to the east of the Andes Mountains (Cordani & Sato, 1999). The South American Platform lies very low in the Earth's mantle, with more than half its total area less than 100 meters above sea level (m-asl). As a result of the general low relief several areas in South America have been exposed to marine transgressions and regressions over the past c. 110 Ma. These episodes of marine transgression drastically affected the extent and distribution of habitat available to obligate freshwater fishes and can explain marine-derived lineages of the Neotropical fish fauna such as freshwater stingrays, drums, anchovies and needlefish. The marine-derived taxa originated in the western Amazonian Pebas wetland in the Miocene (22-10 Ma), most likely from sources in the Caribbean Sea (Lovejoy *et al.*, 2006).

Two principal highland structures of the South American Platform are the Precambrian Guiana Shield in the northeast (including Suriname) and the much larger Brazilian Shield in the southeast; separating the two shields is the Amazon Graben, an old Proterozoic structural downwarp that is 300-1000 km wide and filled with Cambrian sediments up to 7000 m deep (Putzer, 1984). The Amazon Graben exhibits dislocation by a transcurrent fault system such that the northern block (Guiana Shield) is transposed considerably to the west when compared with the southern block (Brazilian Shield). The Guiana Shield stretches from its eastern margin along the Atlantic coast across French Guiana, Suriname, Guyana, and Venezuela, to southeastern Colombia in the west (approximately 2000 km distance). Bounded by the Amazon Graben to its south and the Orinoco River to its north (1000 km distance) and west, the Guiana Shield occupies 2,288,000 km^2 (Hammond, 2005). The name 'Guiana' is believed to be derived from an Amerindian word meaning 'water' or 'many waters' (Hammond, 2005) and the discharge of the 47 medium to large rivers of the shield averages 2,792 km^3 per year or approximately a quarter of South America's total volume of freshwater exported to the oceans (Hammond, 2005). Topography higher than 150 m-asl is largely comprised by the Roraima Group, an aggregation of fluviolacustrine sediments deposited over much of the northern South American Platform during the Proterozoic and subsequently uplifted along with the basement. Portions of this formation resistant to erosion now comprise most of the striking topographic elements of the Guiana Shield, including Mount Roraima, Pico Neblina (with 3014 m-asl South America's highest non-Andean peak) and the Tafelberg Mountain (1026 m-asl) in Suriname.

Nondeformational, epeirogenic uplift of the Guiana Shield has occurred sporadically almost since its formation in the Proterozoic. Since at least the middle Paleozoic, when the region was first exposed at the surface, cycles of uplift and stasis during which erosion occurred have resulted in elevated erosional surfaces (planation surfaces or peneplains) that are now observed throughout the northern interior of South America (Gibbs & Baron, 1993). At lower elevation, these appear as steps or terraces/stages of Roraima Formation sediments, vertically

Fig. 2.3. The loricariid catfish *Harttiella crassicauda* is endemic to Paramaka Creek in Nassau Mountains. *H. crassicauda* is thought to have evolved in situ during the Oligocene (approximately 30 Ma), but nowadays it is threatened with extinction by a bauxite mine with a projected mine-life of 5-10 years (© T. Larsen).

separated from each other by 60-200 m elevation. At higher elevations, collections of peaks can be identified that share similar elevations (e.g. Brownsberg, Lely and Nassau mountains). In Suriname, four planation surfaces have been identified (King *et al.*, 1964) and these were assigned tentative ages by analysis of pollen in buried Coastal Plain sediments that were derived from erosion of the old planation surfaces (Wijmstra, 1971): Early Tertiary (Paleocene), Late Tertiary I (Oligocene), Late Tertiary II (Pliocene) and Quaternary (Holocene). Thus it is possible to associate bauxitization of duricrusts in the Early Tertiary peneplain (in the relatively dry Eocene, 40 Ma) and subsequent uplift and erosion of the surface surrounding the bauxite duricrust in the Oligocene (30 Ma) with the origins of bauxite plateaus (e.g. Nassau Mountains) and interpret these mountain plateaus as remnants of the Early Tertiary peneplain standing high above the lower, rolling landscape of later planations. The uplift and consequent barrier formation of the Nassau Plateau suggest a minimum age of 30 Ma for relictual endemic fishes of Nassau Mountains (e.g. *Harttiella crassicauda*; Covain *et al.*, 2012; Fig. 2.3); a similar phenomenon can explain the presence of relictual species (*Corymbophanes* spp. and *Lithogenes villosus*) above the Kaieteur Falls in Guyana (Lujan & Armbruster, 2011).

The history of the river basins of South America is a dynamic and complex one, and evidence indicates that many of the paleo-fluvial predecessors of modern drainages were substantially different from the rivers seen today (Lundberg *et al.*, 1998). Prior to the Andean uplift in the Neogene, the Guiana and Brazilian shields were South America's major uplands and likely the continent's most concentrated regions of high-gradient lotic habitat. Lujan & Armbruster (2011) note the occurrence of relictual basal loricariids on the Brazilian (Delturinae) and Guiana (*Lithogenes*) shields and speculate that the rheophilic loricariids originated on the shields and subsequently spread through the rest of northern South America and southern Central America.

One of the largest drainages of the central Guiana Shield during most of the Tertiary was the proto-Berbice, a northeast-flowing river that was derived in the Late Cretaceous to Paleogene from the endorheic Maracanata Lake in the Takutu Graben rift valley in south Guyana and exited to the Atlantic between the modern towns of New Amsterdam, Guyana, and Nickerie, Suriname, i.e. approximately at the location of the mouth of the modern Corantijn River (Sinha, 1968; Lujan & Armbruster, 2011). The proto-Berbice system not only included the modern Essequibo River, but also the Corantijn River, which may explain the occurrence in Corantijn River of Amazonian taxa (e.g. *Phractocephalus, Pterodoras, Amblydoras, Brachyrhamdia, Petulanos, Satanoperca*) that are absent from the other Surinamese river basins. Separation of the Corantijn basin from the proto-Berbice was probably only recently accomplished, i.e. in the Pliocene/Pleistocene when most of the breakup of the proto-Berbice took place (N. Lujan, pers. communication). Morphology of the rheophilic loricariid genus *Lithoxus*, a Guiana Shield endemic restricted to upland habitats, suggests a division into a western, proto-Berbice subgenus (*Lithoxus*, 2 spp), and an eastern Atlantic Coastal subgenus (*Paralithoxus*, 5 spp) (Boeseman, 1982; Lujan, 2008). The North Rupununi Savannas occupy the modern Maracanata depression and form a shallow continental divide between the Essequibo and the Branco/Negro rivers that is regularly flooded in the wet season and thus forms a lentic connection (the Rupununi Portal; De Souza *et al.*, 2012) between the Amazon River system and the Essequibo (Fig. 2.1). The western Guiana Shield features one of the largest and most notable river captures events in the Neotropics: that of the ongoing piracy of the Upper Orinoco River by the Negro River via the southwest flowing Casiquiare Canal (Winemiller *et al.*, 2008; Fig. 2.1), a relatively recent (Pleistocene-Holocene) phenomenon.

The eastern lobe of the Guiana Shield, including Suriname and French Guiana, is drained to the south by tributaries of the eastward flowing Amazon River and to the north by Atlantic coastal rivers (including the rivers of Suriname). The birth of the (eastern) proto-Amazon River can be dated back at least to the opening of the equatorial seaway 112 Ma that finally linked the North and South Atlantic Ocean (Maisey, 2000). The eastern proto-Amazon was much smaller than the modern Amazonas-Solimões system, but its mouth was located approximately coincident with its modern delta. For over 100 My following the breakup of Western Gondwana, upper (Solimões) and lower (Amazonas) portions of the modern Amazon Basin were separated by the Purús Arch, a continental divide within the Amazon Graben located near the mouth of the Purús River. The modern Amazon drainage system was assembled through a series of Andean tectonic shifts during the Late Miocene (11-7 Ma; Figueiredo *et al.*, 2009); these events not only shifted the prevailing slope of the Andes back-arc basin eastward and caused Andean-derived watercourses to breach the Purús Arch, but also isolated the trans-Andean Magdalena and Maracaibo drainage basins and thus resulted in the modern distinct Amazon and Orinoco drainages.

When considering areal relationships among Guiana Shield fishes based only upon hydrologic history (Lujan & Armbruster, 2011; their figure 13.5), the Surinamese freshwater fish fauna is part of the Atlantic Coast drainages and most closely related to the Lower Amazon drainages. The Atlantic Coast and Lower Amazon drainages (grouped together as North East Atlantic Coast) are in turn closely related to the proto-Berbice drainages (including the Corantijn River). Proto-Berbice and NE Atlantic Coast drainages are separated from the proto-Orinoco drainages of the western Guiana Shield by the Purús Arch that once separated the eastern proto-Amazon from the western Solimões-Orinoco system.

Lowland portions of the proto-Amazon tributaries drained the southern slopes of the eastern Guiana Shield. A series of ridges with peaks in the range of 400 to 1,000 m, the Wassari, Acarai, Grens-gebergte and Tumuc-Humac (Toemoek-Hoemak) Mountains, form a continental divide within the eastern Guiana Shield, separating south-flowing Amazon tributaries and north-flowing Atlantic Coastal drainages. Headwaters of respective northern and southern rivers interdigitate across these highlands, rendering them potentially porous to fish dispersal, e.g. the hypothesized Upper Marowijne connection (Cardoso & Montoya-Burgos, 2009; Fig. 2.1). A molecular phylogeographic study of *Guyanancistrus brevispinis* populations (or species?) from the Corantijn to the Oyapock and in northern tributaries of the Amazon provided strong support that this species invaded the Atlantic Coastal river system via the Jari River, a south-flowing tributary of the Amazon, via headwater interdigitation and stream capture with the north-flowing Marowijne River (Cardoso & Montoya-Burgos, 2009). Subsequent dispersal among coastal rivers in Suriname and French Guiana occurred principally by temporary connections between adjacent rivers during periods of lower sea level (see below). During high sea level intervals, the isolated populations would have diverged leading to the observed allopatric populations/species (i.e. the sea level fluctuation (SLF) hypothesis of diversification; Cardoso & Montoya-Burgos, 2009). Similarly, Nijssen (1970) suggested a seasonal portal between the Upper Corantijn River (Sipaliwini River) and the Paru do Oeste River (Amazon River basin) across the potentially flooded Sipaliwini-Paru Savanna to explain the distribution of *Corydoras bondi* / *C. coppenamensis* although a proto-Berbice corridor might provide a better explanation (Lujan & Armbruster, 2011).

Exchange of fishes between Atlantic coastal drainages and the eastern Amazon basin may also be accomplished via a coastal marine corridor with reduced salinity due to westward deflected Amazon River discharge (Amazon River freshwater plume; Hu *et al.*, 2004) and coastal confluences during times of low sea level and expanded coastal plains (Lower Oyapock / Coastal Corridor connection, Fig. 2.1). The availability of the Eastern Atlantic Corridor as a means of distribution was first suggested by Vari (1988) to explain the range of *Curimata cyprinoides*, a lowland species widely distributed throughout Atlantic Coast drainages from the

Orinoco to the Amazon. The coastal corridor may also explain the distribution of *Hypostomus plecostomus* and *H. watwata*, two Coastal Plain species found in estuaries that may use the low-gradient streams in the Coastal Plain and near-shore marine habitats to move between drainages along the Atlantic Coastal Corridor (Eigenmann, 1912; Boeseman, 1968), *Hoplosternum littorale* (Mol, 1994), *Parotocinclus britskii* (Schaefer & Provenzano, 1993) and several Serrasalminae species (Jégu & Keith, 1999). The species of *Brachyplatystoma* are also possible coastal dispersers if the pelagic young go in the diluted surface waters especially in times of heavy rainfall (J. Lundberg, pers. communication).

Aridity and marine incursions have a similar effect on rivers and riverine fishes – that of reducing and isolating habitats over a broad geographic range. The two are also correlated in their response to global cycles of glaciations (e.g. Bennett, 1990). In general, warmer, interglacial climates correspond to higher sea levels, more extensive marine incursions, and higher levels of precipitation. Cooler, glacial periods result in reduced precipitation, retreat of the sea, expansion of the Coastal Plain, and incision of river channels. Marrenga and Ruleman (2008) suggest that during the last glacial period the Suriname River was a braided river streaming through a savanna landscape in a wide and shallow valley filled with sediments originating from severe erosion of higher parts and that no rapids were present in the river (thus offering opportunities of expansion of fish populations; Cardoso & Montoya-Burgos, 2009). Geological and biogeographical evidence indicates that the climate of South America was much drier in the recent past than it is today and, within the overall trend of late Pleistocene aridity, the period from approximately 21,000-13.000 BP was the driest (Van der Hammen & Hooghiemstra, 2000). Terrestrial vegetation throughout much of the Guianas was of an open savanna or grassland type, with rainforests limited to highland refugia and riparian margins. Although these refugia do not appear to be very useful in explaining freshwater fish distribution, Renno *et al.* (1990, 1991) found molecular evidence of the existence of an eastern and western Pleistocene refuge of *Leporinus friderici* populations in French Guiana from which this species has more recently expanded its range.

CHAPTER THREE

FAUNAL COMPOSITION

The freshwater fish fauna of Suriname is readily recognized as part of the Guiana Shield fish Fauna (e.g. the Guianas ecoregion of Abell *et al.*, 2008) and the Guiana Shield fish fauna (Vari *et al.*, 2009) in turn is part of the Neotropical ichthyofauna. The Neotropical zoogeographical realm has the greatest number of primary and secondary freshwater fish families (Berra, 2001). The Neotropics also have the highest percentage of endemism of freshwater fish families, with 69% of the families found nowhere else (Berra, 2001). Freshwater fishes from the Neotropics belong to relatively few higher-level taxa (e.g. 17 orders, with dominance of the three otophysan orders Characiformes, Siluriformes and Gymnotiformes), each of which is characterized by a distinct suite of morphological traits. However, despite its relatively poor diversity at higher taxonomic levels, the Neotropical ichthyofauna is extremely diverse at lower taxonomic levels. Estimates of total Neotropical fish richness range from between 5000 and 8000 species (Lundberg *et al.*, 2000; Reis *et al.*, 2003) and the pace of the description of 'new' species is still accelerating. Many species are very small, even miniatures (Weitzman & Vari, 1988). This disproportionate distribution of taxonomic categories, with many lower taxa and few higher taxa, has resulted from a lengthy history of geographical isolation and in situ diversification (Lundberg, 1998).

The great majority (>97%) of Neotropical freshwater fish species are members of the three otophysan orders (75%) and the non-otophysan clades of Cyprinodontiformes and Cichlidae. These primary and secondary freshwater fish taxa, which have little or no tolerance for salt water and very poor capacities for dispersal over marine barriers, trace their origins to before the Early Cretaceous separation of Africa and South America (110 Ma; Myers, 1949, 1966; chapter 2). With the exception of a handful of recent anthropogenic transplants there are few examples of fishes from other continents that have naturally established themselves in cis-Andean South American waters (Hrbek *et al.*, 2007). The only fish taxa that appear to have successfully joined Neotropical communities during the whole of the Cenozoic are certain groups of marine origin (Lovejoy *et al.*, 1998), with only the potamotrygonid stingrays attaining moderate level of diversity (>25 spp.).

Data derived from the literature supplemented by examination of specimens in collections show that 487 species of fish live in the fresh and brackish inland waters of Suriname, with 400 of these restricted to fresh waters (Table 2). These 487 species represent 16 orders and 64 families. Orders with the largest numbers of freshwater species in the Surinamese inland fish fauna are the Siluriformes (165 species), Characiformes (150 species), Perciformes (40 species), Gymnotiformes (21 species) and Cyprinodontiformes (22 species). At the family level, the

Table 2. List of the fresh- and brackish-water fishes of Suriname with occurrences in river systems (T = type locality, S = confirmed by specialist). COR = Corantijn River, NIC = Nickerie River, COP = Coppename River, SAR = Saramacca River, SUR = Suriname River, COM = Commewijne River, MAR = Marowijne River, TLS = Type Locality 'Suriname'. Listing includes freshwater species (F), brackish-water species (B) and marine species that seasonally penetrate into the lower reaches of rivers (M); F(B) denotes a species that occurs in both fresh and brackish water, but not in sea. Introduced species not native to Suriname are marked with an asterisk; a question mark is added if the introduction is considered questionable. (Modified after Mol *et al.*, 2012).

Taxa	Habitat	COR	NIC	COP	SAR	SUR	COM	MAR	TLS
ORDER: PRISTIFORMES									
FAMILY PRISTIDAE									
Pristis perotteti	MB					X		X	
ORDER: MYLIOBATIFORMES									
FAMILY: POTAMOTRYGONIDAE									
Potamotrygon boesemani	F	T							
Potamotrygon marinae	F							T	
Potamotrygon orbignyi	F	X	X	X		S		S	
ORDER: ELOPIFORMES									
FAMILY: MEGALOPIDAE									
Megalops atlanticus	FBM	X			X	X	X	X	
FAMILY: ELOPIDAE									
Elops saurus	BM		X			X	X		
ORDER: CLUPEIFORMES									
FAMILY: CLUPEIDAE									
Harengula jaguana	FBM			X		S			
Opisthonema oglinum	BM		X			X		X	
Rhinosardinia amazonica	FBM	X		X		S			
FAMILY: ENGRAULIDAE									
Anchoa spinifer	FBM	X	X	S		S		S	
Anchovia clupeoides	BM							X	
Anchovia surinamensis	FB	X			S	X		S	T
Anchoviella brevirostris	BM	X				X			
Anchoviella cayennensis	B			X				X	
Anchoviella guianensis	FB	X						S	
Anchoviella lepidentostole	FBM	X		S		X		X	T
Anchoviella sp.	F							S	
Cetengraulis edentulus	BM	S		X		X			
Lycengraulis batesii	FB	X		S		X		S	
Lycengraulis grossidens	FBM	X				X		X	
Pterengraulis atherinoides	FB	X		S		X		S	T
FAMILY: PRISTIGASTERIDAE									
Odontognathus mucronatus	FBM	S		X		X		X	
Pellona flavipinnis	FB	X				X			
Pellona harroweri	MB	S		X		X			
ORDER: CHARACIFORMES									
FAMILY: PARODONTIDAE									
Parodon guyanensis	F	S		X	S	X		X	

Table 2. Continued.

Taxa	Habitat	COR	NIC	COP	SAR	SUR	COM	MAR	TLS
FAMILY: CURIMATIDAE									
Curimata cyprinoides	F	S	S		X	S	X	S	T
Curimatopsis crypticus	F	X				X	X?	X	
Cyphocharax biocellatus	F							T	
Cyphocharax helleri	F	S		X		T	X	S	
Cyphocharax microcephalus	F	S	S		S	S		X	T
Cyphocharax punctatus	F							T	
Cyphocharax spilurus	F	S	S	X	X	S	S	S	
Steindachnerina varii	F					S		T	
FAMILY: PROCHILODONTIDAE									
Prochilodus rubrotaeniatus	F	S	X	X	X	S	S	X	
Semaprochilodus varii	F							T	
FAMILY: ANOSTOMIDAE									
Anostomus anostomus	F	S		S	X	S	X	X?	
Anostomus brevior	F						X	X	
Anostomus ternetzi	F							X	
Hypomasticus despaxi	F							T	
Hypomasticus megalepis	F	S		S	x			X	
Leporinus apollo	F	S	X	T		S			
Leporinus arcus	F	S							
Leporinus fasciatus	F	X	X	S	X	X	X	X	T
Leporinus friderici	F	S	X	S	X	X	X	S	T
Leporinus gossei	F							T	
Leporinus granti	F	S						S	
Leporinus lebaili	F							T	
Leporinus maculatus	F	S	S	S	S	S	S	S	
Leporinus nijsseni	F	S	S	S	S	T		X	
Petulanos plicatus	F	S							
Petulanos spiloclistron	F		T						
Schizodon fasciatus	F	S	X		X	X	X	X	
FAMILY: CHILODONTIDAE									
Caenotropus labyrinthicus	F	S				X			
Caenotropus maculosus	F	S				S		S	
Chilodus punctatus	F		S				S		
Chilodus zunevei	F					S	X	T	
FAMILY: CRENUCHIDAE									
Characidium pellucidum	F							X	
Characidium zebra	F	X	X	X	X	X	X	X	
Crenuchus spilurus	F	X		X	X	X	S	X	
Melanocharacidium blennioides	F	X			X		X	X	
Melanocharacidium dispilomma	F	X		X		X	X	X	
Microcharacidium eleotrioides	F	X		X	X	S	X	X	
FAMILY: HEMIODONTIDAE									
Argonectes longiceps	F							X	

Table 2. Continued.

Taxa	Habitat	COR	NIC	COP	SAR	SUR	COM	MAR	TLS
Bivibranchia bimaculata	F	T				S		S	
Bivibranchia simulata	F		S	S		S			
Hemiodus argenteus	F	S				S			
Hemiodus huraulti	F							T	
Hemiodus quadrimaculatus	F	X	S	S					
Hemiodus unimaculatus	F	X	X	X	X	X		X	T
Family: Gasteropelecidae									
Carnegiella strigata	F	X	X		X	S	S	S	
Gasteropelecus sternicla	F	X	X	X	X	S	S	S	T
Family: Alestidae									
Chalceus macrolepidotus	F	S	X	X	X	X		X	
Family: Characidae									
Genera Incerta Sedis									
Aphyocharacidium melandetum	F			X				X	
Astyanax bimaculatus	F	X	X	X	X	X	S	X	T
Astyanax validus	F							X	
Bryconops affinis	F	S	X	S	X	X		S	
Bryconops caudomaculatus	F	X		S		X		X	
Bryconops melanurus	F	S	X	S	X	X	X	X	T
Bryconops sp. 'redfins'	F					X		X	
Ctenobrycon spilurus	F	X	X		X	X			T
Hemigrammus bellottii	F			X?		S	S	S	
Hemigrammus boesemani	F		X	X	X	X	S	S	
Hemigrammus guyanensis	F		X?	X		X		X	
Hemigrammus lunatus	F	S				X?			
Hemigrammus ocellifer	F	S				X	S	S	
Hemigrammus aff. *ocellifer*	F					X		X	
Hemigrammus orthus	F	X							
Hemigrammus rodwayi	F					X		S	
Hemigrammus unilineatus	F	X	X		X	S	X	S	
Hyphessobrycon borealis	F							S	
Hyphessobrycon copelandi	F			X					
Hyphessobrycon georgettae	F	T							
Hyphessobrycon cf. *minimus*	F		X	X					
Hyphessobrycon minor	F					X?	X?	X?	
Hyphessobrycon rosaceus	F	S	X	X		S			
Hyphessobrycon roseus	F	X				S		T	
Hyphessobrycon simulatus	F				X	S	X	X	
Hyphessobrycon sp. 'blackstripe'	F							X	
Hyphessobrycon sp. 'redline' *?	F					S			
Jupiaba abramoides	F	X	X	X	X	S	S	S	
Jupiaba keithi	F	X	X			S		T	
Jupiaba maroniensis	F							T	
Jupiaba meunieri	F	X		X		S		X	

Table 2. Continued.

Taxa	Habitat	COR	NIC	COP	SAR	SUR	COM	MAR	TLS
Jupiaba ocellata	F						X?		
Jupiaba pinnata	F	S	X	X	X	X		X	
Jupiaba polylepis	F	S	X	X	S	X	X		
Moenkhausia chrysargyrea	F	X		X		X		X	
Moenkhausia collettii	F	S	X	X	X	X		X	
Moenkhausia georgiae	F	S	S	X		X		S	
Moenkhausia grandisquamis	F	X		X	X	X	X	S	T
Moenkhausia hemigrammoides	F	S	X	X	X	X	T	S	
Moenkhausia inrai	F							S	
Moenkhausia intermedia	F					X?		X	
Moenkhausia lepidura	F	S	X	X				X?	
Moenkhausia moisae	F	X?	X?					T	
Moenkhausia oligolepis	F	X	X	X	X	X	X	X	
Moenkhausia surinamensis	F	X	X	X	T	S		X?	
Paracheirodon axelrodi *	F					X			
Pristella maxillaris	F	X		X	X	X	S	S	
Thayeria ifati	F							T	
SUBFAMILY: IGUANODECTINAE									
Iguanodectes aff. *purusii*	F							X	
Piabucus dentatus	F	X			X	X	X	X	
SUBFAMILY: BRYCONINAE									
Brycon falcatus	F	X	X	X	X	X		X	T
Brycon pesu	F							X	
Triportheus brachipomus	F	S	X	X	X	X	X	S	
SUBFAMILY: SERRASALMINAE									
Acnodon oligacanthus	F					S		S	T
Metynnis altidorsalis	F					T	S	X	
Myleus setiger	F	S	S	S	S	S			
Myloplus planquettei	F							T	
Myloplus rhomboidalis	F	S	S	S	S	S		S	
Myloplus rubripinnis	F	S		X	S	S	X	S	
Myloplus ternetzi	F	S			S	S		S	
Myloplus aff. *ternetzi*	F			S					
Pristobrycon eigenmanni	F		X	S	S	X		S	
Pristobrycon striolatus	F							S	
Pygopristis denticulata	F				X	S	X	S	
Serrasalmus rhombeus	F	S	X	X	S	T	X	S	
Tometes lebaili	F						S	T	
SUBFAMILY: APHYOCHARACINAE									
Aphyocharax erythrurus	F	S							
SUBFAMILY: CHARACINAE									
Charax gibbosus	F	X	X		X	S		X?	T
Charax aff. *pauciradiatus*	F			X?	X	X?	X	X	
Cynopotamus essequibensis	F	X			X	S		S	

Table 2. Continued.

Taxa	Habitat	COR	NIC	COP	SAR	SUR	COM	MAR	TLS
Phenacogaster carteri	F	X							
Phenacogaster aff. *microstictus*	F	X		X		X?	X		
Phenacogaster wayana	F	S		X	X	X		S	
Roeboexodon guyanensis	F	S				S		T	
Roeboides affinis	F	X	S		X	X			
SUBFAMILY: STETHAPRIONINAE									
Brachychalcinus orbicularis	F	S		X		S			
Poptella brevispina	F	S		X	X	S	S	S	
Poptella longipinnis	F	S	T		X?	S		X?	
SUBFAMILY: TETRAGONOPTERINAE									
Tetragonopterus chalceus	F	X		X	X	S	X	X	
Tetragonopterus rarus	F	S						T	
SUBFAMILY: STEVARDIINAE									
Bryconamericus guyanensis	F					X?		X	
Bryconamericus heteresthes	F							X	
Bryconamericus aff. *hyphesson*	F	X						X	
Creagrutus melanzonus	F	X						X	
Hemibrycon surinamensis	F			X	T	X		S	
SUBFAMILY: CHEIRODONTINAE									
Odontostilbe gracilis	F	S						S	
FAMILY: ACESTRORHYNCHIDAE									
Acestrorhynchus falcatus	F	X	X	X	X	S	X	X	T
Acestrorhynchus microlepis	F	X	X	X	X	X	X	X	
FAMILY: CYNODONTIDAE									
Cynodon gibbus	F	X							
Cynodon meionactis	F					X		T	
FAMILY: ERYTHRINIDAE									
Erythrinus erythrinus	F	X	X	X	X	X	S	X	T
Hoplerythrinus unitaeniatus	F	X	X	X	X	S	X	X	
Hoplias aimara	F	X	X	X	X	X	X	X	
Hoplias curupira	F	X	S	S	S	S			
Hoplias malabaricus	F	X		X		X?		X	T
FAMILY: LEBIASINIDAE									
Copella arnoldi	F	X			X	S	S	S	
Nannostomus beckfordi	F	X			X	S	S	X	
Nannostomus bifasciatus	F					T	X	S	
Nannostomus harrisoni *	F					X			
Nannostomus marginatus	F	X				X	S	X	
Pyrrhulina filamentosa	F	S	S	S	S	S	S	S	T
Pyrrhulina stoli	F	X		X		S	S	T	
ORDER: SILURIFORMES									
FAMILY: CETOPSIDAE									
Cetopsidium minutum	F	X							
Cetopsidium orientale	F	S		X?	T	X?		X	

Table 2. Continued.

Taxa	Habitat	COR	NIC	COP	SAR	SUR	COM	MAR	TLS
Cetopsis sp.	F						X		
Helogenes marmoratus	F	S	X	X	X	S	X	S	
FAMILY: ASPREDINIDAE									
Aspredinichthys filamentosus	F(B)					S		S	
Aspredinichthys tibicen	F(B)			S	S	T		S	
Aspredo aspredo	F(B)	X	S	S	S	S		S	
Bunocephalus aloikae	F							T	
Bunocephalus amaurus	F	S	S	S	S	S	S	X?	
Bunocephalus coracoideus	F			S					
Bunocephalus verrucosus	F	S				S			
Platystacus cotylephorus	F(B)	X		S	S	S		S	
FAMILY: TRICHOMYCTERIDAE									
Ituglanis amazonicus	F		X	X	X	X		X	
Ituglanis gracilior	F	X							
Ituglanis sp. (Brownsberg)	F				X				
Ochmacanthus aff. *flabelliferus*	F	X	X	X		X	X	X	
Ochmacanthus reinhardtii	F							X	
Trichomycterus aff. *conradi*	F	X?						X	
Trichomycterus guianensis	F	X?							
FAMILY: CALLICHTHYIDAE									
Callichthys callichthys	F	X	X	X	X	X	X	X	T
Corydoras aeneus	F			S	S	S	S	X	
Corydoras baderi	F	S						S	
Corydoras bicolor	F	T				S			
Corydoras boesemani	F	S				T			
Corydoras breei	F	T	X						
Corydoras aff. *breei*	F							S	
Corydoras brevirostris	F	S			S				
Corydoras coppenamensis	F			T					
Corydoras filamentosus	F	T							
Corydoras geoffroy	F					T		S	
Corydoras guianensis	F	X	T	S	S			S	
Corydoras heteromorphus	F		S	T				X?	
Corydoras melanistius	F	S		S	X	S			
Corydoras nanus	F					T		S	
Corydoras oxyrhynchus	F				T	S	S		
Corydoras punctatus	F					T	S		
Corydoras sanchesi	F	S			T				
Corydoras saramaccensis	F				T				
Corydoras sipaliwini	F	T						X?	
Corydoras surinamensis	F	S		T					
Hoplosternum littorale	F	S	X		X	X	X	S	
Megalechis thoracata	F	S	S	S	S	S	S	S	

Table 2. Continued.

Taxa	Habitat	COR	NIC	COP	SAR	SUR	COM	MAR	TLS
FAMILY: LORICARIIDAE									
SUBFAMILY: HYPOPTOPOMATINAE									
Gen.nov. aff. *Parotocinclus* sp.	F							S	
Hypoptopoma guianense	F		T						
Otocinclus mariae	F	S				S		S	
Parotocinclus britskii	F		X	T		S			
SUBFAMILY: LORICARIINAE									
Ctenoloricaria platystoma	F	S		X	S	S		S	T
Farlowella reticulata	F							T	
Farlowella rugosa	F	X?	S					T	
Harttia guianensis	F							S	
Harttia fluminensis	F			T					
Harttia surinamensis	F					T			
Harttiella crassicauda	F							T	
Loricaria cataphracta	F(B)	X?	X?			S	S	T	
Loricaria nickeriensis	F	S	T						
Loricariichthys maculatus	F	S			S	T	S		
Metaloricaria nijsseni	F	T	S	S	S	S			
Metaloricaria paucidens	F							T	
Rineloricaria fallax	F	S							
Rineloricaria aff. *stewarti*	F					S	S	S	
Rineloricaria stewarti	F	S	S	S	S				
Rineloricaria sp.1	F	S	S						
Rineloricaria sp.2	F							S	
SUBFAMILY: HYPOSTOMINAE									
Ancistrus aff. *hoplogenys*	F						S	S	
Ancistrus gr. *leucostictus*	F	S	S					S	
Ancistrus sp. 'reticulate'	F				S		X		
Ancistrus temminckii	F			S	S	T	S	S	
Guyanancistrus brevispinis	F	S	T	S	X	S		S	
Guyanancistrus sp. (Nassau Mountains)	F							S	
? *Hemiancistrus macrops*	F								T
Hemiancistrus medians	F							T	
? *Hemiancistrus megacephalus*	F								T
Hypostomus coppenamensis	F			T					
Hypostomus corantijni	F	T	S						
Hypostomus crassicauda	F	T							
Hypostomus gymnorhynchus	F				S	S	S	S	
? *Hypostomus macrophthalmus*	F	T							
Hypostomus micromaculatus	F					T			
Hypostomus paucimaculatus	F					T			
Hypostomus plecostomus	F	S			S	T	S	S	
? *Hypostomus pseudohemiurus*	F	T							

Table 2. Continued.

Taxa	Habitat	COR	NIC	COP	SAR	SUR	COM	MAR	TLS
Hypostomus saramaccensis	F				T				
Hypostomus taphorni	F	S	X						
Hypostomus watwata	F(B)			S		S	X	S	
Lithoxus gr. *bovallii*	F	S	S						
Lithoxus pallidimaculatus	F				S	T	S	S	
Lithoxus planquettei	F							X	
Lithoxus stocki	F							T	
Lithoxus surinamensis	F					T	S	S	
Lithoxus sp.	F		S						
Panaqolus koko	F							T	
Panaqolus sp.	F						S		
Peckoltia otali	F							T	
Peckoltia sp.	F						S		
Pseudacanthicus fordii	F								T
Pseudacanthicus serratus	F					T		S	
Pseudacanthicus sp.	F	S							
Pseudancistrus barbatus	F							S	
Pseudancistrus corantijniensis	F	T							
Pseudancistrus depressus	F			S		S			T
Pseudancistrus kwinti	F			T					
? *Squaliforma tenuis*	F					T?			
FAMILY: PSEUDOPIMELODIDAE									
Batrochoglanis raninus	F					S		S	
Batrochoglanis villosus	F	S							
Cephalosilurus nigricaudus	F	T				S		S	
Microglanis poecilus	F	S	S	S	X	S	X	X	
Microglanis secundus	F	T	S	S	X	S	X	S	
Pseudopimelodus bufonius	F	S	S			S		S	
FAMILY: HEPTAPTERIDAE									
Brachyrhamdia heteropleura	F	S							
Chasmocranus brevior	F							S	
Chasmocranus longior	F	S	S	S	S	S	X	S	
Chasmocranus surinamensis	F					T			
Heptapterus bleekeri	F					S		T	
Heptapterus tapanahoniensis	F							T	
Imparfinis aff. *stictonotus*	F	X							
Imparfinis hasemani	F							S	
Imparfinis pijpersi	F	T	S			S		S	
Mastiglanis cf. *asopos*	F							S	
Phenacorhamdia tenuis	F							T	
Pimelodella cristata	F	S	S	X	S	S	X	S	
Pimelodella geryi	F	S				S		T	
Pimelodella leptosoma	F							X	
Pimelodella macturki	F	S	S					X	

Table 2. Continued.

Taxa	Habitat	COR	NIC	COP	SAR	SUR	COM	MAR	TLS
Pimelodella megalops	F							S	
Pimelodella procera	F							T	
Rhamdia foina	F							S	
Rhamdia quelen	F	S	S	S	X	S	S	S	
FAMILY: PIMELODIDAE									
Brachyplatystoma filamentosum	F(B)	X		S		S			
Brachyplatystoma rousseauxii	F(B)	X				S			
Brachyplatystoma vaillantii	F(B)	X		S		S			
Hemisorubim platyrhynchos	F	X	X			S		X	
Hypophthalmus marginatus	F(B)	X				S		X	T
Phractocephalus hemioliopterus	F	S							
Pimelabditus moli	F							T	
Pimelodus albofasciatus	F	T	X			X?			
Pimelodus blochii	F(B)	X		S	S	S	X	S	T
Pimelodus ornatus	F	S	X		X	S	X	X	T
Pseudoplatystoma fasciatum	F	S		S	X	S		S	T
Pseudoplatystoma tigrinum	F							X	
FAMILY: ARIIDAE									
Amphiarius phrygiatus	BM							X	T
Amphiarius rugispinis	BM	X				X		X	
Aspistor quadriscutis	FBM							X	T
Bagre bagre	BM	X				X		X	
Bagre marinus	BM	X							
Cathorops arenatus	BM					X			T
Cathorops spixii	BM	X			X	X		X	
Notarius grandicassis	BM			S		X		X	
Sciades couma	FBM	X						X	
Sciades herzbergii	BM	X	X	X		X		X	T
Sciades parkeri	B	X	X			X		X	
Sciades passany	BM	X	X	X				X	
Sciades proops	FB					X		X	
FAMILY: DORADIDAE									
Acanthodoras cataphractus	F	X		S		S	X	X	
Amblydoras affinis	F	X	X						
Doras carinatus	F	X	X	S	X	X		T	
Doras micropoeus	F	X						S	
Platydoras costatus	F	S			S	S	X		
Platydoras sp. 'shallow scutes'	F							S	
Pterodoras aff. *granulosus*	F	X	X						
FAMILY: AUCHENIPTERIDAE									
Ageneiosus inermis	F	X		X	X	T	X	S	
Ageneiosus marmoratus	F	S	X						
Ageneiosus ucayalensis	F	S							
Auchenipterus dentatus	F	S			X	T	X	S	

Table 2. Continued.

Taxa	Habitat	COR	NIC	COP	SAR	SUR	COM	MAR	TLS
Auchenipterus nuchalis	F	X				X?		S	
Centromochlus concolor	F			T					
Centromochlus punctatus	F			X		T		S	
Glanidium leopardum	F							T	
Pseudauchenipterus nodosus	F(B)	X	X		X	S	X	X	
Tatia brunnea	F					T		S	
Tatia gyrina	F	S	S		S	S	S	S	
Tatia intermedia	F	S	S			S			
Trachelyopterus galeatus	F	S	S		X	S	X	X	
ORDER: GYMNOTIFORMES									
FAMILY: GYMNOTIDAE									
Electrophorus electricus	F	X	X	S	X	X		X	T
Gymnotus anguillaris	F					T		S	
Gymnotus carapo	F	X	X	X	X	S	S	X	T
Gymnotus coropinae	F	S	X	X	X	T	S	X	
FAMILY: STERNOPYGIDAE									
Eigenmannia sp. 1	F	X		X			S		
Eigenmannia sp. 2	F	X		X			S	X	
Japigny kirschbaum	F	X						X	
Rhabdolichops jegui	F						X	T	
Sternopygus macrurus	F	X	X	X	X	S	S	X	
FAMILY: RHAMPHICHTHYIDAE									
Gymnorhamphichthys rondoni	F	S	S	S		S			
Rhamphichthys rostratus	F	X			X			X	
FAMILY: HYPOPOMIDAE									
Brachyhypopomus beebei	F	S		S	X	S	S	X	
Brachyhypopomus brevirostris	F	X				S	S	X	
Brachyhypopomus sp. 1	F					S	S		
Brachyhypopomus sp. 2	F						S		
Brachyhypopomus pinnicaudatus	F						X		
Hypopomus artedi	F	X	X	S	X	S	S	X	
Hypopygus lepturus	F	S	S	S	X	S	S	T	
FAMILY: APTERONOTIDAE									
Apteronotus albifrons	F	X	X		X	X	X	X	T
Porotergus gymnotus	F					X?		X	
Sternarchorhynchus galibi	F							T	
ORDER: BATRACHOIDIFORMES									
FAMILY: BATRACHOIDIDAE									
Batrachoides surinamensis	BM	X		S	X	X		X	T
ORDER: MUGILIFORMES									
FAMILY: MUGILIDAE									
Mugil cephalus	FBM	X						X	
Mugil incilis	BM			X		X		X	
Mugil liza	FBM	X						X	

Table 2. Continued.

Taxa	Habitat	COR	NIC	COP	SAR	SUR	COM	MAR	TLS
ORDER: CYPRINODONTIFORMES									
FAMILY: RIVULIDAE									
Kryptolebias marmoratus	F(B)					X?		X	
Kryptolebias sepia	F							T	
Laimosemion agilae	F	S	S	S	S	T	X	S	
Laimosemion breviceps	F						X	X	
Laimosemion frenatus	F							S	
Laimosemion aff. *geayi*	F					X		S	
Rivulus amphoreus	F		S	T				S	
Rivulus gaucheri	F							T	
Rivulus aff. *holmiae*	F	X				S		S	
Rivulus igneus	F			X?		S		X	
Rivulus aff. *lanceolatus*	F	S				S			
Rivulus cf. *lanceolatus*	F	S							
Rivulus lungi	F(B)			X?				X	
Rivulus stagnatus	F	S	S	S	S	S			
FAMILY: POECILIIDAE									
Micropoecilia bifurca	F				S	S	X		
Micropoecilia parae	F(B)		X			S	S	X	
Micropoecilia picta	F(B)	X			X	S	S	X	
Poecilia reticulata	F(B)		X			S			
Poecilia vivipara	F(B)				X	S	S	X	T
Tomeurus gracilis	F	X			X	S	X		
FAMILY: ANABLEPIDAE									
Anableps anableps	F(B)	X				S		X	
Anableps microlepis	F(B)	X		X		X			
ORDER: BELONIFORMES									
FAMILY: BELONIDAE									
Potamorrhaphis guianensis	F	X	S	X	S	S	X	S	
Pseudotylosurus microps	F					S		X	T
Strongylura marina	FMB					X			
FAMILY: HEMIRAMPHIDAE									
Hyporhamphus roberti roberti	BM	X				X		X	
ORDER: GASTEROSTEIFORMES									
FAMILY: SYNGNATHIDAE									
Pseudophallus aff. *brasiliensis*	F	S				S			
ORDER: SYNBRANCHIFORMES									
FAMILY: SYNBRANCHIDAE									
Synbranchus marmoratus	F	X	X	X	X	X	X	X	T
ORDER: PERCIFORMES									
FAMILY: CENTROPOMIDAE									
Centropomus ensiferus	FBM	X				X		X	
Centropomus parallelus	FBM			X		X		X	
Centropomus undecimalis	FBM	X		X		X		X	

Table 2. Continued.

Taxa	Habitat	COR	NIC	COP	SAR	SUR	COM	MAR	TLS
Family: Serranidae									
Epinephelus itajara	BM					X	X	X	
Family: Carangidae									
Caranx hippos	FBM					X		X	
Caranx latus	FBM							X	
Oligoplites saliens	BM					X		X	
Selene vomer	BM					X	X		
Trachinotus cayennensis	BM					X		X	
Family: Lutjanidae									
Lutjanus jocu	BM						X		
Lutjanus synagris	BM						X		
Family: Lobotidae									
Lobotes surinamensis	BM					X			
Family: Gerreidae									
Diapterus rhombeus	BM						X		
Family: Haemulidae									
Genyatremus luteus	BM	X	X	X		X			
Family: Sciaenidae									
Bairdiella ronchus	BM					X	X		
Cynoscion acoupa	FBM	X	X			X		X	
Cynoscion jamaicensis	BM			X	X				
Cynoscion steindachneri	FBM					X		X	
Cynoscion virescens	BM					X		X	T
Lonchurus elegans	BM	T							
Lonchurus lanceolatus	BM	X				X			T
Macrodon ancylodon	BM	X				X		X	T
Micropogonias furnieri	BM					X			
Nebris microps	BM					X		X	T
Pachypops fourcroi	F	S	X	X		X	X	X	T
Pachypops trifilis	F							X?	
Pachyurus schomburgkii	F	X				X		X	
Plagioscion auratus	F	X				S		X	
Plagioscion squamosissimus	F	X			X	S	X	X	
Stellifer microps	BM					X			
Stellifer rastrifer	BM		X			X			
Stellifer stellifer	BM		X			X			T
Family: Polycentridae									
Polycentrus schomburgkii	F	X	X	X	X	X	X	X	T
Family: Cichlidae									
Aequidens paloemeuensis	F	,						T	
Aequidens tetramerus	F	S	S	S	X?	S		S	
Apistogramma ortmanni	F	S							
Apistogramma steindachneri	F	S	S	S	S	S	X?		
Chaetobranchus flavescens	F	S			S	S			

Table 2. Continued.

Taxa	Habitat	COR	NIC	COP	SAR	SUR	COM	MAR	TLS
Cichla monoculus	F	X	X	X					
Cichla ocellaris	F	S	S	X	S	S		X	
Cichlasoma bimaculatum	F	S	S	X	X	S	S	S	T
Cleithracara maronii	F	S	S	X		S	S	T	
Crenicichla albopunctata	F							T	
Crenicichla coppenamensis	F			T	S				
Crenicichla lugubris	F	S							
Crenicichla multispinosa	F					S		T	
Crenicichla nickeriensis	F	S	T						
Crenicichla saxatilis	F		S		X	T	S	S	
Crenicichla sipaliwini	F	T							
Geophagus brachybranchus	F	S	T						
Geophagus brokopondo	F					T			
Geophagus harreri	F							T	
Geophagus surinamensis	F			X	S	S		S	T
Guianacara oelemariensis	F							T	
Guianacara owroewefi	F	X		X	S	S	X	T	
Guianacara sphenozona	F	T							
Krobia guianensis	F	S	S	S	S	S	X		
Krobia itanyi	F						X	T	
Mesonauta guyanae *	F					X			
Nannacara anomala	F	S	S	X	S	S	X	S	
Oreochromis mossambicus *	F(B)	X	X			S	X		
Pterophyllum scalare	F	X			X	X			
Satanoperca leucosticta	F	X	S						
FAMILY: ELEOTRIDAE									
Dormitator lophocephalus	F					T			
Dormitator maculatus	F(B)					S			
Eleotris amblyopsis	F(B)	X				X			T
Eleotris pisonis	F(B)	X?				X	X	X	
Guavina guavina	FBM					X?			T
FAMILY: GOBIIDAE									
Awaous flavus	FB	X				X		X	T
Ctenogobius phenacus	BM	X				X			
Ctenogobius pseudofasciatus	FBM					X			
Ctenogobius thoropsis	FBM					T			
Evorthodus lyricus	FBM	X				X			
Gobioides broussonnetii	FBM	X				X			T?
Gobioides grahamae	FBM	X							
Gobionellus oceanicus	FBM	X	X			X			
FAMILY: EPHIPPIDAE									
Chaetodipterus faber	BM						X		
FAMILY: TRICHIURIDAE									
Trichiurus lepturus	BM					X	X		

Table 2. Continued.

Taxa	Habitat	COR	NIC	COP	SAR	SUR	COM	MAR	TLS
ORDER: PLEURONECTIFORMES									
FAMILY: PARALICHTHYIDAE									
Syacium gunteri	MB							X	
Syacium micrurum	MB					X			
Syacium papillosum	MB	X				X		X	
FAMILY: ACHIRIDAE									
Achirus achirus	FBM	X				X		X	T
Achirus declivis	BM					X	X		
Apionichthys dumerili	FBM	X				X			
Trinectes paulistanus	FBM			X	X	X			
FAMILY: CYNOGLOSSIDAE									
Symphurus plagusia	BM					X		X	
ORDER: TETRAODONTIFORMES									
FAMILY: TETRAODONTIDAE									
Colomesus psittacus	FBM	X	X	X	X	X	X	X	
Sphoeroides testudineus	BM	X		X		X	X	X	
TOTAL NUMBER OF SPECIES	487	278	142	164	146	303	143	319	
TOTAL NUMBER OF ENDEMIC SPECIES PER RIVER		12	3	6	3	7	3	28	

Characidae has the greatest number of freshwater species (86 species), followed by the Loricariidae (60 species), the Cichlidae (30 species), the Callichthyidae (23 species), the Heptapteridae (19 species) and the Anostomidae (17 species). None of the orders, families or even genera of Surinamese freshwater fishes is endemic (restricted in its distribution) to Suriname, but present data indicate that 25 fish species are known solely from locations within Suriname and 93 species (or 23.3% of 400 freshwater species) solely in Surinamese river systems including the Marowijne and Corantijn watersheds which in part extend into French Guiana and Guyana, respectively. Of 487 total species, 187 (38.4 %) were based on type series that originated within Suriname. Continued descriptions of new species from the inland waters of Suriname demonstrate that the present total most likely distinctly underestimates the species-level diversity of the fish fauna. In terms of species per square kilometer, the presently documented richness of the Surinamese freshwater fish fauna is comparable to, or even greater than, the richness of fish faunas of several other tropical South American countries (Mol *et al.*, 2012; also see Abell *et al.*, 2008).

The checklist of Surinamese freshwater fishes by Mol *et al.* (2012) includes 481 species of fishes from the fresh and brackish waters of Suriname and this number represents a 50% increase in documented total of species in these habitats over the last two decades (Ouboter & Mol, 1993). Here, 6 species are added to the list of Mol *et al.* (2012): *Bryconops* sp. 'redfins', *Hemigrammus* aff. *ocellifer*, *Hyphessobrycon* sp. 'blackstripe', *Cetopsis* sp., *Ageneiosis ucayalensis*, and *Laimosemion* aff. *geayi* (Table 2). A major factor in this increase in number of species are the many species that have been described during that interval (e.g. *Potamotrygon marinae, P. boesemani, Cyphocharax biocellatus, Leporinus apollo, Hyphessobrycon borealis, Myloplus planquettei, Tometes lebaili, Tetragonopterus rarus, Bryconamericus guyanensis, Cetopsidium orientale, Harttia guianensis, Harttia fluminensis, Panaqolus koko, Peckoltia otali, Pseudancistrus corantijniensis, P. kwinti, Pimelabditus moli, Japigny kirschbaum, Rhabdolichops jegui, Sternarchorhynchus galibi, Kryptolebias sepia* and *Rivulus gaucheri*), many of which originated in ichthyologically relatively poorly sampled regions of Suriname. Other of the Surinamese fishes known to be part of the ichthyofauna of the country were described from samples collected at various times, sometimes long ago, but which were unidentified or misidentified in museum collections (e.g. *Phenacogaster wayana, Hoplias curupira*). Yet other of the species contributing to this increased diversity represent new records within Suriname of species previously known to occur in Guyana and/or French Guiana. Future collecting efforts in poorly sampled regions and habitats and critical analysis of the many problematic genera will undoubtedly reveal that dozens of freshwater fish species in Surinamese rivers are as-of-yet unknown to science (e.g. Cardoso and Montoya-Burgos, 2009; also see Table 2). The great unknown is the degree to which future studies will add to the list of species that are presently known to occur in Suriname.

CHAPTER FOUR

ECOLOGY

Climate and Tropical Freshwater Ecosystems

Flowing (lotic) waters of streams and rivers are the most characteristic freshwater ecosystems of Suriname. Standing (lentic) waters are mainly represented by wetlands in the Coastal Plain, a single small natural lake (Nanni Lake, Corantijn River watershed) in northwestern Suriname, and the hydroelectric Brokopondo Reservoir (Ouboter, 1993). Lewis (2008) shows that generalizations about tropical streams and rivers are most easily approached through the influences of climate. Suriname is located within the migration area of the Intertropical Convergence Zone, where the Hadley circulation favors heavy seasonal rainfall. The ITCZ low pressure zone or 'meteorological equator' is not aligned with the geographical equator. Its position shows seasonal variation loosely correlated with the path of the sun (often with a notable lag of a few weeks) and differs from one continent to the other. Movement of the ITCZ is the main cause of annual rainfall cycles in Suriname with a long rainy season from late April to middle of August, a long dry season from middle of August to early December, a short rainy season from early December to early February, and a short dry season from early February to late April (with the short rainy and short dry seasons more outspoken in the Coastal Plain than in the Interior of the country) (Amatali, 1993). The long-term mean annual rainfall varies from 1,450 mm in the coastal district Coronie to 3,000 mm at Tafelberg Mountain in the center of the country (Amatali, 1993). Approximately 67% of the precipitation in Suriname originates from evapotranspiration of the rainforest while only about 33% is derived from the Atlantic Ocean (Amatali, 1993). Irregular variations in the annual rainfall are caused by fluctuations in the pressure gradient of the equatorial Pacific (the 'southern oscillation'), with extremes referred to as 'El Niño Southern Oscillation' (ENSO) conditions. El Niño events correspond to unusually dry conditions from December to February in northeastern South America, including Suriname (e.g. Mol *et al.*, 2000), while conditions in La Niña years are for the most part opposite and may include flooding in some locations. Hydrographic seasonality typifies Surinamese streams and rivers (e.g. Crul & Reyrink, 1980). In large rivers, the intra-annual variation in discharge and water level is largely predictable, but hydrographs of small streams often show numerous spikes associated with local thunderstorms and a sudden rise in water level of several meters is not unusual. For example, the Mindrineti River, a tributary of the Saramacca River, can rise 6 m overnight and overflow its banks in May (Fig. 4.11a), but then two months later cease to flow, being reduced to a series of interconnected pools (Fig. 4.33; pers. observations). In 'normal' years, Surinamese rivers typically show well-defined seasonality in depth and velocity of flow, water chemistry, and presumably metabolic rates, based primarily on hydrology alone rather than hydrology in conjunction with temperature (Crul & Reyrink, 1980), as would be more typical of temperate latitudes. Seasonality in discharge results in seasonal variation in concentrations of dissolved and suspended solids, organic matter, and nutrients, often in an order of magnitude. The fishes respond to the changing conditions in the rivers by longitudinal migrations up and down the river and lateral migrations moving between the main channel and the inundated floodplain forest and small tributaries. Most Neotropical freshwater fishes reproduce at the beginning of (or during) the wet season (Lowe-McConnell, 1964, 1987; Mol, 1996a), thus ensuring favorable environmental conditions (space, food, reduced predation) for their offspring. The high-water season is also the main feeding period for most fish species, with food extremely limited in the dry season (Lowe-McConnell, 1964; Goulding, 1980), except for piscivorous fishes which may find prey fishes crowded in drying pools in the main channel of streams (Winemiller, 1989). In the dry season fishes often have empty stomachs (Lowe-McConnell, 1964; Arrington *et al.*, 2002) and they then live off fat stores in various parts of the body (e.g. along the intestines) that they have developed in the rainy season (Lowe-McConnell, 1964; Junk, 1985).

As elsewhere in the humid tropics, water temperatures in Suriname are high at low elevation (range about 25-29 °C, mean 27 °C at 0 m-asl) and slightly lower at high elevation (18-24 °C, mean 21 °C at 1000 m-asl). The surface water temperature of the Marowijne River showed a gradual rise from 24 °C in its headwaters (Litani) to 31 °C near the river mouth, and a diurnal variation of only 2 °C that was very small when compared to the daily variation in air temperature of 9 °C (Geijskes, 1942). Small source creeks (at 300 m-asl), protected from the sun by a closed forest canopy, showed a relatively constant low temperature of 22 °C corresponding to the temperature of the soil through which they flowed (Geijskes, 1942). Because mean water temperatures are high, dissolved oxygen levels at saturation are substantially lower for tropical waters (7.3-8.1 mg L^{-1}) than for temperate waters (7.5-14 mg L^{-1}) (Lewis, 2008). Tropical waters as a whole have a low oxygen reserve and a high potential oxygen demand for a given amount of organic loading, and thus they are vulnerable to organic pollution.

White-Water, Black-Water, and Clear-Water Rivers

The Sioli (1950) classification of Amazonian rivers is mainly based on optical characteristics and distinguishes between white-water rivers of high turbidity, black-water rivers with high humic and fulvic acid content, and clear-water rivers that are neither turbid with suspended sediments nor colored by humic compounds (Fig. 4.1).

In tropical rivers, transport and concentration of total dissolved solids is loosely related to that of suspended solids, with an overall ratio of suspended to dissolved solids of about 5:1 (Lewis, 2008). Inorganic dissolved solids (mostly salts; measured as electric conductivity) are much less significant ecologically than sus-

Fig. 4.1. The meeting of a clear- and a black-water stream (© W. Kolvoort).

pended solids, and they can be generally equated with availability of nutrients such as phosphorus and silicon. In tropical streams and rivers, dissolved phosphorus and nitrogen are usually present in quantities sufficient to support moderate to high biomass of autotrophs even under pristine conditions (Lewis, 1995). The ecological significance of suspended solids is related to the interception of light (i.e. the penetration depth for 1% photosynthetically available radiation (PAR)): photosynthesis is virtually impossible over most of a stream channel at concentrations exceeding 100 mg suspended solids per liter. Topography (slope), lithology (rock type) and (anthropogenic) disturbance are the dominant controls on suspended and dissolved solids in tropical streams. Concentrations of dissolved solids range from a few mg L^{-1} in wet areas of low gradient and resistant lithology (e.g. clear-water streams draining the weathered Precambrian Guiana Shield in Suriname) to 1000 mg L^{-1} on high gradients with readily eroded lithologies. White-water streams originating in the geologically young Andes Mountains (i.e. in 12% of the Amazon drainage area) produce 80-97% of both dissolved and suspended solids in the Amazon Basin (Lewis, 2008). Stallard (1985) distinguished between weathering-limited denudation associated with steep relief and thin soils and transport-limited denudation associated with low relief and thick soils, and identified the shields as transport-limited regions that approach the limits of ion-depletion from crystalline parent material. Undisturbed rivers draining the Interior of Suriname usually have clear water (some suspended sediment may be present for a few days after heavy rains) and only streams impacted by anthropogenic disturbance (often small-scale gold mining; Mol & Ouboter, 2004) show high turbidity throughout the year.

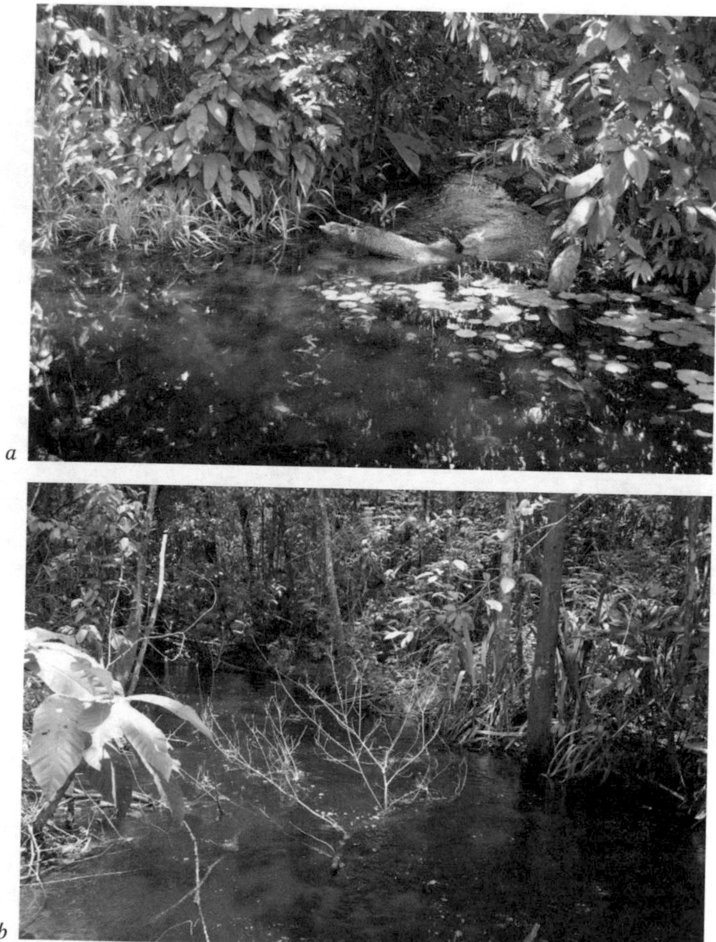

Fig. 4.2. (a) A small black-water creek in the Savanna Belt of Suriname. (b) Upper Coesewijne River.

Black waters owe their striking reddish-brown color (that appears black in deep water) (Fig. 4.2) both to the absence of inorganic turbidity and the impressive absorption of short wavelengths of light by humic acids within an otherwise clear water column; they typically have a low pH and are found in areas with low relief and podzol soils (Lewis, 1995). The best known black-water river is the very large Rio Negro (Goulding *et al.*, 1988), but the Surinamese black-water streams originating in the Savanna Belt are mostly small (e.g. Para River and tributary creeks, Coesewijne River, Marataka River). Most black-water streams in Amazonia and Suriname are associated with white sandy podzol soils consisting of several meters bleached quartz sand below the surface humus layer and followed by an accumulation zone that may be endurated or cemented by a thin ortstein or pan (Klinge, 1967; Goulding *et al.*, 1988). The endurated pan is related to lateral percolation of rainwater which carries away the organic matter migrating downward through the podzol profile and in this way the mobile Dissolved Organic Carbon

Fig. 4.3. Aquatic macrophytes in black-water streams. (a) Para River with luxuriant aquatic macrophyte vegetation, including *Marsilea polycarpa*, *Cabomba aquatica*, *Pistia stratiotes*, *Salvinia auriculata*, *Azolla caroliniana*, *Eichhornia crassipes*, *Utricularia* spp, and many others. (b, c). *Cabomba* stands in a small black-water creek (b and c © W. Kolvoort).

Fig. 4.4. Fishes from black-water streams. (a) Datrafisi *Crenicichla saxatilis*. (b) A small school of *Hyphessobrycon* sp. 'redline' in Coropina Creek (© W. Kolvoort).

passes into the creeks. The black-water streams of Suriname have a high humic acid concentration (3-3.5 mg L^{-1}), no measurable hardness, low pH (4.3-5.2), low electrolyte content (34-38 µS cm^{-1}), low dissolved oxygen concentration (1.0-2.8 mg L^{-1}) and clear water (Secchi transparency 80-90 cm) (Haripersad-Makhanlal & Ouboter, 1993). They often have abundant aquatic macrophyte vegetation (e.g. Fig. 4.3) that offers hiding places for the many small-sized fish species from larger predatory fishes (*Hoplias malabaricus*, *Hoplerythrinus unitaeniatus*, *Crenicichla saxatilis*). Due to the absorption of short wavelengths of light, the underwater environment of black-water streams has little color contrast and poor vision, comparable to what humans would experience in a (red) fog (Fig. 4.3b, Fig. 4.4a). The nutrient deficiency of the soils along the shores of the Rio Negro and the absence of some large food fishes (e.g. *Prochilodus*; Goulding *et al.*, 1988) makes this river known by Amerindians as the 'River of Hunger'. Surinamese black-water streams are also best known for their small-sized fishes that are popular in the aquarium hobby (Fig. 4.4b); among these are the exotic cardinal tetra (*Paracheirodon axelrodi*) and the freshwater angelfish (*Pterophyllum scalare*) that are both native to the Rio Negro. Other characteristic Surinamese black-water fishes include merkikwikwi *Acanthodoras cataphractus*, plarplari *Ageneiosus inermis*, noya *Trachelyopterus galeatus*, dyaki *Rhamdia quelen*, katrina kwikwi *Megalechis thoracata*, pencilfishes (*Nannostomus* spp), splashing tetras (*Copella arnoldi*, *Pyrrhulina filamentosa*), hatchet fishes (*Carnegiella strigata*, *Gasteropelecus sternicla*), *Curimatopsis crypticus*, makasriba *Curimata cyprinoides*, *Crenuchus spilurus*, *Pristella maxillaris*, *Hemigrammus ocellifer*, *Pygopristis denticulata*, *Metynnis altidorsalis*, *Hoplias malabaricus*, *Hoplerythrinus unitaeniatus*, several cichlids (e.g. *Apistogramma steindachneri*, *Chaetobranchus flavescens*, *Cleithracara maronii*, *Crenicichla saxatilis* (Fig. 4.4a), *Krobia guianensis*, and *Nannacara anomala*), *Polycentrus schomburgkii*, *Micropoecilia bifurca*, *Laimosemion agilae*, and several gymnotiform knifefishes (e.g. *Gymnotus carapo*, *G. coropinae*, *Brachyhypopomus* spp, *Hypopygus lepturus*). The fish fauna of a small black-water stream like Para River can include more than 60 fish species (Ouboter & Mol, 1993).

The River Continuum and Flood Pulse Concepts and Rapid Complexes

The river continuum concept (Vannote *et al.*, 1980) can help to understand the ecology of Surinamese river systems. The river is described as a continuum of mosaics of intergrading population aggregates from low-order headwaters to the high-order reaches in the river mouth, with the constituent populations adjusting to a continuous gradient of physical conditions (e.g. stream width, depth, velocity, temperature, and sediment load) along the length of a river. Headwater streams are influenced strongly by the riparian forest vegetation which reduces autotrophic production by shading and contributes large amounts of allochthonous detritus. The coarse particulate organic matter (e.g. fallen leaves) is broken down by 'shredder' organisms in a system that is largely heterotrophic (Production/Respiration < 1). The shredders utilize coarse particulate organic matter with a significant dependence on the associated microbial biomass and produce fine particulate organic matter. This is carried downstream, where it is filtered from transport or gathered from the sediments and then processed by 'collectors'. Like shredders, collectors depend on the microbial biomass associated with the particles (primarily on the surface) and products of microbial metabolism for their nutrition. As stream order increases and the stream widens, primary production increases (P/R > 1), and the shredders are replaced by grazers and scrapers, living alongside collectors. Scrapers are adapted primarily for shearing attached algae from surfaces. The dominance of scrapers follows shifts in primary production, being maximized in midsized rivers. In high-order reaches near the river mouth, primary production may often be limited by depth and turbidity, and such light attenuated systems would again be characterized by P/R < 1 and dominance of collectors. Downstream communities are thus fashioned to capitalize on upstream processing inefficiencies, and both upstream inefficiency (leakage) and downstream adjustments seem predictable.

The longitudinal river continuum concept was developed for small temperate rivers and restricted to permanent lotic habitats. Junk *et al.* (1989) pointed out the importance of lateral exchange between the river and its floodplain, i.e. the seasonal inundation of large floodplains (i.e. the flood pulse concept). The larger the floodplain, the greater the ratio of periodically lentic to lotic areas, resulting in adaptations of biota distinct from those in systems dominated by stable lotic or lentic habitats. The river is compared to a highway used by fishes to gain access to adult feeding areas, spawning grounds and nurseries in the adjoining floodplain forest or as a refuge during low-water periods. The fisheries in the main channel are thus mainly based on production derived from the floodplain habitats. The importance of river floodplains to tropical fish populations has been shown by Welcomme (1979) and Goulding (1980).

The rivers in Suriname do not have extensive floodplains (see below), but are characterized instead by the many large rapid complexes (Zonneveld, 1972; see below) which are mostly absent in temperate rivers. These rapid complexes, although relatively restricted in area compared to the reaches between the rapid complexes, can be important for the river system as a whole, not only because of their fish diversity (with many loricariid, serrasalmine, and other species restricted to the rapid habitat), but also in terms of primary production (e.g. stands of Podostemaceae) and secondary production (aquatic insects, snails), especially where extensive stands of aquatic macrophytes, floating meadows and large floodplain forests do not occur in the middle and upstream reaches of Surinamese rivers.

River Reaches in the Interior

In the seven main rivers of Suriname (Table 3) it is possible to distinguish an upper section, a lower freshwater section and the estuary with brackish water. The upper section of the river is located upstream of the first major rapid complex (Sur. sula) in the Interior of the country, i.e. the surface area of Suriname where the Precambrian Shield is not covered with Tertiary/Quaternary sediments. This section may be further divided in middle reaches characterized by large rapid-complexes and headwater reaches where the river is smaller, shallow and often meandering extensively. The upstream limit of the lower freshwater section of the river is formed by the first rapid complex, but the downstream limit is less clear because salt intrusion (e.g. the 300 mg Cl L^{-1} limit) and silt content both shift with the tides and, more important, with season (Amatali, 1993). In the Marowijne River, the 300 mg Cl L^{-1} limit is located at km 37 and km 59 in the high-water and low-water season, respectively, in the Coppename River at km 31 and km 83, and in the Corantijn at km 40 and km 82 (Table 3). The lower freshwater section corresponds loosely with the Coastal Plain and Savanna Belt (Fig. 1.1). The downstream limit of the estuary can be subjectively defined as the zero-point (km 0) or

Table 3. Hydrological characteristics of the seven main Surinamese rivers (data from Amatali, 1993). Since the construction of a dam at Afobaka (km 194) in 1964 the Suriname River is a regulated river and discharge and salt intrusion depend on the discharge at Afobaka. Distances along the length of a river are measured along the thalweg in kilometers from its mouth: river Km numbers begin at zero, designated as the river mouth at the 15 m depth contour at low-water spring tide, and increase further upstream.

Variable	Corantijn	Nickerie	Coppename	Saramacca	Suriname	Commewijne	Marowijne
Catchment area (km^2)	67,600	10,100	21,700	9,000	16,500	6,600	68,700
Mean discharge at outfall ($m^3 s^{-1}$)	1,572	178	500	225	426	120	1,785
Maximum limit of salt-intrusion (300 mg Cl L^{-1}) during low flow in the dry season	Km 82	Km 110	Km 83	Km 89	Km 90	Km 150	Km 59
Minimum limit of salt intrusion (300 mg Cl L^{-1}) during peak flow in the rainy season	Km 40	Km 28	Km 31	Km 37	Km 54	Km 55	Km 37
Location of the first major rapid complex	Km 235	Km 240	Km170	Km 285	Km 194	-	Km 115
Tidal range at outfall (m)	2	2	2	-	1.8	1.9	2
Tidal volume (1,000,000 m^3)	300	10	75	50	125	40	200
Maximum depth along the thalweg (m)	30 (km 112)	17 (km 73)	22 (km 82)	24 (km 145)	17 (km 95)	31 (km 71)	21 (km 47)
Silt content (g L^{-1}) upstream of the minimum limit of salt intrusion	0.02-4	0.03-45	0.02-1.3	0.003-1.3	0.001-1.5	0.07-20	0.01-0.08
Sediment discharge (1,000,000 ton $year^{-1}$)	1.2	0.1	0.25	0.13	0.25	0.06	1.3

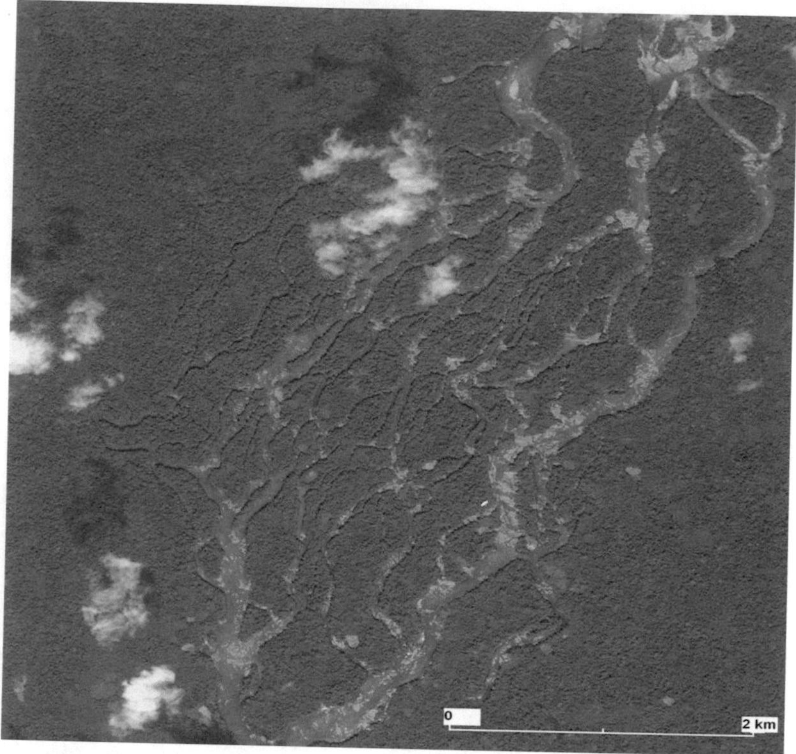

Fig. 4.5. Satellite image of a large rapid complex in the Middle Tapanahony River, Marowijne River system (N 4° 0' 45", W 54° 49' 11").

Fig. 4.6. The high-energy environment of rapids. Blanche Marie Falls, Upper Nickerie River.

river mouth, where the water depth is 15 m at low-water spring tide (Amatali, 1993). The upstream limit of the estuary, the 300 mg Cl L^{-1} limit of salt intrusion, shifts with the season.

The upper sections of Surinamese rivers are characterized by the many waterfalls and rapids (Sur. sula) that often occur in special patterns, in combination with other rapids and islands enclosed between branches of the river that has split itself in many anatomizing branches in areas called rapid or sula complexes by Zonneveld (1972) (Figs. 4.5, 4.6). Local people often use names for individual falls or rapid complexes (e.g. Marrenga & Ruleman, 2008). Up- and downstream of the rapid complex the river is relatively deep and flows in not more than one definitely limited bed of a certain width in nearly level reaches (giving the river a stair-like character). In the rapid complex, however, the total width may be more than three, four, sometimes ten times the proper river width and the branches may enclose islands measuring several kilometers in length. Aerial photographs often reveal directional correlations between rapids indicating the presence of resistant layers, dikes, or veins (Zonneveld, 1972). The rapid complex has some superficial resemblance with a braided river (Fig. 4.5), but in classical braided streams the river bed and the islands consist of unconsolidated sediments, with the braiding itself closely related to the presence of much sediment, the erodibility of the riverbanks, and the heterogeneity of the transported material (Leopold *et al.*, 1964), whereas in rapid complexes many islands consist of hard rock or are even hilly like the terrain on both sides of the river outside the rapid sector. The rapids show a relatively scarcity of erosional features as indicated, for example, by the extensive growth of Podostemaceae plants (Sur. kumalunyanyan) on rock ledges in the rapids, which would be destroyed if the water would carry many abrasives. The tropical rainforest effectively protects the soils against erosion, leaving the river with very little sediments for abrasive action. Although the Surinamese rivers do transport some sandy sediments (sandbanks and sandy beaches can be observed in many places), the poverty of abrasive material (sand) of the clear-water rivers, that do not even carry silt in suspension, is obvious. Sandbars do creep, but this is more like rolling a sandbag than sandpapering and there is no external friction (Marrenga & Ruleman, 2008). This explains why the rivers in the Interior of Suriname, with the exception of the headwater reaches, have no true meanders (and oxbow lakes), and do not form substantial flood plains (Zonneveld, 1972; Lewis *et al.*, 1995). Thus, the course of the rivers in the Interior is structurally controlled by the nature of the weathering rock surface and the stream channel tends to follow troughs that coincide with joints and faults in the rock; instead of smooth meanders the rivers have straight stretches ending in often irregular and angular bends. The headwater reaches of Surinamese rivers can have both extensive meandering in narrow, sediment-filled valleys and large areas of associated flooded forest (e.g. headwaters of the Koetari, Oelemarie and Paloemeu rivers.

The rapids are impressive high-energy ecosystems, but Marrenga and Ruleman (2008) point out that the big waves and hissing waters, suggestive of strong

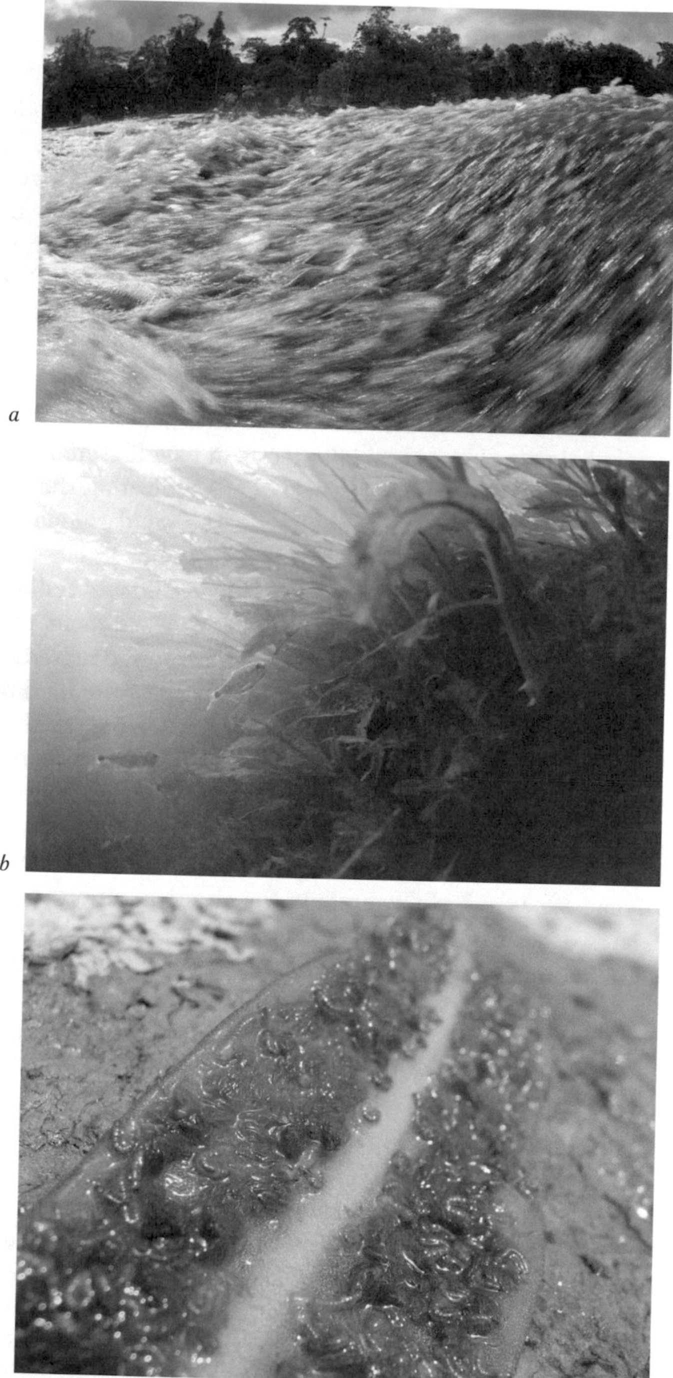

Fig. 4.7. (a) Strong currents in rapids of the Upper Suriname River. (b) In the same rapids, small tetras find refuge from the current behind boulders overgrown with Podostemaceae macrophytes. (c) Aquatic insects living in Podostemaceae stands are important food for rapid-dwelling fishes; here Diptera larvae on the underside of a Podostemaceae leaf (a and b © W. Kolvoort).

a

b

Fig. 4.8. Two endemic loricariid catfishes feeding on periphyton in Raleighvallen rapids, Coppename River. (a) Juvenile *Pseudancistrus kwinti*. (b) *Hypostomus coppenamensis* (© W. Kolvoort).

currents (Fig. 4.7a), can be misleading. Slow currents in shallow water look as impressive as swift currents in deep water. Typical currents in the Upper Suriname River were 0.6 m s^{-1} (or 2 km h^{-1}), while currents in the rapids were up to 2.8 m s^{-1} (10 km h^{-1}) (Marrenga & Ruleman, 2008). Still it takes a great deal of energy for fishes to maintain position in swift waters, and most inhabitants of rapids have special mechanisms for avoiding or withstanding the current. Many fishes find refuge in calmer backwaters, behind rocks (Fig. 4.7b), within Podostemaceae vegetation or on and beneath the streambed. The Podostemaceae stands may also have an important invertebrate fauna (Odinetz Collart *et al.*, 1996; Fig. 4.7c) on which many rapid-dwelling fishes feed (Horeau *et al.*, 1998). Loricariid catfish of the rapids (e.g. *Harttia* spp, *Lithoxus* spp, *Pseudancistrus* spp.) have a flattened body and their mouth modified into a ventral sucking disc by means of which

Fig. 4.9. *Characidium zebra* in Raleighvallen rapids, Coppename River (© W. Kolvoort).

Fig. 4.10. The electric eel *Electrophorus electricus*, hiding between rocks in Raleighvallen rapids, Coppename River (© W. Kolvoort).

they are able to attach themselves so tightly to the substrate that they sometimes cannot be pulled free without doing damage. Movement is by short 'hops' between which they attach themselves to the substrate so that the strong current cannot pull them free as they rest (they breathe by taking water into the gill chamber and expelling it only through the gill openings). With their rasping teeth they can scrape periphyton algae (Power, 1984) and detritus (Horeau *et al.*, 1998) from the substrate (Fig. 4.8). However, other fishes of the rapids have fusiform bodies (*Parodon guyanensis*, *Hemiodus quadrimaculatus*, and *Characidium zebra*; Fig. 4.9) or even deep, laterally compressed bodies, e.g. *Brachychalcinus orbicularis*, *Jupiaba* spp., and several vegetarian Serrasalminae (*Myleus* spp., *Myloplus* spp., *Acnodon oligacanthus*, and *Tometes lebaili*). The electric eel *Electrophorus electricus* (Fig. 4.10) and anyumara *Hoplias aimara* (Fig. 8.2) are two large pisci-

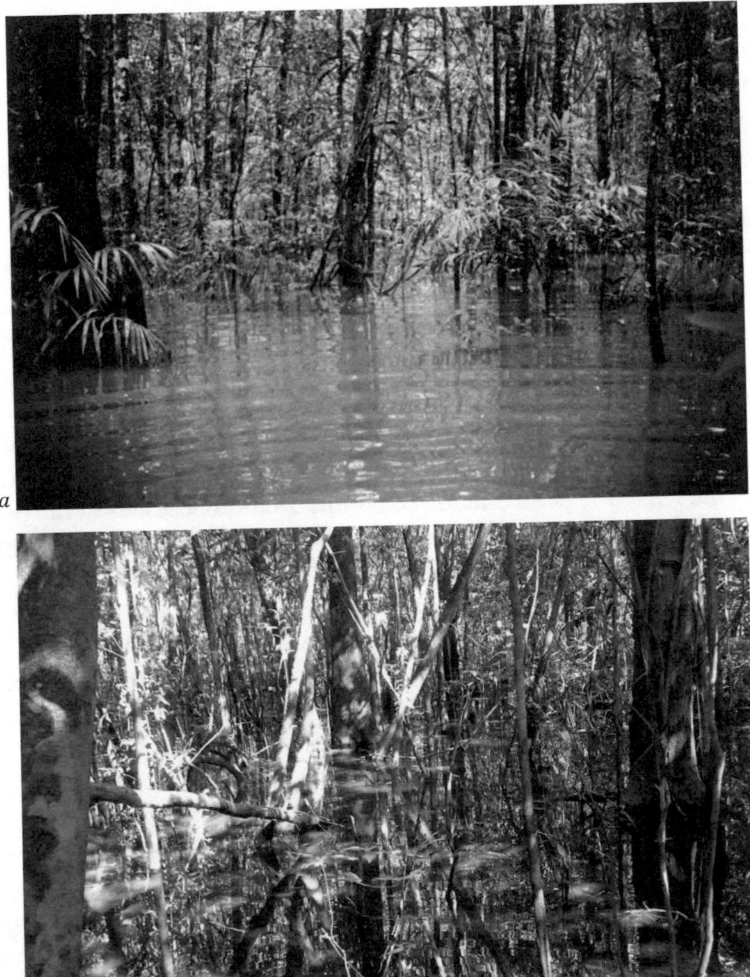

Fig. 4.11. The flooded forest. (a) Mindrineti River, Saramacca River system. (b) Upper Corantijn River (© P. Willink).

vores that are often found in rapids. Rapid complexes in Surinamese rivers are inhabited by many fish species that do not occur to any extent outside the rapid habitat. The rapids have an important primary (Podostemaceae stands, periphytic algae) and secondary (aquatic invertebrates, fishes) production. In the Amazon, large rapids such as the Teotônio cataract in the Rio Madeira are important sites of fisheries that exploit upstream migrating catfishes (Goulding, 1981).

In Suriname, the relatively deep and sluggish reaches between rapid complexes do not have extensive stands of aquatic macrophytes or large floodplains with forest that are inundated in the high-water season. In the high-water season, a strip of 500 m to a few km width of flooded forest was observed on the banks of the upper Nickerie, Coppename and Mindrineti rivers (pers. observations; Fig. 4.11). However, this relatively narrow band of floodplain forest is important for reproduction and feeding of riverine fishes (Junk *et al.*, 1989; Goulding, 1980; Vari, 1982). In the Amazon, approximately 50% of the carbon in adult commercial fishes is derived from the forest, while C4-macrophytes (grasses of floating meadows) are relatively unimportant as carbon source (Forsberg *et al.*, 1993). I observed a small group of *Leporinus nijsseni* in full breeding colors and with ripe gonads entering the flooded forest along the Nickerie River and at the same location I collected a large dyaki catfish (*Rhamdia quelen*) with a bulging stomach due to the presence of a large earthworm. Macrophyte rafts that originate from macrophyte stands that grow along the river banks play an important role in long-distance dispersal of Amazonian fishes (Schiesari *et al.*, 2003); in Suriname, such rafts of grasses or water hyacinth *Eichhornia crassipes* are most often observed in the lower reaches of rivers. The fish fauna of the reaches between rapid complexes includes many large species that are often important food fishes to the people of the Interior. In the Amazon, the diet of these large food fishes was studied by Goulding (1980). The fish communities of the deep channel habitat in Suriname are not well studied (Vari, 1982) and may hold surprises (e.g. a community of planktivorous electric knifefishes was recently discovered in the deep channel of the Orinoco River; Lundberg *et al.*, 1987), but include the large pimelodid catfishes lalaw *Brachyplatystoma filamentosum*, spigrikati *Pseudoplatystoma fasciatum* and *P. tigrinum*, mototyar *Phractocephalus hemioliopterus*, and *Hemisorubim platyrhynchos*. The detritivores kuri (kwi-)mata *Prochilodus rubrotaeniatus* and (in the Marowijne River) *Semaprochilodus varii* are known for their large-scale upriver spawning migrations during which they sound like an outboard engine (Lowe-McConnell, 1987). Interestingly, Araujo-Lima *et al.* (1986) showed that these detritivorous characoids have stable carbon-isotope signatures consistent with diets based on direct or indirect use of carbon from phytoplankton (and not from vascular plants). In Suriname, local people notice the migrating *Prochilodus* schools by the noice they make and then catch the kwimata in shallow water after they enter flooded areas (often the low banks of inner river bends) to spawn. Piranhas *Serrasalmus rhombeus*, *Pristobrycon eigenmanni* and (in the Marowijne River) *Pristobrycon striolatus* also live in the main channel of rivers, but they also move inshore to feed. Moroko *Brycon falcatus*, sardin *Triportheus brachipomus*, tukunari *Cichla ocellaris*, anyumara *Hoplias aimara*, freshwater stingrays or libaspari (*Potamotrygon* spp; mainly on sandy bottoms) and many smaller anostomids, characids and cichlids live close to the shore. The large gymnotiform knifefish *Rhamphichthys rostratus* and stonkubi (*Pachypops fourcroi*, *Pachyurus schomburgkii*) also seem mainly restricted to large rivers of the Interior. Even small-sized fishes can occur in special habitats of the main river channel, e.g. *Corydoras* species or the miniature loricariid catfish *Parotocinclus britskii* which was collected in large numbers on a sandy beach of the Upper Coppename River (Mol *et al.*, 2006). In the Amazon, sandy river beaches may have a specialized fish fauna (Ibarra & Stewart, 1989; Jepsen, 1997; Zuanon *et al.*, 2006; Pereira *et al.*, 2007).

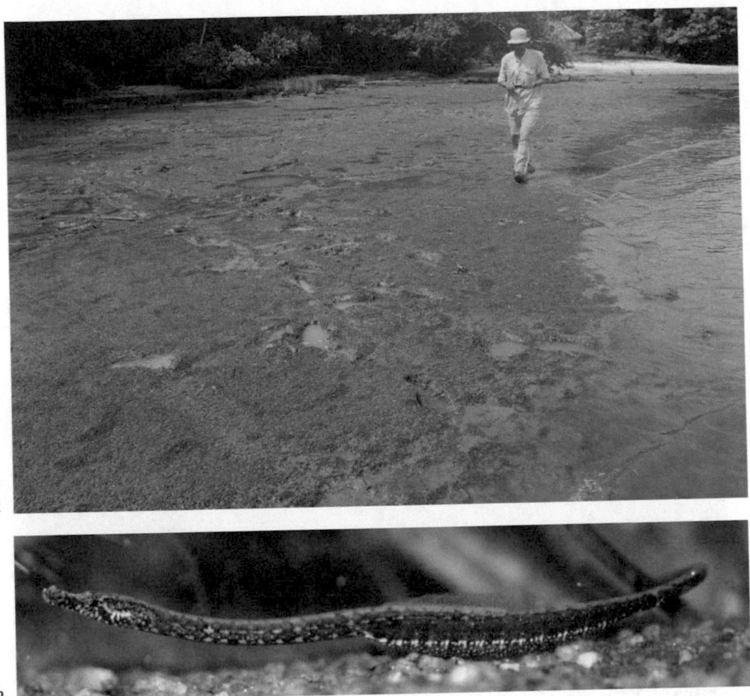

a

b

Fig. 4.12. (a) In the low-water season a meadow of the dwarf Amazonian sword plant *Helanthium tenellum* is exposed at ebb tide; collection site of the miniature freshwater pipefish *Pseudophallus* aff. *brasiliensis* (Sand Landing, Corantijn River; km 155). (b) Live male *Pseudophallus* aff. *brasiliensis* with ripe brood pouch in a field aquarium.

Lower Freshwater Reaches in the Coastal Plain

The lower freshwater reaches of the Surinamese rivers (i.e. downstream of the first rapids and upstream of the estuary) are the deepest reaches of the rivers (Table 3), but shallow beaches and mid-river sandbanks also occur in these reaches. A beach in the Lower Corantijn River near the village Apoera (km 155) with a mixed sand/silt substrate and a stand of dwarf Amazonian sword plant *Helanthium tenellum* (Fig. 4.12a) featured a fish community including a small freshwater pipefish *Pseudophallus* aff. *brasiliensis* (Fig. 4.12b), the slender guppy *Tomeurus gracilis*, the eleotrid *Eleotris pisonis* and the goby *Evorthodus lyricus* (Mol, 2012). A mid-river sandbank in the Lower Suriname River (km 105; Fig. 4.13a) featured abundant aquatic macrophyte vegetation (Fig. 4.13b), large numbers of anchovy larvae, a few larvae of the halfbeak *Hyporhamphus*, and gobiid and eleotrid species (Fig. 4.13c). In the depositional environment of the Coastal Plain, the lower freshwater reaches often show extensive meandering (especially the smaller rivers; Fig. 1.1). Tributaries that originate in the Savanna Belt have black water (e.g. Para, Coesewijne, Marataka, Kaboeri). The large Corantijn and Marowijne rivers have large, stable river islands in their lower reaches and the Suriname River has extensive stands of aquatic macrophytes (*Cabomba*, *Mayaca*;

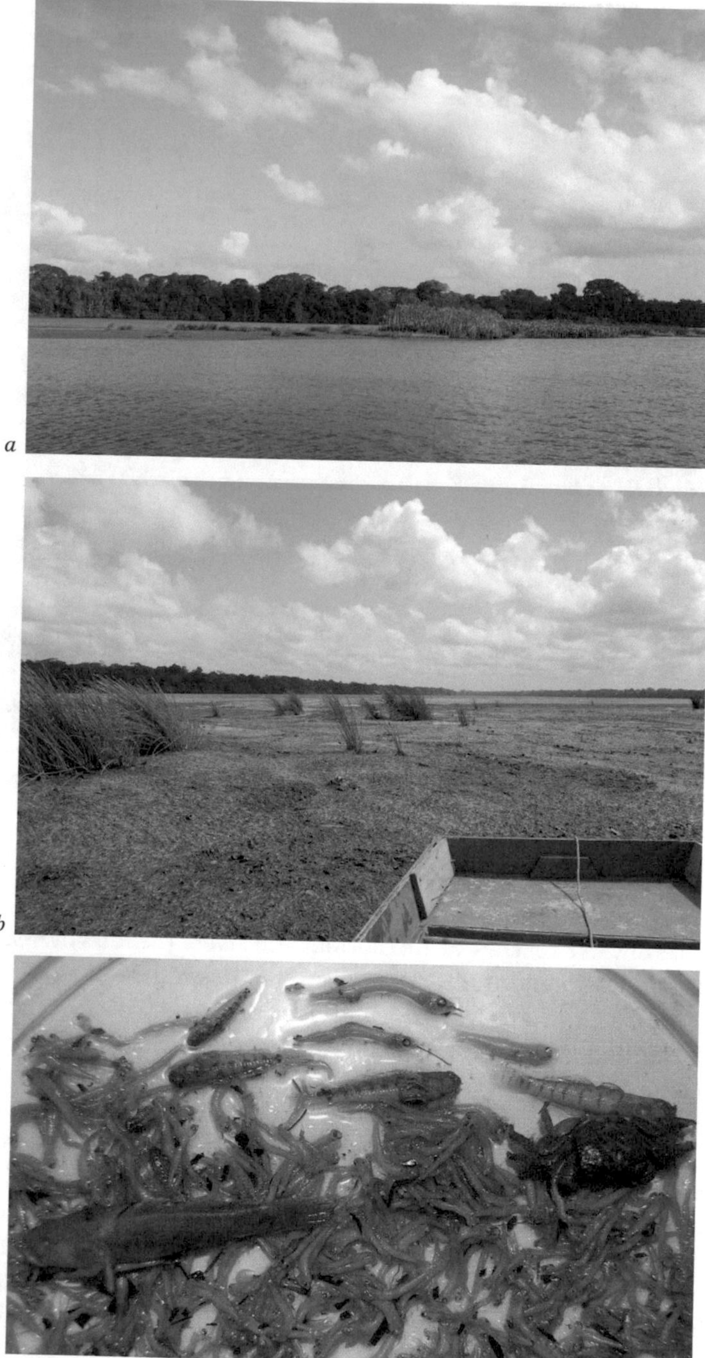

Fig. 4.13. (a) A mid-river, vegetated sandbank near Overbridge is exposed at ebb tide, Lower Suriname River (km 105). (b) Abundant aquatic macrophytes are exposed during the ebb tide. (c) Catch with a small seine net at the Overbridge mid-river bank, including eleotrid and gobiid fishes and early life stages of anchovies (Engraulidae) and the halfbeak *Hyporhamphus* (Hemiramphidae).

Fig. 4.14. Stands of aquatic macrophytes in the lower freshwater section of Suriname River. (a) *Mayaca* sp. with small characins. (b) *Cabomba* sp. (© W. Kolvoort).

a

b

Fig. 4.15. (a) A mud bank in the lower freshwater section of Corantijn River (km 105) is exposed at ebb tide; a fisherman is inspecting for holes of *Gobioides* cf. *broussonnetii*. (b) The goby *Gobioides* cf. *broussonnetii*.

Fig 4.14), that are preferred (spawning) habitat of small characins and even marine anchovies. Upstream reaches in this river section have a sandy bottom, while downstream reaches have a mixed or muddy bottom substrate (Fig. 4.15). The fish communities of the lower freshwater reaches include the three goliath catfishes *Brachyplatystoma filamentosum*, *B. rousseauxii*, and *B. vaillantii* (Fig. 4.16), the ariid kumakuma *Sciades couma*, the pelagic catfishes plarplari *Ageneiosus inermis* and kwasimama *Hypophthalmus marginatus* and the armored catfish warawara *Hypostomus plecostomus*. Two piranha species, *Serrasalmus rhombeus* and the smaller *Pristobrycon eigenmanni*, and two kubi species, *Plagioscion auratus* and

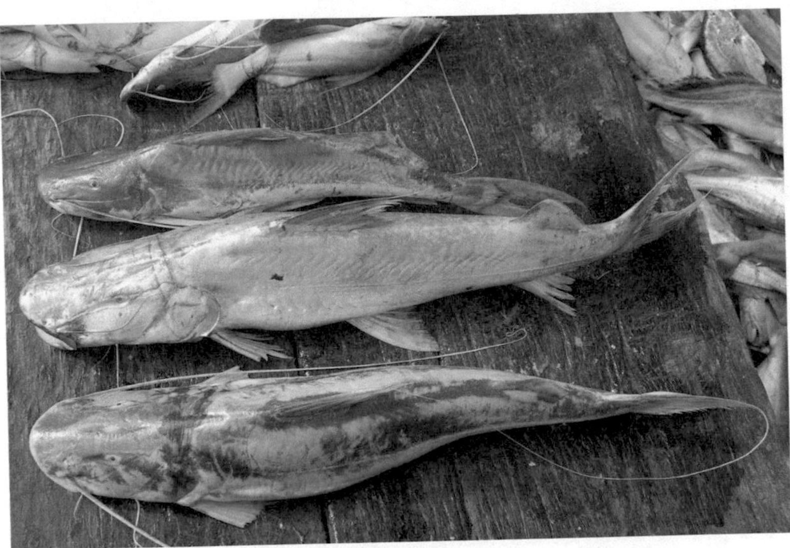

Fig. 4.16. The three species of goliath catfish (Pimelodidae) of Suriname: *Brachyplatystoma vaillantii* (top), the silvery *Brachyplatystoma rousseauxii* (middle) and lalaw *Brachyplatystoma filamentosum* (Lower Corantijn River).

Fig. 4.17. A large kubi (*Plagioscion*) caught by river seine net fisheries on Lower Corantijn River.

P. squamosissimus (Fig. 4.17), are also common. Several diadromous fishes migrate between the estuary or even the ocean and the lower freshwater reaches of the rivers, for example tarpon (Sur. trapun) *Megalops atlanticus* (spawns in sea; Crabtree *et al.*, 1992), several anchovies Engraulidae (spawn in freshwater; Vari, 1982; pers. observations) and botromanki *Pseudauchenipterus nodosus* (spawns in freshwater tributaries; pers. observations in Mindrineti River).

a

b

Fig. 4.18. (a) Wind action in the large open area of the river estuaries can result in large waves (Corantijn River Estuary). (b) Strong tidal currents are used in estuarine shrimp fisheries (Nickerie River Estuary).

The Estuaries

The estuary has a downstream, seaward limit that can be arbitrarily defined as the river mouth (km 0) or the 15 m water depth contour at low-water spring tide (Amatali, 1993), but defining the upper limit in terms of salt intrusion (e.g. 300 mg Cl L^{-1}) can be difficult as this limit shifts with the tide and season. In large rivers like the Amazon it even produces an anomaly as there is no significant penetration of seawater into the estuary, while mixing of river and seawater is still evident 1000 km seaward of the river mouth (i.e. the Amazon freshwater plume,

Fig. 4.19. Mangrove (*Rhizophora* sp.) vegetation along the shores of the Corantijn Estuary.

which also influences coastal waters off Suriname; Hu *et al.*, 2004). Most estuaries are exceptionally new ecosystems in geological and evolutionary terms (having only existed after the last glaciation for approximately the past 6,000 years), which has significant consequences for their colonization, diversity and ecology. The tidal range of the Surinamese estuaries is small (≤ 2 m; Table 3). The estuaries of Surinamese rivers have turbid water and show longitudinal (river-ocean), temporal (tidal, seasonal), and comparatively minor vertical (water depth) variation in salinity. The role of salinity in structuring the fish assemblages of the Caeté River estuary in northeastern Brazil was shown by Barletta *et al.* (2005) and Barletta-Bergan *et al.* (2002). The most important aspect of salinity variation in estuaries is that salinity at any one point varies considerably over time. With every tidal cycle, a fish could potentially be exposed to near fresh water at low tide and near marine conditions at high tide. This variability results in harsh conditions, especially in the mid-estuary area (the salinity range decreasing considerably towards both the freshwater-head and marine-mouth ends). Wind action on the large, open area near the mouth of the estuary (Fig. 4.18) can result in big waves. River and tidal currents can be strong (> 1 m s^{-1}) especially in mid-channel where the friction of the estuary banks does not slow down the water speed. Mangrove forests grow on the banks of the estuaries (Fig. 4.19) and all along the coast of Suriname (in eastern Suriname interrupted by a few sandy beaches where sea turtles nest).

Perhaps the most evocative feature of the Surinamese estuaries are the extensive intertidal mudflats composed of soft, fine grey silt. The fine sediment is difficult to colonize as it has no points of attachment for sessile organisms, while organisms intolerant of mechanical disturbance would not be able to deal with the mobility of the substratum. Fine suspended particles can clog delicate feeding and respiratory structures and the sediment makes locomotion difficult. However,

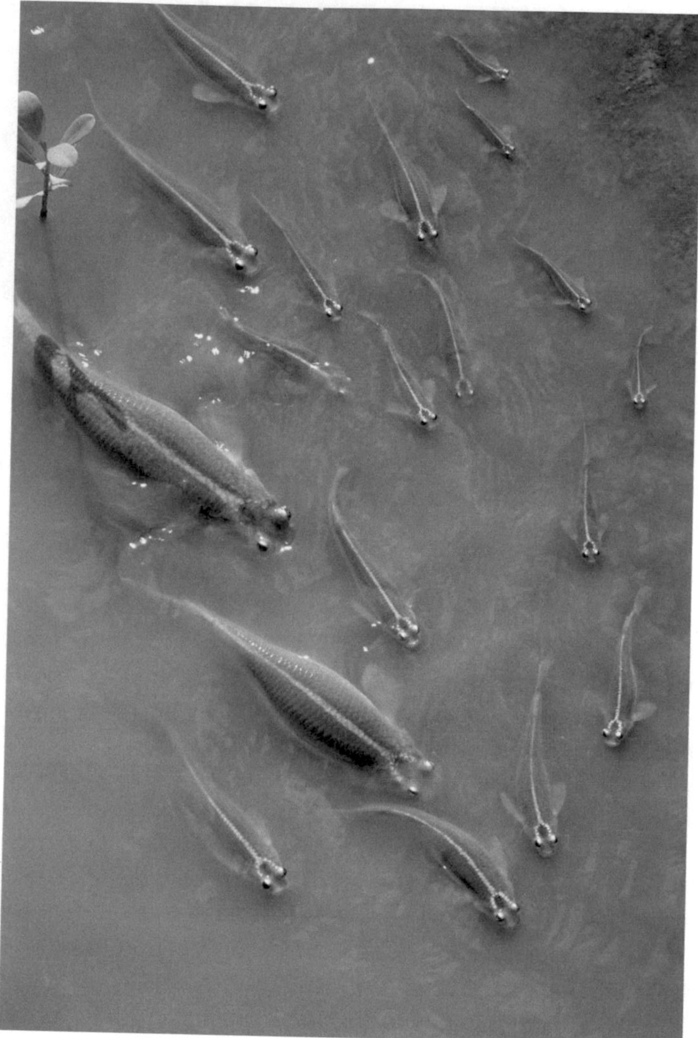

Fig. 4.20. A small group of kutai or four-eyes (*Anableps* sp) in the Suriname River Estuary (© W. Kolvoort).

large fine-grained mudflats only tend to form in the middle reaches of the estuary and sediments near the mouth and the head of estuaries tend to be much coarser as tidal and river flows are great here and thus prevent settlement of finer particles (Dyer, 1997). Estuarine fishes have adapted to the high suspended sediment concentrations and turbidity of the water, which in turn may provide shelter from (visual) predators. The 'four-eyed' fish or kutai *Anableps* (Fig. 4.20) always swims near shore at the water surface with the upper part of the eye exposed to the air. The lens is elliptical with the longer axis refracting light from below the water surface and the short axis light from above the surface; light can thereby be focused on a divided retina for simultaneous aquatic and aerial vision (Bone & Moore, 2008). The four-eyes are difficult to catch as they can see one approaching

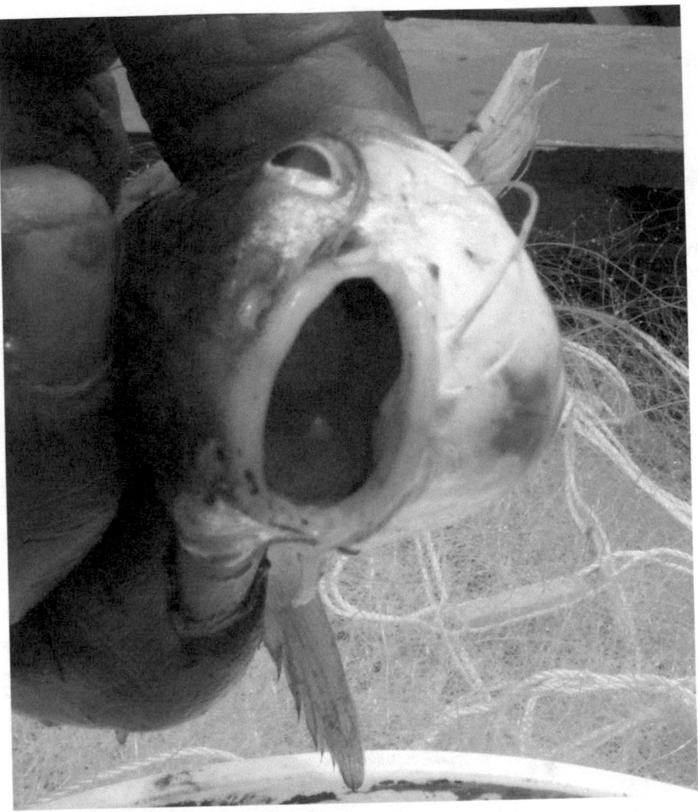

Fig. 4.21. A male of the ariid catfish *Sciades herzbergii* brooding a few large eggs in its mouth (Bigi Pan Lagoon, Northwest Suriname).

from a great distance. Many estuarine fishes protect their eggs from deposition of fine sediments on their surface (which would prevent gas exchange and development) either by carrying the eggs with them in the mouth (tilapia *Oreochromis mossambicus*, ariid catfishes; Fig. 4.21) or attached to their body (aspredinid catfishes) or by retention of the eggs within the body of the female (e.g. vivipary in *Anableps* and poeciliids).

Dissolved oxygen levels tend to be normal at the head and mouth of an estuary, but show a notable dip in the mid estuary (i.e. the so-called oxygen sag). Many estuaries worldwide have been favored places of human settlement (e.g. the city of Paramaribo on the left (west) bank of the Suriname River) and human activity such as waste disposal can stimulate excessive removal of oxygen. In extreme cases (e.g. the Thames; Attrill, 1998), estuarine waters can become depleted of oxygen with dramatic consequences for the ecology of the system and, for example, prevent migration of fish species to their freshwater breeding grounds. Estuarine food webs are powered by detritus from marine, freshwater, terrestrial and estuarine sources, although stable-isotope studies show that terrestrial carbon (e.g. from mangrove forests) is relatively unimportant (Loneragan *et al.*, 1997). Like estuaries worldwide, the Surinamese estuaries are exceptionally important

Fig. 4.22. The large pristigasterid *Pellona flavipinnis* caught in the Corantijn River Estuary.

Fig. 4.23. A Guiana dolphin (*Sotalia guianensis*) in the Suriname River Estuary (© W. Kolvoort).

nursery grounds for many marine fish species as demonstrated by large catches of juvenile sciaenids (Sur. tri). The importance of (upper) estuaries for larval and juvenile fishes is demonstrated by studies in the Cayenne (Tito de Morais & Tito de Morais, 1994) and Caeté (Barletta-Bergan et al., 2002) estuaries in French Guiana and northeast Brazil, respectively. In Suriname, estuarine fish communities include among others 13 ariid, 4 aspredinid, 1 auchenipterid (*Pseudauchenipterus nodosus*), 5 pimelodid and 1 loricariid (*Hypostomus watwata*) catfishes, tarpon *Megalops atlanticus*, ladyfish *Elops saurus*, 17 clupeiform species (Fig. 4.22),

2 species of kutai or four-eyes (*Anableps*), lompu *Batrachoides surinamensis*, 3 mullets (*Mugil* spp), 3 snook species (*Centropomus* spp), 13 sciaenids, 12 gobioid and 8 pleuronectiform species, 2 puffers (*Colomesus psittacus* and *Sphoeroides testudineus*), and two large, IUCN-redlisted, critically endangered fishes, the jewfish or goliath grouper (*Epinephelus itajara*) and the largetooth sawfish (*Pristis perotteti*) (Table 2). The small Guiana dolphins (*Sotalia guianensis*) feed (and breed) in the Suriname River estuary; they are easy to observe and thus very popular with tourists (Fig. 4.23). Allowing for differences in sampling methods and intensity, the fish faunas of estuaries in Suriname (pers. observations) and those of the turbid Cayenne (Tito de Morais & Tito de Morais, 1994), Sinnamary (Boujard & Rojas-Beltrán, 1988) and Caeté (Barletta *et al.*, 2005) estuaries in northeastern South America showed similarities in species number and composition, which

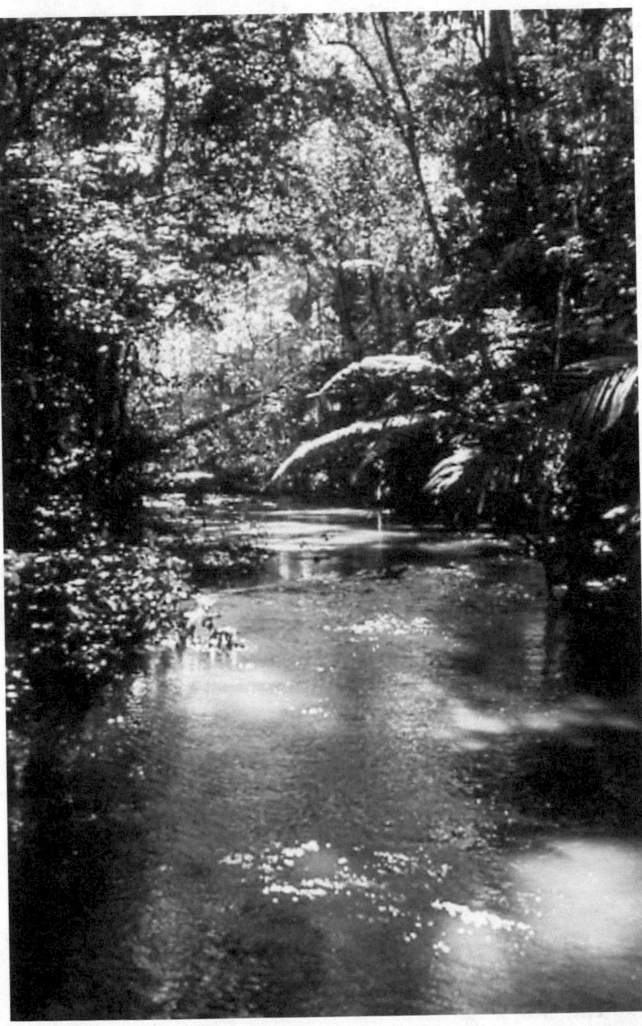

Fig. 4.24. A small shaded rainforest creek in southern Suriname (Upper Corantijn River).

Fig. 4.25. (a) Current in a rainforest creek with clear water and sandy bottom. (b) Watrabebe tree (*Pterocarpus officinalis*) on the bank of a small rainforest creek (© W. Kolvoort).

can be attributed to similar environmental conditions and to close proximity of the sites.

Rainforest Creeks

In the Interior of Suriname, small rainforest creeks up to 10-15 m width are largely shielded from sunlight under a closed canopy of rainforest trees (Fig. 4.24). In lowland creeks the water temperature is about 25-27.5 °C and varies very little during the day or year (± 1 °C; Geijskes, 1942); in mountain streams the water tem-

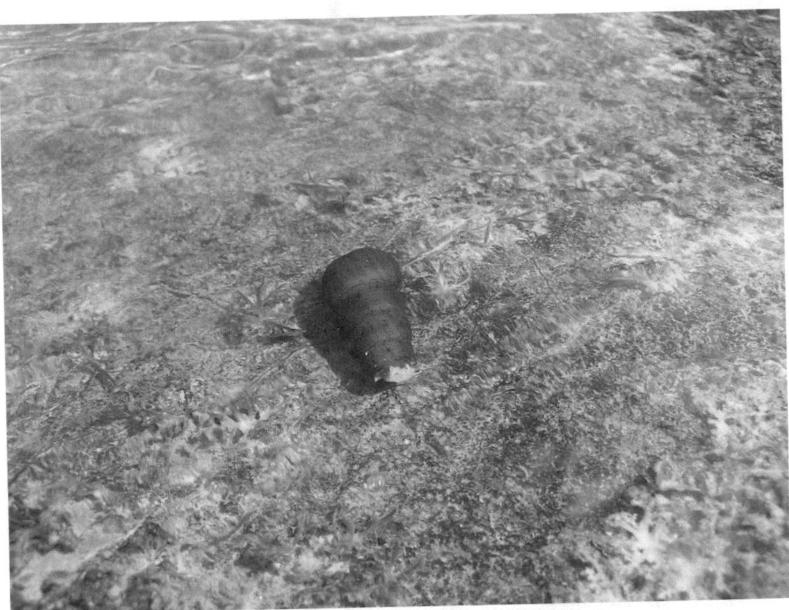

Fig. 4.26. Freshwater snail (*Doryssa* sp., Thiaridae) with eroded apical whorls, Upper Corantijn River.

perature is even lower (22 °C; Mol *et al.*, 2007b). The water is transparent clear (Fig.4.25ab), and occasionally slightly brown by humic acids. The water is extremely poor in dissolved minerals with the smallest tributaries having a slightly higher conductivity (20-30 μS cm^{-1}) than the larger streams and rivers (<20 μS cm^{-1}) (pers. observations; also see Walker, 1995): somehow, solutes are lost on the way. Bivalves and snails with calcareous cases are rare or show shells with missing apical whorls (Fig. 4.26). The callichthyinid catfish *Megalechis thoracata* is apparently not negatively affected by the low Ca and Mg concentrations in the water and seems to store calcium in its bony armor (Mol *et al.*, 1999).

In Suriname, small mountain streams do not occur at altitudes where forest cannot grow and they differ from lowland rainforest streams mainly in current velocity, bottom substrate (bedrock, boulders, pebbles and gravel in mountain streams; mainly sand and mud in lowland streams) and the presence of small waterfalls (Fig. 4.27). The fish faunas of mountain streams in Suriname are not well-known, apparently relatively poor in species, but may include highly endemic species (e.g. Paramaka Creek in Nassau Mountains; Mol *et al.*, 2007b). Fish species that are known to occur at altitudes >500 m include *Rivulus* spp, *Synbranchus marmoratus*, *Ituglanis* spp, *Callichthys callichthys*, and several small loricariid catfish (*Guyanancistrus* spp, *Harttiella crassicauda*, *Lithoxus* spp; Fig. 2.3).

In lowland streams, the low-order, upper courses (headwaters) originate between relatively steep slopes and declivity does not allow for sinuosity. The small, shallow water course is continually dammed by obstacles such as fallen trees and tree roots, with regions of running and stagnating water alternating. The streambed is sandy (Fittkau, 1967; Walker, 1995). Muddy stream valley bottoms

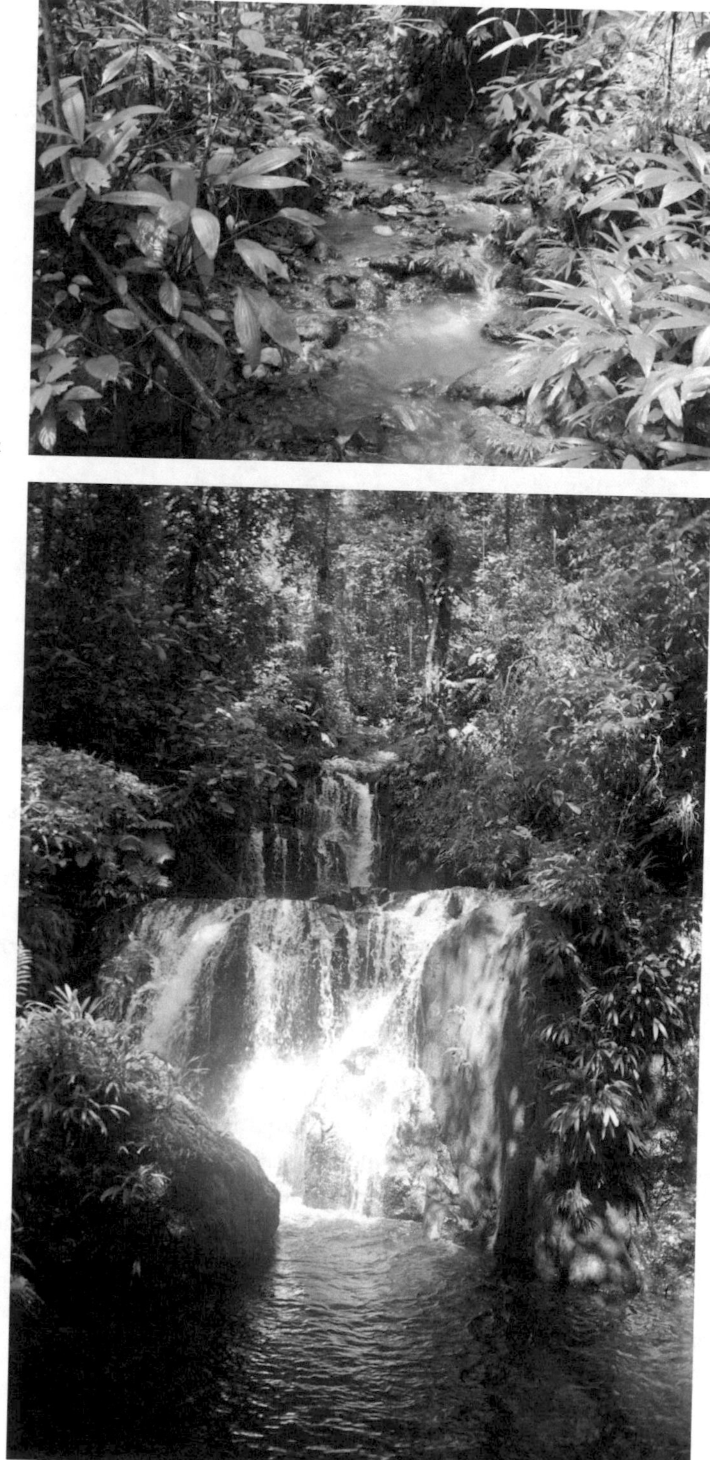

a

b

Fig. 4.27 (a-b). Mountain creeks in Suriname.

Fig. 4.28. Bubble nest of katrina kwikwi (*Megalechis thoracata*) under the undercut bank in a dry season pool of Maikaboe Creek, Saramacca River System.

develop further down with diminishing channel slope and get progressively wider as the streams increase in size and sinuosity, and relatively regular meanders are forming (Fittkau, 1967). The substrate of the stream bottom itself, however, is usually sand. The outer margin of the meanders is an erosion zone with steep banks and occasional deep recesses with overhanging tree root mats, while gentle backflows along the inner margin may build up sand and leaf-litter banks (Walker, 1995). The undercut banks with dense overhanging root mats are used as refuges by small caimans (*Caiman crocodilus*) and electric eel (*Electrophorus electricus*), and provide nesting habitat for katrina kwikwi (*Megalechis thoracata*) (Fig. 4.28). Small crevices in steep banks are often occupied by *Synbranchus marmoratus*, particularly juveniles (Walker, 1995). In the lower courses the bed of the stream is muddy and below the medium high water mark of the main river system, and accordingly is annually flooded by the waters of the main river, which fills the wide valley of the tributary stream. Only during the low-water period this part of the river represents a tributary creek, while during the high-water season it represents a section of the main river, the warm water of which covers the cooler water of the creek (Fittkau, 1967). In the rainforest of the Amazon, the light reaching the forest floor is 1% or less of the light reaching the canopy (Sioli, 1975). From this observation it appears that autotrophy in such streams is negligible. The only flowering plants are the emergent, semi-aquatic *Thurnia sphaerocephala* (Fig. 4.29a) and, in spots where some sunlight reaches the stream, the likewise semi-aquatic *Tonia fluviatilis* (Fig. 4.29b). Lower aquatic plants include periphytic diatoms (e.g. *Eunotia* spp, *Navicula* spp) and filamentous red algae (e.g. *Batrachospermum* spp; Fig. 4.30) (Fittkau, 1967; Walker, 1995; Mol *et al.*, 2007b). The rainforest creeks receive organic matter from allochthonous sources, e.g. 4,200 kg ha^{-1} leaf fall (ca 800 leaves m^{-2}) and 1,500 kg ha^{-1} woody debris in Central

a

b

Fig. 4.29. (a) Stands of the emergent macrophyte *Thurnia sphaerocephala* in Upper Paramaka Creek, Nassau Mountains. (b) Stands of *Tonia fluviatilis* (b © W. Kolvoort).

Amazonian rainforest streams (Walker, 1995). Benthic macroinvertebrates appear to depend mainly on allochthonous organic matter (i.e. fallen leaves), presumably because periphytic algae grow poorly in this light and nutrients impoverished habitat. The fauna is essentially associated with submerged forest litter, decomposed primarily by fungi (Walker, 1995). Litter fungus feeding is prevalent for chironomids, Ephemeroptera, Trichoptera and oligochaetes, which constitute the main food of predatory Odonata, shrimps and fish (Walker, 1995). Among

Fig. 4.30. Gelatinous tufts of periphytic filamentous red algae (*Batrachospermum* sp) in Paramaka Creek, Nassau Mountains (© P. Ouboter).

predators, relative size seems to be the major criterion: large Odonata larvae feed on small fish, shrimps and Odonata, and vice versa for all three groups. The longer the potential food chain, the larger the prey spectrum, because top-level predators feed on all lower levels. As litter fungus input is infallible, and assuming species-specific prey vulnerability, there is an enormous redundancy within the foodweb (there are any number of substitutes if one prey is absent) and thus the foodweb is very robust (Walker, 1995). The fact that aquatic litter decomposition does not result in an increase in mineral content (conductivity) of the water in downstream direction (see above) shows that the decomposition fungi are an efficient mineral filter: whatever is released into the water is immediately re-absorbed (Walker, 1995). The food-chains are essentially cyclic, also in relation to the terrestrial rainforest: (1) nutrient uptake by riparian / flooded forest trees which provide most of the litter in the first place, (2) terrestrial adults of aquatic insects move to the land, and (3) terrestrial predators, such as fishing birds and mammals (e.g. otters and the fishing bat *Noctilio leporinus*), removing nutrients to the forest. Walker (1995) points out that while the foodweb is essentially robust, it is also extremely vulnerable as concerns interference with biomass cycling.

Chironomids are a basic component of the community of leaf-litter banks (Walker *et al.*, 1991), which includes many shrimps (*Macrobrachium* spp., *Palaemonetes carteri*, *Euryrhynchus*), crabs (Trichodactylinae, Pseudothelphusinae), tadpoles and small-sized fishes (Henderson & Walker, 1990; Walker, 1994). The leaves are shredded by chironomids through selective removal of the mesophyll, which sustains the unaltered appearance of the leaf, even after it is extensively shredded. The submerged leaf-litter can form deep layers (up to 2 m, pers. obser-

Fig. 4.31. Floating root masses in Kumbu Creek, Brownsberg Mountains, Saramacca River System (© K. Wan Tong You).

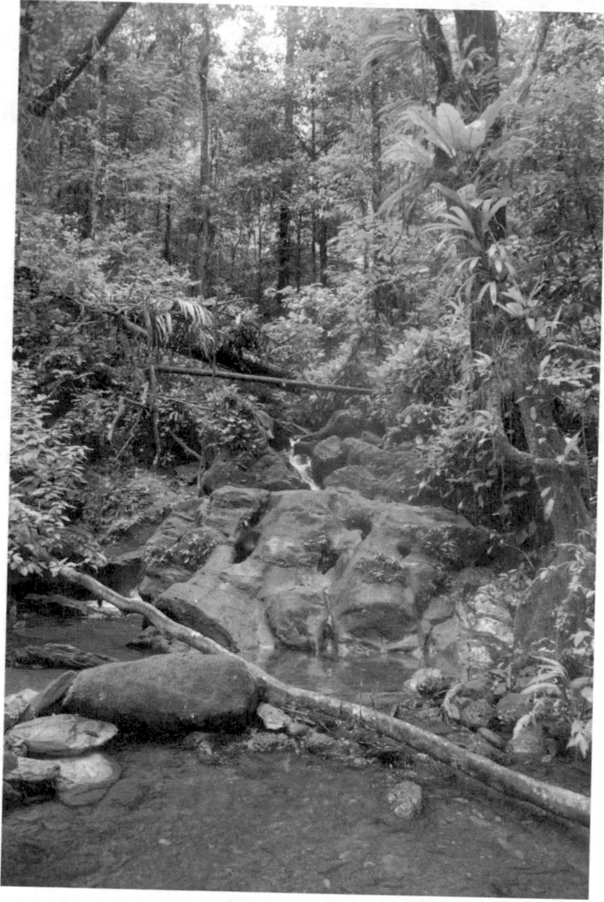

Fig. 4.32. Rocky habitat in small rainforest creek.

a

b

Fig. 4.33. (a) Abundant large woody debris in Mindrineti River, Saramacca River System, as revealed during the low-water season. Large woody debris is an important nesting habitat and shelter against predators for small fishes of rainforest streams (© P. Ouboter). (b) A loricariid catfish (*Hypostomus coppenamensis*) feeding on periphyton on woody debris (Coppename River) (© W. Kolvoort).

vations in a tributary of the Upper Nickerie River, Bakhuis Mountains) that are used as hiding places for small diurnal fishes (Crenuchidae, *Corydoras* spp, Rivulidae, juvenile *Synbranchus marmoratus*, *Polycentrus schomburgkii*, *Nannacara anomala*, *Apistogramma steindachneri*, juvenile *Crenicichla*) and nocturnal fishes (gymnotiform knifefishes, catfishes). Several fish species resemble dead

Fig. 4.34. The water surface habitat. A freshwater needlefish (*Potamorrhaphis guianensis*) at the water surface in a small stream (© W. Kolvoort).

leaves in color and shape (*Polycentrus schomburgkii, Helogenes marmoratus, Bunocephalus* spp, *Hypopygus lepturus*) and thus may find camouflage against visually hunting diurnal predators among fallen leaves on the stream bed (Sazima *et al.*, 2006) or use this crypsis for hunting their prey (*Polycentrus*). Floating root mats (Fig. 4.31), rock (Fig. 4.32) and large woody debris (Fig. 4.33) are also important habitat for fishes of rainforest streams. Rainforest streams in Suriname can have up to 100 species (e.g. 68 species in Maikaboe Creek, Saramacca River System;

Fig. 4.35a. The bottom habitat. (a) A small loricariid catfish (*Rineloricaria fallax*) of tributaries of Corantijn River.

Fig. 4.35b. The bottom habitat. (b) A freshwater stingray (*Potamotrygon* sp.) in a small rainforest creek (© W. Kolvoort).

Mol & Ouboter, 2004), including species that live at the water surface (e.g. *Gasteropelecus sternicla*, *Pyrrhulina filamentosa*, *Copella arnoldi*, *Potamorrhaphis guianensis*; Fig. 4.34), on the stream bed (mainly catfishes, but also stingrays; Fig. 4.35), and in the water column (*Bryconops*, *Charax*, *Hemigrammus*, *Jupiaba*, *Moenkhausia*, curimatids). Fishes are active at night (mainly catfishes and gymnotiform knifefishes) or during the day (mainly cichlids and characiforms) (Lowe-McConnell, 1987). Most fish species are small and appear to be adapted to use allochthonous food sources such as terrestrial insects that fall into the water;

Fig. 4.36. (a) The detritivorous curimatid *Cyphocharax microcephalus*. (b) X-ray photograph of the same specimen showing the long intestines of a detritivorous fish (© R. Vari).

they have either flexible or nonspecialized diets (e.g. Knöppel, 1970). Many species may ingest detritus incidentally with algae or invertebrates that are their main nutrient source, but only the curimatids (*Curimata cyprinoides*, *Cyphocharax* spp) can really be classed as detritivorous (Fig. 4.36). The main piscivorous fishes of rainforest creeks are erythrinids (especially *Hoplias* spp), two *Acestrorhynchus* species and pike cichlids (*Crenicichla*).

Brackish and Freshwater Wetlands of the Coastal Plain

The Coastal Plain of Suriname is a young (Pleistocene-Holocene) flat, lowland landscape consisting of Andes/Amazon-derived clays (Eisma *et al.*, 1991) and beach ridges with sand and shells that are oriented east-west, (semi-) parallel to the coast line. Teunissen (1993) distinguishes four ecological landscapes in the Coastal Plain: (1) brackish wetlands along the coast with mangrove forest, salt water lagoons and brackish herbaceous swamps (10% of the Coastal Plain area), (2) freshwater wetlands consisting of herbaceous swamps (15%), swamp wood and low swamp forest (30%), and high swamp forest (15%), (3) cultivated areas such as agricultural impoldered land, abandoned plantations and mined out lands (10%), and (4) marsh and highland areas covered with marsh forest and high dryland forest (20%). The wetlands have few open water bodies (lagoons, river sections, creeks and man-made canals) and consist mainly of swamp lands or wetland covered with closely packed emergent aquatic plants or forest. The swampland soils are poorly drained and inundated either permanently or at least during the greater part of the year. Generally the soils are covered by a layer of peat or slightly decomposed organic material (which remains soaked with water during average dry seasons). In northern parts of the Coastal Plain (i.e. the Young Coastal Plain), the swamps are often shallow (<1 m) and slow moving so that the characteristic swamp vegetation can be established. Deep swamps (1-3.5 m) dominated by *Eleocharis* sp. are known as relatively narrow fringes along creeks and rivers and in gullies between the clay plateaus of the southern Old Coastal Plain (Teunissen, 1993).

Brackish-water mangrove wetlands are associated with extensive mudflats (alternating with sandy cheniers) that migrate northwestward along the north coast of South America from the mouth of the Amazon River to that of the Orinoco River. Attachment of mud and cheniers to the coast is a continuous process causing a general progradation of the coastline (Augustinus, 1978, 1980; Eisma *et al.*, 1991), and occurred since the uplift of the Andes started to yield abundant fine-grained sediment to the Amazon River. Tidal mudflats are protected against erosion by biofilms of microalgae (mainly motile diatoms) and associated fauna (see Artigas *et al.*, 2003). The coastline is further sheltered from erosion after the rapid establishment of mangrove transforms the unstabilized mudflat into a mangrove forest. The soft substrates of intertidal mudflats had the highest biomass (32-37 g ash-free dry weight m^{-2}, mainly fiddler crabs *Uca maracoani*) around the high tide level, at the border of the mangrove forest, and on sheltered places just outside the *Uca* zone (ca 20 g m^{-2}, mainly Tanaidacea in densities of 6,000-13,000 individuals m^{-2}); biomass was low in the middle and lower parts of the flats (only a few g m^{-2} or less), probably as a result of the instability of the sediment (Swennen *et al.*, 1982). The coastal waters up to 30 m depth may have a low salinity <25‰ at the surface (8-10 m depth), muddy bottoms, and are rich in sediments and nutrients (the latter derived from the Amazon River freshwater plume, terrestrial sources and upwelling generated by the prevailing direction of the winds), but high primary production was mainly observed outside the turbid coastal waters in deeper water (20-60 m) on the continental shelf (Cadée, 1975). The fish fauna of the shallow coastal waters (i.e. the brown-fish zone of Lowe-McConnell, 1962) includes ariid catfishes, rays (*Dasyatis americana, D. guttata* and *Gymnura micrura*) and young fishes of several families, but mainly Sciaenidae. The coastal fishes of French Guiana are illustrated and described by Leopold (2004); also see Cervignon *et al.* (1993).

The white mangrove tree or parwa *Avicennia germinans* dominates the coastal mangrove forests (Fig. 4.37), while two species of red mangrove *Rhizophora mangle* and *R. racemosa* dominate upstream from the estuaries along the rivers (Fig. 4.19). The water in the coastal mangrove forests is rich in nutrients and suspended matter (turbid; Fig. 4.38) and tends to be brackish although its salinity varies greatly, depending on rainfall and the presence of old beach ridges (cheniers) that block tidal inflow of ocean water. Shallow (<1 m), open-water lagoons (Sur. pan)

Fig. 4.37. In a coastal mangrove (*Avicennia germinans*) forest.

Fig. 4.38. Man-made canal with turbid, brackish water in mangrove forest, Saramacca District.

are present in the mangrove forest (Fig. 1.1) and these often have important fisheries for juvenile penaeid shrimps and fishes like tilapia (*Oreochromis mossambicus*), snook (*Centropomus* spp), mullet (*Mugil* spp.) and tarpon (*Megalops atlanticus*). The Bigi Pan Area (530 km^2) in northwestern Suriname is such a brackish wetland known for its extensive mangrove forests and coastal lagoons

a

b

Fig. 4.39. Brackish lagoons in the mangrove forest can sustain important fisheries and function as nurseries for marine penaeid shrimps. (a) Fishermen camps in Bigi Pan Lagoon, Nickerie District. (b) The catch in Bigi Pan Lagoon is dominated by the introduced tilapia (*Oreochromis mossambicus*); note the shallow water depth.

(including the Bigi Pan Lagoon itself with an open-water surface area of 7 km^2; Fig. 4.39). The lagoons are situated 2-5 km behind the coast and cover approximately 10% of the area. Widgeon grass *Ruppia maritima* can form dense stands in the lagoons, while *Cabomba*, *Nymphaea ampla*, *Typha* and *Eleocharis* point to freshwater influence in southern parts of the lagoons. Seawater enters the lagoons via a few small creeks during spring tides. With the tides shrimp (post)larvae

Fig. 4.40. An herbaceous freshwater swamp in the Young Coastal Plain, Saramacca District. (a) Dense stands of the emergent aquatic macrophyte *Typha*. (b) Parts of the swamp with open water often show extensive growth of floating macrophytes (here *Salvinia auriculata* and *Pistia stratiotes*)

(Dumas, 2006) and fish (e.g. *Anableps*; Brenner & Krumme, 2007) migrate via these creeks in and out of the mangroves. In the rainy season, fresh water enters the area from the extensive freshwater swamps to the south and any surplus water will flow to the ocean. During the dry season the salt content of the lagoons may increase through evaporation from 5 g L^{-1} in the wet season to 40 g L^{-1} at the end of the dry season and some of the small lagoons dry up completely (Mol *et al.*, 2000). The fauna of these brackish wetlands is characterized by euryhaline species (Artigas *et al.*, 2003). Fish species of Bigi Pan Lagoon include daguboi or landyan (*Elops saurus*), trapun (*Megalops atlanticus*), wetkati (*Sciades herzbergii*), pani (*Sciades passany*), prasi or kweriman (*Mugil* spp), mangrove rivulus (*Kryptolebias marmoratus*), kutai (*Anableps* spp), molly (*Poecilia vivipara*), snuku

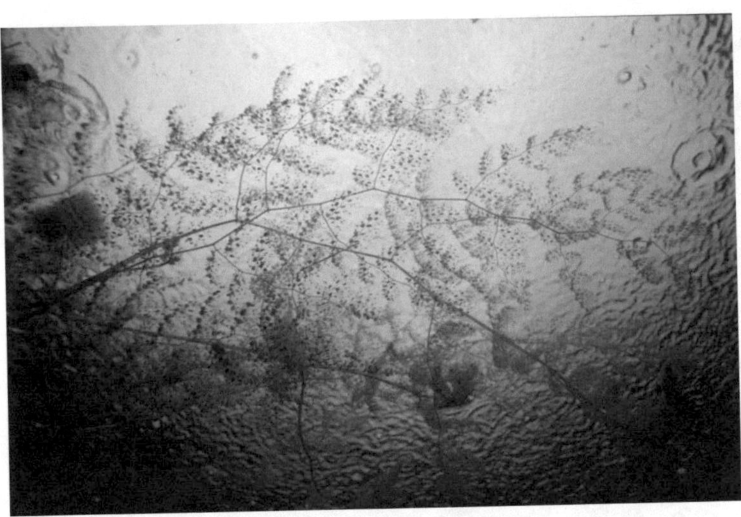

Fig. 4.41. Stand of *Utricularia* during heavy rain shower (© W. Kolvoort).

(*Centropomus* spp), juvenile granmorgu (*Epinephelus itajara*), blakatere (*Cynoscion steindachneri*), tilapia (*Oreochromis mossambicus*), and boki (Achiridae). The spawning grounds of tarpon, mullet, snook, granmorgu, blakatere, and penaeid shrimps are on the adjacent continental shelf, but tilapia, wetkati, mangrove rivulus, and molly reproduce in the lagoons. Fishes of freshwater wetlands (see below) can occur in southern parts of the area that are influenced by freshwater inflow in the wet season. The once diverse fisheries in Bigi Pan Lagoon (Geijskes, 1943) are now dominated by the introduced tilapia (Fig. 4.38b) and, in some years, juvenile *Penaeus subtilis* shrimps and srika crabs *Callinectes bocourti* (Mol *et al.*, 2000). Mangrove forests are known as important nurseries for juvenile penaeid shrimps (Primavera, 1998) and, in Suriname, such a function is best illustrated by the substantial fisheries on juvenile *Penaeus subtilis* in Bigi Pan Lagoon (pers. observations). The habitat function of mangroves for both terrestrial and marine fauna is reviewed by Nagelkerken *et al.* (2008).

To the south of the brackish mangrove wetlands is a zone with herbaceous freshwater swamps. In these swamps, growth of emergent plants (e.g. *Cyperus giganteus, Eleocharis interstincta, Montrichardia arborescens, Typha* sp.; Fig. 4.40a) is often so dense that vast areas of water are entirely hidden from view. The photosynthesizing parts emerge into the air above the water which is thereby in varying degree shaded from sunlight and wind. In unshaded areas the shallow water supports floating (e.g., Lemnaceae, *Salvinia auriculata, Azolla caroliniana, Pistia stratiotes, Eichhornia* spp; Fig. 4.40b), submerged (e.g. *Cabomba* sp., *Utricularia* spp.; Fig.4.41), and rooted (*Nymphaea* spp, *Nymphoides indica, Hydrocotyle umbellata, Ceratopteris* sp) macrophytes (Werkhoven & Peeters, 1993). The root zone of both the floating meadows of the Middle Amazon River (Junk, 1973) and the floating aquatic macrophytes of the Coastal Plain in Suriname (Mol, 1993a) supported an important aquatic invertebrate fauna. The peculiar conditions in

Fig. 4.42. Male pataka *Hoplias malabaricus* guarding a nest in shallow, newly flooded swamp land at the start of the rainy season. Weg Naar Zee, Young Coastal Plain, Suriname.

the water of these herbaceous swamps are primarily due to the rapid decomposition of large quantities of plant material (e.g. Carter & Beadle, 1930). The upper parts of the emergent plants, exposed above the water surface, grow fast and ultimately die and fall into the water where they undergo anaerobic decomposition on the bottom of the swamp. The 'closed canopy' of emergent *Typha* leaves reduces light (and temperature) in the swamp water and for this reason there is little or no photosynthetic production of oxygen in the water, that is also protected from wind stirring. The only source of oxygen is therefore slow diffusion from rather still air, while on the other hand consumption of oxygen by the decomposing organic matter is so rapid that the swamp water may be devoid of oxygen to within a few cm of the surface. It is often impossible to detect oxygen in surface samples except in regions with a rapid current during heavy rains. The water is slightly acid (pH usually 6.0-7.0) with high concentrations of CO_2 and often reducing (the gases CH_4 and H_2S may be formed in the bottom peat layer). Animal life in the interior of these dense *Typha* stands is undoubtedly much less than in the open water at the edge of the swamp, but the fauna is richer than a cursory examination would suggest, even during the dry season when the water is most stagnant and typical swamp conditions are most marked (e.g. Mol, 1993a). The fish fauna of herbaceous swamps in the Young Coastal Plain includes the small characoids *Ctenobrycon spilurus, Hemigrammus boesemani, Hemigrammus unilineatus, Pristella maxillaris, Crenuchus spilurus* and makasriba *Curimata cyprinoides*, the erythrinids stonwalapa or matuli *Erythrinus erythrinus*, walapa *Hoplerythrinus unitaeniatus* and pataka *Hoplias malabaricus*, the catfishes hei-ede-kwikwi *Hoplosternum littorale*, platkop-kwikwi *Callichthys callichthys*, noya *Trachelyopterus galeatus* and *Loricariichthys maculatus*, the cichlids datrafisi

Crenicichla saxatilis, krobia (*Cichlasoma bimaculatum* and *Krobia guianensis*) and *Nannacara anomala*, leaf fish *Polycentrus schomburgkii*, trapun *Megalops atlanticus*, zwampaal *Synbranchus marmoratus*, *Rivulus* spp and poeciliids. Additional fish species occur in the deeper *Eleocharis*-dominated swamps of the Old Coastal Plain (e.g. *Astyanax bimaculatus*, *Hyphessobrycon simulatus*, *Cyphocharax spilurus*, *Curimatopsis crypticus*, pencilfish *Nannostomus beckfordi*, splashing tetra *Copella arnoldi*, *Pyrrhulina filamentosa*, dyaki *Rhamdia quelen*, katrina-kwikwi *Megalechis thoracata*, logologo *Gymnotus carapo*, *Apistogramma steindachneri*, and *Cleithracara maronii*). Fishes of the swamps overcome the lack of dissolved oxygen by breathing atmospheric air from above the water surface (e.g. walapa *Hoplerythrinus unitaeniatus*, stonwalapa *Erythrinus erythrinus*, zwampaal *Synbranchus marmoratus* and kwikwi species Callichthyinae; Carter & Beadle, 1931; Graham, 1997) or extracting oxygen from the thin oxygen-rich surface layer with swollen, highly vascular lips (*Astyanax bimaculatus*; i.e. aquatic surface respiration; Winemiller, 1989). Platkop kwikwi *Callichthys callichthys* (pers. observations) and zwampaal *Synbranchus marmoratus* (Lüling, 1980) can construct burrows to hibernate during unfavorable dry season conditions and migrate overland between drying pools on wet nights. Predatory pataka *Hoplias malabaricus* spawn early in the rainy season in newly flooded shallow swamp land (Prado *et al.*, 2006; Fig. 4.42) to avoid low dissolved oxygen levels that develop later in the wet season (and perhaps to give their young a lead in growth over their prey species that reproduce later in the season). Kwikwi species (Callichthyinae) deposit their eggs in floating bubble nests above the water surface where the eggs can develop in the humid oxygen-rich environment of the bubbles (Mol, 1993b; Fig. 4.43). The distribution of hei-ede-kwikwi (*Hoplosternum littorale*) is largely restricted to the herbaceous swamps of the Young Coastal Plain, while platkop-kwikwi (*Callichthys callichthys*) and katrina-kwikwi (*Megalechis thoracata*) are more common in swamp(forests) of the Old Coastal Plain and also occur in rainforest creeks of the Interior (Fig. 4.27) (Mol, 1994).

Low swamp forest or swamp wood can be dominated by several tree species including watrabebe *Pterocarpus officinalis*, mira-udu *Triplaris surinamensis*, zwamppruim *Chrysobalanus icaco*, zwampzuurzak *Annona glabra*, morisi palms *Mauritia flexuosa* or kofimama *Erythrina glauca* (Teunissen, 1993). High swamp forest in shallow swamps of the northern Young Coastal Plain (Fig. 4.44) is dominated by babun trees *Virola surinamensis* and pina palms *Euterpe oleracea*, and relatively rich in species with many monocotyledons in the undergrowth. Posentri *Hura crepitans* and mataki *Symphonia globulifera* trees may also dominate this forest. In the southern Old Coastal Plain the up to 3.5 m deep swamp forest is relatively poor in species and dominated by watrabiri *Crudia glaberrima* and waratapa *Macrolobium acaciifolium* (Teunissen, 1993). Fish faunas of the swamp forest do not differ much from fish faunas of herbaceous swamps in the Coastal Plain.

Fig. 4.43. Bubble nest of hei-ede kwikwi *Hoplosternum littorale* in a *Typha* swamp at Weg Naar Zee, Suriname. (a) The large (30-40 cm diameter, 10 cm above the water surface) floating bubble nest is made of grass. (b) A nest is frozen and dissected to show the eggs positioned above the water surface and embedded in the oxygen-rich foam.

Fig. 4.44. A high swamp forest in the Young Coastal Plain, Saramacca District.

The Brokopondo Reservoir

Brokopondo Reservoir (officially known as 'Prof.Dr.Ir. W.J. Van Blommestein-meer') is situated in *terra firme* (high dry-land, i.e. not flooded in the wet season) rainforest of the Precambrian Shield in the Suriname River Basin (Van der Heide, 1982). The 54-m high dam was constructed at Afobaka (km 194) and completed in 1964, making Brokopondo Reservoir (1,560 km²) the oldest, large hydroelectric reservoir in tropical rainforest habitat. Prior to the closure of the dam, the Suriname River in the area of the future Brokopondo Reservoir was a low gradient (0.64 m per km), clear-water river with alternation of wide, shallow reaches with rapids and large river islands, and narrow, slowly flowing deeper runs about 300 m in width and 5-7 m deep. The rapids (up to 5 m high) probably presented no biogeographic barrier to dispersal of fishes as most species would have been able to bypass the rapids during the high-water season. Drainage density was high in the area and the river had narrow fringing floodplains without permanent (oxbow) lakes. When the reservoir was finally filled in 1968, some 1160 hill tops formed small islands. The small reservoir catchment (12,550 km²) and low relief in the reservoir area resulted in a long residence time of the water of approximately 28 months and a relatively inefficient energy production (180 MW for a reservoir of 1,560 km²). Van der Heide (1982) described chemical and biological developments in the forming reservoir as determined by (1) the change from turbulence and mixing under riverine conditions to stagnation and stratification in lacustrine conditions, and (2) the decomposition of the drowned rainforest vegetation. Eutrophication, deoxygenation and explosive growth of floating aquatics were characteristic of the filling phase, but currently the water of the reservoir is oligo-

Fig. 4.45. Brokopondo Reservoir with characteristic tree trunks of hard-wood trees still emerging from the water 40 years after impoundment (© W. Kolvoort).

trophic, clear and with high dissolved oxygen concentrations (Mol *et al.*, 2007). Hard-wood trees did not decompose easily and, in 2005, 70% of the reservoir area still had tree trunks emerging from the water (Fig. 4.45). Mol *et al.* (2007) described changes in the fish fauna during 40 years of impoundment and compared the equilibrium fish fauna of the reservoir in 2002-2005 to the fish fauna of the former Suriname River in the Brokopondo Reservoir Area. In 1963, the fish fauna of the Suriname River had 172 species, but only 41 species were collected in the reservoir in 2002-2005. Most fishes from rapids and small rainforest streams did not survive in the reservoir. The fish community of the open water habitat of Brokopondo Reservoir was dominated by three piscivorous species (*Serrasalmus rhombeus, Acestrorhynchus microlepis* and *Cichla ocellaris*; Fig. 4.46ab) and their prey (*Bryconops melanurus* and two *Hemiodus* species; Fig. 4.47), resulting in the food web shown in Fig. 4.48. The inshore fish community was more diverse than the open-water community and was dominated by seven cichlid species (juvenile *Cichla ocellaris, Cichlasoma bimaculatum, Crenicichla multispinosa* and *C. saxatilis, Geophagus surinamensis, Guianacara owroewefi* and *Krobia guianensis*). Interestingly, stunting was observed in several fish species of Brokopondo Reservoir: fishes in Brokopondo Reservoir showed a decreased maximum length, lower size at first maturation, increased batch fecundity and larger oocytes when compared with conspecific riverine populations (Merona *et al.*, 2009; for cichlids in Petit Saut Reservoir French Guiana, also see Ponton & Mérigoux, 2000). Richter and Nijssen (1980) comment on the low fishery potential of Brokopondo Reservoir (piranhas destroying the nets), and, except for sport fishers targeting tukunari (*Cichla ocellaris*), commercial fisheries have indeed not been established in the reservoir.

a

b

Fig. 4.46. Two piscivorous species of Brokopondo Reservoir. (a) The redeye piranha *Serrasalmus rhombeus*. (b) Tukunari *Cichla ocellaris* (© W. Kolvoort).

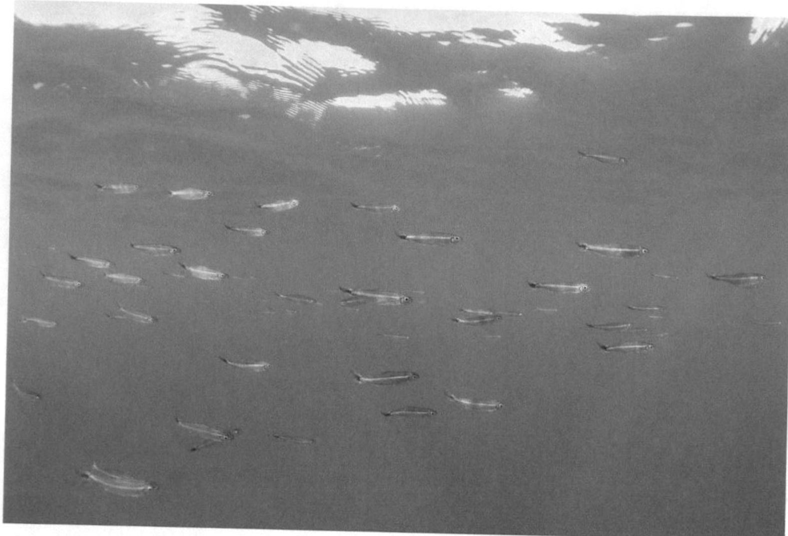

Fig. 4.47. A school of *Bryconops melanurus* in the open water habitat of Brokopondo Reservoir (© W. Kolvoort).

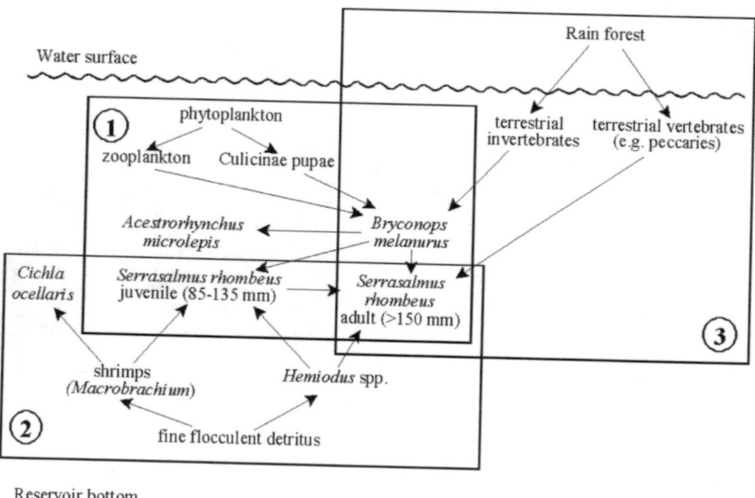

Fig. 4.48. Food web of the open-water habitat of Brokopondo Reservoir, including redeye piranha *Serrasalmus rhombeus*, dagufisi *Acestrorhynchus microlepis*, tukunari *Cichla ocellaris*, nyanga-nyanga *Bryconops melanurus* and dyogu *Hemiodus* spp. (from Mol *et al.*, 2007).

CHAPTER FIVE

HUMAN IMPACTS AND THREATS TO THE FISHES

Healthy fish populations are known to provide important fundamental and demand-derived ecosystem services to human societies (Holmlund & Hammer, 1999). Fish provide directly for human needs as food, medicine, entertainment, and jobs. But most important fishes are part of larger ecosystems and ecosystem function and resilience (ecosystem integrity or health) often rely on the abundance, species composition, and ecological roles of fishes. Indirect fundamental services such as nutrient cycling are often not directly linked to any specific economic market value, but ultimately they are a prerequisite for human existence. In addition, all direct, demand-derived ecosystem services ultimately depend on natural systems and thus on fundamental ecosystem services. To secure the generation of ecosystem services from fish populations it is important to take into account that the fish are embedded in ecosystems and that the integrity of ecosystems in turn is dependent on complex interactions of many chemical, hydrological, biological, energy and habitat structure factors (National Resources Conservation Service, 1998).

Conservation biologists are increasingly realizing that efforts targeted solely at the species level are shortsighted and simplistic, and that to save species, ecosystems and the processes and interactions that occur within them must be protected. Thus Sheldon (1988) argues that conservation efforts at stream fishes should focus on the largest reasonable natural drainages in as many major river systems as possible, i.e. ecosystem management as opposed to single species management.

Human activities have severely affected the integrity of freshwater ecosystems worldwide (Revenga *et al.*, 2005; Dudgeon *et al.*, 2006) and this unfortunately also holds for the inland aquatic systems of Suriname. Habitat loss/modification, including hydroelectric dams, roads, dykes and canals and other hydrological alterations, introduction of nonnative species and pollution all threaten the freshwater fishes of Suriname. Helfman (2007, p. 51) recorded 965 imperiled fish species on the IUCN (2004) red list, of which 656 occurred in fresh water (Revenga *et al.*, 2005). However, assessments of threatened fishes vary greatly among countries and the IUCN red list almost certainly vastly underestimates the real number of endangered fish species. Suriname, with high endemism in its freshwater fish fauna (23.3%; chapter 3; Mol *et al.*, 2012) and at least some of these endemic fish species clearly imperiled (see below), has never developed a red list for its freshwater fishes. In Suriname environmental impacts affecting fishes have mostly been associated with mining and, in the Coastal Plain, with agriculture.

Pollution

In the 1970s toxic pesticides used in mechanized rice culture have caused fish kills, including kwikwi (*Hoplosternum littorale*), sriba (*Astyanax bimaculatus*), krobia (*Cichlasoma bimaculatum*), pataka (*Hoplias malabaricus*), noya (*Trachelyopterus galeatus*) and tilapia (*Oreochromis mossambicus*), in the Nickerie District, northwest Suriname (Vermeer *et al.*, 1974). A large tailing pond of a gold mine in the Mindrineti River catchment area, Brokopondo District, had high concentrations of cyanide and no fish at all (which is very unusual for a water body in Suriname), while a treated-water pond of the same gold mine had fish kills (*Loricariichthys maculatus*, *Astyanax bimaculatus* and *Hoplias* sp.; Fig. 5.1a) probably due to a bloom of the toxic blue-green algae *Microcystis aeruginosa* (Fig. 5.1b). This so-called Harmful Algae Bloom was in turn associated with the transformation of cyanide (used in the recovery of gold from the ore) to nitrate, i.e. eutrophication (J. Mol, pers. observations). Mercury pollution has been associated with artisanal gold mining in eastern Suriname (the greenstone belt) and high mercury concentrations have been detected in freshwater fishes, especially in piranhas (*Serrasalmus rhombeus*) from Brokopondo Reservoir (Mol *et al.*, 2001). The high mercury concentrations pose a threat to humans that consume the fishes (e.g. Cordier *et al.*, 1998), but the fishes themselves are apparently more affected by high suspended sediment concentrations (turbidity) and the accumulation of fine sediment on the stream bed and woody debris that are associated with mining (i.e. habitat degradation, Mol & Ouboter 2004).

Fig. 5.1a. Eutrophication of a treated water pond in a gold mine, Gros Rosebel Area. (a) Dead loricariid catfishes *Loricariichthys maculatus* (© S. Kesarsing).

Fig. 5.1b. Eutrophication of a treated water pond in a gold mine, Gros Rosebel Area. (b) Bluegreen algae *Microcystis aeruginosa* drifting on the water surface.

Habitat Degradation

In the densely populated Coastal Plain of Suriname, wetland habitat degradation is associated with agriculture, bauxite and oil exploitation, and the expansion of the city of Paramaribo. However, Teunissen (1993) concludes that human-associated peat fires destroying large areas of climax swamp forest vegetation constitute the largest human impact on wetlands in the Coastal Plain. In the dry season, large areas of (herbaceous) swamp land are set on fire intentionally to keep these areas accessible to humans or unintentionally during campfires. In extreme dry years associated with the El Nino Southern Oscillation phenomenon, vegetation fires may turn into peat fires which can destroy swamp forest vegetation (e.g. 1963/1964 peat fires reported by Bubberman, 1973).

In northwestern Suriname (Nickerie District), 2000 km^2 of wetlands (herbaceous swamps, swamp forests, marsh forests) have been transformed into rice polders (Teunissen, 1993). In addition, dykes have been built through swamps, not only for purpose of impoldering, but also for water storage, drainage diversion and construction of roads, and where irrigation/drainage canals have been excavated. The dykes and canals change the water regime of the swamps (e.g. Teunissen, 1976). Numerous north-south oriented canals have been constructed in the Buru Swamp (Saramacca District) in order to supply rice fields with irrigation water; some of these canals are more than 10 km long and extend into the brackish mangrove swamps, thus increasing salt penetration in the freshwater

Fig. 5.2. Minimal disturbance of freshwater wetlands during oil extraction with the 'wet operation' technique. Saramacca District, Young Coastal Plain of Suriname.

wetlands. In the dry season, when the swamps dry up and most swamp fishes are found in the canals, the canals (Sur. visgat) are very popular with sport fishers who target mainly hei-ede-kwikwi (*Hoplosternum littorale*).

Since 1980 the national oil company Stateoil (Staatsolie) produces a crude oil from oil fields in the Saramacca District. The first oil fields were impoldered before exploitation was attempted, but recent oil fields are exploited using the so-called 'wet operations' technique in which construction of dykes and canals is not necessary and disturbance of the hydrology of the wetlands is minimal (Fig. 5.2).

In 1941, BHP-Billiton started to mine buried bauxite deposits in the Onverdacht-Lelydorp area (Aleva & Wong, 1998). These mines were situated below groundwater level in freshwater wetlands of the Old Coastal Plain. After exploitation was completed the mine pits filled with groundwater and formed rather deep (>10 m) lakes. Because the soils in the area were mainly pyrite (FeS_2) containing cat-clays, the mine pit lakes were initially vary acid (pH < 4) and also very clear (Secchi transparency >10 m; with aluminum sulfate $Al_2(SO_4)_3$ flocculating humic acids). In Kankantri Lake, for example, aquatic macrophytes (*Cabomba*) were observed to grow from the deep parts of the lakes (6 m) to the surface attesting to the great transparency of the water. The acid mine pit lakes have an ethereal beauty (Fig. 5.3a), but biologically they are rather sterile. However, some fish species still manage to survive under these extreme pH conditions (Fig. 5.3b). As a result of dilution with rain water, the pH of the lakes increases with time to approximately 6-7 and, in later stages, large riverine fishes (e.g. kubi *Plagioscion*, piranha *Serrasalmus rhombeus*, kwasimama *Hypophthalmus marginatus*) can live in the mine pit lakes (K. Wan Tong You, pers. communication).

a

b

Fig. 5.3. (a) Water lilies in an acid mine pit lake, Para District, Old Coastal Plain, Suriname. (b) Krobia cichlids (*Krobia guianensis*) in an acid mine pit lake (© W. Kolvoort)

Shallow swamps in the immediate neighborhood of the bauxite refinery plant at Paranam have been used to dump the NaOH-rich residue of the refinery process, a red, alkaline slurry known as 'red mud'. These confined swamps have thus been changed into shallow, open 'red mud lakes'. Compared to natural swamps the red mud lakes have a high and highly variable pH (pH may change from 5 in the early morning to 9 in the afternoon), a high alkalinity, high dissolved oxygen, high conductivity, and a low humic acid content. Biologically they are more productive than the acid mine pit lakes (Ouboter & De Dijn, 1993). Although typical swamp fishes like kwikwi (*Callichthys callichthys*, *Hoplosternum littorale* and

Fig. 5.4a-c. Legends see next page.

d

e

Fig. 5.4a-e. (a) The hydroelectric Brokopondo Reservoir in the Interior of Suriname, showing skeletons of hardwood trees still standing 40 years after construction of the dam (© W. Kolvoort). (b) Juvenile piranhas (*Serrasalmus rhombeus*) in Brokopondo Reservoir (© W. Kolvoort). (c) A school of the characoid *Bryconops melanurus* in Brokopondo Reservoir (© W. Kolvoort). (d) A large group of peccaries crossing the relatively small Petit Saut Reservoir, French Guiana (© HYDRECO). (e) A dead white-lipped peccary (*Tayassu pecari*) washed ashore at Tonka Island, Brokopondo Reservoir (© P. Ouboter). Peccaries occasionally drown in Brokopondo Reservoir when trying to cross the large reservoir in rough weather.

Megalechis thoracata), noya (*Trachelyopterus galeatus*) and splashing tetra (*Copella arnoldi*) were conspicuously absent from the alkaline red mud 'lakes', many other fish species thrived, e.g. *Leporinus friderici*, *Astyanax bimaculatus*, *Hemigrammus boesemani*, *Charax gibbosus*, *Curimata cyprinoides*, *Cyphocharax microcephalus*, *Hoplerythrinus unitaeniatus*, *Hoplias malabaricus*, *Gymnotus carapo*, *Crenicichla saxatilis*, *Polycentrus schomburgkii*, a pimelodid catfish and a poeciliid (Ouboter & De Dijn, 1993).

In the Interior of Suriname, habitat degradation is mainly associated with bauxite and gold mining. In order to provide energy to the aluminum smelters at Paranam, a dam was constructed across the Suriname River at Afobaka in 1964, resulting in the large hydroelectric Brokopondo Reservoir (approx. 1560 km^2) (Fig. 5.4a). This dam had a major impact on the fish fauna of the middle Suriname River and, in 2005, only 41 fish species were recorded from Brokopondo Reservoir out of a total 172 species that were recorded in 1963-64 from the middle Suriname River and its tributaries in the area of the future reservoir (Mol *et al.*, 2007a). The fish fauna of the open water habitat of Brokopondo Reservoir was dominated by piranhas (*Serrasalmus rhombeus*) (Fig. 5.4b) and large areas of the reservoir had very few species and looked like an aquatic desert (Fig. 5.4c). Forty years after its completion, the reservoir still had an impact on the terrestrial fauna of the adjacent rainforest as groups of peccaries occasionally drowned in the large reservoir when trying to cross it (Fig. 5.4de); the piranhas in Brokopondo Reservoir then fed on the drowned peccaries (Mol *et al.*, 2007a; Fig. 4.48).

There are plans for more dams in the Interior of Suriname. An old plan (1970s) which has recently received renewed interest involves modifications of large portions of the Corantijn River basin resulting in a hydroelectric dam in the Kabalebo River. The drastic alterations of the existing hydrological regime over a significant portion of the river basin will negatively impact the extant fresh- and brackish-water fish communities (Vari, 1982). Twenty fish species endemic to the Marowijne River are threatened by a proposed dam in the Tapanahony River, a major tributary of the Marowijne River (the so-called Tapajai project). This dam would divert water from the Tapanahony River via Jai Creek to the Brokopondo Reservoir (Suriname River system) and thus severely diminish the flow in the Marowijne River with consequent significant changes in aquatic habitats and likely the fish fauna downstream from the reservoir. In addition, the Tapanahony River diversion would also effectively connect the Marowijne and Suriname river systems, each with their own endemic species, and mixing these faunas may well lead to an ecological disaster, i.e. the introduction of Marowijne endemics into the Suriname River system and *vice versa*.

Two small endemic loricariid catfishes of Nassau Mountains, *Harttiella crassicauda* (Fig. 2.3) and *Guyanancistrus* sp., are threatened with extinction by a proposed bauxite mine with a short (5-10 year) mine life.

In Suriname, gold mining is an old practice dating from 1876 (De Vletter & Hakstege, 1998), but the recent (post 1980) gold rush involves both large numbers (25,000-35,000) of 'small-scale' gold miners and at least two multinational companies (one of these with a gold mine in operation since 2004). The largely uncontrolled mining of gold by 'small-scale' miners results in the clearance of large areas of riparian forest along rainforest creeks (Fig. 5.5), changing the aquatic habitat of cool, clear, running water into a series of warm, muddy, sunlit pools. The hydraulic extraction of gold is basically artificial erosion at a very high rate and the increased supply of sediment also changes the instream habitat of the clear-water rainforest streams, particularly with respect to an increased load of suspended

Fig. 5.5. Riparian forest vegetation has been removed by small-scale gold miners and a shaded forest creek with cool, clear, running water has been changed into a series of muddy, sunlit pools. Gros Rosebel area, Saramacca River System.

Fig. 5.6. Large number of noya *Trachelyopterus galeatus* in a turbid dry-season pool in Maikaboe Creek, Mindrineti River, Saramacca River System, Gros Rosebel area. The creek is severely disturbed by small-scale gold miners and, as a result, has turbid water and low fish diversity. Note the albino noya in the center of the photograph.

and deposited fine sediments. The result is a muddy stream with reduced habitat diversity (e.g. streambed structure, leaf litter, and woody debris buried under a thick layer of fine sediments) (Mol & Ouboter, 2004). Fish communities of affected streams show low fish species diversity and changes in community structure (Fig. 5.6). In one of these streams affected by gold mining diurnal, visually orienting fishes (most Characiformes, except *Gasteropelecus sternicla* and *Bryconops*, and cichlids) decreased in numbers, while nocturnal fishes (Gymnotiformes and many catfishes) increased in numbers (Mol & Ouboter, 2004).

The immediate threat to a small, freshwater pipefish in Corantijn River, *Pseudophallus* aff. *brasiliensis* (Fig. 4.12), was removed when the mining company that proposed to construct a jetty in its main habitat left Suriname in 2010, but population densities of this species are low and its survival in Suriname is still precarious (Mol, 2012).

Introduction of Exotic Fish Species

In 1955, the tilapia *Oreochromis mossambicus* was introduced in Suriname by the Fisheries Department for aquaculture purposes (Lijding, 1958). The tilapia was subsequently released intentionally in brackish-water lagoons (e.g. Bigi Pan

Fig. 5.7a. Exotic fish species introduced in Suriname. (a) The landings of Bigi Pan Lagoon fisheries in Northwestern Suriname are dominated by tilapia *Oreochromis mossambicus* since its introduction in 1955 by the Fisheries Department of Suriname.

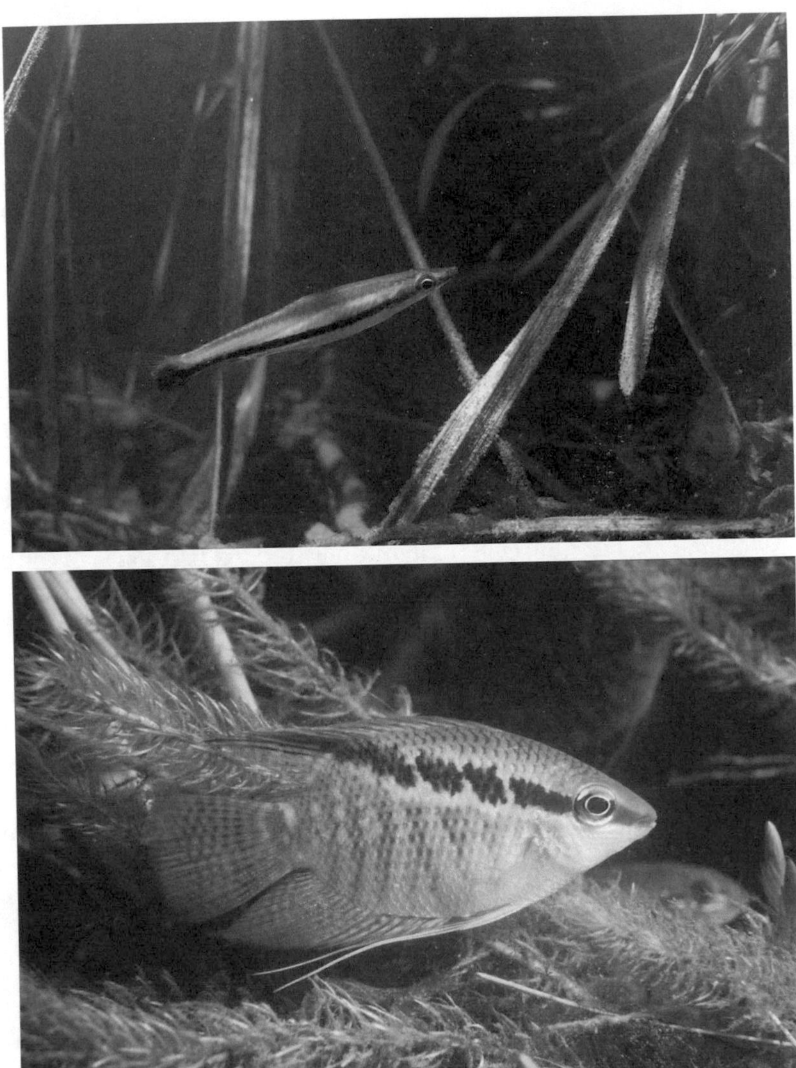

Fig. 5.7b-c. Exotic fish species introduced in Suriname. (b) The elongated pencilfish *Nannostomus harrisoni* that occurs in Para River was introduced in the 1970s by aquarium hobbyists (© W. Kolvoort). (c) The cichlid *Mesonauta guyanae* was also introduced by aquarium hobbyists and occurs in Suriname only in the Para River (© W. Kolvoort).

Lagoon, Nickerie District) and canals. Although the introduction of tilapia probably has not resulted in the extinction of indigenous fish species, it has reduced fish diversity in brackish-water lagoons and canals. In 1943, the fish fauna of Bigi Pan Lagoon was diverse, including two snook (*Centropomus*) species, two *Mugil* species, tarpon (*Megalops atlanticus*), daguboi (*Elops saurus*), granmorgu (*Epinephelus itajara*), two ariid catfish (*Sciades herzbergii* and *Sciades passany*) and, in freshwater parts, kwikwi (*Hoplosternum littorale*) and pataka (*Hoplias malabaricus*) (Geijskes, 1943). Nowadays, the fish catch of Bigi Pan Lagoon is completely dominated by tilapia (Fig. 5.7a), and the fish species mentioned above are either extinct or rare in Bigi Pan Lagoon. In the canals of northern Paramaribo, sport fishers used to organize competitions in catching the largest kwikwi (*Hoplosternum littorale*) (Heyde, 1986), but nowadays the fish fauna of these canals is dominated by tilapia and kwikwi are rarely caught.

In November 2011, an aquarium shop in Paramaribo offered an unusual loricariid catfish for sale and the owner of the aquarium shop claimed to have collected the species in 'a swamp near Paramaribo in the district Saramacca'. This catfish was subsequently identified as *Pterygoplichthys multiradiatus*, an Amazonian species that is not known from Suriname. In February 2012, *Pterygoplichthys multiradiatus* was offered for sale in some numbers as a food fish at the central market of Paramaribo and thus this species may have established a viable population in Suriname.

Ornamental (aquarium) fishes have been imported in Suriname and exported from that country since the 1970s. The trade in ornamental fish in Suriname is poorly documented, but Junk (1984) interestingly reports that 10.1% of the ornamental fishes exported in 1980 from Manaus, Brazil, was shipped to Suriname. Foremost in numbers and commercial value in these exports from Manaus was the cardinal tetra (*Paracheirodon axelrodi*), and this fish from the Rio Negro is now present in the Para River basin (e.g. in Coropina Creek). Other fish species that are only known in Suriname from Para River (close to the international airport) and thus suspect of being accidentally or intentionally introduced in this stream are the pencilfish *Nannostomus harrisoni* (Fig. 5.7b), the flag cichlid *Mesonauta guyanae* (Fig. 5.7c) and the freshwater angelfish *Pterophyllum scalare* (however, in Kaboeri Creek, a tributary of Corantijn River, *P. scalare* is almost certainly native). The impact of these introduced ornamental species on the indigenous fishes of Para River is not known.

CHAPTER SIX

NAMES AND CLASSIFICATION OF FISHES

Fish species often (but not always) have local names, but all formally described species have a scientific or Latin name that is used by the international scientific community. Suriname has many languages and fish species with a country-wide distribution thus may have many local names. On the other hand, fish species that are endemic to (the upper reaches of) a single river system may have a single local name of restricted use or no local name at all (if they are small and little known). Sranan Tongo (S) is a creole language developed by negro slaves and probably the most commonly spoken language in Suriname (besides the official Dutch language). In the spelling of Sranan Tongo this book follows Stichting Volkslectuur Suriname (1995, 3rd ed.). The Carib languages are spoken by a number of indigenous Amerindian peoples in Suriname (e.g. Kalina-caribs, Tiriyo, Wayana) and many local names of plants and animals are derived from a Carib language. The six maroon peoples of Suriname, descendents of escaped slaves, have their own languages, the best known being Saramaka or Saramacca. In the Coastal Plain, East Indian and Javanese (Indonesia) descendants of contract laborers brought in after the abolition of slavery, speak Hindi and Javanese, respectively.

All described species have a single valid scientific name, although this valid name often has many synonyms (for updated valid fish names and synonyms see W. Eschmeyer & Fricke, 2011). The binomial, scientific name, as first proposed by Linnaeus in 1753, is composed of two parts, the first of these is the genus, as in *Homo*, and the second is the species, as in *sapiens*. For example, the redeye piranha or piren (S) is *Serrasalmus rhombeus*. The generic name *Serrasalmus* pertains to a group of closely related species that share a number of common features related to general shape, type of teeth, scalation, fin-ray counts and so on. The specific name *rhombeus* applies only to a single entity that is distinguished from its relatives by a unique combination of characters, often including color/pigmentation pattern. One can speak of the genus *Serrasalmus*, but not of the species *rhombeus*; the specific name is always used in combination with the generic name (thus it is the species *Serrasalmus rhombeus*). The 'Latin' or 'scientific' names of species are always written in italics; the generic name always begins with an upper-case letter and the specific name is always lower case even when it is derived from a proper name (e.g. when constructed after people or places, as in *surinamensis*). When it is obvious which genus we are talking about it is permissible to present the generic name simply as an initial, and so we can say *S. rhombeus*. But this we should do only if it is absolutely clear that the 'S' stands for *Serrasalmus*. After all many different genera of all kinds have names that begin with 'S' and many share the same specific qualifier.

Linnaeus also proposed a formal hierarchical system of classification. The binomial name already implies this: the small category of the 'species' nests within the broader category 'genus'. Related genera (plural of genus) are then grouped together in a family, whose spelling always ends in *-idea*. Worldwide there are approximately 515 families of fishes with living species (Nelson 2006); 64 are represented in fresh and brackish waters of Suriname (Table 2). Large families such as the Characidae and Loricariidae can be subdivided in subfamilies whose spelling ends in *-inae*. A group of similar families is classified in one of the 62 orders of fishes whose spelling ends in *-iformes*. The highest levels of classification pertain to class and phylum. The Class Chondrichthyes contains sharks and rays (including the freshwater sting rays or spari (S) in the family Potamotrygonidae) and the Class Osteichthyes contains the vast majority of bony fishes (Nelson, 2006). All fishes belong to the Phylum Chordata, as do other higher animals including amphibians, reptiles, birds and mammals.

The classification of the redeye piranha can therefore be summarized as follows:

Phylum	Chordata
Class	Osteichthyes
Order	Characiformes
Family	Characidae
Subfamily	Serrasalminae
Genus	*Serrasalmus*
Species	*S. rhombeus*

Characters used to separate species, and often genera, include external (morphological) features such as the number of fin rays, size and number of scales, ratio of various body proportions and color patterns. Internal structures, particularly those pertaining to skeletal elements, are often indicative of relationships at levels of higher classification (family, order, class). A good example are the Weberian ossicles that characterize otophysan fishes (orders Characiformes, Siluriformes, Gymnotiformes and Cypriniformes). In recent years, genetic (molecular) methodology (electrophoresis and DNA analysis) has become more sophisticated and widely used as a supplemental tool for elucidating relationships or recognizing distinct species (see for example Covain *et al.*, 2012).

Why do the scientific names of fishes sometimes change? There are four primary reasons that systematists change names of organisms: (1) 'splitting' what was considered to be a single species into two (or more) (see for example Reis *et al.*, 2005 for splitting of katrina kwikwi *Megalechis thoracata*); (2) 'lumping' two species that were considered distinct into one; (3) changes in classification (e.g. a species is hypothesized to belong in a different genus); and (4) an earlier name is discovered and becomes the valid name by the Principle of Priority (the correct name is the senior synonym; other, later names are junior synonyms).

Many fish species previously unknown to science have been discovered in fresh waters of Suriname over the past few decades and this is a process that continues to this day. When a 'new' (i.e. not formally described) fish is discovered it is given a scientific name by the researcher who formally publishes a detailed description in a recognized scientific journal. Scientific names are frequently descriptive. For example, *rhombeus* refers to the rhomboid body shape of piranhas. New fishes are sometimes named after the locality where they were first collected; for example, *Corydoras saramaccensis* (Saramacca River), *Jupiaba maroniensis* (Marowijne River) and *Loricaria nickeriensis* (Nickerie River). Specific names based on geographic locations often end in *-ensis*. In a third category, fishes named after people have spellings that end in either *-i* (males) or *-ae* (females).

Important scientific specimens such as type specimens used in the description of a 'new' species are generally stored in collections where they serve as vouchers to document identification in published scientific research. Collections are similar to libraries in many aspects: specimens are filed in an orderly and retrievable fashion (usually following the classification system of Linnaeus). Historically most collections of fishes have been preserved in formalin, and then transferred to alcohol for permanent storage (see chapter 7). Fishes collected in Suriname are stored in the National Zoological Collection of Suriname (NZCS, Paramaribo) and in large museums located in major cities of nations that played important roles in the exploration of the world in earlier times or have developed more recently, including the Academy of Natural Sciences Philadelphia (ANSP, Philadelphia), Field Museum of Natural History (FMNH, Chicago), Muséum d'histoire naturelle de la Ville de Genève (MHNG, Geneva), Muséum National d'Histoire Naturelle (MNHN, Paris), Netherlands Centre for Biodiversity Naturalis (NCB Naturalis, Leiden; a fusion of the Zoological Museum Amsterdam [ZMA] and the Rijks Museum voor Natuurlijke Historie in Leiden [RMNH]), and National Museum of Natural History, Smithsonian Institution (USNM, Washington, D.C.).

CHAPTER SEVEN

PRESERVING FISHES

Sometimes it may be desirable to preserve fish specimens, particularly if identification by a specialist is required. Also small, unusual or rare fishes can be kept as curios or as teaching aids. The recommended method of preservation is the same as that used by museum biologists. The main ingredient is full-strength formaldehyde (formalin), which can be obtained from a pharmacy, university laboratory or museum. It is a dangerous chemical and care must be taken to avoid breathing the fumes (where possible use a fumehood). This chemical should always be handled with gloves and it must be kept out of reach of children.

The preserving solution is made by diluting one part of full-strength formalin with nine parts of water. The fish should be fully immersed in the solution. If larger than 15 cm a slit along the side of the belly will facilitate preservation of the internal organs. For long-term storage it is best to transfer the specimen to a 70 per cent solution of ethyl alcohol after the fish is fixed in formalin (that is after two or three weeks).

Anyone sending specimens to a museum or other institution should first contact appropriate staff for detailed instructions on packing and shipping. Freezing is another option, which avoids messy chemicals and is a good short-term storage, particularly if the specimen can be hand-delivered to local authorities (National Zoological Collection Suriname, and Fisheries Department of the Ministry of Agriculture).

CHAPTER EIGHT

DANGEROUS FISHES

In Brazil, injuries to humans caused by freshwater fishes (or other aquatic animals) are common and can be observed in many situations, causing painful processes, necrosis, mutilation, high morbidity and even mortality (Da Silva *et al.*, 2010). Most injuries occurred in fishermen and accidents were mainly associated with stressful work conditions, inattention to basic safety/preventive measures (e.g. working barefooted) and carelessness in handling fish when removing them from a net or hook. Fresh waters in Suriname also have some potentially dangerous fish species. They can be divided into four categories; a fifth category, the Amazonian parasitic candiru catfish (*Vandellia* spp; Trichomycteridae), with their habit of entering the human urethra and feeding on blood (Spotte *et al.*, 2001), does not occur in Suriname.

Biters

Piranhas have a reputation of grisly man-eaters, but most human deaths attributed to piranhas are probably cases of scavenging on drowned or otherwise already dead persons (e.g. the case of two epilepsy patients at the Amerindian village Corneliskondre; Reijenga *et al.*, 1962). Piranhas are usually more dangerous out of the water than in it and most bites occur on shore or in boats when removing a piranha from a gill net or a hook or when a 'loose' piranha is flopping about and snapping its jaws. Amazonian red-belly piranhas (*Pygocentrus nattereri*) are most often accused of attacking humans when in the water (but well-documented cases of attacks are rare), and Amerindians in Guyana are most afraid of this species when entering dry-season pools in the Rupununi Savannas (Lowe-McConnell, 2000). However, in Suriname the redeye piranha *Serrasalmus rhombeus* has also been implicated in recent attacks on bathing humans, notably in two river resorts on the Lower Suriname River and two Amerindian villages on the Wayombo River (Mol, 2006). Redeye piranhas are large (up to 41 cm SL and 3 kg; Willink *et al.*, 2011) and with their razor-sharp, interlocking teeth (Fig. 8.1b) and powerful jaws they can cut out a good piece of flesh with a single bite (Fig. 8.1c), much like how a human takes a bite out of an apple. The piranha attacks in Suriname were associated with special conditions of high juvenile piranha densities in the dry season, high human prey densities, commotion in the water, and spillage of food, fish offal or blood in the water.

People living in the interior of Suriname are often more afraid of the anyumara *Hoplias aimara* than of piranhas. The anyumara is a large, solitary, ambush (sit-and-wait) predator that usually hides motionless under a dead tree trunk or rock

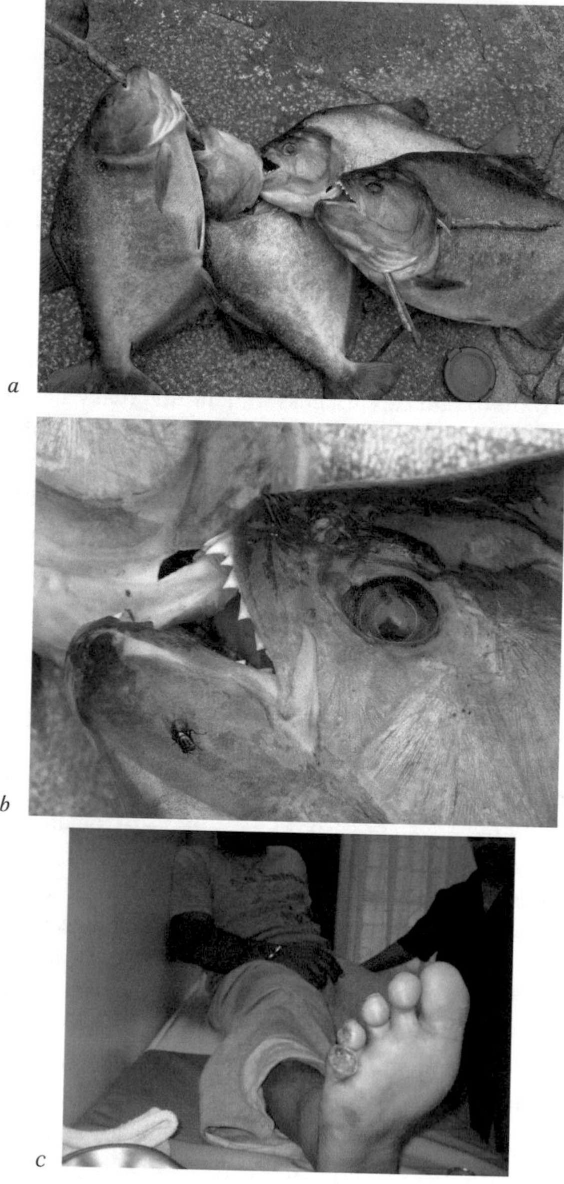

Fig. 8.1. (a) Large redeye piranhas *Serrasalmus rhombeus* are abundant in all Surinamese rivers and Brokopondo Reservoir (photograph shows piranha catch from Upper Coppename River). (b) Detail of the head of *S. rhombeus* showing razor-sharp, interlocking teeth. (c) Injury resulting from a bite of *S. rhombeus* in a recreation park on the Lower Suriname River (© De Ware Tijd).

and waits for prey to come within striking distance. It hunts mainly during the day and can be caught with a large hook (baited with a big chunk of meat or fish) attached to a strong line. When the bait is thrown into the water with a big splash this will draw the attention of the anyumara; if no bite follows it probably means that at that particular spot there is no anyumara present within striking distance

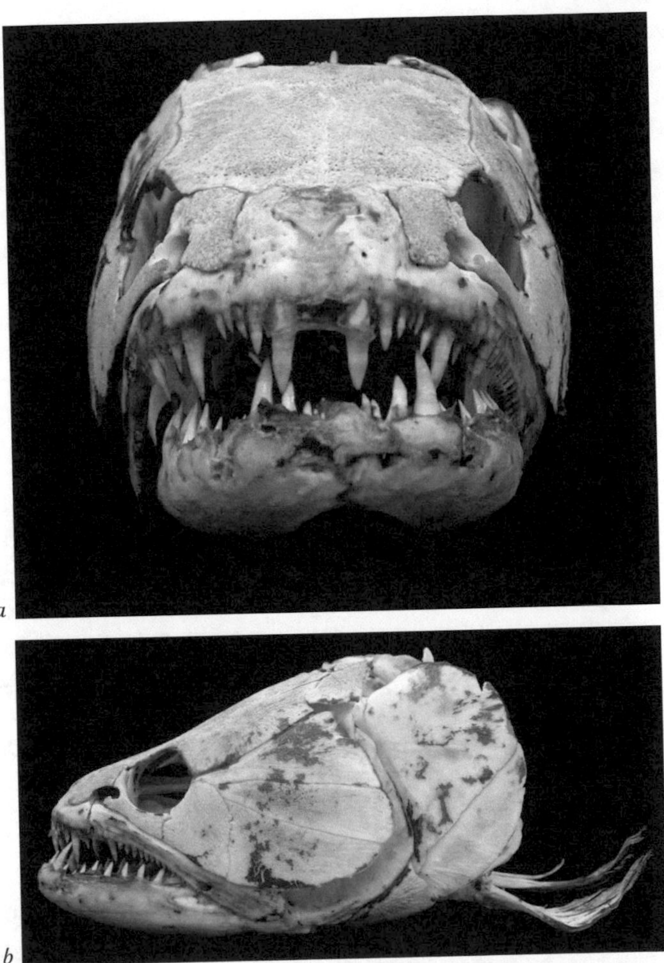

Fig. 8.2. Frontal (a) and lateral (b) view of the skull of the anyumara *Hoplias aimara* (© P. Willink).

and thus one moves a short distance along to stream in order to try again. Taking into account the hunting strategy of this formidable predator it can be dangerous to jump in a stream, especially when the stream is not often visited by humans. The anyumara has sharp, cone-shaped teeth (Fig. 8.2) which are used for tearing flesh (while turning its body, much like a crocodile); the anyumara is not able to cut out a chunk of flesh like a piranha and the resulting wound is very nasty. The smaller pataka *Hoplias malabaricus* can also inflict painful bites (e.g. when 'gropping' for fish in shallow, muddy dry-season pools, pers. observation) (Da Silva *et al.*, 2010).

Some *Cetopsis* species from the Amazon have voracious feeding habits attacking not only carrion, but also live fishes in gill-nets (Burgess, 1989), and on occasion humans (Goulding, 1989: 185). Their smooth, streamlined bodies enable them to penetrate wounds or ripped or torn flesh, becoming wedged in as they continuously chew away flesh of netted fish (usually large pimelodid catfishes) (Burgess,

1989). However, the *Cetopsis* species that was recently (February 2012) discovered in the Upper Commewijne River (see p. 362) did not seem to be aggressive toward its collectors.

We usually think of sharks as a marine hazard rather than a freshwater one, but some shark species (e.g. the bull shark *Carcharhinus leucas*) frequent the lower brackish-water sections of large, tropical rivers, sometimes penetrating well into fresh water. Sharks are caught by estuarine fishers (drifting gill nets; Mario IJspol, Fisheries Department, personal communication), but there are no known cases of shark attacks in Suriname (for example no swimmers have been attacked during the yearly swim marathon from Domburg to Paramaribo, Suriname River).

Stingers

Injuries to humans by stinging catfishes and stingrays are purely defensive; no report has been made of an offensive strike (such as a shark attack). The wounds have a traumatic (puncture) component and a toxic (envenomation) component. The puncture component is like a stiletto-type knife wound often inflicted on the lower leg (waders) or arm (fishermen). Rare puncture injuries to the thorax or abdomen can cause serious injuries and death. The venoms associated with the spines appear to be primarily pain-inducing, since the venom is defensive in nature: the fish is trying to hurt what is attacking it, so the venom is painful and drives away the predator (compare this to a snake, which has venom intended to kill a prey item). Envenomation causes intense pain that is out of proportion to the apparent injury. Information on fish venoms is relatively sparse (Church & Hodgson, 2002). There is often confusion concerning the terms sting, spine, and barb. The sting properly refers to the entire structure: the spine, its integumentary sheath, and the venom glands. The term spine properly refers to the rigid surface of the sting, which is made of bone/dentin. The barbs are the backwards facing serrations associated with the lateral aspect of the spine. The barbs facilitate the tearing of the spine's integumentary sheath and the broadening of the victim's wound. Barbs also work like a backwards pointing fish hook and make disengagement more time consuming and traumatic.

Many catfish have sharp serrated bony spines on the dorsal and pectoral fins which can be locked into place when the catfish is threatened (Lowe-McConnell, 1987, p.282); these are used in defense against predators. Many species of catfishes also have venom glands associated with these spines (Wright, 2009), and stings of some ariids, heptapterids and pimelodids are reported to be very painful. Venom glands were apparently absent in auchenipterids and aspredinids, while in doradids they were morphologically different from those in other catfish (Wright, 2009). While catfish-induced injuries are generally characterized by the pain associated with envenomation, the stings in some species are sufficiently long and sharp to cause severe penetrating trauma. The chemical nature of catfish venoms is poorly known though the loss of toxicity seen when these venoms are subjected to common denaturing agents suggests that proteins constitute the major toxic

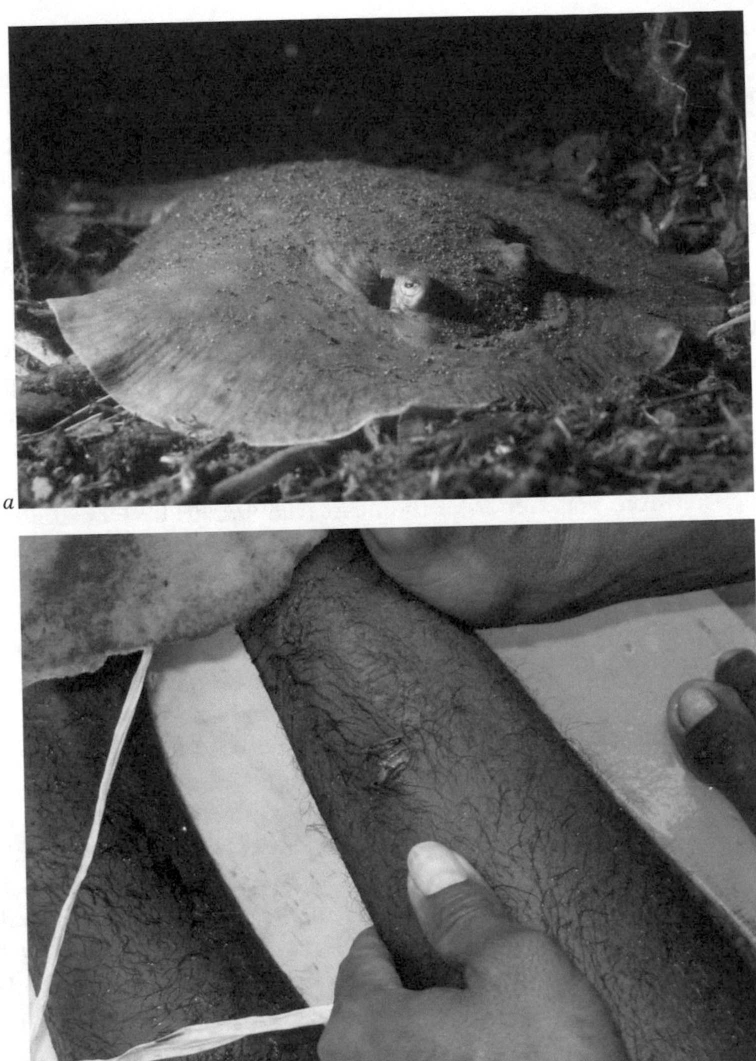

Fig. 8.3. (a) Freshwater stingray (*Potamotrygon* cf. *orbignyi*) on the sandy bottom of a shallow stream (© W. Kolvoort). (b) Injury from the sting of a freshwater stingray.

component of the secretions. A cytolytic activity due to pore formation in cell membranes may explain the 'pain-producing' characteristic of the venoms. The pain is usually localized around the wound, and lasts anywhere between 30 minutes to 2 days. Secondary infections are very common (Da Silva *et al.*, 2010), and they are often bacterial (Junqueira *et al.*, 2006). In 2009, Dr H. Yang (neurologist, then at the Academic Hospital Paramaribo; pers. comm.) reported a case of a fisherman who stepped with his bare foot on a wetkati catfish (*Sciades herzbergii*) in a coastal brackish-water lagoon and was stung by one of its pectoral/dorsal spines; the patient showed symptoms much like the Guillain-Barré syndrome and four weeks after the incident he was still suffering from weakness in the muscles of both arms and legs. When such symptoms increase in intensity until the muscles cannot be used at all and the patient is almost totally paralyzed, the disorder is life-threatening. It is possible that a secondary bacterial infection could have caused the 'Guillain-Barré' syndrome symptoms in the case of the Surinamese fisherman. Haddad *et al.* (2008) report the death of a Brazilian fisherman caused by myocardial perforation from an ariid sting.

Stingrays, both freshwater (Potamotrygonidae; Fig. 8.3a) and marine species, are the most common group of fish involved in human envenomations (Meyer, 1997). Injuries from freshwater stingrays (Fig. 8.3b) are extremely common in South American countries (Haddad *et al.*, 2004) including Suriname, where these fish are plentiful and often come in contact with local people. For example, the word 'Sipaliwini' in Sipaliwini River, Upper Corantijn River System, renowned for its many stingrays, is translated from the Carib language as 'water/river of stingrays'. However, death from a stingray wound is rare (Meyer, 1997), and fatalities typically result from penetration of the heart (the famous case of the death of Steve Irwin in Australia) and abdomen. Freshwater stingrays frequent shallow water, where they often lie camouflaged on the bottom (Fig. 8.3a, Fig. 4.35) or hidden under the sand with only eyes, spiracles and tail exposed. The poisonous spine(s) (usually one but up to four) at the base of the long, thin tail is used to defend against attack by predators. The spine, which can reach up to 37 cm in some marine species, is bone-hard, sharp, and pointed on the end. The sides are retroserrated, so the spine is extracted from human tissue only with difficulty and damage. Sting victims are typically either fishermen who remove stingrays from nets or hooks or waders who step on a stingray, so that the fish thrusts its tail upwards and forward, driving the spine into the foot or leg. Shoes and clothing do not offer adequate protection (penetration of heavy rubber or leather boots has been reported). The traumatic component of the stingray injury should be regarded as an injury caused by a serrated, stiletto-type knife entering the body. Envenomation occurs when the integumentary sheath surrounding the barb ruptures on penetration. All stingray venoms are very similar. They contain serotonin, 5-nucleotidase, and phosphodiesterase. The latter two enzymes are responsible for the necrosis and tissue breakdown seen in stingray envenomations; serotonin is the cause of inexorable pain in the region of the injury. These actions will continue unabated if left untreated. Since the serotonin in stingray venoms produces severe and immediate onset of local pain, any sting that is relatively free of pain indicates that no actual envenomation occurred and the 'lucky' victim endured a 'dry' sting. This may be due to one or more of several reasons: the sheath was previously ruptured, releasing its venom store; the sheath failed to penetrate the wound; the sheath failed to rupture, so the venom remained contained; or, the spine had been broken off previously. But for those people who receive a dose of venom along with the physical trauma of being hit, the tissue necrosis and subsequent secondary bacterial infection that occurs as a result is extremely difficult to treat; and many months and several courses of intravenous antibiotics may be necessary. Stings to the legs should be treated, as well, by several weeks (or perhaps months) of bed rest to help prevent exacerbation of the necrosis and bacterial infection occasioned by the dependent position in which legs are kept when the victim stands upright or walks. Suggested treatment of stingray injuries include (1) washing the wound with clean tap water (to remove the toxin from the wound), (2) applying warm water soaks to the injured body part, as hot as can be tolerated without scalding, for 30-60 minutes (hot-water immersion relieves the venom-provoked pain in less than 1 min), (3) adjunctive local and/or systemic anesthesia if needed, (4) irrigation of wounds with sterile saline to remove toxin; debridement of necrotic tissue (there is variation in recommendation of these vigorous local measures by emergency physicians), (5) soft tissue radiographs or MRI to locate foreign body (part of the ray's barb) remaining in the wound, and (6) antibiotic treatment of acute wound infections directed against tetanus, *Vibrio* and *Aeromonas* (Meyer, 1997).

Electro Shockers

All gymnotiform knifefishes generate electric pulses, but the electric output is generally weak (fractions of a volt). The only species that can produce really powerful, even lethal discharges is the electric eel *Electrophorus electricus* ('sidderaal', 'stroomfisi', 'maisi' or 'praké'): up to 600 volts, 1 ampere, in large specimens of 2-2.5 m (Fig. 8.4). These strong discharges are used for stunning prey or deterring predators and can be deadly for an adult human (electrocution death is due to current flow; the level of current that is fatal in humans is roughly 0.75A). The shocks of the electric eel are only dangerous at close distance to the fish because the water of the streams where it lives is very poor in dissolved ions and thus not very conductive to electricity. I am not aware of any deaths from electric eels in Suriname, but such deaths are apparently quite common in Amazonia (about one

Fig. 8.4. A large (>2 m) electric eel *Electrophorus electricus* caught with a fyke net in a small forest creek near Raleighvallen, Upper Coppename River. Large electric eels can generate powerful shocks of up to 600 volts (1 ampere) (© National Geographic).

death per year in the area of Iquitos, Peru; T. Roberts, pers. communication). Multiple shocks can cause respiratory or heart failure and people have been known to drown in shallow water after a stunning jolt. Touching a large eel (or touching it with a conducting metal (e.g. a cutlass)) can knock an adult human unconscious for several minutes (J. Montoya, pers. communication) and result in drowning if he or she falls unconscious in the water. An interesting early account of the powerful shocks of the electric eel was given by Houttuyn (1764, p.111-120; his *Gymnotus tremulens* or 'beef-aal').

Poisonous Fishes

The only threat in this category are the puffer fishes or bosrokoman (Tetraodontidae) (Fig. 8.5). The 'flesh' (especially the viscera and gonads) of some puffers, including the two estuarine species of Suriname (*Colomesus psittacus* and *Sphoeroides testudineus*), contains the alkaloid poison tetraodotoxin (tetrodotoxin), produced by the fish, which can be fatal. In at least some species, the gonads at spawning time contain the highest concentration of this poison; none occurs in the muscle. Although puffer fishes are eaten in Japan (where they are

Fig. 8.5. A bosrokoman or pufferfish *Colomesus psittacus* (Tetraodontidae) from the estuary of the Nickerie River. Puffer fish contain the lethal tetraodotoxin in their viscera and gonads and should not be eaten.

known as 'fugu') when specially prepared by licensed chefs, they are extremely dangerous and specimens from Surinamese estuarine waters should never be eaten. In Japan, some people die each year after eating fugu that was not well prepared. A case in which a 46-year old man from Nieuw Nickerie died of tetraodotoxin poisoning after eating a bosrokoman *Colomesus psittacus* is described by De Kom *et al.* (2001).

CHAPTER NINE

FISH COLORS

Colors in fishes include fixed colors present throughout a phase of life, and chromatophore-influenced colors which change with the emotional state of the fish and with light intensity. A single species of fish often exhibits a large range of color patterns depending on several factors, including age, sex, geographical location, environmental surroundings (especially water clarity), behavioral mood and stress. The last factor includes the often very different patterns that are flashed out of the water when a fish is captured. Early life stages (post-larvae, juveniles) of Surinamese freshwater fish may look very different from the adult fish, as shown for example in *Triportheus brachipomus* (Géry, 1977, p. 337), katrina kwikwi *Megalechis thoracata* (Fig. 9.1), plarplari *Ageneiosus inermis* (Fig. 9.2), makasriba *Curimata cyprinoides* (Fig. 9.3) and redeye piranha *Serrasalmus rhombeus* (Fig. 9.4). Ponton and Mérigoux (2001) illustrate pigmentation patterns of early stages of the fishes of the Sinnamary River in French Guiana (many of these fishes also occur in Suriname). Good examples of color patterns that differ between the

a

b

Fig. 9.1. Juvenile (a) and adult (b) of katrina kwikwi *Megalechis thoracata*. Juvenile shows early stage of formation of bony plates along the lateral line (a © M. Littmann, b © M. Sabaj Pérez).

Fig. 9.2. Early life stage (a) and adult (b) of plarplari *Ageneiosus inermis* (© M. Sabaj Pérez).

Fig. 9.3. Early life stage (a) and adult (b) of makasriba *Curimata cyprinoides*. Note the black blotch in the dorsal fin of the early life stage (a © R. Vari, b © W. Bronaugh).

Fig. 9.4. Juvenile (a) and adult (b) redeye piranha *Serrasalmus rhombeus* (© P. Willink).

sexes are presented by *Curimatopsis crypticus* (Fig. 9.5) and by many cichlids. In the goldeneye dwarf acara *Nannacara anomala*, it is very easy to tell the males and females apart during breeding, as the males will be in full colors of green with a yellow lower jaw area and some blue in the fins and the females will develop a checkerboard appearance on the upper half of the body from the horizontal band up. Males try to dominate an area containing several females and challenge unfamiliar males, with the defeated male folding its fins and changing color from lustrous green to dull grey and the color of the eyes from golden to black (Hurd, 1997). In Trinidad, coloration of male guppies (*Poecilia reticulata*) was determined by both sexual selection for bright colors to increase mating success and natural selection for subdued colors to reduce predation risk (Endler, 1983, his figure 1). In the related guppy *Micropoecilia parae* males exhibit one of the most complex color polymorphisms known to occur within populations, whereas females are monomorphic; five color morphs have been described (Lindholm *et al.*, 2004; Fig. 9.6): the 'immaculate' morph has the males grey-brown like the female (but lacking the gestation spot on the belly), the 'parae' morph has a colorful tail stripe and vertical bars, while in the 'melanzona' type morphs the males are bright blue, red or yellow-green with double horizontal black stripes. Different color

Fig. 9.5. Female (a) and male (b) of *Curimatopsis crypticus* (© M. Sabaj Pérez).

morphs can occur syntopically in the same shallow ditch (pers. observations), but the frequency of occurrence of the five color morphs may be the result of the combined forces of sexual and natural (predation) selection (Lindholm *et al.*, 2004). Pencilfishes (*Nannostomus* spp.) show very different color patterns during the day and at night. The diurnal pattern of horizontal stripes is only weakly present and supplanted by two or three dark comma-shaped blotches at night (Weitzman & Cobb, 1975). The nocturnal pattern is also shown during short periods in courtship displays. In the turbid water of rainforest streams disturbed by small-scale gold miners, fishes often look pale with colors and pigmentation suppressed (Mol & Ouboter, 2004).

Obviously it is not possible to cover in this book the full range of colors of fish species and thus, with few exceptions of preserved specimens, the colors shown are the 'normal' or average ones displayed by live fish in their natural habitat. Permanent colors include the 'poster' colors of many territorial reef fishes, but also include the shining neon colors of small forest species such as cardinal tetras and the redline tetra of Para River (Fig. 9.7). In dark-water streams, black pigmentation patterns surrounded or supported by white, yellow, orange or red show up most clearly (Fig. 9.8). These pigmentation patterns are usually found as vertical, longitudinal or diagonal stripes on the body, in caudal and humeral spots, or on

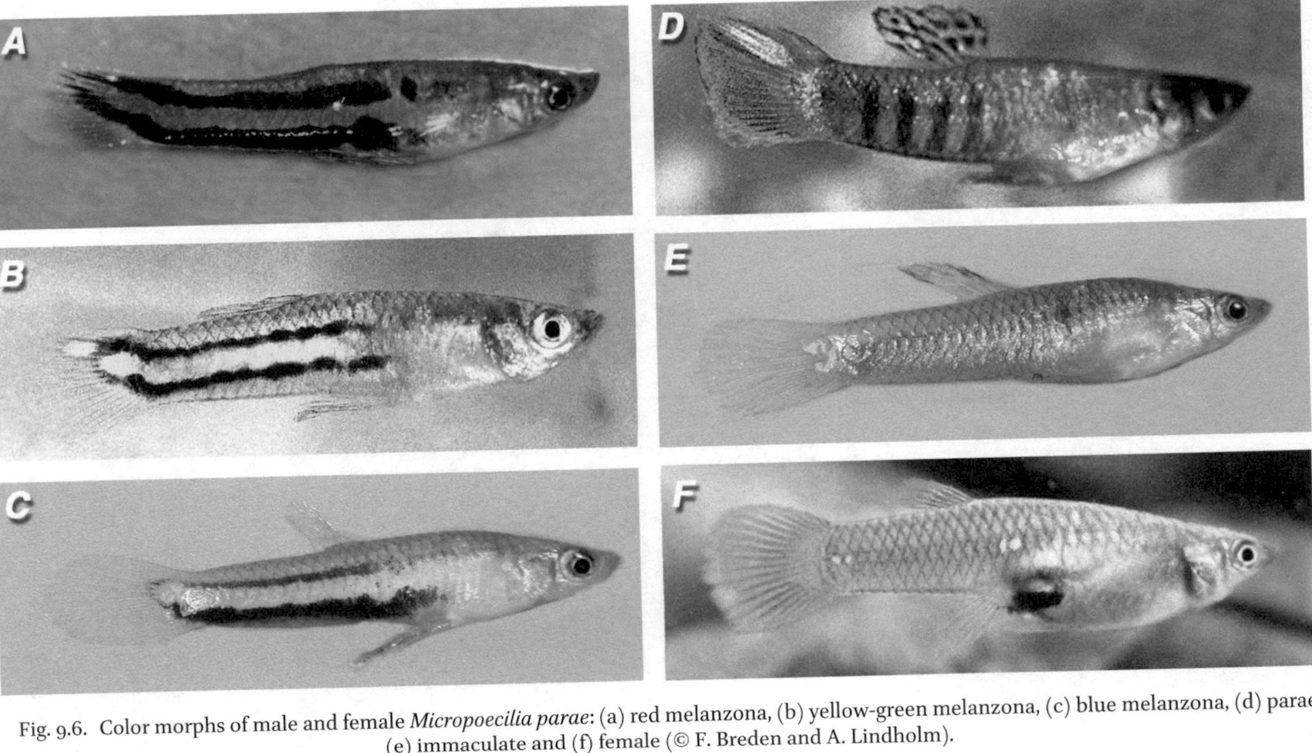

Fig. 9.6. Color morphs of male and female *Micropoecilia parae*: (a) red melanzona, (b) yellow-green melanzona, (c) blue melanzona, (d) parae, (e) immaculate and (f) female (© F. Breden and A. Lindholm).

Fig. 9.7. Shining neon colors of *Hyphessobrycon* sp. 'redline' in Para River (© W. Kolvoort).

Fig. 9.8. Black caudal spots show up clearly in these tetras of the black-water Para River (© W. Kolvoort).

the caudal, dorsal or adipose fin, and are often an important aid in the identification of Surinamese freshwater fishes (e.g. characoids, cichlids, *Corydoras*). In tributaries of the black-water Para River, small characoids schooling in midwater often show very similar pigmentation patterns and also occur in mixed species bands. Selection here is likely for uniformity in appearance, with any individual differences making it easier for predators to pick out an individual. This can help explain the many cases of mimicry, with fishes in different genera being similar in appearance (compare for example *Pristella maxillaris*, *Hyphessobrycon simulatus*, and *Moenkhausia hemigrammoides*). The vertical black stripe through the eye in cichlids (*Pterophyllum scalare*, *Guianacara* spp, *Cleithracara maronii*) or the black 'mask' in many *Corydoras* species can conceal the eye, while black blotches or ocelli on the tail(fin) can direct the attention of a predator to parts of the body that are less vital than the head and eyes, thus allowing a potential prey to survive the attack of a predator. Color patterns and markings come to have behavioral clues, as shown by aquarium experiments (collated by Keenleyside, 1979). In the small characoid *Pristella maxillaris*, the conspicuous black patch on the dorsal fin is jerked rapidly when the fish is alarmed, leading other *Pristella* to follow the school (Keenleyside, 1955). In many cichlids (e.g. *Pterophyllum*, *Mesonauta*) the pelvic fins become conspicuous in color and are jerked in characteristic fashion when 'calling' young to follow the parent. The caudal ocellus developed in many cichlids (e.g. *Cichla ocellaris*, *Mesonauta guyanae*, *Crenicichla* spp) may help the orientation of the young to the parent. Small marine cleaner fishes (*Labroides dimidiatus*) advertise their ectoparasite removing activity with a conspicuous color pattern (including a dark longitudinal band) that allows them to safely clean large predatory fish that would otherwise eat small fish such as these. In the Amazon, cleaning of large predatory *Hoplias* has recently been observed in *Anostomus ternetzi* (Lucanus, 2009) and in juvenile *Platydoras costatus* (Carvalho *et al.*, 2003). Both juvenile *P. costatus* and *A. ternetzi* show a characteristic coloration with longitudinal dark stripes; the two species also occur in Suriname.

CHAPTER TEN

EXTERNAL ANATOMY

This book is designed as an identification and reference guide. It relies mainly on visual comparison of photographs supplemented with features, often related to color pattern, that are useful in recognizing species and distinguishing species from close relatives. Technical jargon, pertaining to various body parts of fishes, is kept at a minimum, but is often unavoidable. The accompanying drawings (Figs. 10.1, 10.2 and 10.3) will help explain some terms and the reader is referred to the 'Glossary of technical terms' for definitions of other terms.

When attempting to identify an unknown fish, particular attention should be paid to overall shape including the head and snout, the number of dorsal fins (presence of adipose fin), fin shape (especially the caudal fin), presence/number or absence of spines, scale number (along the lateral line), and presence/absence and form of teeth.

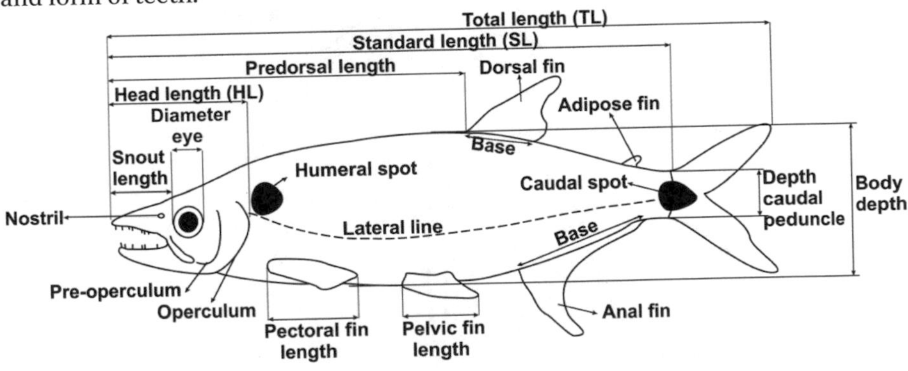

Fig. 10.1a. Standard measurements taken on characiform fishes. Scales are counted in the horizontal row along the lateral line (when present), and sometimes also above and below the lateral line.

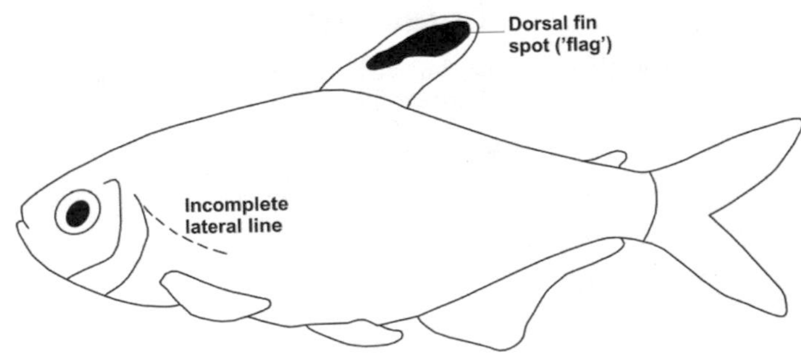

Fig. 10.1b. A small characiform fish with an incomplete lateral line and a spot ('flag') on the dorsal fin.

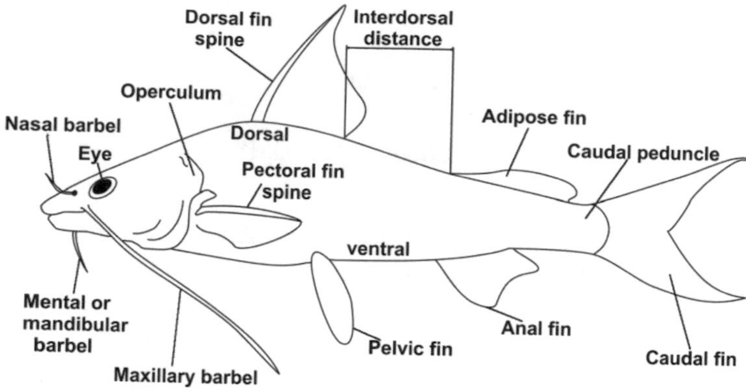

Fig. 10.2a. External anatomy of a catfish (Siluriformes).

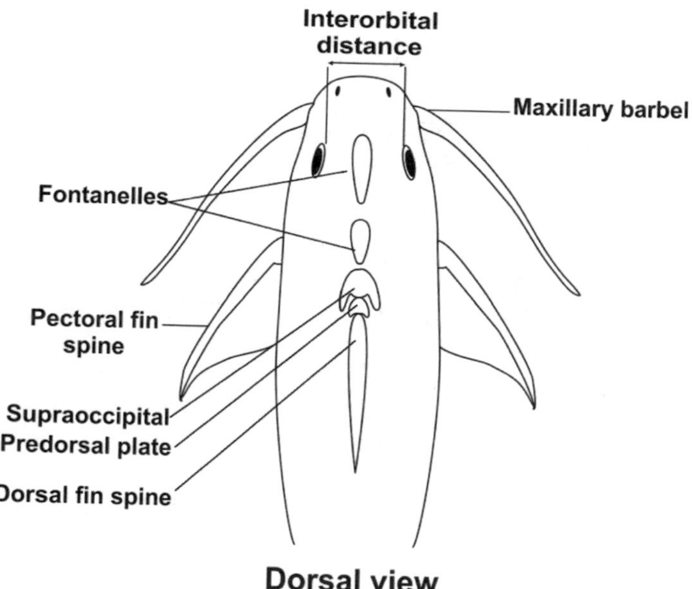

Fig. 10.2b. Dorsal view of a catfish showing fontanelles, supraoccipital and predorsal plate (after Burgess, 1989).

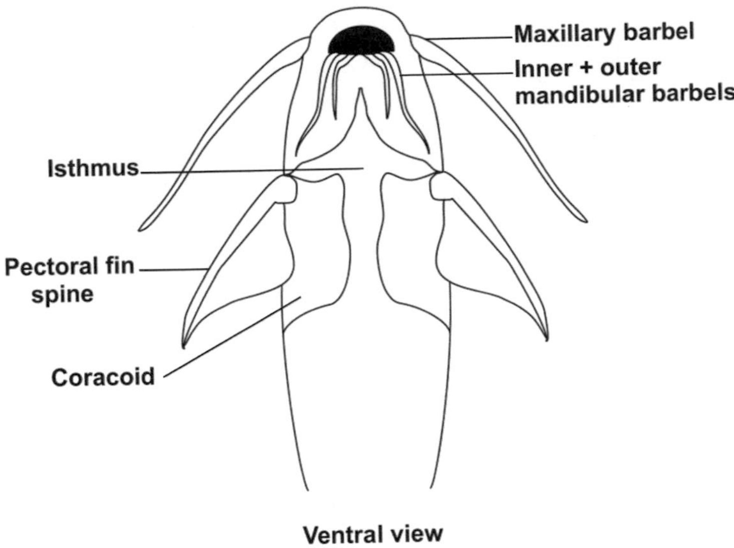

Fig. 10.2c. Ventral view of a catfish showing isthmus, coracoids and maxillary and mandibular barbels (after Burgess, 1989).

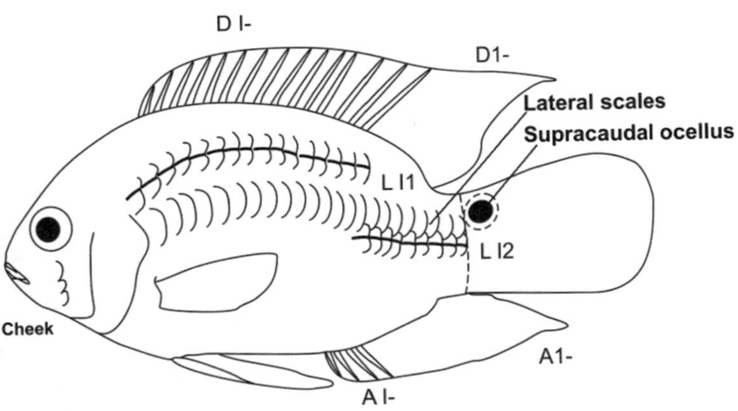

Fig. 10.3. Standard counts taken on a cichlid fish (after Kullander & Nijssen, 1989). Ll1 = upper lateral line scales, Ll2 = lower lateral line scales, cheek = horizontal scale series below the orbit, AI- anal fin spines (roman numerals), A1- = anal fin (soft) rays (arabic numerals), DI- = dorsal fin spines (roman numerals), D1- = dorsal fin (soft) rays (arabic numerals). Lateral scales are counted in the horizontal series next above that including the lower lateral line.

CHAPTER ELEVEN

KEY TO THE (SUB)FAMILIES

1-a Gill openings (4-7) not covered by bony operculum; body extremely depressed (flattened dorso-ventrally)
2

1-b Gills covered by operculum with a single, lateral opening or a single small gill opening as slit or pore under the head
3

2-a Snout produced in a long flat blade with teeth on each side; marine species, sporadically in estuaries
Pristidae (*Pristis perotteti*) (p. 134)

2-b Snout not produced in a long flat blade with lateral teeth; freshwater
Potamotrygonidae (p. 138)

3-a Both eyes on the same side of the body; body compressed laterally
4 (Pleuronectiformes, flatfishes)

3-b Eyes not at the same side of the body; body form variable
6

4-a Caudal fin continuous with dorsal and anal fins
Cynoglossidae (p. 800)

4-b Caudal fin distinct from other fins
5

5-a Both eyes situated on the left side of the body; edge of pre-operculum clearly visible
Paralichthyidae (p. 788)

5-b Both eyes situated on the right side of the body; edge of pre-operculum hidden under the skin, poorly visible
Achiridae (p. 794)

6-a Absence of a dorsal fin
7 (Gymnotiformes or Neotropical knifefishes)

6-b Dorsal fin present
11

7-a Head dorso-ventrally depressed, body cylindrical in cross section
Gymnotidae (p. 574)

7-b Head and body laterally compressed
8

8-a Distinct caudal fin present
Apteronotidae (p. 602)
8-b Absence of well-defined caudal fin
9

9-a Tubular, elongated snout
Rhamphichthyidae (p. 590)
9-b Snout rounded and only moderately elongated
10

10-a Teeth present at both jaws; opening of anus positioned in the anterior part of the head
Sternopygidae (p. 582)
10-b Teeth missing on lower jaws; opening anus positioned in the posterior part of the head
Hypopomidae (p. 594)

11-a Body eel-like, without scales; pelvic and pectoral fins absent; dorsal and anal fins reduced to a rayless ridge; single ventral opening to the gills
Synbranchidae (*Synbranchus marmoratus*) (p. 662)
11-b Body elongate and encased in bony plates that form rings around the body; two gill openings; no pelvic fins (and, in *Pseudophallus*, no anal fin); males with a ventral brood pouch
Syngnathidae (*Pseudophallus* aff. *brasiliensis*) (p. 658)
11-c Body not eel-like or encased in bony rings; two gill openings
12

12-a Body skin without scales or body with bony plates ('scutes')
13 (Siluriformes or catfishes)
12-b Scales covering at least part of the body
21

13-a Body more or less covered with bony plates
14
13-b Body skin naked, without bony plates
15

14-a A single row of spiny bony plates along the lateral line on each side of the body
Doradidae (p. 544)
14-b Two rows of bony plates cover each side of the body
Callichthyidae (p. 382)
14-c Several (>2) rows of bony plates cover each side of the body
Loricariidae (p. 412)

15-a Broad depressed head and body followed by slender caudal peduncle (banjo shape)
Aspredinidae (p. 366)

15-b Variable body shape
16

16-a Distance between anterior and posterior nasal openings larger than two times the diameter of the eye
Ariidae (p. 528)

16-b Distance between the anterior and posterior nasal openings less than two times the diameter of the eye
17

17-a Presence of a patch of hooked odontodes on the interopercle, visible in ventral or ventrolateral aspect of the head
Trichomycteridae (p. 376)

17-b Absence of odontodes on the interopercle
18

18-a Anal fin long (20-50 fin rays); absence of dorsal and pectoral fin spines; small size (<20 cm)
Cetopsidae (p. 358)

18-b Anal fin short (except in *Hypophthalmus* which is >20 cm); pectoral and/or dorsal fin spines present
19

19-a Base of adipose fin much smaller than the base of the anal fin
Auchenipteridae (p. 554)

19-b Base of adipose fin equal to or larger than base of the anal fin
20

20-a Eyes small, without free orbital margin; large mouth with short barbels not reaching the origin of the dorsal fins; small to medium-sized species
Pseudopimelodidae (p. 478, p. 484)

20-b Eyes with free orbital rim; large barbels reaching or surpassing the origin of the dorsal fin; usually medium to large sized species (>20 cm SL)
Pimelodidae (p. 478, p. 512)

20-c Orbital rim free or not; adipose fin well developed; barbels long or short; small sized, usually 20 cm SL or less (species of *Pimelodella* and *Rhamdia* can exceed); the family is easily diagnosed by means of anatomical features, but lacks unique externally visible characters
Heptapteridae (p. 478, p. 492)

21-a Opening to the gill chamber of equal size or smaller than the base of the pectoral fin
22

21-b Opening to the gill chamber larger than the base of the pectoral fin
23

22-a Absence of pelvic fins; each jaw with only two large teeth
Tetraodontidae (p. 804)

22-b Presence of pelvic fins; numerous teeth in each jaw
Batrachoididae (*Batrachoides surinamensis*) (p. 608)

23-a Both jaws elongated in a beak; body slender, elongated
Belonidae (p. 648)

23-b Lower jaw elongated, upper jaw much shorter, not elongated; body slender, elongated
Hemiramphidae (*Hyporhamphus roberti*) (p. 654)

23-c Jaws not elongated in beak; body form variable
24

24-a Presence of at least two spines (unbranched, unsegmented, stiff rays) in the dorsal fin or one spine in the anal fin; dorsal fin often composed of two parts (one part with spines, and a second part with segmented, often branched rays) that may or may not be separated from each other
25

24-b Not more than one spine present in the dorsal fin and no spines in the anal fin; dorsal fin always singular (not taking into account the presence of an adipose fin)
37

25-a Lateral line continuous on caudal fin
26

25-b Lateral line absent or not continuing on caudal fin
27

26-a Less than three spines in the anal fin; the two parts of the dorsal fin united or very close
Sciaenidae (p. 698)

26-b Three spines in the anal fin; the two parts of the dorsal fin well separated by free space
Centropomidae (p. 666)

27-a The two parts of the dorsal fin well separated; 4 spines in the anterior part of the dorsal fin
Mugilidae (p. 610)

27-b The two parts of the dorsal fin united or very close; >4 spines in the anterior part of the dorsal fin
28

28-a Caudal fin absent; body very elongate and laterally compressed
Trichiuridae (*Trichiurus lepturus*) (p. 786)

28-b Caudal fin rounded or square
29

28-c Caudal fin forked or quarter-moon shaped
34

29-a A single non-branched ray in the anal fin; this ray is often soft
30
29-b Several spines in the anal fin
31

30-a Pelvic fins always well separated from each other
Eleotridae (p. 766)
30-b Pelvic fins united, usually forming an adhesive or sucking disc
Gobiidae (p. 774)

31-a Lateral line generally interrupted; single nostril on each side
Cichlidae (p. 720)
31-b Lateral line continuous or absent; two pairs of nostrils
32

32-a More than 6 spines in the anal fin; absence of a lateral line
Polycentridae (*Polycentrus schomburgkii*) (p. 716)
32-b Less than 6 spines in the anal fin; lateral line present
33

33-a Opercle with three spines, central spine largest, with one above and one below it (the two smaller spines may be difficult to see)
Serranidae (*Epinephelus itajara*) (p. 672)
33-b Opercle without spines; mouth large, slightly oblique; dorsal fin with 12 spines; rounded lobes of posterior dorsal, caudal and anal fins giving fish appearance of having three tail fins
Lobotidae (*Lobotes surinamensis*) (p. 690)
33-c Opercle without spines; mouth small, terminal and protractile; dorsal fin with 9 spines
Ephippidae (*Chaetodipterus faber*) (p. 784)

34-a More than 14 soft, branched rays in the anal fin; presence of bony scutes on the (posterior part of) the lateral line in subfamily Caranginae
Carangidae (p. 676)
34-b Less than 14 soft, branched rays in the anal fin; never bony scutes on the caudal peduncle
35

35-a 9 spines in the dorsal fin; mouth highly protrusible
Gerreidae (*Diapterus rhombeus*) (p. 692)
35-b More than 9 spines in the dorsal fin; mouth little or not protrusible
36

36-a Presence of small teeth on palatines and usually vomer; often enlarged canine teeth on jaws
Lutjanidae (p. 684)

36-b Absence of teeth on palatines and vomer; canine teeth rarely present on jaws
Haemulidae (*Genyatremus luteus*) (p. 694)

37-a Body strongly laterally compressed with a well-developed (sharp) ventral keel (hatchet form)
38
37-b Body not with sharp ventral keel
39

38-a Keel with saw-like teeth and short anal fin (<30 rays)
Clupeidae (p. 146)
38-b Keel with saw-like teeth and long anal fin (>30 rays)
Pristigasteridae (p. 164)
38-c Keel smooth
48 (Characiformes)

39-a Presence of an adipose fin posterior to the dorsal fin
48 (Characiformes)
39-b Absence of an adipose fin
40

40-a Caudal fin forked
41
40-b Caudal fin rounded or emarginate
45

41-a Mouth width < eye diameter; superior lobe of caudal fin larger than inferior lobe:
48 (Characiformes)
41-b Mouth width > eye diameter; two lobes of the caudal fin equal in size
42

42-a Presence of a lateral line
43
42-b Absence of a lateral line
44

43-a Last ray of the dorsal fin elongated in a filament; body strongly compressed
Megalopidae (*Megalops atlanticus*) (p. 142)
43-b Dorsal fin with last ray not elongated; body rounded, little compressed
Elopidae (*Elops saurus*) (p. 144)

44-a Ventral keel with saw-like teeth; upper and lower jaws of equal length; short anal fin (<30 rays)
Clupeidae (p. 146)

44-b Ventral keel with saw-like teeth; upper and lower jaws of equal length; long anal fin (>30 rays)
Pristigasteridae (p. 164)
44-c Ventral keel smooth; upper jaw much longer than lower jaw resulting in a 'nose' (the tip of the snout extending beyond the lower jaw)
Engraulidae (p. 152)

45-a Prominent eyes (elevated above top of head) that are horizontally divided into upper and lower portions giving two pupils on each side
Anablepidae (p. 642)
45-b Eyes simple and not prominent
46

46-a Mouth large with upper jaw extending beyond the centre of the pupil
48 (Characiformes)
46-b Mouth small, upper jaw not extending beyond the centre of the pupil
47

47-a Anal fin of male transformed into a copulatory organ (gonopodium); first 3 rays of anal fin of female unbranched
Poeciliidae (p. 632)
47-b Anal fin males not modified; first 3 rays of the anal fin branched
Rivulidae (p. 616)

48-a Body strongly compressed with ventral keel (hatchet-shaped); large pectoral fins
Gasteropelecidae (p. 224)
48-b Body not hatchet-shaped
49

49-a Absence of an adipose fin
50
49-b Presence of an adipose fin
51

50-a Mouth large, often with strong canine teeth; caudal fin rounded; medium-sized to large species (SL > 10 cm)
Erythrinidae (p. 342)
50-b Mouth very small; caudal fin forked; small species (SL < 10 cm)
Lebiasinidae (p. 350)

51-a Presence of teeth
52
51-b Total absence of teeth
Curimatidae (p. 170)

52-a Teeth not (or very little) movable, always on the jaws
53

52-b Teeth weakly implanted (movable) on the lips
61

53-a Dorsal scales two times as large as ventral scales
Alestidae (*Chalceus macrolepidotus*) (p. 228)
53-b Dorsal and ventral scales equal in size
54

54-a Anal fin of medium to long size (at least 3 unbranched rays and more than 11 branched rays)
55
54-b Anal fin short with 2 or 3 unbranched rays followed by less than 11 branched rays
58

55-a Presence of serrae ('spines') on the belly, at least in the post-pelvic region; always a spine in front of the dorsal fin
Serrasalminae (p. 288)
55-b Absence of spines on the belly; predorsal spine usually missing
56

56-a Predatory fishes with caniniform teeth (except in *Phenacogaster* and the scale-eaters *Roeboides* and *Roeboexodon*; see 57-c); maxillary elongated and with teeth
57
56-b Omnivorous fishes with small (more or less conical) teeth; maxillary rarely elongated and without teeth
63 (Characidae)

57-a Large fishes with elongate and strongly compressed body with ventral keel; mouth very oblique and superior; relatively expanded pectoral fins, reduced ventral fins and long anal fin
Cynodontidae (p. 338)
57-b Body very elongate without ventral keel, pike-like, with long snout and small scales; short, falciform caudal fin; origin of dorsal fin much nearer to caudal base than to the tip of the snout
Acestrorhynchidae (p. 334)
57-c Body not very elongate, without ventral keel and with a short snout; top of the head horizontal and dorsal profile often humped; long anal fin; three genera without conical canine-like teeth are the scale-eaters *Roeboexodon guyanensis* (dorsal profile without hump) and *Roeboides affinis* with an outer series of mammilla-like teeth pointing forward and outward, and the omnivorous *Phenacogaster* with small tricuspid to conical teeth (and only very slightly humped dorsal profile)
Characinae (p. 306)

58-a Pectoral fins well developed with 3-5 unbranched rays (except in *Crenuchus* which can be recognized by its oval black spot on the lower part of the caudal peduncle); usually small (< 10 cm) bottom fishes
59
58-b Pectoral fins less developed with only a single unbranched ray; species of medium size
60

59-a Terminal mouth with numerous conic or tricuspid teeth on both jaws; the upper jaw has the teeth organized in one row, in the lower jaw the teeth usually form two rows
Crenuchidae (p. 208)
59-b Inferior mouth with four (rarely two) premaxillary teeth with a horizontal multicuspid cutting border; gill membranes joined together and free of scaly isthmus
Parodontidae (*Parodon guyanensis*) (p. 168)

60-a Teeth in one row of 3 or 4 firmly fixed on each (side of the) premaxillary or dentary; the mouth is small, distant from the anterior orbital rim and may be terminal, (rarely) inferior or superior (in *Anostomus*)
Anostomidae (p. 184)
60-b More than 4 compressed teeth not firmly fixed on premaxilla; mouth usually inferior; most members possess a round, midlateral, body spot and a longitudinal stripe on the (lower) lobe(s) of the caudal fin
Hemiodontidae (p. 216)

61-a Presence of a spine in front of the dorsal fin; numerous, very small teeth set on the lips forming a comb; mouth protrusible, forming a suction disc
Prochilodontidae (p. 180)
61-b Absence of predorsal spine; teeth less numerous and poorly visible
62

62-a Mouth inferior and extremely protrusible; absence of black spots on the scales
Bivibranchia (Hemiodontidae) (p. 216)
62-b Mouth terminal and not protrusible; presence of a black spot on each scale
Chilodontidae (p. 202)

63-a Teeth of lower jaw in two rows (inner row composed of a pair of small conical teeth); premaxillary teeth in upper jaw usually in three rows with presence of larger teeth in the inner row
Bryconinae (p. 282)
63-b Teeth of lower jaw in one row; premaxillary teeth very rarely in three rows
64

64-a Postorbital bone very large, teeth conical or tricuspid, single series of well-developed teeth on the premaxillary, mandible and maxilla; lateral line incomplete (9-11 perforated scales), absence of caudal spot; caudal fin colored red in life

Aphyocharacinae (*Aphyocharax erythrurus*) (p. 304)

64-b Presence of pseudotympanum (which looks like a brilliant humeral spot); a single tooth series on the premaxilla with teeth perfectly aligned and similar in shape and cusp number; lateral line not complete, but rather long with 10-14 perforated scales, reaching to level of dorsal fin; symmetrical black caudal spot present; black pigmentation at the base of the anal fin; caudal fin hyaline

Cheirodontinae (*Odontostilbe gracilis*) (p. 332)

64-c Postorbital bone not very developed, pseudotympanum absent, teeth variable

65

65-a A predorsal spine present, body deep and compressed, usually with a ventral keel

Stethaprioninae (p. 316)

65-b Predorsal spine absent

66

66-a Elongate body with a ventral keel; basally-contracted, multicuspid teeth, outer premaxillary row of teeth weak or even absent; posterior end of maxilla not extending to the eye; dorsal fin origin posterior to the middle of the body and behind the origin of the long anal fin

Iguanodectinae (p. 278)

66-b Body usually not very elongate, body depth rarely more than 4 in standard length, except in *Bryconops*; without ventral keel, dorsal fin usually at mid body, usually two premaxillary rows of teeth

67

67-a Body deep and lozenge-shaped with flat preventral area and long anal fin base; start of lateral line descending sharply like a staircase (*T. chalceus*) or presence of dark longitudinal zigzag stripes on lateral side of body (*T. rarus*)

Tetragonopterinae (p. 320)

67-b Body elongate (body depth > 3.5 times in standard length; 2.8 times in *Hemibrycon surinamensis*), upper jaw at equal level or below the pupil of the eye

Stevardiinae (p. 324)

67-b Not as described above

Genera Incertae Sedis in Characidae (p. 230)

CHAPTER TWELVE

SYSTEMATIC ACCOUNT OF THE SPECIES

FAMILY PRISTIDAE (SAWFISHES)

Sawfish are a family of rays, characterized by a long, toothy extension of the snout (rostrum; Fig. 12.1). Several species can grow to approximately 7 meters. The rostrum is covered with motion- and electro-sensitive pores that allow sawfish to detect movement and even heartbeats of prey hiding under the ocean floor. The rostrum serves as a digging tool to unearth buried crustaceans. When a suitable prey swims by, the normally lethargic sawfish springs from the bottom and slashes at it with its saw. This generally stuns or injures the prey sufficiently for the sawfish to devour it. Sawfish also defend themselves with their rostrum, against predators such as sharks and intruding divers. The "teeth" protruding from the rostrum are not real teeth, but modified tooth-like structures called denticles. The body and head of a sawfish are flat, and they spend most of their time lying on the sea floor. Like rays, sawfish's mouth and nostrils are on their flat undersides. The mouth is lined with small, dome-shaped teeth for eating small fish and crustaceans; sometimes the fish swallows them whole. Sawfish breathe with two spiracles just behind the eyes that draw water to the gills. The skin is covered with tiny dermal denticles that give the fish a rough texture. They are not to be confused with sawsharks (order Pristiophoriformes), which have a similar appearance but a pair of long barbels about halfway along the snout and the gill slits on the side of the heads rather than underneath the head (sawsharks are not known from Suriname). All species of sawfish are considered critically endangered by the IUCN.

Pristis perotteti Müller & Henle, 1841
Local name(s): krari(n), zaagvis (largetooth sawfish)

Diagnostic characteristics: snout extremely elongated as a flat, narrow, and firm blade, edges of which are armed with a single series of large tooth-like structures (up to 20 on each side); pectoral fins barely enlarged, not extending forward to the level of the mouth, terminating rearward anterior to origin of pelvic fins; eyes and spiracles on top of head, spiracles well posterior to eyes, mouth transverse and straight with numerous teeth in pavement pattern forming bands along jaws, nostrils well in front of the mouth; color more or less uniformly brown
Type locality: Senegal, West Africa, in fresh water
Distribution: cosmopolitan (in Suriname, known from the Marowijne and Suriname estuaries)
Size: 700 cm
Habitat: sandy or muddy bottoms of shallow coastal waters, estuaries and river mouths; large adults can also be found in fresh water
Position in the water column: bottom
Diet: bottom living invertebrates and fishes

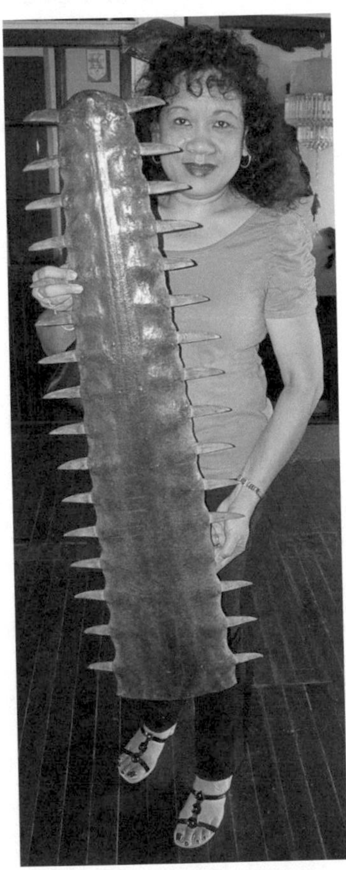

Fig. 12.1. Toothed rostrum of a largetooth sawfish *Pristis perotteti* from the Suriname River Estuary.

Reproduction: the largetooth sawfish is estimated to mate once every two years, with an average litter of around eight. Males and females show sexual dimorphism in the number of rostral teeth, with males averaging 18 (16-20) teeth on each side of the rostrum and females only 16 (14-18) (Thorson, 1976). They mature very slowly; it is estimated that they do not reach sexual maturity until they are 3.5 to 4 m long and 10 to 12 years old. They reproduce at lower rates than most fish. This makes the animals especially slow to recover from overfishing. Females are viviparous, bearing live pups, whose semi-hardened rostrum is covered with a membrane. This membrane prevents the pup from injuring its mother during birth and it eventually disintegrates and falls off.
Remarks: a nocturnal species, usually sleeping during the day and hunting at night, with undeveloped eyes, also related to their muddy habitat. The rostrum is the main sensory device. The largetooth sawfish is a critically endangered species according to the IUCN.

Pristis perotetti, dorsal view of the toothed rostrum of a large specimen, Suriname River Estuary

FAMILY POTAMOTRYGONIDAE (NEOTROPICAL RIVER STINGRAYS)

The family Potamotrygonidae comprises four genera (*Potamotrygon*, *Paratrygon*, *Plesiotrygon* and *Heliotrygon*) of Neotropical freshwater stingrays and in Suriname is represented by three *Potamotrygon* species. The Potamotrygonidae is the only living chondrichthyan family restricted to freshwater habitats. Potamotrygonid stingrays are clearly monophyletic, sharing unique morphological and physiological specializations, including a pelvis with a greatly expanded anterior median process (pre-pelvic process), blood with low concentrations of urea, and reduction of the rectal gland (Thorson *et al.*, 1983a). They are generally medium to large sized batoids, ranging from about 25 cm in disc width or length to well over 100 cm in adults of some species. Species of *Potamotrygon* have moderately stout and short tails, usually shorter than disc length, whereas *Paratrygon* and *Plesiotrygon* have slender, filiform or whiplike tails. The recently described genus *Heliotrygon* has a highly circular disc and an extremely reduced sting (Carvalho & Lovejoy, 2011). The dorsal surface of the disc and tail are usually covered with many denticles, thorns and tubercles. The caudal sting (or serrated spine) is a rigid dermal derivative, located on the dorsal surface of the tail, containing small lateral serrations directed toward its base and an acute distal tip. The sting is well developed and located posteriorly in both *Plesiotrygon* and *Potamotrygon*. The stings contain longitudinal grooves to conduct venom produced in special glands at their bases, and are continuously worn, shed and replaced; up to four stings may be present in one individual. The disc is usually slightly longer than wide and covers most of the pelvic fins posteriorly. Dorsal and caudal fins are absent, but membranous skin folds (finfolds), with rudimentary internal radial elements, occur in *Potamotrygon* on both upper and lower tail midlines posterior to caudal stings. The eyes are moderately large in *Potamotrygon*. Oral teeth are small with short, single cusps (more prominent in adult males), in usually less than 50 rows in either jaw, and set in quincunx. Most potamotrygonid species have colorful dorsal arrangements, including spots of various dimensions, ocelli, reticulate patterns, and vermiform markings, which are generally species-specific, and grey, brown, reddish-brown or black background coloration. Potamotrygonids are ovoviviparous (aplacentally viviparous), and the developing embryos are nourished by uterine milk secreted by trophonemata (Thorson *et al.*, 1983b). Gestation may be restricted to certain stations or occur throughout the year, and the number of young produced in each gestation varies among species, but is usually from two to seven. Potamotrygonids occur only in South American rivers that drain into the Atlantic Ocean or Caribbean Sea. They are conspicuously absent, however, from the São Francisco basin in northeastern Brazil, rivers that drain into the Atlantic from the Atlantic rainforest of northeastern and southeastern Brazil, the Upper Paráná basin, and rivers south of the La Plata River in Argentina. In Suriname they are apparently absent from the Saramacca River. Generally, most potamotrygonid species have distributions restricted to a single basin or river system, with only a few species present in more than one basin (e.g. *Potamotrygon orbignyi*). This high degree of endemism has led recent workers to express concern that some species may be endangered or are at least clearly vulnerable at present (two species are cited on the IUCN Red List as "data deficient"). Potamotrygonids are generally not consumed as food, but are commercialized in increasing quantity by the aquarium trade, and are only seldom bred in captivity for commercial purposes (Ross & Schäfer, 1999; cf. Carvalho, 2001). Potamotrygonids are much maligned and feared because of their venomous caudal stings (chapter 8), but pose little or no threat if not stepped on or directly interfered with.

1-a Dorsal region of disc light or dark sometimes with vague dark spots or a reticulated pattern

Potamotrygon orbignyi

1-b Dorsal region of disc dark with clear yellowish or orange-red spots

2

2-a Spots ocellated, deep orange to red of irregular form, encircled by a relatively broad black ring; Corantijn River

Potamotrygon boesemani

2-b About 40 wide circular yellowish spots, themselves formed by smaller patches; Marowijne River

Potamotrygon marinae

Potamotrygon boesemani Rosa, Carvalho & Wanderley, 2008
Local name(s): liba-spari

Diagnostic characteristics: spots ocellated, deep orange to red of irregular form, encircled by a relatively broad black ring
Type locality: Matapi Creek, ca. 1 kilometer from Corantijn River, Nickerie District, Suriname, 5°00'N, 57°16'W
Distribution: Corantijn River drainage, Suriname
Size: 43 cm disc width
Habitat: mainly found in the main river channel, clear water, but also in smaller tributary streams over sandy substrate
Position in the water column: bottom
Diet: diet and hunting tactics of potamotrygonid rays are described by Silva & Uieda (2007) and Garrone-Neto & Sazima (2009), respectively
Reproduction:
Remarks: nothing is known about its biology.

Potamotrygon marinae Deynat, 2006
Local name(s): liba-spari

Diagnostic characteristics: about 40 wide circular yellowish spots, themselves formed by smaller patches
Type locality: Maroni, grand Inini, French Guiana
Distribution: Marowijne (Maroni) River drainage and Oyapock River drainage
Size: 41 cm disc width
Habitat: apparently mainly found in the main river channel
Position in the water column: bottom
Diet:
Reproduction:
Remarks: nothing is known about its biology.

Potamotrygon orbignyi (Castelnau, 1855)
Local name(s): liba-spari

Diagnostic characteristics: dorsum light or dark brown sometimes with vague dark spots or a reticulated pattern
Type locality: Rio Tocantins, Brazil
Distribution: Widespread in South America
Size: 35 cm disc width
Habitat: clear-water rivers of the Interior (also in Brokopondo Reservoir)
Position in the water column: bottom
Diet:
Reproduction:
Remarks: this species was identified as *P. histrix* (e.g. in Planquette *et al.*, 1996), but later identified as *P. orbignyi* (Rosa *et al.*, 2008).

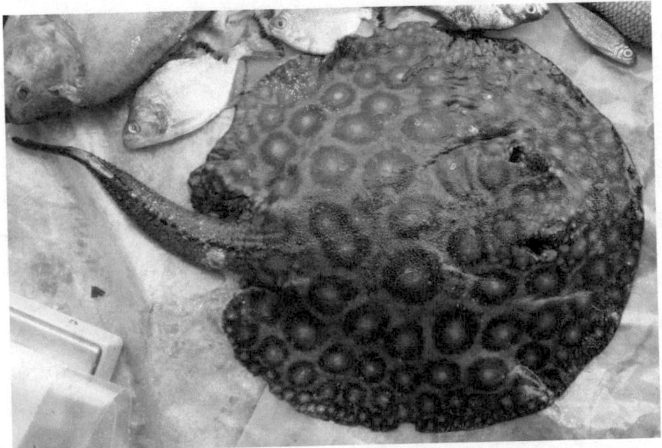

Potamotrygon boesemani (live), Corantijn River (km 160) (© Ross Smith)

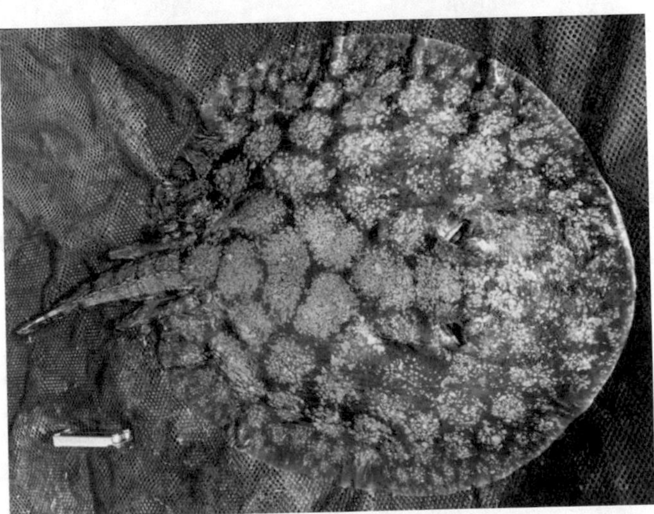

Potamotrygon marinae (live), 400 mm disc width, Lawa River (Marowijne River), ANSP187098 (© M. Sabaj Pérez)

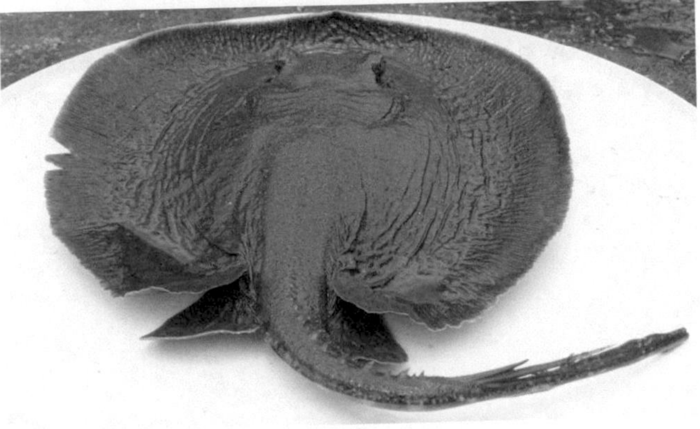

Potamotrygon cf. *orbignyi* (live), Upper Coppename River

FAMILY MEGALOPIDAE (TARPONS)

The Megalopidae consists of two species of large, silvery bodied fishes generally referred to as tarpon. Tarpon can be distinguished from similar looking fishes by the presence of a filamentous last ray of the single dorsal fin, a large mouth, in which the maxilla extends at least to the posterior margin of the orbit, large scales covering the body, and a large gular plate. Tarpon are found only in near-shore habitats in tropical and warm temperate waters. Both species can exist in freshwater and are often found in river mouths and estuaries. Tarpon are generally not considered to be a valuable food fish, but are important game fish through much of their distribution.

MEGALOPS ATLANTICUS Valenciennes, 1847
Local name(s): trapun, kofun

Diagnostic characteristics: a large, moderately elongate, silvery fish with a highly compressed body and large scales; a greatly prolonged final ray of the dorsal fin; mouth large and oblique, upturned, with lower jaw prominently projecting
Type locality: Antilles (Guadeloupe, Martinique Island, Santo Domingo, Puerto Rico)
Distribution: tropical Atlantic and associated estuaries, lagoons, and rivers
Size: up to 250 cm TL (120 kg; males are smaller and rarely reach 50 kg)
Habitat: juveniles live in freshwater habitats of lower (coastal) reaches of rivers and coastal swamps, adults spawn in sea (Crabtree *et al.*, 1992)
Position in the water column: mid-water
Diet: small fishes, shrimps and crabs
Reproduction: the female spawns millions of very small eggs (0.6-0.75 mm diameter) at sea; the transparent leptocephalus larvae (approx. 5 cm TL) move inshore into the mangroves (Crabtree *et al.*, 1992)
Remarks: the tarpon are facultative air-breathers and at least the juveniles can survive in water with low dissolved oxygen concentrations (e.g. swamps) (Geiger *et al.*, 2000); they are relatively long-lived (age >50 years) and very popular with sport fishers.

Megalops atlanticus (live), Lower Suriname River

FAMILY ELOPIDAE (TENPOUNDERS OR LADYFISHES)

Elopidae is a family of ray-finned fishes containing the single genus *Elops* with 6 species They are commonly known as ladyfishes or tenpounders. They have a long fossil record, and are easily distinguished from other fishes by the presence of an additional set of bones in the throat. The body is fusiform (tapering spindle shape) and oval in cross-section; being slightly laterally compressed, the eyes are large and partially covered with fatty (adipose) eyelids. They are related to the order of eels, which shows in their larvae (highly compressed, ribbon-like and transparent leptocephali, looking very similar to those of eels). After initial growth the leptocephali shrink and then metamorphise into the adult form. The ladyfish are coastal dwelling fish found throughout the tropical and sub-tropical regions. Spawning takes place at sea and the larvae then move inland, entering brackish waters. Their food is smaller fish and crustaceans (shrimp).

ELOPS SAURUS Linnaeus, 1766
Local name(s): daguboi, landyan (ladyfish)

Diagnostic characteristics: body elongate, fusiform, moderately compressed, very small scales; eye large; mouth large, nearly horizontal and almost terminal
Type locality: Carolina, U.S.A.; Jamaica
Distribution: Western Atlantic
Size: 90 cm TL (most are <50 cm TL; they seldom reach a weight of 10 pounds or 4.5 kg)
Habitat: brackish water lagoons and estuaries
Position in the water column: mid-water
Diet: shrimps and small fishes
Reproduction: spawns at sea, with leptocephalus larvae migrating inshore to estuarine waters
Remarks: mainly caught in shallow water of estuaries and lagoons (e.g. Bigi Pan Lagoon) in mangrove forest.

Elops saurus (live), Lower Commewijne River

FAMILY CLUPEIDAE (HERRINGS)

The family Clupeidae includes fishes known as herrings, shads, sardines, pilchards, and sprats. Clupeids are generally fusiform fishes with a strongly compressed body and typically a single series of scutes running along the ventral midline of the body (forming a sharp keel with saw-like teeth). Fins are without spines; anal fin with less than 28 fin rays; pelvic fins inserting below dorsal-fin base; there is no adipose fin or lateral line canal system on the body. The mouth is usually terminal and articulation of lower jaw is always anterior to the vertical through the middle of the eye. Clupeids are especially abundant in temperate coastal waters, whereas anchovies (Engraulidae) are more abundant in tropical waters. Most clupeids are schooling, coastal marine species. Some species venture into estuaries and freshwaters and a few are found primarily, if not exclusively, in inland waters. Many of the coastal marine species are important food fishes. However, none of the species listed here for Suriname are significant food fishes.

1-a Last dorsal-fin ray filamentous; supramaxillae (hypomaxillae) absent; upper part of maxilla without sharp, backward pointing spine
Opisthonema oglinum
1-b Last dorsal-fin ray normal
2

2-a Hind-border of gill opening with two fleshy outgrowths; presence of small toothed supramaxillae (hypomaxillae) between posterior tip of premaxilla and expanded blade of maxilla; absence of backward pointing spine near base of maxilla; medium-sized species (common to 12 cm SL, but up to 20 cm SL)
Harengula jaguana
2-b Hind-border of gill opening evenly rounded, without two fleshy outgrowths; absence of toothed supramaxillae; upper part of maxilla at about level of eye with sharp, backward pointing spine; a small species (up to 8 cm SL)
Rhinosardinia amazonica

Harengula jaguana Poey, 1865
Local name(s): – (Scaled herring)

Diagnostic characteristics: hind-border of gill opening with two fleshy outgrowths; presence of toothed supramaxillae (hypomaxillae); absence of backward pointing spine near base of maxilla; body fusiform, moderately deep, compressed, body depth usually 34% or more of standard length (over 40% of standard length in large individuals), and greater than head length; gill rakers fine, usually 32-39 on lower limb of first arch; pelvic fins with 8 rays; color of back and upper sides blue/

This page is intentionally left blank.

black, with faint longitudinal streaks, a dark spot behind gill cover, faint or conspicuous; fins hyaline, but tips of caudal often dusky
Type locality: no types known
Distribution: Western Atlantic, probably throughout the area, northward to New Jersey and southward to southern Brazil
Size: maximum 22 cm standard length; commonly to 12 cm SL
Habitat: pelagic and demersal in coastal waters over sand and mud bottoms, often near estuaries and sometimes in hypersaline lagoons; can be abundant in nearshore areas, estuaries and bays
Position in the water column: pelagic
Diet: plankton? (fine gill rakers)
Reproduction: spawns at night with some evidence of intermittent or spasmodic spawning; most mature at age-1 (80 to 130 mm), all by age-2; size at first maturity 78 to 85 mm standard length; estimated fecundity 5,563 to 52,753 eggs (based on 22 females, 85 to 163 mm standard length); relative fecundity 323 to 807 eggs/g; eggs pelagic, transparent, spherical, 1.55 to 1.85 mm in diameter
Remarks:

OPISTHONEMA OGLINUM Lesueur, 1818
Local name(s): – (Atlantic thread herring)

Diagnostic characteristics: body fusiform, moderately compressed; abdomen with 32 to 36 scutes forming a distinct keel; no hypomaxilla; gill rakers fine and numerous, increasing with size of fish up to 12 cm standard length, then stable at 28 to 46 gill rakers; posterior margin of gill chamber with 2 fleshy lobes; dorsal fin slightly anterior to centre point of body, its last ray filamentous; anal fin short and placed well posterior to vertical through posterior base of dorsal fin; 7 branched pelvic-fin rays, pelvic-fin origin inserted at point about at vertical through middle of dorsal-fin base; color: dorsum and upper sides blue-green, sometimes with dark horizontal lines, lower sides and abdomen silvery; dark spot on side posterior to gill cover, about equal in size to diameter of pupil (followed in some specimens by one or more irregular lines of smaller spots); margin of dorsal fin and its filament dusky; caudal fin with black tips
Type locality: Newport, Rhode Island
Distribution: Western Atlantic
Size: maximum 38 cm standard length, commonly to 20 cm SL
Habitat: mainly a marine species that probably does not enter estuarine, low-salinity water.
Position in the water column: pelagic
Diet: adults feed on small fishes, crabs, and shrimps; juveniles on planktonic organisms.
Reproduction: between April and June in the Guianas; estimated fecundity 13,638 to 67,888 eggs; relative fecundity 471 to 746 eggs/g; eggs pelagic, 1.08 to 1.31 mm in diameter.
Remarks:

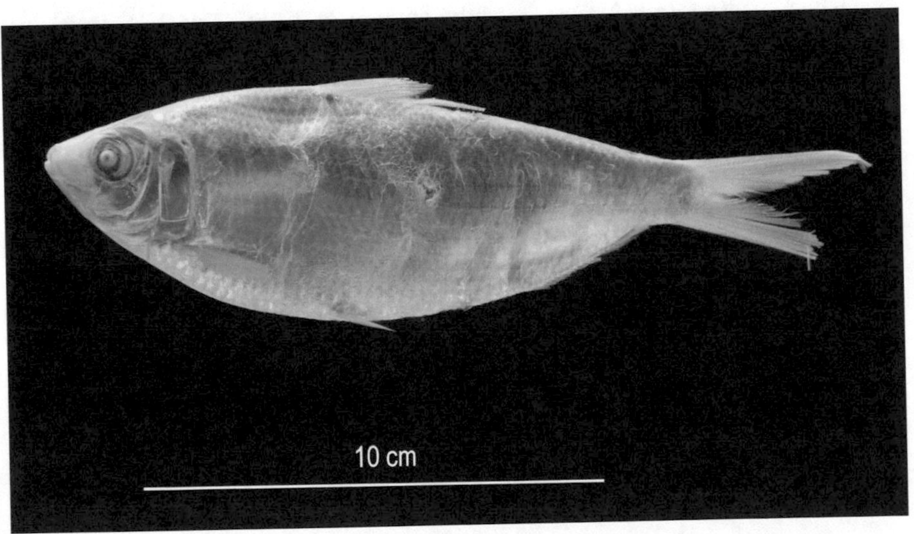

Opisthonema oglinum (alcohol), Suriname River Estuary, RMNH26235 (© Naturalis)

Rhinosardinia amazonica (Steindachner, 1879)
Local name(s): – (Amazon spinejaw sprat)

Diagnostic characteristics: upper part of maxilla at about level of eye with sharp, backward pointing spine, absence of elongated dorsal fin rays; a relatively small species
Type locality: Bahia, Brazil
Distribution: along the Atlantic coast from Orinoco River to northeastern Brazil: Brazil, French Guiana, Guyana, Suriname and Venezuela; in Suriname, known from the Corantijn Estuary (Vari, 1982)
Size: maximum size to 8 cm standard length, commonly to 5 cm SL
Habitat: freshwaters of rivers, but also enters estuaries
Position in the water column: pelagic; presumably schooling
Diet: zooplankton
Reproduction:
Remarks: at present only known from the Lower Corantijn River (Vari, 1982), but probably present in other Surinamese rivers as well.

Rhinosardinia amazonica (alcohol), scale bar = 5 mm, INHS49009 (© M. Sabaj Pérez)

FAMILY ENGRAULIDAE (ANCHOVIES)

Small to moderate-sized silvery fishes (to 30 cm standard length, commonly 8 to 15 cm standard length), usually with fusiform, subcylindrical bodies, but sometimes quite strongly compressed; no scutes present along the abdomen in New World species (except for plate-like scute at pelvic-fin bases). Snout prominent, blunt, and projecting beyond the tip of the lower jaw. Lower jaw almost always long and slender and characteristically 'underslung'; its articulation extending well behind the eye. Typically with 2 supramaxillae. Maxilla (posterior tip of upper jaw) short and rounded or long and pointed; jaw teeth usually small or minute, but sometimes absent (*Cetengraulis*) or canine-like (*Lycengraulis*). Eyes large, with adipose eyelid completely covering the eyes. Most species feed on plankton (a few by filter-feeding), but large species are piscivorous. In Suriname, the anchovies are best known from the river estuaries, but they are known to migrate upriver to freshwater habitats to spawn. A single, undescribed *Anchoviella* species is known to occur in the upper reaches of the Marowijne River (i.e. upstream of the rapids); this would be a freshwater species.

1-a Anal-fin origin anterior to vertical at dorsal-fin origin; pectoral fins long, reaching posteriorly beyond pelvic-fin base
Pterengraulis atherinoides
1-b Anal-fin origin equal with or posterior to vertical at dorsal-fin origin; pectoral fins short, not reaching posteriorly beyond pelvic-fin base
2

2-a Teeth on lower jaw small and evenly spaced or absent
3
2-b Teeth on lower jaw enlarged and canine-like (*Lycengraulis*)
9

3-a Lower gill rakers on first arch less than 45
4
3-b Lower gill rakers on first arch more than or equal to 45
10

4-a Maxilla long, tip pointed, reaching onto or beyond preoperculum (*Anchoa*)
Anchoa spinifer
4-b Maxilla short, tip blunt, not reaching or just reaching anterior margin of preoperculum (*Anchoviella*)
5

5-a Total length < 5 cm; freshwater upstream of first rapids; Upper Marowijne River

Anchoviella sp.

5-b Total length > 5 cm; marine, estuarine or freshwater downstream of first rapids

6

6-a Maxilla longer, reaching to anterior margin of preoperculum

Anchoviella lepidentostole

6-b Maxilla failing to reach anterior margin of preoperculum by 1/3 or ½ pupil diameter

7

7-a Snout very short, projecting only slightly beyond lower jaw; lower jaw symphysis almost at tip of snout

Anchoviella brevirostris

7-b Snout longer, projecting beyond lower jaw; lower jaw symphysis more posterior, not at tip of snout

8

8-a Less than 16 rays in the anal fin

Anchoviella cayennensis

8-b 16-18 rays in the anal fin

Anchoviella guianensis

9-a More than 15 gill rakers on the lower part of the first arch

Lycengraulis grossidens

9-b 12-15 gill rakers on the lower part of the first arch

Lycengraulis batesii

10-a Branchiostegal membrane broadly joined across isthmus; 8 branchiostegal rays

Cetengraulis edentulus

10-b Branchiostegal membrane not broadly joined across isthmus; 9 or more branchiostegal rays

11

11-a Maxilla short, not extending beyond end of second supramaxilla; anal fin moderate with 20-25 rays, its origin about at vertical through middle of dorsal-fin base

Anchovia surinamensis

11-b Maxilla moderate, extending beyond end of second supramaxilla; anal fin long (28-35 rays), its origin at vertical through anteriormost dorsal-fin rays

Anchovia clupeoides

ANCHOA SPINIFER (Valenciennes, 1848)
Local name(s): sardin, krafana

Diagnostic characteristics: posterior margin of gill cover with small, triangular projection on suboperculum, long anal fin with 36-40 rays
Type locality: Cayenne, French Guiana
Distribution: South and Central America: French Guiana, Guyana, Panama, Suriname and Venezuela and eastern Pacific from Guatemala to Ecuador and possibly Peru
Size: 16-20 cm TL (maximum 24 cm TL)
Habitat: shallow coastal waters, in estuaries, and may migrate upstream in the rivers for up to 30 km (Vari, 1982)
Position in the water column: mid-water
Diet: small fishes and shrimps
Reproduction:
Remarks:

ANCHOVIA CLUPEOIDES (Swainson, 1839)
Local name(s): sardin, krafana

Diagnostic characteristics: long anal fin (28-35 rays) inserted at a vertical through the anteriormost rays of the dorsal fin (i.e. at the same level as the dorsal fin)
Type locality: Pernambuco coast, Brazil
Distribution: Western Atlantic
Size: 17 cm TL (maximum 24 cm TL)
Habitat: coastal waters, estuaries, and may migrate upriver into fresh water
Position in the water column: mid-water
Diet: plankton
Reproduction: batch fecundity 3,500-28,000 eggs/female
Remarks:

ANCHOVIA SURINAMENSIS (Bleeker, 1865)
Local name(s): sardin, krafana

Diagnostic characteristics: anal fin moderate (20-25 rays), its origin at vertical through middle of dorsal-finbase
Type locality: Suriname
Distribution: Estuaries and lower parts of rivers; Trinidad to Amazon River, northeastern South America: Brazil, French Guiana, Guyana, Suriname, Trinidad and Tobago
Size: 8 cm TL (max 15 cm TL)
Habitat: coastal waters, estuaries, and may migrate far upriver (150 km; Vari, 1982) into fresh water
Position in the water column: mid-water

Anchovia surinamensis (alcohol), scale bar = 1 cm, ANSP189252 (© M. Sabaj Pérez)

Diet:
Reproduction:
Remarks:

ANCHOVIELLA BREVIROSTRIS (Günther, 1868)
Local name(s): sardin, krafana

Diagnostic characteristics: short anal fin (15-17 branched rays), its origin at the vertical through the posteriormost dorsal-fin rays, snout rounded and very short, projecting only slightly beyond lower jaw
Type locality: Caxoeira, Rio Paraguacu, Bahia, Brazil
Distribution: Atlantic estuaries and river mouths: Brazil, French Guiana, Guyana, Suriname and Venezuela.
Size: 7 cm TL (maximum 9 cm TL)
Habitat: brackish, marine (pelagic up to a depth of 50 m), but may migrate upriver into freshwater (Vari, 1982)
Position in the water column: mid-water
Diet:
Reproduction: one 7.2 cm female contained 20,000 eggs
Remarks:

ANCHOVIELLA CAYENNENSIS (Puyo, 1946)
Local name(s): sardin, krafana

Diagnostic characteristics: snout longer, projecting beyond lower jaw, short anal fin (16 rays) inserted at the vertical through the posteriormost rays of the dorsal fin, axial scale of pectoral fin reaching beyond midpoint of fin
Type locality: Cayenne River at Macouria, about 4 kilometers from its mouth, French Guiana
Distribution: Atlantic coastal river mouths: Brazil, French Guiana, Suriname and Trinidad and Tobago
Size: 9 cm TL (maximum 12 cm TL)
Habitat: best known from estuaries
Position in the water column: mid-water
Diet:
Reproduction:
Remarks:

ANCHOVIELLA GUIANENSIS (Eigenmann, 1912)
Local name(s): sardin, krafana

Diagnostic characteristics: long snout projecting beyond lower jaw, short anal fin (16-18 rays), axial scale of pectoral fin reaching only to about midpoint of fin
Type locality: Bartica rocks, Guyana

Anchoviella guianensis (live) (© Institut National de la Recherche Agronomique – P.Y. Le Bail [INRA-Le Bail])

Distribution: Caribbean to Amazon River basin: Brazil, Colombia, French Guiana, Guyana, Suriname and Venezuela
Size: 6 cm TL (maximum 9 cm TL)
Habitat: occurs in low salinity, brackish water, but predominantly in fresh water
Position in the water column: mid-water
Diet:
Reproduction: one female contained about 2,000 eggs
Remarks:

ANCHOVIELLA LEPIDENTOSTOLE (Fowler, 1911)
Local name(s): sardin, krafana

Diagnostic characteristics: snout prominent, but bluntly rounded; maxilla long, reaching to anterior margin of preoperculum; 22-25 rays in anal fin,
Type locality: Suriname
Distribution: Atlantic and South America, estuaries and entering rivers: Brazil, French Guiana, Guyana, Suriname and Venezuela
Size: 9 cm TL (maximum 11 cm TL)
Habitat: coastal waters and estuaries; occurs mainly in river mouths and estuaries but also offshore in open water up to depths of 50 m; may also migrate upriver (Vari 1982)
Position in the water column: mid-water
Diet: plankton
Reproduction: a female of 7.7 cm SL contained about 20,000 eggs
Remarks:

ANCHOVIELLA sp.
Local name(s): –

Diagnostic characteristics:
Type locality:
Distribution: Upper Marowijne River (upstream of the village Anapaike)
Size: 3.5 cm SL
Habitat: freshwater, upper reaches Marowijne River
Position in the water column:
Diet:
Reproduction:
Remarks: this is the only anchovy species of Suriname that is known to occur in the Interior, i.e. upstream of the first rapids; it was collected in shallow water near the shore over a sandy bottom substrate.

Anchoviella sp. (alcohol), scale bar = 5 mm, Lawa River (Marowijne River), ANSP189234 (© M. Sabaj Pérez)

CETENGRAULIS EDENTULUS (Cuvier, 1829)
Local name(s): –

Diagnostic characteristics: large species with short and pointed snout; gill rakers fine and very numerous (>45 on first gill arch); branchiostegal membrane broad, expanding posteriorly to cover the isthmus
Type locality: Rio de Janeiro, Brazil [original locality Jamaica]
Distribution: Western Atlantic
Size: 17 cm TL
Habitat: pelagic in shallow coastal waters, often forming large schools, but also entering estuaries (tolerating salinities of 10.3-31 ppt)
Position in the water column: mid-water
Diet: phyto- and zooplankton
Reproduction: relative fecundity 585 eggs/g and minimum size of maturity for females about 10 cm; eggs hatch in 20-24 hours.
Remarks:

LYCENGRAULIS BATESII (Günther, 1868)
Local name(s): sardin, krafana

Diagnostic characteristics: teeth on lower jaw enlarged and canine-like (especially in the middle of the jaw); gill rakers short-stumpy and very few in number (12-15 on the first arch); presence of a longitudinal, silvery, mid-lateral band; presence of a curved line of small, dark spots on the upper part of the operculum
Type locality: Para River, Brazil
Distribution: Amazon and Orinoco River basins, and Guianas: Brazil, Colombia, Guyana, French Guiana, Suriname and Venezuela
Size: common 15-20 cm TL (maximum 30 cm TL)
Habitat: low saline, estuarine waters and in freshwater
Position in the water column: mid-water
Diet: a carnivorous predator feeding on small fishes and shrimps
Reproduction:
Remarks:

LYCENGRAULIS GROSSIDENS (Spix & Agassix, 1829)
Local name(s): sardin, krafana

Diagnostic characteristics: teeth on lower jaw enlarged and canine-like (especially in the middle of the jaw); gill rakers short-stumpy and few in number, but more than 15 on the first arch; absence of a curved line of small, dark spots on the upper part of the operculum
Type locality: Rio de Janeiro, Brazil
Distribution: widespread in South and Central America, mostly marine but ascending rivers

Lycengraulis batesii (alcohol), scale bar = 1 cm, ANSP189251 (© M. Sabaj Pérez)

Size: 20 cm TL (maximum 26 cm TL)
Habitat: mainly in shallow offshore waters, but also entering estuaries and freshwater
Position in the water column: mid-water up to 40 m depth
Diet: a carnivorous predator feeding on small fishes and shrimps
Reproduction:
Remarks:

Pterengraulis atherinoides (Linnaeus, 1766)
Local name(s): sardin, krafana

Diagnostic characteristics: dorsal fin short and positioned far back on the body; anal fin long (28-32 branched rays), its origin anterior of the vertical through the origin of the dorsal fin
Type locality: Suriname
Distribution: Trinidad Island to northeastern Brazil: Brazil, Guyana, French Guiana, Suriname, Trinidad and Tobago and Venezuela
Size: common to 20 cm TL (maximum 30 cm TL)
Habitat: found in estuarine waters of low salinity and in fresh water, reaching some distance inland (at least 17 km up the Suriname River)
Position in the water column: mid-water
Diet: small fishes
Reproduction:
Remarks:

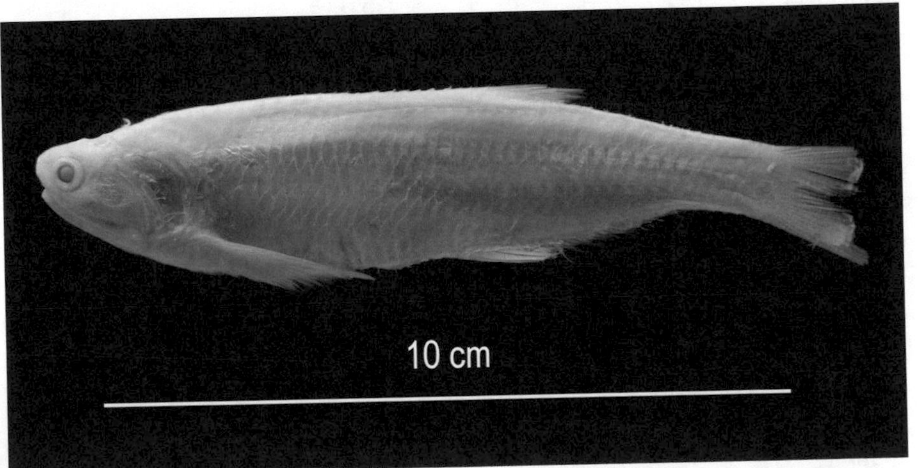

Pterengraulis atherinoides (alcohol), RMNH26301 (© Naturalis)

FAMILY PRISTIGASTERIDAE (LONGFIN HERRINGS OR PELLONAS)

Moderate or sometimes quite large clupeoid fishes (usually to about 20 to 25 cm standard length), but some South American members up to 50 cm standard length (*Pellona*). Body compressed, very deep in some (*Pristigaster*), more elongate in others. Mouth terminal or more often with lower jaw projecting beyond upper (i.e. superior); jaw teeth usually small or minute. A toothed hypomaxillary bone present in some genera. Ventral keel with saw-like teeth present. Long anal fin (>30 rays). Adipose eyelids with broad vertical opening in middle. No spiny rays in the single, short dorsal fin (if present) that is positioned near the midpoint of the body. Until recently, pristigasterines were considered a subfamily of the Clupeidae.

1-a Pelvic fins absent; insertion of anal fin is anterior to the insertion of the dorsal fin; toothed supramaxilla absent; 70-85 anal fin rays

Odontognathus mucronatus

1-b Pelvic fins present; insertion of dorsal fin anterior to insertion of anal fin; toothed supramaxilla present (*Pellona*)

2

2-a 5-7 abdominal spines in the region between the pelvic fins and the anal opening

Pellona harroweri

2-b 8-14 abdominal spines in the region between the pelvic fins and the anal opening

Pellona flavipinnis

Odontognathus mucronatus Lacépède, 1800
Local name(s): –

Diagnostic characteristics: elongate body; abdomen with distinct, but interrupted, keel of scutes (a short gap under pectoral-fin base; abdominal scutes 7 or 8 prepectoral, then a gap, followed by 12 or 13 postpectoral); mouth small, directed upward, lower jaw prominent, maxilla long, reaching to or beyond gill opening; pectoral fins large, pelvic fins absent; dorsal fin small and placed far back on body; anal fin very long, with 70 to 85 rays
Type locality: Cayenne, French Guiana
Distribution: Western Atlantic
Size: 12 cm SL (maximum 16 cm SL)

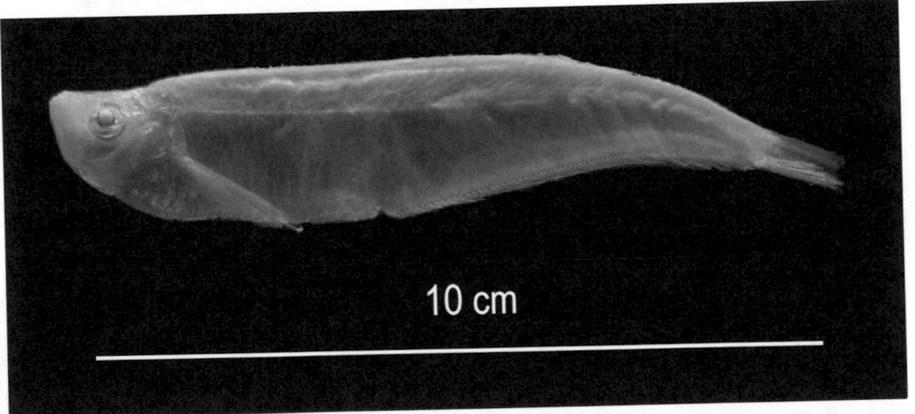

Odontognathus mucronatus (alcohol), Suriname River Estuary, RMNH18235 (© Naturalis)

Habitat: in shallow coastal waters over sand and mud bottoms to depths of about 30 m, but usually occurring much shallower; also close to shore, abundant in estuaries, and ascends rivers.
Position in the water column: mid-water, pelagic
Diet:
Reproduction: it is possible that *O. mucronatus* spawns in rivers.
Remarks:

Pellona flavipinnis (Valenciennes, 1837)
Local name(s): sardin

Diagnostic characteristics: 8-14 abdominal spines in the region between the pelvic fins and the anal opening; a large species with upturned mouth
Type locality: Buenos Aires, Argentina
Distribution: widespread in South America: Argentina, Brazil, Colombia, French Guiana, Guyana, Suriname, Venezuela and Uruguay
Size: 45-55 cm SL
Habitat: mainly a freshwater species, but sometimes caught in estuaries
Position in the water column: mid-water
Diet:
Reproduction:
Remarks:

Pellona harroweri (Fowler, 1917)
Local name(s): sardin

Diagnostic characteristics: 5-7 abdominal spines in the region between the pelvic fins and the anal opening; a much smaller species than *P. flavipinnis*
Type locality: Colon, (Atlantic) Panama
Distribution: Western Atlantic
Size: common to 12 cm SL (maximum 18 cm SL)
Habitat: coastal waters, estuaries, and surrounding areas, over muddy bottoms to a depth of about 35 m (usually less); perhaps not tolerating very low salinities
Position in the water column: mid-water
Diet:
Reproduction:
Remarks:

Pellona flavipinnis (live), Corantijn River Estuary

Pellona flavipinnis (live), Suriname River Estuary

ORDER CHARACIFORMES

The order Characiformes is one of the largest components of the freshwater fish fauna world-wide with nearly 2,000 species now recognized and a key component of Neotropical freshwater ecosystems. The speciose assemblage of New World characiforms (1,700 species) is split into 14 families (Acestrorhynchidae, Anostomidae, Characidae, Chilodontidae, Crenuchidae, Ctenoluciidae, Curimatidae, Cynodontidae, Erythrinidae, Gasteropelecidae, Hemiodontidae, Lebiasinidae, Parodontidae, and Prochilodontidae) (Reis *et al.* (2003), with the Serrasalmidae also recognized as a family by some authors (e.g. Orti *et al.*, 2008). A single Neotropical genus (*Chalceus*) has been assigned to the otherwise African family Alestidae (Zanata & Vari, 2005). See Oliveira *et al.* (2011) for a new phylogenetic framework for the Characiformes based on molecular and morphological analysis and a definition of the family Characidae.

FAMILY PARODONTIDAE (PARONDONTIDS)

Parodontidae are a relatively small family within the Characiformes. Its species are distributed throughout South America and part of Panama, except in some coastal basins, Patagonia and the Amazon channel. Most species do not exceed 15 cm in length. All species have fusiform bodies, no fontanel, and an inferior mouth with a poorly developed, or absent, upper lip. There are commonly four (rarely two) spatulate premaxillary teeth, which have a straight or cusped cutting border. Dentary and maxillary teeth occur in some species. Gill membranes are joined together and free of the scaly isthmus.

PARODON GUYANENSIS Géry, 1959
Local name(s): –

Diagnostic characteristics: mouth inferior, elongate body with a flat and broad breast, paired fins well developed and used to 'walk' on the bottom (especially the pectoral fins, inserted low on the body); black bar on the dorsal fin and a v-shaped zigzag mark along the lateral line
Type locality: Crique Deux-Branches, Upper Mana River, French Guiana
Distribution: Orinoco River basin and coastal rivers of northern South America: French Guiana, Guyana, Suriname and Venezuela
Size: 12 cm TL
Habitat: on rocky substrate in rapids and reaches with strong currents in rivers of the Interior
Position in the water column: bottom
Diet:
Reproduction:
Remarks: apparently most active at dusk or during the night.

Parodon guyanensis (live), scale bar = 1 cm, Lawa River (Marowijne River), ANSP189204 (© M. Sabaj Pérez)

FAMILY CURIMATIDAE (TOOTHLESS CHARACIFORMS)

The family Curimatidae is characterized by the absence in adults of dentition in either jaw. The species of the Curimatidae are broadly distributed across southern Central America and much of tropical and temperate South America. Externally curimatids range from fusiform to deep bodied, slab-sided fishes. Many curimatid species travel in large schools that often constitute a major portion of the fish biomass in both riverine and lacustrine habitats. Curimatids have a number of modifications of the mouth, gill-arches, and digestive tract which allow them to efficiently utilize the flocculent organic matter, microdetritus, microvegetation, and filamentous algae that are common in those habitats across the Neotropics. Some members of the family are known to engage in mass spawning migrations.

1-a Lateral line incomplete in adults; pronounced sexual dimorphism in depth of caudal peduncle (deeper in males) and the form of the caudal fin (deeply forked in females, marginally forked in males)

Curimatopsis crypticus

1-b Lateral line complete in adults; no obvious sexual dimorphism

2

2-a Elaborations of soft tissues of the roof of the mouth limited to three simple, non-fleshy folds; folds not paralleled by series of secondary folds; no lobulate bodies extending ventrally into oral cavity

3

2-b Roof on mouth with three fleshy folds paralleled by series of secondary folds, or with series of lobulate fleshy bodies extending ventrally into oral cavity

7

3-a Caudal peduncle lacking distinct dark midlateral pigmentation (dark midlateral or caudal spot); part of the lobes of the caudal fin covered by scales in larger specimens

Cyphocharax microcephalus

3-b Caudal peduncle with distinct dark pigmentation in form of rounded or horizontally elongate spot

4

4-a Lateral surface of body with series of dark longitudinal stripes along junctions of scales rows; dark pigmentation more intense in larger individuals; greatest body depth equal to distance from tip of snout to posterior margin of fifth or sixth scale of lateral line

Cyphocharax helleri

4-b Lateral surface of body lacking series of longitudinal stripes along junctions of scales rows; greatest body depth equal to distance from tip of snout to posterior margin of second or third scale of lateral line

5

5-a Body without dark midlateral spot in region under dorsal-fin base

Cyphocharax spilurus

5-b Body with dark midlateral spot in region under dorsal-fin base

6

6-a Rounded midlateral dark spot present in region under dorsal-fin base in all but specimens under 2 cm SL; spot ranges from diffuse concentration of dark chromatophores in individuals of 1.8-4.5 cm SL to distinct black spot in larger specimens; no dark midlateral spots present in region between spot under dorsal-fin base and spot at rear of caudal peduncle

Cyphocharax biocellatus

6-b One rounded midlateral very dark spot present in region under dorsal-fin base in all specimens (1.8-4.2 cm SL); spot under dorsal-fin base followed by two or three dark midlateral spots in region between spot located under dorsal-fin base and spot at rear of caudal peduncle

Cyphocharax punctatus

7-a Roof of mouth with three longitudinal folds paralleled by series of secondary folds; absence of dark spot on caudal peduncle

Curimata cyprinoides

7-b Roof of mouth with series of lobulate fleshy bodies extending into oral cavity; presence of elongated spot on caudal peduncle extending to the posterior edge of the caudal fin

Steindachnerina varii

Curimata cyprinoides (Linnaeus, 1766)
Local name(s): makasriba, makafisi

Diagnostic characteristics: a medium-sized species (40-60 scales in the lateral line row) with three longitudinal folds paralleled by series of secondary folds on the roof of mouth; absence of dark spot on caudal peduncle
Type locality: America (Suriname)
Distribution: widespread in South America: Brazil, French Guiana, Guyana, Suriname and Venezuela
Size: 21.3 cm SL
Habitat: small streams and creeks with leaf litter and organic debris on the bottom, but also in coastal freshwater swamps and black-water streams of the Savanna Belt
Position in the water column: mid-water
Diet: detritus
Reproduction: females mature at 13 cm SL; may migrate in schools during the reproductive season
Remarks: very small juveniles have a black spot on the dorsal fin (R. Vari, pers. comm.; Fig. 9.3a).

Curimatopsis crypticus Vari, 1982
Local name(s): –

Diagnostic characteristics: a small species with an incomplete lateral line; on the caudal peduncle is an oval black spot surrounded by bright-orange color
Type locality: stream 2 kilometers east of Lake Amucu, Rupununi District, Guyana, about 3°43'N, 59°25'W
Distribution: widespread in South America: Brazil, French Guiana, Guyana and Suriname
Size: 5 cm SL
Habitat: slow-flowing or standing water of coastal swamps and black-water creeks of the Savanna Belt
Position in the water column: mid-water and near the surface, between aquatic vegetation
Diet:
Reproduction: sexual dimorphism in which mature males have a much deeper caudal peduncle than the females; the caudal fin of the male is less deeply forked than the caudal of the female
Remarks: with 5 cm one of the smallest curimatids.

Curimata cyprinoides (live), Suriname River

Curimatopsis crypticus (live), female, 32.4 mm SL, Para River (Suriname River), ANSP189091 (© M. Sabaj Pérez)

Curimatopsis crypticus (live), male, 28.8 mm SL, Para River (Suriname River), ANSP189091 (© M. Sabaj Pérez)

Cyphocharax biocellatus Vari, Sidlauskas & Le Bail, 2012
Local name(s): –

Diagnostic characteristics: in addition to a black spot on the caudal peduncle, there is a smaller, rounded midlateral dark spot present in region under dorsal-fin base in all but specimens under 2 cm SL; spot ranges from diffuse concentration of dark chromatophores in individuals of 1.8-4.5 cm SL to distinct black spot in larger specimens; no dark midlateral spots present in region between spot under dorsal-fin base and spot at rear of caudal peduncle
Type locality: Upper Marowijne River, Suriname
Distribution: Marowijne River
Size: 6.5 cm SL
Habitat:
Position in the water column:
Diet:
Reproduction:
Remarks:

Cyphocharax helleri (Steindachner, 1910)
Local name(s): –

Diagnostic characteristics: lateral surface of the body with characteristic longitudinal zigzag lines; black spot on the caudal peduncle; mouth subterminal; lateral line with 29-34 scales
Type locality: Upper Suriname River, Suriname
Distribution: widespread in northern South America: Brazil, French Guiana, Guyana, Suriname and Venezuela
Size: 9.6 cm SL
Habitat: in small streams of the Interior with slow current and muddy or sandy bottom
Position in the water column:
Diet: detritus
Reproduction:
Remarks:

Cyphocharax microcephalus (Eigenmann & Eigenmann, 1889)
Local name(s): –

Diagnostic characteristics: a silvery fish without a black spot on the caudal peduncle or body
Type locality: Suriname
Distribution: Atlantic drainages of northern South America: French Guiana (?), Guyana and Suriname
Size: 16.9 cm SL

Cyphocharax biocellatus (alcohol), Lawa River (Marowijne River) (© M. Sabaj Pérez)

Cyphocharax helleri (alcohol), 42.6 mm SL, Lawa River (Marowijne River), ANSP189154 (© M. Sabaj Pérez)

Cyphocharax microcephalus (alcohol), Corantijn River (© R. Vari)

Habitat: coastal swamps, black-water streams in the Savanna Belt and clear-water tributaries of lower freshwater reaches of rivers
Position in the water column:
Diet: detritus
Reproduction:
Remarks: in adults the lobes of the caudal fin may be scaled halfway to the tips of the lobes, much like the caudal squamation in *Curimatella alburna* (an Amazonian species that is not present in Suriname); in collections *C. microcephalus* from Suriname is often misidentified as *Curimatella alburna*. In *C. alburna*, however, the scales on the caudal fin are distinctly smaller than those of the body.

CYPHOCHARAX PUNCTATUS (Vari & Nijssen, 1986)
Local name(s): –

Diagnostic characteristics: a small species with one rounded midlateral very dark spot present in region under dorsal-fin base in all specimens (1.8-4.2 cm SL); spot under dorsal-fin base followed by two or three dark midlateral spots in region between spot located under dorsal-fin base and spot at rear of caudal peduncle
Type locality: Litani River near Kawatop Village, Upper Marowijne River system, Suriname, 3°11'N, 54°12'W
Distribution: Marowijne (Maroni) River system: French Guiana and Suriname
Size: 5 cm TL
Habitat: sandy river banks and small creeks with slow current
Position in the water column:
Diet:
Reproduction:
Remarks:

CYPHOCHARAX SPILURUS (Günther, 1864)
Local name(s): –

Diagnostic characteristics: a relatively small, silvery species with a black spot on the caudal peduncle and 32-36 scales in the lateral line row; mouth subterminal
Type locality: Essequibo River, Guyana
Distribution: widespread in South America: Brazil, French Guiana, Guyana, Suriname and Venezuela
Size: 10.4 cm SL
Habitat: in small streams with sandy or rocky bottom substrate, but also in rapids with strong currents
Position in the water column: mid-water and bottom
Diet: detritus
Reproduction:
Remarks:

Cyphocharax punctatus (alcohol), Marowijne River (© B. Sidlauskas)

Cyphocharax punctatus (live) (© INRA-Le Bail)

Cyphocharax spilurus (alcohol), 87 mm SL, Lawa River (Marowijne River), ANSP189157 (© M. Sabaj Pérez)

Cyphocharax spilurus (live) (© INRA-Le Bail)

STEINDACHNERINA VARII Géry, Planquette & Le Bail, 1991
Local name(s): –

Diagnostic characteristics: a silvery body with a longitudinal, mid-lateral black band on the posterior part of the body (beginning at approximately the vertical through the posterior end of the dorsal fin), continuing on the caudal fin; fins colored yellowish-orange, especially the caudal fin; 41-43 scales in the lateral line row; mouth inferior
Type locality: Crique Awahakiki, Marowijne River, French Guiana
Distribution: French Guiana and Suriname; in Suriname only known from Marowijne and Suriname rivers
Size: 9.4 cm SL
Habitat: in small streams with moderate current and sandy bottom
Position in the water column:
Diet: periphyton algae
Reproduction:
Remarks: not vey abundant; seems to live in small groups.

Steindachnerina varii (alcohol), Marowijne River

Steindachnerina varii (live) (© INRA-Le Bail)

FAMILY PROCHILODONTIDAE (FLANNEL-MOUTH CHARACIFORMS)

Members of the family Prochilodontidae can be readily distinguished from other fishes, except as larvae, by their fleshy lips equipped with two series of numerous, relatively small, falciform or spatulate teeth, movably attached to the lips. Upon protraction these lips form an oral disk encircled by teeth. The two tooth rows in each jaw are variably separated from each other proximate to the symphysis, but converge towards the lateral margins of each jaw. Prochilodontids are moderate to large sized, robust fishes (reaching up to 43 cm TL in *Prochilodus rubrotaeniatus*) with relatively large scales and fleshy lips. The dorsal fin is preceded by a procumbent spine which is bifurcate in *Prochilodus* and *Semaprochilodus*. Prochilodontids are important food fishes that are well-known for their extensive riverine migrations in relation to reproduction (Carolsfeld *et al.*, 2003).

1-a Caudal fin lacking black and yellow stripes, but sometimes with small black spots; presence of dark spot on opercle
Prochilodus rubrotaeniatus
1-b Caudal fin colored with numerous black and yellow horizontal to slightly oblique stripes; opercle lacking a black spot; Marowijne River
Semaprochilodus varii

PROCHILODUS RUBROTAENIATUS Jardine, 1841
Local name(s): kwimata, kurimata

Diagnostic characteristics: mouth subterminal; a dark spot on the operculum; presence of a bifurcated spine in front of the dorsal fin; caudal fin sometimes with small black spots
Type locality: Branco, Negro and Essequibo rivers
Distribution: widespread in South America: Brazil, French Guiana, Guyana, Suriname and Venezuela
Size: 43 cm TL
Habitat: large rivers, mainly upstream of the first rapids
Position in the water column: mid-water and near the bottom
Diet: detritus, algae, and associated micro-organisms (Bowen, 1983)
Reproduction: spawning takes place in the rainy season; relative fecundity is high with 100,000-1,000,000 eggs/kg; (spawning) migrations are known to occur in the Suriname River: at the start of the rainy season *P. rubrotaeniatus* migrates downstream to their spawning habitat, often flooded, swampy areas in inner bends of rivers. Migrating, prespawning females and courting males emit a low-pitched, drumming sound that presumably attracts females (but also fishermen and caimans!). Fertilized eggs, embryos and larvae are found in floodplain habitats,

Prochilodus rubrotaenitus (alcohol), scale bar = 1 cm, ANSP175495 (© M. Sabaj Pérez)

where the juveniles feed on rich invertebrate and algal production. During the falling-water period, massive schools migrate upstream and disperse among large tributaries.
Remarks: *P. rubrotaeniatus* also occurs in Brokopondo Reservoir, but apparently it is no longer present in the Suriname River downstream of the dam (Mol *et al.*, 2007).

Semaprochilodus varii Castro, 1988
Local name(s): péni-kwimata, kurimata

Diagnostic characteristics: caudal fin colored with numerous black and yellow horizontal to slightly oblique stripes; operculum lacking a black spot
Type locality: Marowijne River, about 25 kilometers south of Albina, Suriname
Distribution: only known from Marowijne River basin: French Guiana and Suriname
Size: 28 cm SL
Habitat: main channel of Marowijne River and large tributaries
Position in the water column: probably bottom-feeding
Diet:
Reproduction: upstream directed spawning migrations are known from *Semaprochilodus* species in the Amazon; it is possible that *S. varii* also makes such spawning migrations in the Marowijne River
Remarks: in general, prochilodontid spawning migrations appear to allow adults and early life stages to feed in productive floodplain habitats during the high-water phase, and allow adults and large sub-adults to disperse to other locations within the river basin during the low-water phase (which presumably reduces the impact of resource competition and predation).

Semaprochilodus varii (live), scale bar = 1 cm, Lawa River (Marowijne River), ANSP187435 (© M. Sabaj Pérez)

FAMILY ANOSTOMIDAE (HEADSTANDERS)

The Anostomidae are distinguishable from all other Characiformes by the presence of a unique series of only three or four teeth, or a combination of both, in each premaxillary or dentary, disposed as steps of a stair. The mouth is small, distant from the anterior orbital rim, and may be terminal, subinferior or almost superior. The mouth is terminal or sub-inferior in the genera *Abramites*, *Leporinus*, *Schizodon*, *Leporellus*, *Rhythiodus*, and *Laemolyta*, and elevated, almost superior, in the genera *Gnathodolus*, *Synaptolaemus*, *Sartor*, *Anostomus*, *Anostomoides*, and *Pseudanos*. The body is entirely covered with large scales, with the perforated lateral series varying from 32-34 to 42-44 scales, and the circumpeduncular scales in 12 to 16 series. The branchial opening is small, with the membranes firmly joined to isthmus. An adipose-fin is always present. The anostomids are all fusiform fishes, including species varying from 10 cm (small species from upper tributaries of Orinoco and Amazon system) to 80 cm SL (large species from the main tributaries of the Paraná River and Pantanal of Mato Grosso), and a large number of species of intermediate body size, from mostly South American river basins. Many anostomid species are known by their habit of feeding in an inclined position (i.e. headstanders; Géry, 1977; Santos & Rosa, 1998) and they constitute a significant portion of the fish biomass in the diverse aquatic habitats. By their variable position of mouth and extended digestive tract, they are efficient in utilizing sponges, detritus, insects, and vegetal items like seeds, leaves, and filamentous algae (Santos & Rosa, 1998), all common in the forest habitats of the Neotropics. Some large members of the genera *Leporinus* and *Schizodon* are known to undergo spawning migrations (Carolsfeld *et al.*, 2003).

1-a Mouth inferior, with premaxilla oriented such that long axis of premaxillary teeth points ventrally or slightly posteroventrally (*Hypomasticus*)

2

1-b Mouth subterminal, terminal, supraterminal or superior, but not inferior; premaxillary oriented such that long axis of premaxillary teeth points anteriorly or anteroventrally

3

2-a Lateral surface of body with multiple longitudinal dark stripes

Hypomasticus despaxi

2-b Lateral surface of body with numerous small spots, with spots below lateral line largest and surrounded by ring of eight discrete and evenly spaced dots

Hypomasticus megalepis

3-a Premaxillary teeth, including symphysal tooth, strongly and uniformly bicuspid or multicuspid, with all cusps on each tooth roughly equivalent in size, premaxillary teeth exclusive of tooth furthest from symphysis all approximately of equal size

4

3-b Premaxillary teeth not uniformly bicuspid or multicuspid; each tooth with single large central cusp; size of premaxillary teeth progressively decreasing in stepwise fashion posteriorly from symphysis

9

4-a Body with four irregular vertical dark blotches and dark spots at base of median caudal-fin rays; all dentary teeth strongly multicuspid giving overall jagged appearance to distal margin; exposed portion of dentary teeth wider than high; mouth terminal in specimens >9 cm SL, slightly supraterminal in individuals 4-9 cm SL, and distinctly supraterminal in juveniles <4 cm SL

Schizodon fasciatus

4-b Dark pigmentation on body not as in 4a; dentary teeth either uniformly with two or more cusps or with symphysal tooth having truncate margin; exposed portion of dentary teeth higher than wide; mouth superior in individuals of all sizes

5

5-a Dark pigmentation on body consisting of numerous longitudinal stripes (*Anostomus*)

6

5-b Dark pigmentation on body consisting of one or more dark spots centered on lateral-line scale row and dark transverse markings across dorsal surface of body (*Petulanos*)

8

6-a Pale median middorsal stripe extending from supraoccipital crest to dorsal-fin origin; rows of light spots with intervening thin dark stripes present on anterior surface of body in region dorsal to thicker dark stripe along lateral-line scale row; three branchiostegal rays

Anostomus ternetzi

6-b Dark median middorsal stripe extending from supraoccipital crest to dorsal-fin origin; single dark stripe midway present between middorsal stripe and that on lateral-line scale row; four branchiostegal rays

7

7-a Greatest body depth 24.0-27.1% of SL (mean 25.8%); fins hyaline; Marowijne River

Anostomus brevior

7-b Greatest body depth 20.1-23.0% of SL (mean 21.5%); fins bright-red

Anostomus anostomus

8-a Transverse bands on body broad, typically two scales wide; 40-42 lateral-line scales; spots on lateral-line scale row typically rounded

Petulanos spiloclistron

8-b Transverse bands on body thin, typically one scale wide; 38-39 lateral-line scales; spots on lateral-line scale row typically horizontally elongate
Petulanos plicatus

9-a Dark pigmentation on body consisting of series of vertical bands with intervening lighter regions
10
9-b Dark pigmentation on body not as in 9a (specimens may have dark longitudinal bands, various patterns of spots, etc)
11

10-a Dark pigmentation on body consisting of alternating wide and narrow transverse bars; eight (rarely nine) branched pelvic-fin rays
Leporinus maculatus
10-b Dark pigmentation on body in adults consisting of ten regularly spaced transverse bars beginning with bar situated on opercle; nine branched pelvic-fin rays
Leporinus fasciatus

11-a Multiple dark longitudinal stripes present on body, with stripe along lateral-line scale row widest and most intense
Leporinus arcus
11-b Dark pigmentation on body not as in 11a
12

12-a Shallow bodied, with maximum body depth 19-25% of SL; four spots centered along lateral line scale row, one discrete dark spot present ventral to seventh, eighth or ninth scale of lateral-line row; series of smaller dark spots variably present dorsal and ventral to lateral-line scale row
Leporinus apollo
12-b Relatively deep bodied, with maximum body depth 28-38% of SL
13

13-a Dark vertical bar on lateral surface of body outlining entire posterior margin of opercle to pectoral-fin insertion; four additional dark spots present along lateral line scale row in individuals of all sizes and additional scattered spots present in juveniles
Leporinus lebaili
13-b No discrete dark pigmentation outlining posterior margin of opercle dorsal to pectoral-fin insertion; other dark pigmentation patterns variably present on body
14

14-a Body with multiple dark spots along lateral-line scale row; body dorsal to lateral line scale row with scattered discrete dark spots; additional spots variably present ventral to lateral-line scale row
15

14-b Primary dark pigmentation on body located along lateral-line scale row either as series of one or more distinct spots or as partial lateral stripe; body dorsal to lateral-line scale row unpigmented or with transverse bars, but without scattered discrete dark spots as in 14a; series of smaller dark spots variably present posterior to opercle and ventral to lateral line
16

15-a 12 circumpeduncular scales; juveniles and adults with series of discrete spots ventral to lateral line scale row, spots fading somewhat ontogenetically
Leporinus nijsseni
15-b 16 circumpeduncular scales; juveniles with series of discrete spots ventral to lateral line scale row, specimens of 6.5 cm SL or larger often with spots ventral to lateral line united into diffuse dark band
Leporinus granti

16-a Four transverse scales above lateral-line scale row (counted between the first ray of the dorsal fin and the lateral line); 36-37 scales in lateral-line row
Leporinus gossei
16-b Five transverse scales above lateral-line scale row (counted between the first ray of the dorsal fin and the lateral line); 36-39 scales in lateral-line row
Leporinus friderici

Anostomus anostomus (Linnaeus, 1758)
Local name(s): –

Diagnostic characteristics: body with three longitudinal black stripes, fins reddish; greatest body depth 20.1-23.0% of SL (mean 21.5%)
Type locality: South America (probably Suriname)
Distribution: Amazon and Orinoco River basins: Brazil, Guyana, Peru and Suriname; in Suriname known to occur in Corantijn, Coppename, Saramacca and Suriname rivers
Size: 16 cm TL
Habitat:
Position in the water column:
Diet:
Reproduction:
Remarks: a well-known and popular aquarium fish.

Anostomus brevior Géry, 1961
Local name(s): –

Diagnostic characteristics: body with three longitudinal black stripes, but fins not reddish and greatest body depth 24.0-27.1% of SL (mean 25.8%)
Type locality: Río Camopi River, a tributary of Oyapock River, French Guiana
Distribution: rivers of French Guiana; in Suriname only known from Marowijne (Maroni) River
Size: 12 cm SL
Habitat: shallow water habitats in clear-water rivers and streams of the Interior
Position in the water column: they swim in characteristic head-down position near the sandy bottom substrate
Diet: algae (diatoms)
Reproduction:
Remarks:

Anostomus ternetzi Fernández-Yépez, 1949
Local name(s): –

Diagnostic characteristics: only two dark longitudinal bands; the back dark with rows of light spots; fins transparent
Type locality: Palital, Estado Guárico, Venezuela
Distribution: widespread in South America: Brazil, French Guiana, Guyana, Suriname and Venezuela; in Suriname only known from the Marowijne and Corantijn rivers
Size: 12 cm SL
Habitat: clear-water rivers of the Interior
Position in the water column:

Anostomus anostomus (live), 72.5 mm SL, ANSP180172 (© M. Sabaj Pérez)

Anostomus brevior (live), scale bar = 1 cm, Lawa River (Marowijne River), FMNHexANSP189141 (© M. Sabaj Pérez)

Anostomus ternetzi (live) (© INRA-Le Bail)

Diet: omnivorous, eating both plants and invertebrates (both aquatic and terrestrial) (Santos & Rosa, 1998)
Reproduction:
Remarks: *A. ternetzi* lives in the same habitat as *Hypomasticus despaxi*, a species with an almost identical black-striped color pattern. In the Rio Teles Pires (Brazilian Shield), *A. ternetzi* (or a species close to it) was observed picking parasites of the scales and fins of a large *Hoplias aimara* (Lucanus, 2009, p. 149).

Hypomasticus despaxi (Puyo, 1943)
Local name(s): –

Diagnostic characteristics: mouth inferior, lateral surface of body with multiple longitudinal dark stripes
Type locality: Upper Marowijne River
Distribution: coastal rivers of French Guiana, from Maroni to Oyapock River; in Suriname only known from Marowijne (Maroni) River
Size: 16 cm TL
Habitat: clear-water streams and rivers of the Interior
Position in the water column: near the bottom
Diet:
Reproduction:
Remarks:

Hypomasticus megalepis (Günther, 1863)
Local name(s): –

Diagnostic characteristics: mouth inferior, lateral surface of body with numerous small spots, with spots below lateral line largest
Type locality: Essequibo River, Guyana
Distribution: Essequibo River, Guyana, and Coppename and Saramacca rivers, Suriname
Size: 7 cm SL
Habitat: clear-water streams and rivers of the Interior
Position in the water column:
Diet:
Reproduction:
Remarks:

Anostomus ternetzi (live), Corantijn River (© P. Willink)

Hypomasticus despaxi (live), Marowijne River

Hypomasticus despaxi (live), scale bar = 1 cm, Lawa River (Marowijne River), ANSP189010 (© M. Sabaj Pérez)

Hypomasticus megalepis (live), 68.3 mm SL, AUM37999 (© M. Sabaj Pérez)

Leporinus apollo Sidlauskas, Mol & Vari, 2011
Local name(s): –

Diagnostic characteristics: slender, shallow bodied species, with maximum body depth 19-25% of SL; four spots centered along lateral line scale row, one discrete dark spot present ventral to seventh, eighth or ninth scale of lateral-line row; series of smaller dark spots variably present dorsal and ventral to lateral-line scale row; the presence of a series of 9-14 dark bars across the dorsal surface of the body
Type locality: Coppename River, Sidonkrutu Rapids, sand island and channel (04°31'51"N, 56°30'56"W), Suriname
Distribution: only known from Coppename, Corantijn and Suriname rivers, Suriname
Size: 14 cm SL
Habitat: clear-water rivers of the Interior
Position in the water column:
Diet:
Reproduction:
Remarks:

Leporinus arcus Eigenmann, 1912
Local name(s): –

Diagnostic characteristics: multiple dark longitudinal stripes present on the body, with stripe along lateral-line scale row widest and most intense; mouth terminal
Type locality: Tukeit, Guyana
Distribution: widespread in South America: Guyana, Suriname and Venezuela; in Suriname, only known from Upper Corantijn River
Size: 40 cm TL
Habitat: clear-water rivers of the Interior
Position in the water column:
Diet:
Reproduction:
Remarks:

Leporinus fasciatus (Bloch, 1794)
Local name(s): kwana

Diagnostic characteristics: ten regularly spaced transverse black bars beginning with bar situated on opercle; nine branched pelvic-fin rays
Type locality: unknown locality [probably Suriname].
Distribution: Amazon River basin; countries Brazil, French Guiana, Peru, Suriname; occurs in all rivers in Suriname and also in Brokopondo Reservoir
Size: 30 cm SL
Habitat: clear-water rivers and streams of the Interior

Leporinus apollo (alcohol), Coppename River (© M. Littmann)

Leporinus arcus (live) (© W. Kolvoort)

Leporinus fasciatus (live), scale bar = 1 cm, Lawa River (Marowijne River, ANSP189158 (© M. Sabaj Pérez)

Position in the water column: mid-water and near the bottom
Diet: omnivorous, eating both fruits and small fishes
Reproduction:
Remarks:

LEPORINUS FRIDERICI (Bloch, 1794)
Local name(s): waraku, abonkia

Diagnostic characteristics: body pigmentation of adults consisting of 3 or 4 spots (including a caudal spot); five transverse scales above lateral-line scale row; 36-39 scales in lateral-line row
Type locality: Suriname
Distribution: Suriname and Amazon River basin; Brazil, French Guiana, Guyana, Suriname, Trinidad and Tobago
Size: 35 cm SL
Habitat:
Position in the water column:
Diet: omnivores, feeding mainly on fruits and allochthonous invertebrates (Boujard *et al.*, 1990)
Reproduction: females spawn 100,000-200,000 eggs; in aquaculture, newly hatched larvae are very small (3 mm) and grow, with a low survival rate of 10%, in 3 months to 8-cm juveniles; they can reach an age of 4 years and a weight of 2.3 kg
Remarks: Renno *et al.* (1990, 1991) found molecular evidence of the existence of an eastern and western Pleistocene refuge of *Leporinus friderici* populations in French Guiana, from which this species has more recently expanded its range.

LEPORINUS GOSSEI Géry, Planquette & Le Bail, 1991
Local name(s): –

Diagnostic characteristics: body pigmentation of adults resembling that of *L. friderici*, consisting of 3 or 4 spots (including a caudal spot), but only four transverse scales above lateral-line scale row; 36-37 scales in lateral-line row
Type locality: Crique Balaté, Lower Marowijne (Maroni) River, French Guiana
Distribution: Marowijne River basin; French Guiana, Suriname
Size: 25 cm SL
Habitat: mainly in clear-water rivers of the Interior
Position in the water column:
Diet:
Reproduction:
Remarks: according to Planquette *et al.* (1996) *L. gossei* is also present in the Suriname River.

Leporinus friderici (alcohol), 95.2 mm SL, USNM225383 (© T. Britt Griswold)

Leporinus friderici (live) (© INRA-Le Bail)

Leporinus gossei (alcohol), MHNG2700-051 (© B. Sidlauskas)

Leporinus gossei (live) (© INRA-Le Bail)

Leporinus granti Eigenmann, 1912
Local name(s): –

Diagnostic characteristics: presence of a red, oblique post-operculum bar; 16 circumpeduncular scales
Type locality: Maripicru Creek, Guyana
Distribution: all rivers in French Guiana except Oyapock River (including Marowijne/Maroni), Guyana; in Suriname known from Marowijne and Corantijn rivers
Size: 25 cm TL
Habitat:
Position in the water column:
Diet: omnivorous
Reproduction: from December to June
Remarks:

Leporinus lebaili Géry & Planquette, 1983
Local name(s): –

Diagnostic characteristics: a dark vertical bar present on the lateral surface of the body outlining entire posterior margin of opercle to the pectoral-fin insertion; presence of a very characteristic, small, red spot at each side of the mouth, just above the upper jaw
Type locality: Laissé-Dédé Falls, Marowijne (Maroni) River, French Guiana
Distribution: only known from the Marowijne/Maroni and Mana rivers
Size:
Habitat: clear-water rivers of the Interior
Position in the water column:
Diet: omnivorous
Reproduction:
Remarks: occurs syntopically with *L. fasciatus*, *L. granti* and juvenile *L. friderici*.

Leporinus maculatus Muller & Troschel, 1844
Local name(s): –

Diagnostic characteristics: dark pigmentation on body consisting of alternating wide (4) and narrow (3-4) transverse bars; eight (rarely nine) branched pelvic-fin rays
Type locality: Guyana
Distribution: coastal rivers of the Guianas and São Francisco River basin; Brazil, French Guiana, Guyana, Suriname (Corantijn, Nickerie, Coppename, Suriname and Marowijne rivers)
Size: 18 cm TL
Habitat: clear-water rivers of the Interior

Leporinus granti (live) (© P. Willink)

Leporinus granti (live) (© R. Smith)

Leporinus lebaili (live), scale bar = 1 cm, Lawa River (Marowijne River), ANSP189043 (© M. Sabaj Pérez)

Leporinus maculatus (live), 123.4 mm SL, Lawa River (Marowijne River), ANSP189041 (© M. Sabaj Pérez)

Position in the water column:
Diet: invertebrates
Reproduction:
Remarks: although the species name 'maculatus' suggests a spotted species, *L. maculatus* has not a spotted pigmentation pattern, but a pattern with transverse (vertical) bands; *Leporinus pellegrini* Steindachner, 1910 is a junior synonym (Willink & Sidlauskas, 2004).

Leporinus nijsseni Garavello, 1990
Local name(s): –

Diagnostic characteristics: resembles *L. granti*, with a red, oblique post-operculum bar, but with 12 circumpeduncular scales
Type locality: about 27 km south of Village Dam, Sara Creek, Suriname River, Brokopondo district, Suriname.
Distribution: Oyapock River (French Guiana), and Suriname and Nickerie rivers (Suriname)
Size: 17 cm TL
Habitat: clear-water rivers of the Interior
Position in the water column:
Diet:
Reproduction:
Remarks:

Petulanos plicatus (Eigenmann, 1912)
Local name(s): –

Diagnostic characteristics: 4 roundish or slightly horizontally elongate spots on flank; longitudinal lines between rows of scales; about 12, narrow bars (1 scale wide) across back, some extending onto the flanks between the spots; 38-39 lateral line scales
Type locality: Crab Falls, Essequibo River, Guyana
Distribution: Essequibo River basin, Guyana, and Corantijn River, Suriname
Size: 9.5 cm SL
Habitat: clear-water rivers of the Interior
Position in the water column:
Diet:
Reproduction:
Remarks:

Leporinus nijsseni (alcohol), Coppename River, FMNH-SUR2004F21 (© B. Sidlauskas)

Leporinus nijsseni (live) (© INRA-Le Bail)

Leporinus nijsseni (live), Nickerie River (© R. Smith)

Petulanos plicatus (alcohol), ANSP180173 (© B. Sidlauskas)

Petulanos spiloclistron (Winterbottom, 1974)
Local name(s): –

Diagnostic characteristics: pigmentation pattern much like *P. plicatus*, but transverse bands on body broad, typically two scales wide, and spots on lateral-line scale row typically rounded; 40-42 lateral-line scales
Type locality: Falawatra River at rapids, 5 kilometers south-southwest of Stondansi Falls, Nickerie River system, Suriname
Distribution: only known from the Nickerie River, Suriname
Size: 16 cm SL
Habitat:
Position in the water column:
Diet:
Reproduction:
Remarks:

Schizodon fasciatus Spix & Agassiz, 1829
Local name(s): nyamsifisi

Diagnostic characteristics: body coloration with four irregular vertical dark bands plus a dark caudal spot; all dentary teeth strongly multicuspid forming a characteristic, continuous crenulate cutting border
Type locality: Rivers of Brazil
Distribution: Upper Amazon River and rivers of French Guiana and Suriname: Brazil, French Guiana, Peru, Suriname and Venezuela; in Suriname known to occur in the Corantijn, Nickerie, Saramacca and Marowijne rivers
Size: 40 cm TL
Habitat: rivers and medium-sized streams of the Interior
Position in the water column:
Diet: strictly herbivorous
Reproduction:
Remarks: Amazonian *Schizodon* species are known for their long-distance river migrations.

Petulanos spiloclistron (alcohol), Nickerie River, RMNH37463 (© Naturalis)

Schizodon fasciatus (live), Lower Nickerie River

FAMILY CHILODONTIDAE (SPOTTED HEADSTANDERS)

The Chilodontidae can be distinguished from other members of the Characiformes by the presence of a single series of relatively small teeth movably attached to the lips of the upper jaw and, in most species, to the lower jaw together with a sixth lateral-line scale distinctly smaller than the other scales in that series. Chilodins share with certain anostomids the faculty of swimming head down (Fig.12.2). Species of *Chilodus* are exported from various locations for the aquarium trade in which they are known as headstanders.

1-a Branched anal-fin rays typically 10 or 11, rarely 9; mouth terminal or slightly superior; dorsal fin with series of dark spots on posterior rays; absence of a longitudinal dark band along the lateral line

2

1-b Branched anal-fin rays 6-8; mouth subterminal; dorsal fin with dark pigmentation on distal portion of anterior rays, but lacking dark spots on remaining portion of fin; presence of a longitudinal dark band along the lateral line

3

2-a Body with evenly distributed conspicuous dark spots which are ventrally equally prominent as dorsally, spots also being present around the pectoral fin; no trace of a longitudinal stripe on the flanks; Marowijne River

Chilodus zunevei

2-b Pigmentation of body variable, dorsally more prominent than ventrally, absent around the pectoral fin; longitudinal stripe on the flanks rarely absent, but sometimes faded or showing a zigzag pattern (in such cases the spots on the body are also faded and then confined to the margin of the scales); Nickerie River

Chilodus punctatus

3-a Lateral-line scales, not including terminal elongate scale, typically 27, rarely 28; most individuals with teeth in lower jaw; distal portions of anterior dorsal-fin rays with distinct patch of dark pigmentation; distinct midlateral stripe extending from snout to base of caudal fin; Corantijn and Marowijne rivers

Caenotropus maculosus

3-b Lateral-line scales, not including terminal elongate scale, 28-32, typically 29 or 30; teeth absent in lower jaw; distal portions of anterior dorsal-fin rays sometimes dusky but not with distinct patch of dark pigmentation; diffuse midlateral stripe extending from snout to base of caudal fin, most often with rotund dark blotch along stripe slightly anterior of vertical through origin of dorsal fin; Suriname and Saramacca rivers

Caenotropus labyrinthicus

Fig. 12.2. The spotted headstander *Chilodus zunevei* swimming characteristically with its head down (© W. Kolvoort).

Caenotropus labyrinthicus (Kner, 1858)
Local name(s): –

Diagnostic characteristics: 28-32, typically 29 or 30, lateral-line scales, not including terminal elongate scale; teeth absent in lower jaw; distal portions of anterior dorsal-fin rays sometimes dusky but not with distinct patch of dark pigmentation
Type locality: Rio Branco and mouth of River Negro, Brazil
Distribution: widespread in South America: Bolivia, Brazil, Colombia, Ecuador, Guyana, Peru, Suriname and Venezuela; in Suriname known from the Suriname and Saramacca rivers
Size: 15.2 cm SL
Habitat:
Position in the water column:
Diet:
Reproduction:
Remarks: a headstander, feeding with body in an oblique (45°) position, head down near the bottom substrate.

Caenotropus maculosus (Eigenmann, 1912)
Local name(s): –

Diagnostic characteristics: typically 27, rarely 28, lateral-line scales, not including terminal elongate scale; most individuals with teeth in lower jaw; distal portions of anterior dorsal-fin rays with distinct patch of dark pigmentation
Type locality: Creek below Potaro Landing, Guyana
Distribution: widespread in South America: French Guiana, Guyana, Suriname and Venezuela; in Suriname known from the Marowijne (Maroni) and Corantijn rivers
Size: 10.8 cm SL
Habitat: in streams with fast flowing water or rapids
Position in the water column:
Diet: probably omnivorous, feeding on small invertebrates
Reproduction:
Remarks: a headstander, feeding with body in an oblique (45°) position, head down near the bottom substrate.

Chilodus punctatus Müller & Troschel, 1844
Local name(s): – (spotted headstander)

Diagnostic characteristics: pigmentation of body (dark spots) variable, dorsally more prominent than ventrally, absent around the pectoral fin; longitudinal stripe on the flanks rarely absent, but sometimes faded or showing a zigzag pattern (in such cases the spots on the body are also faded and then confined to the margin of the scales

Caenotropus labyrinthicus (alcohol) (© T. Britt Griswold)

Caenotropus maculosus (live), 70 mm SL, Litani River (Marowijne River)ANSP189147 (© M. Sabaj Pérez)

Chilodus punctatus (live) (© P. Willink)

Type locality: Lake Amuku, Guyana
Distribution: widespread in South America: Brazil, Colombia, Ecuador, Guyana, Peru and Suriname; in Suriname only known from Nickerie River
Size: 7.9 cm SL
Habitat:
Position in the water column:
Diet:
Reproduction:
Remarks: a headstander, feeding with body in an oblique (45°) position, head down near the bottom substrate.

CHILODUS ZUNEVEI Puyo, 1946
Local name(s): –

Diagnostic characteristics: body with evenly distributed conspicuous dark spots which are ventrally equally prominent as dorsally, spots also being present around the pectoral fin; no trace of a longitudinal stripe on the flanks
Type locality: creek flowing into Litany River, Upper Maroni (Marowijne) River, Plateaux de Guyanes region, French Guiana
Distribution: Marowijne, Sinnamary, Approuague, and Oyapock River basins, French Guiana
Size: 7.7 cm SL
Habitat: small streams and rivers
Position in the water column: probably near the bottom
Diet: periphyton algae, but also aquatic insect larvae
Reproduction: eggs are deposited on bottom substrate or vegetation; eggs hatch in three days
Remarks: a headstander, feeding with body in an oblique (45°) position, head down near the bottom substrate; this species is said to produce metallic sounds when feeding or when captured.

Chilodus zunevei (alcohol), MHNG2608-040 (© B. Sidlauskas)

Chilodus zunevei (live) (© INRA-Le Bail)

FAMILY CRENUCHIDAE (SOUTH AMERICAN DARTERS)

The Crenuchidae are relatively small (usually less than 10 cm SL) fishes. The current composition of the family has been established relatively recently (Buckup, 1998); the diversity of the subfamily Characidiinae was reviewed by Buckup (1993). Crenuchids are diagnosed by the presence of paired foramina located in the frontal bones, posterodorsally to the orbits. These foramina are more easily recognizable in the three members of the subfamily Crenuchinae, where they are relatively large and associated with a distinct depression in the frontals which is located immediately in front of the foramina (Alexander, 1963; Bossy *et al.*, 1965). These foramina, however, are very small in members of the Characidiinae, which encompass the majority of crenuchids. Fortunately, most characidiin fishes can be easily recognized by their general external appearance. They are easily distinguished from most other characiform fishes, such as the ubiquitous Characidae, by the low number of anal-fin rays (fewer than 14 rays). The Characidiinae have conic or tricuspid teeth on both jaws. In the upper jaw they are organized in a single row. In the lower jaw they usually form two rows (the inner one may be absent in some species).

1-a Presence of a large, oval black spot on the lower portion of the caudal peduncle; upper jaw reaching beyond the center of the eye

Crenuchus spilurus

1-b If a black spot is present on the caudal peduncle, it is small and positioned symmetrically at the base of the caudal fin

2

2-a Pectoral-fin rays 10 or fewer (small species reaching no more than 3 cm SL); lateral line incomplete

Microcharacidium eleotrioides

2-b Pectoral-fin rays 11 or more; lateral line complete

3

3-a Supraorbital present; absence of black stripes on the pelvic, anal and caudal fins; presence of a small black spot on the caudal peduncle at the base of the caudal fin (*Characidium*)

4

3-b Supraorbital absent; presence of black stripes on the pelvic, anal and caudal fins; absence of a small black spot on the caudal peduncle at the base of the caudal fin (*Melanocharacidium*)

5

4-a Body transparent with black dots; absence of a midlateral longitudinal black stripe

Characidium pellucidum

4-b Body yellowish with transverse black stripes; presence of a midlateral longitudinal black stripe

Characidium zebra

5-a Absence of well-defined black spots above the eye; pectoral fins uniformly grey; 14 circumpeduncular scales; absence of a black stripe between the eye and the mouth

Melanocharacidium blennioides

5-b Presence of black spots above the eye; pectoral fins yellowish with 3-4 black stripes; 12 circumpeduncular scales; presence of at least one black stripe between the eye and the mouth

Melanocharacidium dispilomma

A little known species, *Melanocharacidium* cf. *melanopteron*, was collected in the Upper Coppename River (Mol *et al.*, 2006); more as yet undescribed Crenuchidae may be present in Suriname, especially in the Interior of the country.

Crenuchus spilurus Günther, 1863
Local name(s): –

Diagnostic characteristics: presence of a large, oval black spot on the lower portion of the caudal peduncle; long dorsal fin with 10-11 rays; lateral line not complete; upper jaw reaching beyond the center of the eye
Type locality: Essequibo, Guyana
Distribution: Orinoco and Amazon River basins, and coastal rivers in the Guianas: Brazil, Colombia, French Guiana, Guyana, Peru, Suriname and Venezuela
Size: 5.7 cm TL
Habitat: coastal freshwater swamps and black-water streams of the Savanna Belt
Position in the water column:
Diet: the large mouth may indicate a predatory habit, but it feeds mostly on aquatic insect larvae and zooplankton
Reproduction: male generally larger and with more brilliant colors than female; in the aquarium the male is known to defend and fan the spawned eggs and after hatching the male continues to defend the newborn for a few days
Remarks: presence of paired foramina located in the frontal bones, posterodorsally to the orbits; although known for a long time (Alexander, 1963) the function of these foramina is still not known.

Characidium pellucidum Eigenmann, 1909
Local name(s): –

Diagnostic characteristics: absence of black stripes on the pelvic, anal and caudal fins; presence of a small black spot on the caudal peduncle at the base of the caudal fin; body transparent with black dots; absence of a midlateral longitudinal black stripe
Type locality: Gluck Island, Guyana
Distribution: Essequibo and Marowijne (Maroni) River basins: French Guiana, Guyana and Suriname
Size: 3.4 cm SL
Habitat: on sandy bottom substrate in streams of the Interior with moderate to fast current
Position in the water column: bottom
Diet:
Reproduction:
Remarks: a small, transparent species that is, in Suriname, only known from the Marowijne River.

Crenuchus spilurus (live) (© P. Willink)

CHARACIDIUM ZEBRA Eigenmann, 1909
Local name(s): –

Diagnostic characteristics: absence of black stripes on the pelvic, anal and caudal fins; presence of a small black spot on the caudal peduncle at the base of the caudal fin; body yellowish with transverse black stripes; presence of a midlateral longitudinal black stripe
Type locality: Maripicru Creek, Ireng River, Guyana
Distribution: Amazon and Essequibo River basins and coastal basins of Guiana Shield: Argentina (?), Brazil, French Guiana, Guyana and Suriname
Size: 4.9 cm SL
Habitat: small streams and rivers of the Interior on sandy bottom in moderate to fast currents
Position in the water column: bottom
Diet: insect larvae
Reproduction: the female may spawn approximately 150 eggs which hatch after 30-40 hours
Remarks: the intensity of coloration varies with the environment, especially the bottom substrate.

MELANOCHARACIDIUM BLENNIOIDES (Eigenmann, 1909)
Local name(s): –

Diagnostic characteristics: presence of black stripes on the pelvic, anal and caudal fins; absence of a small black spot on the caudal peduncle at the base of the caudal fin; absence of well-defined black spots above the eye; pectoral fins uniformly grey; 14 circumpeduncular scales; absence of a black stripe between the eye and the mouth; body and head mostly colored dark brown; all fins with a distal transparent margin
Type locality: Erukin River, tributary of Potaro River, Guyana
Distribution: Essequibo River basin, coastal streams of northern South America: Guyana, Suriname, French Guiana and Venezuela; in Suriname, known from the Corantijn, Saramacca, Commewijne and Marowijne rivers
Size: 6-7 cm TL
Habitat: bottom and on large woody debris
Position in the water column: bottom
Diet:
Reproduction:
Remarks:

Characidium zebra (live), Nickerie River (© R. Smith)

Melanocharacidium blennioides (live) (© INRA-Le Bail)

Melanocharacidium cf. *melanopteron* (alcohol), Coppename River, (© P. Willink)

Melanocharacidium dispilomma Buckup, 1993
Local name(s): –

Diagnostic characteristics: presence of black stripes on the pelvic, anal and caudal fins; absence of a small black spot on the caudal peduncle at the base of the caudal fin; presence of two black spots above the eye and at least one black stripe between the eye and the mouth; pectoral fins yellowish with 3-4 black stripes; 12 circumpeduncular scales;
Type locality: Cachoeira Morena, about 2°10'S, 59°30'W, Rio Uatuma, Estado do amazonas, Brazil
Distribution: Northern South America: Brazil, French Guiana, Guyana (inferred), Suriname and Venezuela
Size: 5 cm TL
Habitat: creeks and rivers of the Interior (upstream of the first rapids) with moderate or strong current and rock bottom substrate
Position in the water column: bottom
Diet:
Reproduction:
Remarks: less abundant than *M. blennioides* with which it lives in sympatry; its coloration resembles that of a juvenile *M. blennioides*. Its name '*dispilomma*' refers to the two supra-orbital spots on the head.

Microcharacidium eleotrioides (Géry, 1960)
Local name(s): –

Diagnostic characteristics: incomplete lateral line; 10 or fewer pectoral-fin rays (a very small species reaching no more than 2 cm SL); body transparent yellowish with a narrow, longitudinal black band and approximately 8 broad transversal black bands
Type locality: Little brook between St. Patawa and St. Grand Bacou, Middle-Mana, French Guiana, between 4°N-5°N, 53°W-54°W
Distribution: coastal streams of French Guiana and Suriname; in Suriname, it is known from the Coppename, Saramacca, Suriname. Commewijne and Marowijne rivers
Size: 2.1 cm SL
Habitat: small creeks
Position in the water column: bottom
Diet:
Reproduction:
Remarks:

Melanocharacidium dispilomma (live), Nickerie River (© R. Smith)

Melanocharacidium dispilomma (live) (© INRA-Le Bail)

Microcharacidium eleotrioides (live) (© INRA-Le Bail)

FAMILY HEMIODONTIDAE (HEMIODONTIDS)

Hemiodontids are swift swimmers with a fusiform and streamlined body (7-30 cm SL). An adipose, well-developed eyelid, a suprapectoral sulcus, and 9 to 11 branched pelvic-fin rays are some of the synapomorphies uniting all the species within the Hemiodontidae. Most of its members can be distinguished from other Characiformes by the possession of a round, midlateral, body spot on the vertical through the posterior portion of the dorsal-fin base, and a longitudinal stripe on the lower lobe of the caudal fin. Hemiodontids are mainly benthic, feeding on sand, detritus, mud, filamentous algae, higher plants, larvae of chironomids (Diptera), corixids (Heteroptera) and Ephemeroptera, and fish droppings. *Argonectes* and *Bivibranchia* are the only characiform fishes having protractile upper jaws (especially pronounced in *Bivibranchia*) and *Bivibranchia* also has apomorphies in the roof of the mouth, gill arches and gill chamber, used to sort food particles taken with sand (Eigenmann, 1912; Vari, 1985). They form small or large schools (Vari, 1985), and most inhabit open waters in lakes and large rivers (Roberts, 1972), while others live in much smaller environments, like forest creeks. In Suriname, hemiodontids are abundant in Brokopondo Reservoir.

1-a Body coloration with four transverse black bands
2
1-b Presence of a midlateral black spot immediately posterior of the dorsal fin
3

2-a 50-52 scales in the lateral-line row
Hemiodus huraulti
2-b 42-45 scales in the lateral-line row
Hemiodus quadrimaculatus

3-a Teeth with three cusps; adipose lid completely covering the pupil; Marowijne River
Argonectes longiceps
3-b Teeth multicuspid; adipose lid not entirely covering pupil (with vertical slit at pupil)
4

4-a Gill membrane partly united to isthmus; mouth strongly protractile; fontanel narrow; inner face of gill-rakers with numerous tubercles
5
4-b Gill membrane free from isthmus; mouth not very protractile; fontanel broad; gill-rakers normal
6

5-a 59-63 scales in lateral-line row
Bivibranchia bimaculata
5-b 46-50 scales in lateral-line row
Bivibranchia simulata

6-a Scales of back only a little smaller than those of abdominal region; >80 (120-125) lateral-line scales
Hemiodus argenteus
6-b Scales of back much smaller than those of abdomen; ratio of apparent height of scales below and above lateral line 1.8-2.2 in adult; 60-72 lateral-line scales
Hemiodus unimaculatus

Argonectes longiceps (Kner, 1858)
Local name(s): –

Diagnostic characteristics: presence of a midlateral black spot immediately posterior of the dorsal fin; teeth with three cusps; adipose lid completely covering the pupil; mouth partially protractile; 83-90 scales in the lateral line row; presence of a dark band along the posterior margin of the operculum
Type locality: Rio Icanno (Icana River), Brazil
Distribution: widespread in South America: Brazil, French Guiana and Suriname; in Suriname, only known from Marowijne (Maroni) River
Size: 24.5 cm SL
Habitat: in strong currents over a sandy bottom, immediately downstream the rapids
Position in the water column:
Diet:
Reproduction:
Remarks: apparently rare.

Bivibranchia bimaculata Vari, 1985
Local name(s): –

Diagnostic characteristics: presence of a midlateral black spot immediately posterior of the dorsal fin; teeth multicuspid; adipose lid not entirely covering pupil (with vertical slit at pupil); gill membrane partly united to isthmus; mouth strongly protractile; 59-63 scales in lateral-line row
Type locality: Corantijn River, Suriname
Distribution: French Guiana, Guyana and Suriname; in Suriname known from the Corantijn, Suriname, and Marowijne (Maroni) River basins
Size: 12.9 cm SL
Habitat: larger streams and rivers of the Interior over sandy or rocky bottom substrates
Position in the water column: bottom
Diet: feeding on benthic invertebrates
Reproduction:
Remarks: *B. bimaculata* and *B. simulata* occur sympatrically in the Suriname River.

Bivibranchia simulata Géry, Planquette & Le Bail, 1991
Local name(s): –

Diagnostic characteristics: presence of a midlateral black spot immediately posterior of the dorsal fin; teeth multicuspid; adipose lid not entirely covering pupil (with vertical slit at pupil); gill membrane partly united to isthmus; mouth strongly protractile; 46-50 scales in lateral-line row

Argonectes longiceps (alcohol), scale bar = 1 cm, Lawa River (Marowijne River), ANSP189151 (© M. Sabaj Pérez)

Bivibranchia bimaculata (live), 111 mm SL, Lawa River (Marowijne River), ANSP189149 (© M. Sabaj Pérez)

Bivibranchia simulata (alcohol), Coppename River

Bivibranchia silumata (live) (© INRA-Le Bail)

Type locality: Panacupelu, Moutouci Fall, Wilapaleya, French Guiana (Oyapock River)
Distribution: Brazil, Suriname and French Guiana; in Suriname known from the Coppename, Nickerie and Suriname River basins:
Size: 13.5 cm SL
Habitat: larger streams and rivers of the Interior over sandy or rocky bottom substrates
Position in the water column: bottom
Diet: feeding on benthic invertebrates
Reproduction:
Remarks: this species seems less abundant than *B. bimaculata.*

HEMIODUS ARGENTEUS Pellegrin, 1909
Local name(s): dyogu

Diagnostic characteristics: presence of a midlateral black spot immediately posterior of the dorsal fin; mouth not very protractile; scales of back only a little smaller than those of abdominal region; >80 (120-125) lateral-line scales
Type locality: Rio Orinoco, Venezuela
Distribution: widespread in South America: Brazil, Guyana, Suriname and Venezuela
Size: 23.7 cm SL
Habitat: rivers and larger streams of the Interior; also present in Brokopondo Reservoir
Position in the water column: bottom
Diet: detritus, periphyton and mud
Reproduction:
Remarks:

HEMIODUS HURAULTI (Géry, 1964)
Local name(s): stonfisi

Diagnostic characteristics: body coloration with four transverse black bands; 50-52 scales in the lateral-line row
Type locality: Upper Maroni River, near Litani Falls, French Guiana
Distribution: Marowijne (Maroni) and Mana River basins: French Guiana and Suriname
Size: 11.4 cm SL
Habitat: in rapids and strong currents in rivers of the Interior
Position in the water column: bottom
Diet:
Reproduction:
Remarks: this species occurs sympatrically with *Leporinus maculatus* which has more or less the same pigmentation pattern of four transverse black bands (but *L. maculatus* has narrow bands in between the four broad bands).

Hemiodus huraulti (live) (© INRA-Le Bail)

Hemiodus quadrimaculatus Pellegrin, 1909
Local name(s): –

Diagnostic characteristics: body coloration with four transverse black bands (exactly like *H. huraulti*), but with 42-45 scales in the lateral-line row
Type locality: Riviere Camopi, French Guiana
Distribution: Northern South America: Brazil, French Guiana, Guyana and Suriname
Size: 13.1 cm SL
Habitat: rapids and strong currents in rivers of the Interior
Position in the water column: bottom
Diet:
Reproduction:
Remarks: apparently replaces *H. huraulti* in Surinamese rivers to the west of the Marowijne River.

Hemiodus unimaculatus Bloch, 1794
Local name(s): dyogu

Diagnostic characteristics: presence of a midlateral black spot immediately posterior of the dorsal fin; mouth not very protractile; scales of back much smaller than those of abdomen; ratio of apparent height of scales below and above lateral line 1.8-2.2 in adult; 60-72 lateral-line scales
Type locality: Brazil, Suriname
Distribution: widespread in South America: Brazil, French Guiana, Guyana, Peru and Suriname
Size: 21.5 cm SL
Habitat: rivers and larger streams of the Interior; also present in Brokopondo Reservoir
Position in the water column: bottom
Diet: detritus, periphyton and mud
Reproduction: spawns in the rainy season
Remarks:

Hemiodus quadrimaculatus (live), Nickerie River (© R. Smith)

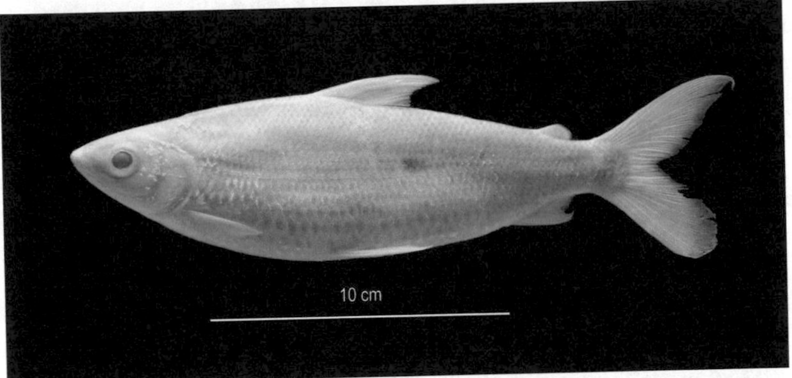

Hemiodus unimaculatus (alcohol), Suriname River, RMNH36116 (© Naturalis)

Hemiodus unimaculatus (live) (© INRA-Le Bail)

FAMILY GASTEROPELECIDAE (FRESHWATER HATCHETFISHES)

Fishes of the family Gasteropelecidae have the frontal bone longitudinally corrugated, bearing a strong longitudinal ridge. The posttemporal and supracleithrum are fused into a single bone. The pelvic fins and associated bones are minute. They have an enlarged, strongly convex muscular pectoral girdle region, consisting of greatly expanded coracoids fused to a single fan-shaped and corrugated median bone. Their lateral line extends ventroposteriorly to approach the anterior termination of the anal fin, or they have 0 to 2 or 3 scales behind head and one or a very few scales on the tail-fin base. The dorsal fin has 10 to 17 rays and the anal fin 22 to 44 rays. An adipose fin is present in the larger species, but absent in the smaller species. All are capable of jumping relatively long or high distances out of the water by use of their modified, elongate pectoral fin rays and heavily-muscled, enlarged pectoral girdle. See Wiest (1995) and Weitzman & Palmer (1996) for discussions of their jumping (flying) abilities.

1-a Presence of an adipose fin; body color uniformly silvery
Gasteropelecus sternicla
1-b Absence of an adipose fin; body marbled with broad bands extending from the keel to the dorsum
Carnegiella strigata

CARNEGIELLA STRIGATA (Günther, 1864)
Local name(s): banketman (marbled hatchetfish)

Diagnostic characteristics: body hatchet-shaped, marbled with broad bands extending from the keel to the dorsum, large pectoral fins, absence of an adipose fin
Type locality: no type locality known
Distribution: Amazon River basin, south of Amazon, and Caquetá River in Colombia: Brazil, Colombia, Guyana, Peru and Suriname; all rivers of Suriname
Size: 3.5 cm SL
Habitat: freshwater, Savanna Belt and Interior, mainly small streams
Position in the water column: surface
Diet: small (allochthonous) invertebrates
Reproduction: eggs deposited on aquatic vegetation or left on the bottom substrate, hatching follows after 36 h at 25°C
Remarks:

Carnegiella strigata (alcohol), scale bar = 5 mm, INHS49173 (© M. Sabaj Pérez)

Gasteropelecus sternicla (Linnaeus, 1758)
Local name(s): banketman (silver or common hatchetfish)

Diagnostic characteristics: body color uniformly silvery with a thin longitudinal, black band and a black stripe at the base of the anal fin, large pectoral fins, hatchet-shaped body with horizontally straight back, presence of an adipose fin;
Type locality: Suriname
Distribution: Peruvian Amazon, middle Amazon, Guianas and Venezuela: French Guiana, Guyana, Peru and Venezuela; all rivers of Suriname
Size: 4 cm SL
Habitat: freshwater, Savanna Belt and Interior, both small streams and large rivers
Position in the water column: surface
Diet: small (allochthonous) invertebrates
Reproduction: group spawning, females slightly larger than males
Remarks: can escape from its enemies by jumping out of the water (10 m), but they do not actually fly by flapping their pectoral fins; the pectoral fins are used in the take off mechanism for moving the hatchetfish from the water into the air, but they are not used by the fish while airborne (Wiest, 1995).

Gasteropelecus sternicla (alcohol), scale bar = 5 mm, Para River (Suriname River), ANSP189193 (© M. Sabaj Pérez)

FAMILY ALESTIDAE ('AFRICAN' TETRAS)

Zanata and Vari (2005) shifted the Neotropical genus *Chalceus* with 5 species into the family Alestidae or African tetras (Géry, 1977), which thus becomes the only trans-Atlantic family-level group within the Characiformes (with a minimum age of 90-112 Ma). The approximately 105 species in the Alestidae range from the diminutive species of the West African genus *Virilia*, some of which mature at 18.8 mm standard length, to the tigerfish, *Hydrocynus goliath* which at 1320 mm standard length achieves the greatest length of any characiform (Weitzman & Vari, 1998). Such a dramatic range in body size, 18.8–1320 mm, a 70x size range, is unique to members of the Alestidae among the families of the Characiformes (Zanata & Vari, 2005). The monophyly of the Alestidae is supported by 6 synapomorphies: (1) contralateral premaxillae with medial interdigitations present, (2) two functional rows of teeth present on premaxilla (further developed into three rows of teeth in *Chalceus*), (3) area of contact of ectopterygoid with palatine narrow, (4) ligamentous attachment of ectopterygoid to neurocranium absent, and (6) small ossification associated with first dorsal-fin proximal radial absent (Zanata & Vari, 2005). The genus *Chalceus* is characterized by large scales above the (low) lateral line (which has more numerous and smaller perforated scales than the series above the lateral line), and three rows of teeth on the upper jaw.

CHALCEUS MACROLEPIDOTUS Cuvier, 1818
Local name(s): alampiya

Diagnostic characteristics: lateral scales above the lateral line (situated low on the body, 33-37 scales) very large, while those below the lateral line (including lateral line scales) small; fins pink to bright red, eye yellow, body silvery; anal fin short; mouth terminal with three rows of teeth on the upper jaw
Type locality: Brazil
Distribution: Negro and Orinoco River basins, and coastal rivers of northern South America: Brazil, Colombia, French Guiana, Guyana, Suriname and Venezuela
Size: 24.5 cm SL
Habitat: upper reaches of rivers and large streams of the Interior, upstream of the first rapids
Position in the water column: surface and mid-water
Diet: carnivorous, mainly aquatic insects but also feeding on small fishes
Reproduction: in aquarium, up to 2000 eggs are scattered among vegetation
Remarks: popular, both as a food fish and in the aquarium hobby (but needs large space).

Chalceus macrolepidotus (live), Nickerie River (© R. Smith)

FAMILY CHARACIDAE – GENERA INCERTAE SEDIS (CHARACINS, TETRAS)

The eighty eight characid genera listed in Reis *et al.* (2003) as Incertae Sedis constitute a largely heterogeneous assemblage of small to large-sized fishes. Most of these genera had been included in the subfamily Tetragonopterinae (see Géry, 1977). However, considering the lack of evidence that this subfamily constitutes a monophyletic assemblage (e.g. Weitzman & Malabarba, 1998), Reis *et al.* (2003) prefer instead to emphasize the fact that interrelationships among the Characidae are poorly known, and only consider subfamilies for which some evidence of monophyly is available. The species in Suriname are smaller than 15 cm TL; the slender *Bryconops* are known as nyanga-nyanga, while others (especially in the genera *Astyanax*, *Jupiaba* and *Moenkhausia*) are known as sriba.

1-a Presence of a broad, black band at the base of the caudal fin
Moenkhausia oligolepis
1-b Absence of a broad, black band at the base of the caudal fin
2

2-a Lower caudal lobe longer than the upper lobe; presence of a black band that begins just in front of the dorsal fin and extends on the lower lobe of the caudal fin; first rays of the anal fin and the extremity of the dorsal fin black (Marowijne)
Thayeria ifati
2-b Lobes of the caudal fin of equal size; absence of an oblique black band
3

3-a Presence of a black spot on the caudal peduncle
27
3-b Absence of a black spot on the caudal peduncle
4

4-a Presence of a black spot on the dorsal fin; species of small size <5 cm in standard length
21
4-b Absence of a black spot on the dorsal fin; variable size
5

5-a Presence of a black spot or margin on the caudal fin
6
5-b Absence of black markings on the caudal fin
11

6-a Anal fin straight; extremity of the upper jaw positioned behind the anterior margin of the pupil; maxilla very arched at its meeting with the premaxilla (very characteristic for the genus *Bryconops*)

9

6-b Anal fin falciform; extremity of the upper jaw not reaching the anterior margin of the pupil; presence of scales on the caudal fin

7

7-a Distal part of both caudal fin lobes black (except their tip)

Moenkhausia intermedia

7-b Black markings on caudal fin not symmetrically distributed

8

8-a Upper caudal lobe usually darker than the lower lobe

Moenkhausia lepidura

8-b Base of caudal fin and distal part of the upper caudal lobe transparent; median part of the upper caudal lobe black (Marowijne)

Moenkhausia inrai

9-a The two lobes of the caudal fin rounded and equally dark; central part of the adipose fin white

Bryconops affinis

9-b Fins reddish, caudal fin red with a white posterior margin

Bryconops sp. 'redfins'

9-c Upper lobe of the caudal fin more dark than the lower lobe; adipose fin generally colored yellow-orange to red

10

10-a Median rays of the caudal fin white, upper caudal lobe with superior black spot; snout rounded, with the tip at the same level as the horizontal of the pupil; maxilla short, its tip not reaching suture between 2nd and 3rd circumorbitals; 28-35 branched anal rays

Bryconops caudomaculatus

10-b Median rays of the caudal fin black, the black spot reaches the extremity of the upper caudal lobe; snout pointed, with the tip at the same level as the upper margin of the pupil; maxillary long, its tip usually reaching suture between 2nd and 3rd circumorbitals; 23-27 branched anal rays

Bryconops melanurus

11-a Body brilliant rose and fins red; large humeral spot (measuring half of the body depth) (Upper Marowijne and Suriname rivers)

Hyphessobrycon roseus

11-b Humeral spot, if present, small, measuring less than half of the body depth

12

12-a Body with a brilliant, red longitudinal line, a red eye, and a small humeral spot (Para River)

Hyphessobrycon sp. 'redline'

12-b Body with a broad, black longitudinal stripe extending on the caudal fin and a thin red line immediately above the black stripe, a red eye and no humeral spot (Paloemeu River, Upper Marowijne River System)

Hyphessobrycon sp. 'blackstripe'

12-c Body with brilliant blue-green dorsal iridescence and bright-red ventral parts, no humeral spot (the cardinal tetra, introduced by aquarium hobbyists) (Para River)

Paracheirodon axelrodi

12-d Body grey, humeral spot absent or small, measuring less than half of the body depth, longitudinal line, when present, thin and black

13

13-a Elongated body, body depth more than 2.5 times in standard length; maximum size <5 cm SL

14

13-b Body deep, body depth less than 2.5 times in standard length; maximum size >5 cm SL

17

14-a Absence of a humeral spot; lateral line complete; presence of a thin, black, curved margin at the base of the caudal fin

Aphyocharacidium melandetum

14-b Presence of a humeral spot (sometimes difficult to see); absence of black, curved margin at the posterior end of the caudal peduncle

15

15-a Presence of a horizontal black bar through the eye; lateral line complete

Moenkhausia collettii

15-b Absence of a black horizontal bar through the eye; lateral line not complete

16

16-a Two fine, black lines at the base of the anal fin; humeral spot conspicuous, vertically ovate

Hemigrammus bellottii

16-b Presence of a dark, lateral stripe, heaviest behind the origin of the anal, but not reaching the base of the caudal; single black stripe at posterior base of anal fin, not continuous at the base of the first seven rays; body depth 3.75 in standard length

Hemigrammus orthus

16-c Absence of black stripe(s) at the base of anal fin, lateral stripe a narrow, dark line overlaid with silvery (or very faint); humeral spot faint or absent; body depth 2.75 in standard length

Hemigrammus lunatus

17-a Presence of a pre-ventral spine; humeral spot with ocellus

18

17-b Absence of pre-ventral spine; humeral spot with or without ocellus

19

18-a 42-46 scales in the lateral line

Jupiaba polylepis

18-b 33-36 scales in the lateral line (Commewijne River?)

Jupiaba ocellata

19-a Eye with orange apex; presence of a longitudinal, broad, silvery band; when adpressed first rays of dorsal fin falling distinctly short of adipose fin; presence of a small rounded humeral spot

Moenkhausia grandisquamis

19-b Absence of broad, silvery band; when adpressed, first rays of dorsal fin reach (or almost reach) adipose fin

20

20-a Apex of the eye colored red; first rays of the dorsal fin reach to or even over adipose fin when adpressed; humeral spot not clear

Moenkhausia chrysargyrea

20-b Presence of a thin, black longitudinal line along the lateral line; first rays of dorsal fin almost reach the adipose fin; large, elongated, horizontal oval humeral spot

Moenkhausia browni

21-a Presence of a black spot or black band on the anterior part of the anal fin

22

21-b Absence of black markings on the anal fin

24

22-a Presence of an oval spot on the anterior part of both the anal and pelvic fins

Pristella maxillaris

22-b Presence of a black band on the anterior part of the anal fin; absence of black markings on the pelvic fins

23

23-a Lateral line complete; dorsal fin spot not reaching the posterior margin of the fin

Moenkhausia hemigrammoides

23-b Lateral line incomplete; black spot on dorsal fin reaches the posterior margin of the fin

Hemigrammus unilineatus

24-a Presence of a conspicuous humeral spot

25

24-b Absence of a conspicuous humeral spot (pseudotympanum may be present)

26

25-a Humeral spot vertically elongated; presence of a vertical, black band through the eye

Hyphessobrycon simulatus

25-b Humeral spot rounded; absence of vertical, black band through the eye

Hyphessobrycon copelandi

26-a Both body and fins rosy colored; very small in size (< 2 cm SL); only 15 anal rays; only known from Paru/Sipaliwini Savanna

Hyphessobrycon georgettae

26-b Only fins colored bright rosy-red; pseudotympanum conspicuous, even in adults; 26-27 anal rays (Corantijn, Nickerie, Coppename)

Hyphessobrycon rosaceus

27-a Body elongated (body depth >2.8 times in SL); lateral line incomplete; small species < 5 cm SL

28

27-b Deep body (body depth <2.8 times in SL); lateral line complete; species > 5 cm SL

32

28-a Apex of the eye colored red or rose

29

28-b Eye with uniform color

31

29-a Absence of humeral spot; absence of scales on the caudal fin (Marowijne)

Hyphessobrycon borealis

29-b Presence of one or two humeral spots (sometimes faint); presence of scales on the caudal fin

30

30-a Presence of a longitudinal band; absence of an iridescent spot on upper part of caudal peduncle

Hemigrammus guyanensis

30-b Presence of longitudinal black band on the posterior half of the body (from the vertical through the insertion of the dorsal fin to the caudal spot and a iridescent spot on the upper part of the caudal peduncle

Hemigrammus aff. *ocellifer*

30-c Absence of longitudinal band; presence of iridescent spot on upper part of caudal peduncle

Hemigrammus ocellifer

31-a Caudal spot occupying entire height of caudal peduncle and not continuous with lateral band; presence of black markings at the base of the anal fin; body depth 3.3-3.5 in standard length

Hemigrammus boesemani

31-b Caudal spot occupying entire height of caudal peduncle and continuous with diffuse, thin black lateral stripe; presence of black markings at base of anal fin; very slender, elongate species with body depth 4.3 in standard length
Hyphessobrycon cf. *minimus*
31-c Caudal spot not occupying the entire height of the caudal peduncle; black markings along anal fin base less conspicuous; body depth 2.75-3.2 in standard length
Hemigrammus rodwayi

32-a Caudal spot horizontally elongated, its anterior margin reaching to the level of the adipose fin; humeral spot horizontally elongated (oval)
33
32-b Caudal spot not elongated horizontally, not reaching to the level of the adipose fin; humeral spot not elongated horizontally
34

33-a Margins of scales black forming a longitudinal zigzag line pattern; apex of eye pale red (Marowijne)
Astyanax validus
33-b Absence of longitudinal zigzag pattern; apex of eye bright red
Astyanax bimaculatus

34-a Anal fin very long (>36 rays); presence of ctenoid (rough) scales at the base of the pelvic fins (a species of the Coastal Plain)
Ctenobrycon spilurus
34-b Anal fin short (<36 rays); absence of ctenoid scales at the base of the pelvic fins
35

35-a Caudal spot very elongated vertically
Jupiaba keithi
35-b Caudal spot rounded
36

36-a Apex of the eye red
37
36-b Eye uniformly colored (without red apex)
39

37-a Origin of anal fin positioned posterior to dorsal fin; absence of pre-pelvic spines; scales on the caudal fin
Moenkhausia georgiae
37-b Origin of anal fin about midway the dorsal fin; presence of pre-pelvic spines; absence of scales on the caudal fin
38

38-a Adpressed pelvic fins reach anal fin; apex eye brown-red; 5-6 scales below the lateral line; head short, 3.6-3.9 times in SL (Marowijne)
Jupiaba maroniensis
38-b Adpressed pelvic fins do not reach to origin of anal fin; apex eye bright-red; 6.5-8 scales below the lateral line; head long, 3.4-3.5 times in SL
Jupiaba meunieri

39-a Presence of pre-pelvic spines; absence of scales at the base of the caudal fin
40
39-b Absence of pre-pelvic spines, presence of scales at the base of the caudal fin
41

40-a Caudal peduncle deep, vertically elongated; 45-50 scales in lateral line; absence of chevron (<) marks along lateral line in preserved specimens
Jupiaba abramoides
40-b Caudal peduncle horizontally elongated; 33-39 scales in lateral line; presence of chevron (<) marks along lateral line in preserved specimens
Jupiaba pinnata

41-a Less than 41 scales in the lateral line; caudal spot oval
Moenkhausia surinamensis
41-b More than 40 scales in the lateral line; caudal spot rounded
Moenkhausia moisae

APHYOCHARACIDIUM MELANDETUM (Eigenmann, 1912)
Local name(s): –

Diagnostic characteristics: small species with terminal or slightly superior mouth with two rows of teeth on the lower jaw; absence of a humeral or caudal spot; lateral line complete (33-35 scales); presence of a thin, black, curved margin at the posterior end of the caudal peduncle
Type locality: Guyana
Distribution: Essequibo River basin, Guyana; in Suriname known only from the Marowijne (Planquette *et al.*, 1996) and Coppename rivers
Size: 3.5 cm SL
Habitat: in the upper reaches of the rivers
Position in the water column:
Diet:
Reproduction:
Remarks: apparently very rare, with only few specimens collected; nothing is known about its biology.

Aphyocharacidium melandetum (alcohol), Coppename River

Aphyocharacidium melandetum (live) (© INRA-Le Bail)

Astyanax bimaculatus (Linnaeus, 1758)
Local name(s): sriba

Diagnostic characteristics: caudal spot horizontally elongated, its anterior margin reaching to the level of the adipose fin; humeral spot horizontally elongated (oval); caudal and anal fins with orange-yellow color; complete lateral line (37-41 scales); absence of pre-pelvic spines
Type locality: South America (Suriname drainages)
Distribution: Panama to the Amazon basin: Argentina, Brazil, Colombia, Ecuador, French Guiana, Guyana, Panama, Peru, Suriname, Trinidad and Tobago and Venezuela
Size: 15 cm TL
Habitat: an abundant species, in rivers and forest creeks of the Interior, coastal swamps, and black-water streams of the Savanna Belt
Position in the water column: mid-water
Diet: zooplankton and detritus and plant matter
Reproduction: in the rainy season, the reproductive season, males develop sexual hooklets on the rays of the pelvic and anal fins, while their colors become more intense; eggs hatch in 24-36 hours and larvae start to feed after 5 days when the yolk-sac is resorbed
Remarks: a robust tetra popular with aquarium hobbyists.

Astyanax validus Géry, Planquette & Le Bail 1991
Local name(s): sriba

Diagnostic characteristics: margins of scales black forming a longitudinal zigzag line pattern (in preserved specimens); apex of eye pale red; caudal spot elongated, its anterior margin reaching to the level of the adipose fin; humeral spot horizontally elongated (oval); caudal and anal fins with orange-yellow color, but less intense compared with *A. bimaculatus*; complete lateral line; absence of pre-pelvic spines
Type locality: Blanche creek, Comté River, French Guiana
Distribution: French Guiana, including Marowijne river; in Suriname only known from the Marowijne River
Size: 15 cm TL
Habitat: in upper reaches of the Marowijne River and tributary creeks
Position in the water column: mid-water
Diet:
Reproduction: in the reproductive season males develop sexual hooklets on the rays of the pelvic and anal fins
Remarks: in French Guiana, *A. validus* is known from large specimens (>8.9 cm SL) as compared with *A. bimaculatus*, and Planquette *et al.* (1996) point out that the two taxa (*A. validus* and *A. bimaculatus*) are possibly just one species, with *A. validus* a geographic variety of *A. bimaculatus*.

Astyanax bimaculatus (live) (© INRA-Le Bail)

Astyanax bimaculatus (live), Nickerie River (© R. Smith)

Astyanax validus (live) (© INRA-Le Bail)

Bryconops affinis (Günther, 1864)
Local name(s): nyanga-nyanga

Diagnostic characteristics: an elongate species with a complete lateral line and a straight anal fin (22-26 rays), its insertion just posterior to the dorsal fin; extremity of the upper jaw positioned behind the anterior margin of the pupil; the two lobes of the caudal fin rounded and equally dark; central part of the adipose fin white
Type locality: Guyana
Distribution: coastal streams of the Guiana Shield: French Guiana, Guyana and Suriname
Size: 12 cm TL
Habitat: in streams and rivers of the Interior, mainly in water with moderate or strong current
Position in the water column: surface
Diet: allochthonous invertebrates
Reproduction:
Remarks:

Bryconops caudomaculatus (Günther, 1864)
Local name(s): nyanga-nyanga

Diagnostic characteristics: upper lobe of the caudal fin more dark than the lower lobe; adipose fin generally colored pale yellow-orange; median rays of the caudal fin white, upper caudal lobe with superior black spot; insertion anal fin about midway under dorsal fin; snout rounded, with the tip at the same level as the horizontal of the pupil; maxilla short, its tip not reaching suture between 2nd and 3rd circumorbitals; 28-35 branched anal rays
Type locality: South America [probably Guyana]
Distribution: coastal streams of the Guiana Shield, Orinoco and Amazon River basins: Brazil, French Guiana, Guyana, Suriname and Venezuela
Size: 14 cm TL
Habitat: streams of the Interior with moderate to strong current and sandy or rocky bottom substrate
Position in the water column: surface
Diet: omnivorous
Reproduction: sexual hooklets on the anal fin rays of males clearly visible with a loupe
Remarks:

Bryconops affinis (live) (© INRA-Le Bail)

Bryconops caudomaculatus (live) (© P. Willink)

Bryconops caudomaculatus (live) (© W. Kolvoort)

Bryconops melanurus (Bloch, 1794)
Local name(s): nyanga-nyanga

Diagnostic characteristics: upper lobe of the caudal fin more dark than the lower lobe; adipose fin generally colored red; median rays of the caudal fin black, the black spot reaches the extremity of the upper caudal lobe; insertion anal fin about midway under dorsal fin; snout pointed, with the tip at the same level as the upper margin of the pupil; maxillary long, its tip usually reaching suture between 2nd and 3rd circumorbitals; 23-27 branched anal rays
Type locality: Suriname
Distribution: coastal streams of the Guiana Shield: Guyana, French Guiana and Suriname
Size: 12 cm SL
Habitat: very abundant in rivers and smaller streams of the Interior and in Brokopondo Reservoir
Position in the water column: surface
Diet: omnivorous, also feeding on allochthonous insects
Reproduction:
Remarks: in Brokopondo Reservoir, this species is abundant, but individuals are smaller than those of riverine populations; in the reservoir it also seems to reproduce at small size (i.e. stunting). A fourth *Bryconops* species in Suriname has red fins with the posterior edge of the caudal fin white; it is known from the Suriname and Paloemeu rivers.

Ctenobrycon spilurus (Valenciennes, 1850)
Local name(s): sriba

Diagnostic characteristics: *Astyanax*-like fish with a compressed body, a complete lateral line, a round caudal spot and a small humeral spot; eye small and anal fin very long (39-47 rays); presence of ctenoid (rough) scales at the base of the pelvic fins
Type locality: Suriname
Distribution: Orinoco River basin and coastal river basins of northern South America: French Guiana, Guyana, Suriname; in Suriname only known from coastal swamps
Size: 8 cm SL
Habitat: standing water of coastal freshwater swamps
Position in the water column: mid-water
Diet: zooplankton and other small aquatic invertebrates
Reproduction: an adult female has approximately 2000 eggs that hatch in 50-70 hours; the larvae start feeding after three days
Remarks:

Bryconops melanurus (live), Lawa River (Marowijne River), ANSP189268 (© M. Sabaj Pérez)

Bryconops melanurus (live), Nickerie River (© R. Smith)

Bryconops sp. 'redfins' (live), Paloemeu River (© P. Naskrecki)

Ctenobrycon spilurus (live) (© INRA-Le Bail)

Hemigrammus bellottii (Steindachner, 1882)
Local name(s): –

Diagnostic characteristics: a very small *Hemigrammus* species with only few scales on the caudal fin, lacking a spot on the caudal peduncle, but with a humeral spot (often difficult to see) and two fine, black lines at the base of the anal fin; lateral line not complete
Type locality: Tabatinga (Solimoes River, Amazonas, Brazil)
Distribution: Solimões and Negro River basins and Maroni River basin: Brazil and French Guiana; in Suriname present in Lower Marowijne River (Planquette *et al.*, 1996) and possibly Corantijn River
Size: 2.6 cm SL
Habitat:
Position in the water column:
Diet: small invertebrates, especially aquatic insect larvae
Reproduction:
Remarks: apparently rare.

Hemigrammus boesemani Géry, 1959
Local name(s): –

Diagnostic characteristics: small, elongate species (body depth 3.3-3.5 in standard length) with a caudal spot occupying the entire height of caudal peduncle, no humeral spot and an inconspicuous lateral band; presence of characteristic black markings at the base of the anal fin
Type locality: creek near Sinnamary, northern French Guiana
Distribution: rivers in French Guiana; Suriname; Upper Amazon basin of Peru; in Suriname known from Saramacca, Suriname, Commewijne and Marowijne rivers
Size: 2.7 cm SL
Habitat: present in both small streams (especially in the lower reaches of the rivers) and coastal swamps
Position in the water column: mid-water
Diet:
Reproduction:
Remarks: often present in large numbers.

Hemigrammus guyanensis Géry, 1959
Local name(s): –

Diagnostic characteristics: presence of a circular humeral spot and a black spot on the caudal peduncle; presence of scales on the caudal fin; presence of a longitudinal band; absence of an iridescent spot on upper part of caudal peduncle (compare *H. ocellifer*)
Type locality: Sable creek, Upper Mana River, French Guiana

Hemigrammus boesemani (live) (© INRA-Le Bail)

Hemigrammus guyanensis (live) (© INRA-Le Bail)

Distribution: Maroni (Marowijne), Mana, Approuague and Oyapock rivers, French Guiana; in Suriname also known to occur in the Coppename and Suriname rivers
Size: 3.5 cm SL
Habitat: rivers and tributary streams of the Interior (upstream of the first rapids), not in coastal swamps
Position in the water column: mid-water
Diet:
Reproduction: males with sexual hooklets on the first 4 rays of the anal fin
Remarks: apparently not very abundant.

Hemigrammus lunatus Durbin in Eigenmann, 1918
Local name(s): –

Diagnostic characteristics: humeral spot very faint and small (often lacking), horizontally elongate, no caudal spot; body depth 2.75 in standard length; all fins hyaline; 19-26 anal rays
Type locality: Amazon River basin
Distribution: Amazon, Paraguay and Suriname River basins: Brazil, Peru and Suriname; in Suriname known from Corantijn and (possibly) Suriname rivers
Size: 5 cm SL
Habitat:
Position in the water column: mid-water
Diet:
Reproduction:
Remarks: not well known and apparently not very abundant.

Hemigrammus ocellifer (Steindachner, 1882)
Local name(s): – (head-and-tail light tetra)

Diagnostic characteristics: presence of an iridescent spot on upper part of caudal peduncle (above the black spot) and a bright red eye ('head-and-tail' lights); two faint humeral spots
Type locality: Vila Bela, now Parintins, and Codajas, Amazon basin, Brazil
Distribution: rivers of Guyana, Suriname, French Guiana, and Amazon basin in Peru and Brazil
Size: 4.4 cm TL
Habitat: in standing water of coastal swamps and black-water streams of the Savanna Belt, but also present in the Interior (e.g. Upper Corantijn River)
Position in the water column: mid-water
Diet: omnivorous, especially allochthonous invertebrates (ants)
Reproduction: reproduction (in groups) takes place in captivity (aquarium), the eggs are adhesive and hatch in about 3 hours.
Remarks: this species can be abundant in coastal freshwater habitats; popular with aquarium hobbyists. A related species with a conspicuous midlateral black stripe on the posterior part of the body, *Hemigrammus* aff. *ocellifer*, is known from the Paloemeu and Suriname rivers.

Hemigrammus ocellifer (live), Sipaliwini River (Corantijn River) (© P. Willink)

Hemigrammus aff. *ocellifer* (live), Paloemeu River (© K. Wan Tong You)

Hemigrammus orthus Durbin, 1909
Local name(s): –

Diagnostic characteristics: a very small *Hemigrammus* species close to *H. bellottii* (Géry, 1977) with a diffuse, round, or somewhat elongate humeral spot; a dark lateral stripe, heaviest behind the origin of the anal, but not reaching the base of the caudal; a thin black line at the base of the posterior anal fin rays, not continuous with that at the base of the first seven rays; dorsal, caudal, first seven rays of anal, and first two or three rays of ventrals dusky; scales of the back dusky, each often bearing a single black spot
Type locality: Tukeit, Guyana
Distribution: Essequibo River in Guyana; lower Tapajós River, Brazil; in Suriname, only known from the Upper Corantijn River
Size: 2.7 cm TL
Habitat: small forest creeks in the Interior
Position in the water column: mid-water
Diet:
Reproduction:
Remarks:

Hemigrammus rodwayi Durbin, 1909
Local name(s): –

Diagnostic characteristics: very close to *H. boesemani*, but with the caudal spot not occupying the entire height of the caudal peduncle and black markings along anal fin base not very conspicuous; body depth 2.75-3.2 in standard length
Type locality: Georgetown Trenches, Botanic garden, Aruka and Barima rivers, Guyana
Distribution: rivers of Guyana, Suriname, French Guiana and Amazon River basin: Brazil, French Guiana, Guyana, Peru and Suriname
Size: 5.3 cm TL
Habitat: mainly a coastal species
Position in the water column:
Diet:
Reproduction:
Remarks:

Hemigrammus unilineatus (Gill, 1858)
Local name(s): –

Diagnostic characteristics: a small *Hemigrammus* species (lateral line incomplete, caudal fin scaled) with a black spot on the dorsal fin that reaches the posterior margin of the fin and the first rays of the anal fin colored black
Type locality: Trinidad Island, West Indies

Hemigrammus orthus (alcohol), Sipaliwini River (Corantijn River) (© P. Willink)

Hemigrammus rodwayi (live) (© INRA-Le Bail)

Hemigrammus unilineatus (live) (© INRA-Le Bail)

Distribution: river basins of Trinidad Island, coastal river basins of Venezuela, and of Guyana, Suriname and French Guiana; Guaporé and Amazon River basins: Brazil, French Guiana, Guyana, Peru, Suriname, Trinidad and Tobago, and Venezuela; in Suriname known from the Corantijn, Saramacca, Suriname, Corantijn and Marowijne rivers
Size: 5.3 cm TL
Habitat: coastal freshwater swamps, black-water streams of the Savanna Belt and forest creeks in the Interior
Position in the water column: mid-water
Diet: small invertebrates
Reproduction: males develop sexual hooklets on the first rays of the anal fin; the females produce 200-500 eggs that hatch in about 60 hours
Remarks: can be easily confused with *Moenkhausia hemigrammoides*, a slightly larger species, but *M. hemigrammoides* has a complete lateral line; *H. unilineatus* is known in the aquarium hobby since 1910 (imports from Brazil).

HYPHESSOBRYCON BOREALIS Zarske, Le Bail & Géry, 2006
Local name(s): –

Diagnostic characteristics: a small, elongate *Hyphessobrycon* species (incomplete lateral line, absence of scales on the caudal fin) lacking a humeral spot, but with a round caudal spot, not extending the whole depth of the caudal peduncle and only a little on the basal parts of the middle rays of the caudal fin; superior margin of eye colored red; a black, sometimes green, shiny band extends on the mid-axis of the body from the gill cover to the black caudal spot (in preserved animals this band is fine, starting at the dorsal fin)
Type locality: Crique St. Anne, Mana River, French Guiana
Distribution: French Guiana, including Marowijne (Maroni) River
Size: 2.8 cm SL
Habitat: lives in the same habitat as *Hemigrammus guyanensis*, rivers and tributary streams of the Interior (upstream of the first rapids)
Position in the water column: mid-water
Diet:
Reproduction:
Remarks: identified as *H.* aff. *sovichthys* in Planquette *et al.* (1996); there is no similar species of the genera *Hyphessobrycon* or *Hemigrammus* in Suriname because of the absence of the humeral spot.

HYPHESSOBRYCON COPELANDI Durbin in Eigenmann, 1908
Local name(s): –

Diagnostic characteristics: an *Hyphessobrycon* species with a black spot on the dorsal fin (a 'flag') and a rounded humeral spot; absence of vertical, black band through the eye and absence of a black band on the first rays of the anal fin
Type locality: Tabatinga, Amazon River basin, Brazil

Hyphessobrycon borealis (live) (© INRA-Le Bail)

Hyphessobrycon copelandi (live) (© INRA-Le Bail)

Distribution: Upper Solimões, Mana and Approuague River basins: Brazil and French Guiana; in Suriname only known from Coppename and Marowijne rivers
Size: 3.5 cm SL
Habitat: rivers of the Interior, upstream of the first rapids, and small forest creeks
Position in the water column: mid-water
Diet:
Reproduction:
Remarks:

HYPHESSOBRYCON GEORGETTAE Géry, 1961
Local name(s): –

Diagnostic characteristics: a very small *Hyphessobrycon* species with a black spot on the dorsal fin (a 'flag'), but lacking a humeral spot and of very small size; both body and fins rosy colored; anal fin with 15-17 branched rays
Type locality: tributary of Sipaliwini River (Upper Corantijn River), a swamp creek in Paru Savanna, Suriname
Distribution: only known from its type locality, Upper Corantijn River, Suriname
Size: 2.2 cm SL
Habitat: pool in a small stream, a tributary of Vier Gebroeders Creek, in the Paru (Sipaliwini) Savanna with standing water and aquatic macrophytes
Position in the water column: mid-water
Diet: probably small invertebrates
Reproduction:
Remarks:

HYPHESSOBRYCON cf. *MINIMUS* Durbin, 1909
Local name(s): –

Diagnostic characteristics: small, elongate species with thin, black lateral stripe (somewhat diffused in humeral region), an intense black, round caudal spot, a black stripe at the base of the anal fin, and lacking a humeral spot; fins without distinct markings
Type locality: Cane Grove Corner [mouth of Mahaica River, Guyana]
Distribution: Mahaica River basin, Guyana; in Suriname, known from Upper Coppename and Nickerie rivers
Size: 2.1 cm TL
Habitat: both main river channel and tributaries (e.g. Adampada Creek)
Position in the water column: mid-water
Diet:
Reproduction:
Remarks: smaller and more elongate (body depth 4.3 in SL in *H.* cf. *minimus* versus 3.3-3.5 in *H. boesemani*) than *Hemigrammus boesemani.*

Hyphessobrycon georgettae (alcohol), Sipaliwini River (Corantijn River), ZMA103270 (© Naturalis)

Hyphessobrycon cf. *minimus* (alcohol), Coppename River

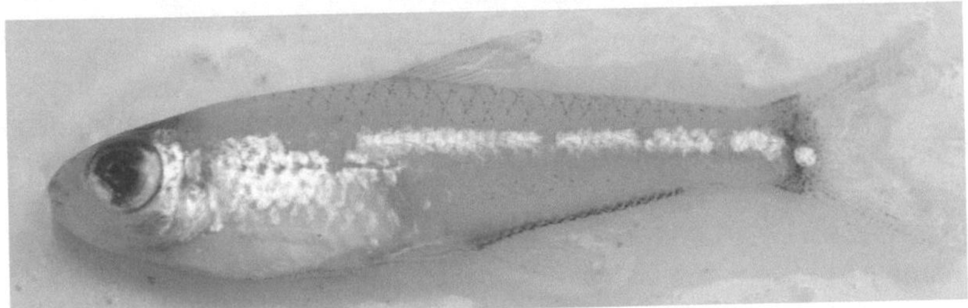

Hyphessobrycon cf. *minimus* (live), Coppename River (© R. Smith)

Hyphessobrycon rosaceus Durbin, 1909
Local name(s): –

Diagnostic characteristics: another *Hyphessobrycon* species with a black spot on the dorsal fin (a 'flag'), and lacking a humeral spot; only the fins are colored bright rosy-red; a pseudotympanum is conspicuous, even in adults; 26-27 anal rays
Type locality: Gluck Island, Essequibo River, Guyana (a river island at about 6°00' to 6°05'N, 58°36'W)
Distribution: Essequibo, Corantijn and Suriname River basins: Guyana and Suriname; in Suriname known from the Corantijn, Nickerie, Coppename and Suriname rivers
Size: 3.4 cm SL
Habitat: rivers and streams of the Interior, upstream of the first rapids
Position in the water column: mid-water, often collected near the shore in shallow water with emergent macrophytes
Diet:
Reproduction:
Remarks:

Hyphessobrycon roseus (Géry, 1960)
Local name(s): –

Diagnostic characteristics: absence of black markings on the dorsal and caudal fins and on the caudal peduncle; body brilliant rose and fins red; presence of a very large oval humeral spot (measuring half of the body depth)
Type locality: creeks near Gaa Kaba, Maroni, French Guiana
Distribution: Maroni (Marowijne), Suriname and Oyapock River basins
Size: 1.9 cm SL
Habitat: small forest creeks of the Upper Marowijne and Upper Suriname
Position in the water column: probably mid-water
Diet:
Reproduction:
Remarks: locally abundant; closely-related species are very popular in the aquarium hobby.

Hyphessobrycon simulatus (Géry, 1960)
Local name(s): reditere-sriba

Diagnostic characteristics: presence of a conspicuous vertically-elongated humeral spot and a black 'flag' on the dorsal fin, but no black markings on the anal fin; presence of a vertical, black band through the eye; caudal fin bright-red
Type locality: Kourou River (coast), French Guiana
Distribution: Maroni, Mana, Sinnamary, Kourou, Comté, Approuague and Oyapock River basins, French Guiana; in Suriname known from Saramacca, Suriname, Commewijne and Marowijne rivers

Hyphessobrycon rosaceus (live), Sipaliwini River (Corantijn River) (© P. Willink)

Hyphessobrycon roseus (live), Paloemeu River (© K. Wan Tong You)

Hyphessobrycon simulatus (live), 30.5 mm SL, Para River (Suriname River), ANSP189416 (© M. Sabaj Pérez)

Size: 3 cm SL
Habitat: coastal freshwater swamps and black-water streams of the Savanna Belt
Position in the water column: mid-water
Diet: omnivorous
Reproduction:
Remarks: *Pseudopristella simulata* is a synonym; the species looks very similar to *Hyphessobrycon copelandi*, but differs from the latter in the vertically elongate form of its humeral spot (against a circular spot in *H. copelandi*; *H. copelandi* is also restricted to the Interior while *H. simulatus* is more a coastal species.

HYPHESSOBRYCON sp. 'blackstripe'
Local name(s): -

Diagnostic characteristics: body with a broad black longitudinal band asymmetrical on the lower half of the body and extending from the eye to the posterior edge of the caudal fin; presence of a thin red line above the black band; eye red; no humeral spot; size < 5 cm SL (Paloemeu River)
Type locality: probably a still undescribed species
Distribution: in Suriname only known from the Upper Paloemeu River, but may be present in other tributaries of the Upper Marowijne River System (P.Y. Le Bail, personal communication)
Size: 3.5 cm SL
Habitat: the species was collected in a tributary of the Upper Paloemeu River (approximately 300 m above sea level), both up- and downstream of a 60-m high waterfall
Position in the water column: mid-water
Diet:
Reproduction:
Remarks: *Hyphessobrycon* sp.'blackstripe' may be close to the flag tetra *H. heterorhabdus* of the middle and lower Amazon.

HYPHESSOBRYCON sp. 'redline'
Local name(s): –

Diagnostic characteristics: body with a brilliant, red, longitudinal line, a red eye, and a small humeral spot; maximum size <5 cm SL (Para River)
Type locality: probably a still undescribed species
Distribution: in Suriname known only from Para River and other black-water tributaries of the Suriname River
Size: 3.5 cm SL
Habitat: black-water creeks of the Savanna Belt
Position in the water column: mid-water
Diet:
Reproduction:

Hyphessobrycon sp. 'blackstripe' (live), Paloemeu River (© P. Naskrecki)

Hyphessobrycon sp. 'redline' (live), Para River (Suriname River), ANSP189414 (© M. Sabaj Pérez)

Remarks: possibly an introduced species; a very similar species is known from the Essequibo River, Guyana. This species was identified erroneously as *H. heterorhabdus* in Ouboter & Mol (1993).

Jupiaba abramoides (Eigenmann, 1909)
Local name(s): sriba

Diagnostic characteristics: presence of pre-pelvic spines (diagnostic of the genus *Jupiaba*); absence of scales at the base of the caudal fin; complete lateral line; presence of a black, circular spot on the caudal peduncle and a faint vertically elongated humeral spot; caudal peduncle deep, vertically elongated; 45-50 scales in lateral line; absence of chevron (<) marks along lateral line in preserved specimens; color of live specimens almost green
Type locality: Tumatumari, Potaro River, Guyana
Distribution: Orinoco River basin and coastal drainages of the Guianas: French Guiana, Guyana, Suriname and Venezuela
Size: 12 cm SL
Habitat: small forest creeks of the Interior
Position in the water column: mid-water
Diet:
Reproduction:
Remarks: locally abundant.

Jupiaba keithi (Géry, Planquette & Le Bail, 1996)
Local name(s): –

Diagnostic characteristics: a *Jupiaba* species (with characteristic pre-pelvic spines) with a characteristically vertically elongated caudal spot and a faint humeral spot
Type locality: Crique Balaté, Maroni, French Guiana
Distribution: coastal rivers of French Guiana and Suriname
Size: 10 cm SL
Habitat: upper reaches of the rivers and streams of the Interior upstream of the first rapids where it seems to prefer calm water (counter currents)
Position in the water column: mid-water
Diet:
Reproduction:
Remarks: sometimes quite abundant.

Jupiaba maroniensis (Géry, Planquette & Le Bail, 1996)
Local name(s): –

Diagnostic characteristics: a small *Jupiaba* species (presence of pre-pelvic spines) with a circular black spot on the caudal peduncle; adpressed pelvic fins reach anal fin; apex eye brown-red; 5-6 scales below the lateral line; head short, 3.6-3.9 times in SL (Marowijne)

Jupiaba abramoides (live) (© P. Willink)

Jupiaba keithi (live) (© INRA-Le Bail)

Jupiaba maroniensis (alcohol), 42 mm SL, Lawa River (Marowijne River), ANSP189611 (© M. Sabaj Pérez)

Type locality: Antecume Pata village at confluence Litani and Marouini, Maroni, French Guiana
Distribution: Marowijne (Maroni) River, French Guiana and Suriname
Size: 6 cm SL (TL?)
Habitat: upper reaches of Marowijne River and tributaries with moderate to strong currents
Position in the water column: mid-water
Diet:
Reproduction:
Remarks: looks very similar to *J. meunieri*, which also occurs in the Marowijne River; *J. meunieri* is, however, more colorful (e.g. bright-red eye) than *J. maroniensis*.

Jupiaba meunieri (Géry, Planquette & Le Bail, 1996)
Local name(s): –

Diagnostic characteristics: a *Jupiaba* species (presence of pre-pelvic spines) with a black spot on the caudal peduncle and the apex of the eye bright-red; adpressed pelvic fins do not reach to origin of anal fin;; 6.5-8 scales below the lateral line; head long, 3.4-3.5 times in SL
Type locality: Approuague River, Arrataye at Saut Japigny, French Guiana
Distribution: coastal rivers of French Guiana and Suriname
Size: 9 cm SL
Habitat: upper reaches of rivers of the Interior in rapids and tributary creeks with fast flowing water
Position in the water column: mid-water
Diet: mainly herbivorous
Reproduction:
Remarks: it occurs in the same biotope as *Moenkhausia georgiae* (a species that looks very similar, but lacks the pre-pelvic spines and has a scaled caudal fin).

Jupiaba ocellata (Géry, Planquette & Le Bail 1996)
Local name(s): –

Diagnostic characteristics: a *Jupiaba* (presence of pre-pelvic spines); humeral spot with ocellus; 33-36 scales in the lateral line
Type locality: Oyapock River between Maripa Fall and Camopi, French Guiana
Distribution: coastal rivers of northeastern South America; in Suriname possibly present in Commewijne River
Size: 12 cm TL
Habitat:
Position in the water column: mid-water
Diet:
Reproduction:
Remarks: the occurrence in Suriname of *J. ocellata* is in need of confirmation.

Jupiaba maroniensis (live) (© INRA-Le Bail)

Jupiaba meunieri (live) (© INRA-Le Bail)

Jupiaba ocellata (live) (© INRA-Le Bail)

Jupiaba pinnata (Eigenmann, 1909)
Local name(s): –

Diagnostic characteristics: a *Jupiaba* species (presence of pre-pelvic spines); caudal peduncle horizontally elongated; 33-39 scales in lateral line; presence of conspicuous chevron (<) marks along lateral line in preserved specimens
Type locality: Amatuk, Lower Potaro River, Guyana
Distribution: coastal rivers of Guyana and Suriname
Size: 6 cm SL
Habitat:
Position in the water column: mid-water
Diet:
Reproduction:
Remarks:

Jupiaba polylepis (Günther, 1864)
Local name(s): –

Diagnostic characteristics: a *Jupiaba* (presence of pre-pelvic spines); humeral spot with ocellus; 42-46 scales in the lateral line
Type locality: Guyana
Distribution: Paru de Oeste, Xingu, Tocantins, and Araguaia River basins, Brazil; coastal rivers in Guyana and Suriname: Brazil, Guyana and Suriname; in Suriname present in all rivers except the Commewijne and Marowijne rivers
Size: 6.1 cm SL
Habitat:
Position in the water column: mid-water
Diet:
Reproduction:
Remarks: a common species.

Moenkhausia browni (Eigenmann, 1909)
Local name(s): –

Diagnostic characteristics: a *Moenkhausia* species (complete lateral line, caudal fin partly scaled) with a thin, black longitudinal line along the lateral line and a large, elongated, horizontal oval humeral spot; first rays of dorsal fin almost reach the adipose fin
Type locality: Atuataima Falls, Potaro River, Guyana
Distribution: Potaro River basin, Guyana; in Suriname, only known from Coppename River
Size: 8.2 cm TL
Habitat:
Position in the water column:
Diet:

Jupiaba pinnata (alcohol), Coppename River (© P. Willink)

Jupiaba polylepis (alcohol), Coppename River (© P. Willink)

Jupiaba polylepis (live), Nickerie River (© R. Smith)

Reproduction:
Remarks: its occurrence in Suriname needs confirmation.

MOENKHAUSIA CHRYSARGYREA (Günther, 1864)
Local name(s): –

Diagnostic characteristics: apex of the eye colored red; dorsal fin inserted anteriorly of mid-body position, its first rays often reach to or even over the adipose fin when adpressed; humeral spot not very clear (especially in large specimens), no spot on the caudal peduncle
Type locality: Essequibo River, Guyana
Distribution: Guianas coastal rivers and Amazon River basin: Brazil, French Guiana, Guyana and Venezuela; in Suriname, known from Corantijn, Saramacca, Suriname, Commewijne and Marowijne rivers
Size: 10 cm SL
Habitat: rather abundant in small streams, especially in water with slow current (pools)
Position in the water column:
Diet:
Reproduction:
Remarks: *M. grandisquamis* has a lower number of scales below the lateral line (5 versus 6-7), a longitudinal broad silvery band above the lateral line, and a short dorsal fin (not reaching the adipose fin when adpressed). Juvenile *M. chrysargyrea* are more elongated and difficult to distinguish from *M. collettii*.

MOENKHAUSIA COLLETTII (Steindachner, 1882)
Local name(s): –

Diagnostic characteristics: an elongated, small *Moenkhausia* species lacking a caudal spot; humeral spot on 3rd, 4th and 5th scales of the (complete) lateral line; presence of a characteristic horizontal black bar through the eye; caudal, anal, dorsal and especially adipose fins colored orange; presence of a fine black line along the base of the anal fin
Type locality: Hyavary; Obidos, Amazon River
Distribution: Amazon River basin from Peru to the Guianas: Brazil, Colombia, French Guiana, Guyana, Peru, Suriname and Venezuela; in Suriname probably present in all rivers
Size: 5 cm SL
Habitat: rivers and small tributary streams of the Interior, upstream of the first rapids, often on muddy substrates, plant detritus or leaf litter
Position in the water column: mid-water
Diet:
Reproduction: males have tiny sexual hooklets on the first rays of the anal fin
Remarks: this species has a wide geographic distribution.

Moenkhausia chrysargyrea (live), Sipaliwini River (Corantijn River) (© P. Willink)

Moenkhausia collettii (live) (© P. Willink)

Moenkhausia georgiae Géry, 1965
Local name(s): –

Diagnostic characteristics: a widely distributed, rather small *Moenkhausia* species with the apex of the eye red and a black spot on the caudal peduncle; humeral spot(s) poorly visible in live specimens; presence of a mid-lateral broad silvery band; the head is short and the eye large (2.2-2.5 times in the head length); the origin of anal fin positioned posterior to dorsal fin; dorsal fin when adpressed falling distinctly short of adipose fin; the scales are large (32-34 in the lateral line row)
Type locality: between Saut-Chien and Saut-Topi-Topi, middle Mana River, French Guiana
Distribution: several river basins: French Guiana and Suriname
Size: 6.9 cm SL
Habitat: in rapids and streams with moderate to strong currents
Position in the water column: mid-water
Diet: allochthonous invertebrates
Reproduction:
Remarks: *M. georgiae* may be distinguished from the similar *Tetragonopterus chalceus* by the form of the lateral line (descends steeply, step-wise after the gill cover in *T. chalceus*) and the color of the eye (transparent in *T. chalceus*); the main differences with *Jupiaba meunieri* is the presence of the pre-pelvic spines in *J. meunieri* and the larger number of lateral line scales in *J. meunieri* (36-38).

Moenkhausia grandisquamis (Müller & Troschel, 1845)
Local name(s): –

Diagnostic characteristics: eye with orange apex; absence of a caudal spot, but presence of a well-defined rounded humeral spot; presence of a longitudinal, broad, silvery band; when adpressed first rays of dorsal fin falling distinctly short of adipose fin; large scales (32-34 in lateral line row)
Type locality: Suriname
Distribution: Amazon, Orinoco and coastal River drainages of the Guianas: Brazil, French Guiana, Guyana, Suriname and Venezuela; in Suriname, known from the Corantijn, Saramacca, Suriname, Commewijne and Marowijne rivers, but possibly present in all rivers; also present in Brokopondo Reservoir
Size: 6.4 cm SL
Habitat: in relatively fast running water of rivers and small streams of the Interior, but also in pools with standing water
Position in the water column: mid-water and near the surface
Diet: probably small allochthonous invertebrates
Reproduction:
Remarks: may escape from aquatic predators by spectacular jumps out of the water. A common species.

Moenkhausia georgiae (live) (© INRA-Le Bail)

Moenkhausia grandisquamis (alcohol), scale bar = 1 cm, Lawa River (Marowijne River), ANSP189606 (© M. Sabaj Pérez)

Moenkhausia grandisquamis (live) (© INRA-Le Bail)

Moenkhausia hemigrammoides Géry, 1965
Local name(s): –

Diagnostic characteristics: a relatively small *Moenkhausia* species (lateral line complete) with a 'flag' on the dorsal fin (the black fin spot not reaching the posterior margin of the dorsal fin) and black coloration on the first rays of the anal fin (much like *Hemigrammus unilineatus*); there is no caudal spot and a small humeral spot (covering two scales of the lateral line)
Type locality: Weyne, Matoekasie creek, on the road Albina-Moengo, Cottica River basin, Suriname
Distribution: coastal rivers of French Guiana and Suriname
Size: 4 cm SL
Habitat:
Position in the water column:
Diet:
Reproduction:
Remarks: this species looks much like *Hemigrammus unilineatus* ('*hemigrammoides*') but can be distinguished from *H. unilineatus* by its complete lateral line (incomplete in all *Hemigrammus* species). The two species can occur syntopically.

Moenkhausia inrai Géry, 1992
Local name(s): –

Diagnostic characteristics: no spot on the caudal peduncle and the base of the caudal fin and distal part of the upper caudal lobe transparent, but the median part of the upper caudal lobe black; two humeral spots that are often difficult to see; apex of the eye bright-red (transparent in *M. intermedia* and *M. lepidura* two other *Moenkhausia* species with black markings on the caudal fin)
Type locality: "Crique Roche au-dessus du Saut Grand Canori, Approuague", French Guiana
Distribution: Maroni (Marowijne) and Approuague River basins, French Guiana
Size: 10 cm TL
Habitat: mainly in the main channel of the Marowijne River
Position in the water column: mid-water
Diet:
Reproduction:
Remarks:

Moenkhausia hemigrammoides (live) (© P. Willink)

Moenkhausia inrai (live) (© INRA-Le Bail)

Moenkhausia inrai (live) (© INRA-Le Bail)

MOENKHAUSIA INTERMEDIA Eigenmann, 1908
Local name(s): –

Diagnostic characteristics: an elongated *Moenkhausia* species lacking a black spot on the caudal peduncle, but the distal part of both caudal fin lobes black (the proximal part of the upper lobe is colored orange-yellow); presence of a silvery longitudinal band, especially well developed in the median part of the fish (i.e. below the dorsal fin); the eye is colored light-yellow and the adipose fin is orange
Type locality: Tabatinga, Amazonas, Brazil
Distribution: Amazon, Orinoco, La Plata, Approuague, Maroni and Mana River basin
Size: 8 cm SL
Habitat: rapids and small streams with moderate to string currents over sandy or rocky substrates
Position in the water column: mid-water
Diet: mainly small invertebrates, but essentially omnivorous
Reproduction: fecundity approximately 8000 eggs
Remarks: is capable of escaping from aquatic predators by jumping out of the water.

MOENKHAUSIA LEPIDURA (Kner, 1858)
Local name(s): –

Diagnostic characteristics: an elongated *Moenkhausia* species close to *M. intermedia*, but with the upper caudal lobe usually darker than the lower lobe, the proximal part of the upper lobe red, and lacking a longitudinal silvery band; eye color light yellow
Type locality: Rio Juapore, Rondonia, Brazil
Distribution: Amazon, Orinoco and coastal rivers in Guyana and Suriname: Brazil, Guyana, Peru, Suriname, and Venezuela; in Suriname known from the Corantijn, Nickerie and Coppename rivers
Size: 8.4 cm SL
Habitat: rivers and tributary streams of the Interior
Position in the water column: mid-water
Diet:
Reproduction:
Remarks:

Moenkhausia intermedia (live), 39.5 mm SL, Lawa River (Marowijne River), ANSP189622 (© M. Sabaj Pérez)

Moenkhausia lepidura (live), Sipaliwini River (Corantijn River) (© P. Willink)

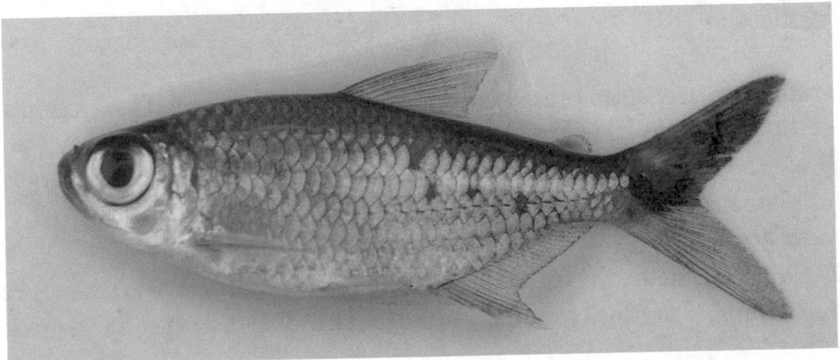

Moenkhausia lepidura (live), Nickerie River (© R. Smith)

Moenkhausia moisae Géry, Planquette & Le Bail, 1995
Local name(s): –

Diagnostic characteristics: a *Moenkhausia* species, close to *M. surinamensis*, with small scales (> 40 in the lateral line, generally 41-47) and a round black spot on the caudal peduncle; the humeral spot is vertically elongated (most pronounced in its center) and surrounded by a lighter area ('ocellus'), much like in *Jupiaba polylepis*; sometimes a dark spot is present on the gill cover
Type locality: Balaté Creek, Lower Marowijne (Maroni) River, French Guiana
Distribution: Marowijne (Maroni) and Mana River basins, French Guiana and Suriname
Size: 9.5 cm SL
Habitat: in streams with alternatively moderate currents over sandy bottom and slow currents over muddy substrate
Position in the water column:
Diet:
Reproduction: in French Guiana, sexual hooklets have been observed on the anal fin in January, pointing to reproduction in March
Remarks: *M. moisae* is very close to *M. surinamensis* and only the number of scales of the lateral line row (36-40 in *M. surinamensis*) and, to a lesser extent, the form of the humeral spot (horizontally elongated in *M. surinamensis*) can be used to distinguish between the two species.

Moenkhausia oligolepis (Günther, 1864)
Local name(s): –

Diagnostic characteristics: presence of a characteristic broad, black band at the base of the caudal fin (this broad black band is bordered anteriorly and posteriorly by white-yellow transversal bands; the apex of the eye is colored bright red; there are few scales in the lateral line row (29-32)
Type locality: Guyana
Distribution: Venezuela, the Guianas, and the Amazon River basin: Brazil, French Guiana, Guyana, Peru, Suriname, and Venezuela; in Suriname probably present in all rivers
Size: 10 cm SL
Habitat: in rivers and small tributaries of the Interior, mainly in shallow water
Position in the water column: near the surface
Diet:
Reproduction:
Remarks: well known in the aquarium hobby for a long time. A common species.

Moenkhausia moisae (live) (© INRA-Le Bail)

Moenkhausia oligolepis (live), scale bar = 1 cm, Lawa River (Marowijne River), ANSP189616 (© M. Sabaj Pérez)

Moenkhausia surinamensis Géry, 1965
Local name(s): –

Diagnostic characteristics: very similar to *M. moisae*, but with less scales (<41) in the lateral line row and an oval caudal spot
Type locality: Browns Creek, km 114 from Paramaribo on the Paramaribo-Dam railroad, Saramacca River, Suriname
Distribution: Suriname, Saramacca, Oyapock, Approuague, Compté, and Sinnamary River basins: Brazil, French Guiana and Suriname
Size: 10 cm SL
Habitat: in rapids and small streams with moderate to strong currents
Position in the water column: mid-water
Diet:
Reproduction:
Remarks: similar in coloration to *Jupiaba abramoides*, but *M. surinamensis* lacks the pre-pelvic spines of *Jupiaba* and has the caudal fin partly scaled.

Paracheirodon axelrodi (Schultz, 1956)
Local name(s): (cardinal tetra)

Diagnostic characteristics: the famous cardinal tetra; body with brilliant blue-green dorsal iridescence and bright-red ventral parts, no humeral spot
Type locality: stream near Tomar, Rio Negro, near Porto Velho, Brazil
Distribution: Upper Orinoco and Negro River basins: Brazil, Colombia and Venezuela (introduced in Suriname, Coropina Creek, Para River, Suriname River system)
Size: 2.5 cm SL
Habitat: black-water streams
Position in the water column: mid-water
Diet: in its natural habitat, the middle Rio Negro, it feeds on microcrustacea (cladocerans and copepods) and chironomid larvae (Walker, 2004)
Reproduction:
Remarks: this is the gorgeous cardinal tetra of the Rio Negro; it was introduced by aquarium hobbyists into Para River in the 1970s.

Pristella maxillaris (Ulrey, 1894)
Local name(s): reditere sriba

Diagnostic characteristics: presence of an oval spot on the anterior part of both the anal and pelvic fins; black spot ('flag') on the dorsal fin; a small rounded humeral spot with the size of the pupil; the caudal fin is colored bright-red
Type locality: Brazil
Distribution: Amazon, Orinoco and coastal river drainages of the Guianas: Brazil, French Guiana, Guyana and Venezueła

Moenkhausia surinamensis (live) (© INRA-Le Bail)

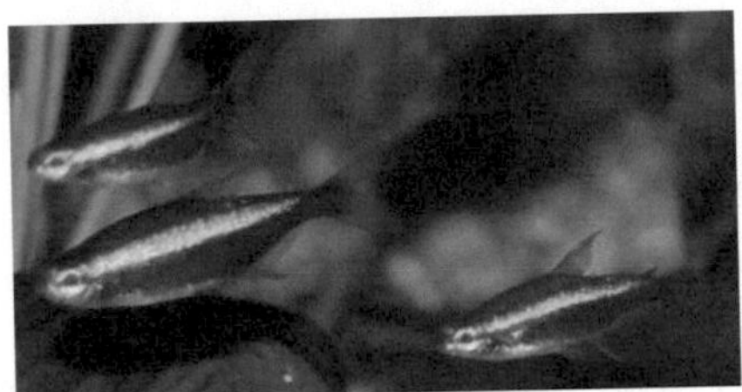

Paracheirodon axelrodi (live), Para River (Suriname River System)

Pristella maxillaris (live), Para River (Suriname River), ANSP189415 (© M. Sabaj Pérez)

Size: 4.5 cm TL
Habitat: coastal swamps and black-water streams of the Savanna Belt among aquatic macrophyte vegetation
Position in the water column: mid-water
Diet: omnivorous
Reproduction: reproduction easily accomplished in the aquarium; about 300-400 eggs are spawned; hatching in about 3 days; the newborn are relatively easy to keep alive
Remarks: very popular and known since 1924 in the aquarium hobby; in Suriname, it lives in coastal swamps and black-water streams of the Savanna Belt with other small species like *Hyphessobrycon simulatus* and *Hemigrammus unilineatus*.

Thayeria ifati Géry, 1959
Local name(s): (penguin or hockey-stick tetra)

Diagnostic characteristics: lower caudal lobe longer than the upper lobe; presence of a black band that begins just in front of the dorsal fin and extends on the lower lobe of the caudal fin; first rays of the anal fin and the extremity of the dorsal fin black
Type locality: “Gaa Kaba”, Marowijne (Maroni) River, north of the 4th parallel, French Guiana
Distribution: Marowijne (Maroni) and Approuague River basins, French Guiana and Suriname
Size: 3.5 cm SL
Habitat: lives in schools in habitats with slow water currents of the upper reaches of the Marowijne River and its tributaries (e.g. Oelemarie River)
Position in the water column: mid-water
Diet:
Reproduction: the related *T. boehlkei* reproduces in the aquarium, spawning about 1000 eggs among aquatic vegetation
Remarks: the well-known penguin or hockey stick tetra of the aquarium hobby; this species swims with its head up at an angle of about 30°); the black band prolonged on the lower caudal lobe associated with its oblique position in the water column breaks up the contours of the fish, which is thus no longer recognizable as a fish (a so-called disruptive color pattern).

Thayeria ifati (alcohol), Oelemarie River (Marowijne River), RMNH37482 (© Naturalis)

Thayeria ifati (live) (© INRA-Le Bail)

FAMILY CHARACIDAE – SUBFAMILY IGUANODECTINAE

The Iguanodectinae is a small characid subfamily composed of 11 valid species in two genera: *Iguanodectes* and *Piabucus*. Fishes of this subfamily are characterized by elongated bodies, with basally-contracted, multicuspid teeth, gill-membranes united and free from the isthmus, the posterior end of the maxilla not extending to the eye, the dorsal-fin origin generally posterior to the middle of the body, and anal fin long. The genus *Piabucus* is distinguished from *Iguanodectes* by the presence of a long pectoral fin, and a well-developed pectoral keel. Little information is known on their ecology. Some species are used as ornamental fishes.

1-a Medium-sized species (up to 15 cm) with a long pectoral fin and a well-developed keel

Piabucus dentatus

1-b Small species (7 cm) with short pectoral fins and lacking a well-developed keel

Iguanodectes aff. *purusii*

IGUANODECTES aff. *PURUSII* (Steindachner, 1908)

Local name(s): –

Diagnostic characteristics: elongate body, lacking large pectoral fins and a ventral keel; anal fin with 28-34 branched rays, 66-71 lateral line scales; presence of dark lateral band overlaid by a brilliant band
Type locality: Rio Purus, Brazil
Distribution: Amazon, Guaporé, Madira, Pastaza and Purus River basins; in Suriname, known only from Marowijne River
Size: 6 cm SL
Habitat:
Position in the water column: surface
Diet: probably primarily herbivorous, feeding occasionally on allochthonous insects (Knöppel, 1970)
Reproduction:
Remarks: very little is known about the ecology of *Iguanodectes* species.

Iguanodectes aff. *purusii* (alcohol) (© INRA-Le Bail)

Piabucus dentatus (Koelreuter, 1763)
Local name(s): –

Diagnostic characteristics: very elongate body with ventral keel, large pectoral fins and a long anal fin; mouth superior with pluricuspid, flat incisiform teeth; posterior end of maxilla not extending to the eye; dorsal fin origin behind midbody and slightly behind the origin of the anal fin; silvery color, with a brilliant longitudinal band overlaid on the lateral line, sometimes prolonged posteriorly with an elongated black spot on the caudal peduncle
Type locality: Brazil
Distribution: northern coastal drainages of South America: Brazil, French Guiana, Guyana, Suriname and Venezuela
Size: 12.9 cm SL
Habitat: mainly the lower freshwater reaches of rivers (i.e. downstream of the first rapids) and medium-sized tributaries with muddy bottom substrate, but not very common
Position in the water column: surface
Diet:
Reproduction:
Remarks: one of the oldest known South American fishes (it is the *Piabuca* of Marcgravius, 17th century); may escape from predatory fishes by jumping out of the water much like hatchet fishes.

Piabucus dentatus (live) (© INRA-Le Bail)

FAMILY CHARACIDAE – SUBFAMILY BRYCONINAE

The ±40 species of the subfamily Bryconinae are medium to large-sized characid fishes, reaching a maximum standard length from 12 cm (*Brycon pesu*) to about 70 cm (*Brycon orbygnianus* and *B. amazonicus*). The Bryconinae could not be diagnosed by a unique set of characters, but the combination of three (rarely four) rows of teeth on the premaxilla, the presence of larger teeth in the inner row of the premaxilla, and the presence of a symphysal tooth behind the main dentary tooth row are uncommon among the remaining Characidae (Fig. 12.3). Species in the genus *Triportheus* have adopted a life near the surface and, thanks to a ventral keel and large pectoral fins positioned high on the body, like that of hatchetfishes (Gasteropelecidae), they are able to jump several meters above the water surface, apparently to escape from predators (without real flight). The Bryconinae occurs from southern Mexico to Panama, across the trans-Andean South American river basins from northern Peru to the Maracaibo system in Venezuela, and in cis-Andean South America in all major river drainages and most Atlantic and Caribbean coastal river basins. Members of the genus *Brycon* are important food fishes throughout Central and South America. Species of the genus are important in commercial fisheries in many river systems and are also being cultivated in several South American countries. *Brycon* species are omnivorous, relying more heavily on allochthonous items, such as fallen fruits, seeds, and insects (e.g. Goulding, 1980). Some species are known to perform long-range reproductive migrations (Carolsfeld *et al.*, 2003; Goulding, 1980).

1-a Presence of a ventral keel; pectoral fins positioned high on the body and very large

Triportheus brachipomus

1-b Absence of a ventral keel; pectoral fins not very developed

2

2-a Presence of a broad, V-shaped or crescent shaped black band at the base of the caudal fin; black band at the base of the anal fin; presence of a humeral black spot; adipose fin hyaline

Brycon falcatus

2-b Absence of a broad, V-shaped band at the base of the caudal fin; caudal fin with some small blackish dots, more or less arranged in two vertical rows; absence of black pigmentation at the base of the anal fin and no humeral spot; adipose fin very black; Marowijne River

Brycon pesu

Fig. 12.3. Teeth of moroko *Brycon falcatus* (© B. Chernoff).

Brycon falcatus Müller & Troschel, 1844
Local name(s): moroko
Diagnostic characteristics: presence of a broad, V-shaped or crescent shaped black band at the base of the caudal fin; black band at the base of the anal fin; presence of a humeral black spot (size about equal to the pupil of the eye); adipose fin hyaline; a large mouth with three rows of teeth on the upper jaw; lateral line situated low on the body
Type locality: Guyana, Suriname
Distribution: rivers northern South America and Amazon and Orinoco River basins: Brazil, Colombia, French Guiana, Guyana, Peru, Suriname and Venezuela; in Suriname, known from most if not all rivers
Size: 37 cm TL (1 kg)
Habitat: rivers and large streams of the Interior
Position in the water column: mid-water
Diet: omnivorous
Reproduction:
Remarks: a well known food fish in the Interior of Suriname.

Brycon pesu Müller & Troschel, 1845
Local name(s): abongoni

Diagnostic characteristics: absence of a broad, V-shaped band at the base of the caudal fin; caudal fin with some small blackish dots, more or less arranged in two vertical rows; absence of black pigmentation at the base of the anal fin and no humeral spot; adipose fin characteristically black; this small species looks much like a *Bryconops* species, silvery with a straight back, but with a large mouth
Type locality: Guyana
Distribution: Amazon River basin and Orinoco River basin: Brazil, Ecuador, French Guiana, Guyana, Peru, Suriname and Venezuela; in Suriname only known from the Marowijne River
Size: 12 cm SL
Habitat: Marowijne River upstream of the first rapids and some large tributaries; mainly in moderate to strong currents over sandy or rocky bottom substrates
Position in the water column: mid-water
Diet: omnivorous
Reproduction:
Remarks:

Triportheus brachipomus (Valenciennes, 1850)
Local name(s): sardin

Diagnostic characteristics: a medium-sized, silvery, laterally compressed species with a conspicuous ventral keel; pectoral fins positioned high and very large; upturned mouth and large scales (33-38 in the lateral line row)
Type locality: Mana, French Guyana, 5°44'N, 53°54'W

Brycon falcatus (alcohol), ANSP161212 (© M. Sabaj Pérez)

Brycon falcatus (live), Coppename River (© B. Chernoff)

Brycon pesu (live), Lawa River (Marowijne River) (© P. Willink)

Triportheus brachipomus, adult (live) (© INRA-Le Bail)

Distribution: Brazil, Suriname, Venezuela and the Guianas
Size: 30 cm TL
Habitat: in slow water currents in both rivers of the Interior and small tributary creeks
Position in the water column: surface
Diet: omnivorous, feeding on both fruits, seeds and invertebrates that float on the water surface (Almeida, 1984)
Reproduction: in the rainy season; post-larval *Triportheus* have their pectorals colored black and a black band at the base of the anal fin (see photograph in Géry, 1977, p. 337,) giving them a very different appearance as compared to the plain-silvery adults
Remarks: adapted to living near the water surface; with their large wing-like pectoral fins they are able to jump several meters a few decimeter above the water surface, apparently to escape from aquatic predators.

Triportheus brachipomus, juvenile (alcohol), scale bar = 5 mm, Lawa River (Marowijne River), ANSP189619 (© M. Sabaj Pérez)

FAMILY CHARACIDAE – SUBFAMILY SERRASALMINAE (PIRANHAS AND PACUS)

Serrasalminae have a deep, laterally compressed body with a series of mid-ventral abdominal serrae ('spines') and, except in *Colossoma*, *Piaractus*, and *Mylossoma*, an anteriorly-directed spine just before the dorsal-fin insertion. Names of genera follow Orti *et al.* (2008). Some species (piranhas) possess only one row of teeth on each jaw. These teeth are interlocking, sharp and pointed; tricuspid in *Pygocentrus*, *Pristobrycon* and *Serrasalmus*, pentacuspid in *Pygopristis*. Teeth are mammiliform and pointed out of the mouth in *Catoprion*. All other species have two rows of teeth on the upper jaw and often a pair of conical teeth just behind the main row of the lower jaw. In *Acnodon*, *Colossoma*, *Piaractus*, *Metynnis*, *Myleus*, *Myloplus*, *Mylossoma*, and *Utiaritichthys*, teeth are molariform, heavily attached to the jaw, and mainly used to grind fruits and seeds. In *Mylesinus*, *Ossubtus* and *Tometes*, teeth are incisiform and tricuspid, weakly attached to the jaw and mainly used to cut leaves. During the breeding period, *Acnodon*, *Metynnis*, *Myleus*, *Mylesinus*, *Myloplus*, *Tometes*, and *Utiaritichthys* exhibit sexual dimorphism in the form of a supplementary lobe of the anal fin, dorsal fin rays elongated into long filaments or a red pattern on the body. The distribution of Serrasalminae is strictly Neotropical and their presence everywhere else in the world is the result of introduction. Serrasalmine species occur in all freshwater biotopes, except in very small forest creeks and benthic areas of deep rivers. *Tometes*, *Mylesinus* and some *Myleus* species are among the most important food fishes for Amerindian people of the Guiana Shield. *Metynnis* species are recognized as aquarium fishes. Victims of their bad reputation, piranhas, mainly *Serrasalmus* and *Pygocentrus* species, are caught and dried to be sold as souvenirs. Piranhas are usually more dangerous out of the water than in it and most bites occur on shore or in boats when removing a piranha from a gillnet or hook or when a "loose" piranha is flopping about and snapping its jaws. However, under some conditions piranhas can be dangerous to human bathers (Mol, 2009; chapter 8).

1-a Upper jaw straight (except in *Pygopristis denticulata*); a single row of close-set, interlocking sharp teeth on the upper jaw forming a continuous saw (piranhas)

2

1-b Upper jaw showing a sharp angle between the premaxillary and maxillary; two rows of teeth on the anterior part of the upper jaw, the inner row molariform (pacus)

5

2-a Absence of black bands on the caudal peduncle or base of the caudal fin; adipose fin long, interdorsal distance (distance between the adipose and dorsal fins) >2 in the base of the dorsal fin; teeth usually with 5 cusps (pentacuspid); a small species (20 cm TL)

Pygopristis denticulata

2-b Presence of a black spot or a black band on the caudal peduncle or base of the caudal fin (at least in juveniles <20 cm TL); adipose fin short, interdorsal distance <2 times in the base of the dorsal fin; teeth with no more than 3 cusps

3

3-a Presence of a black spot on the posterior part of the caudal peduncle and a 'fluorescent' spot on the apex of the caudal peduncle; absence of a spine posterior of the anus; Marowijne River

Pristobrycon striolatus

3-b Presence of a black band at the base of the caudal fin (not on the caudal peduncle), absence of a 'fluorescent' spot on the apex of the caudal peduncle; presence of a spine posterior of the anus (at the insertion of the anal fin)

4

4-a Eye transparent, large humeral spot present; juveniles (spotted) lacking a black band on the posterior margin of the caudal fin

Pristobrycon eigenmanni

4-b Eye bright-red; humeral spot absent; juveniles (spotted), with transparent eyes, a black band along the margin of the anal fin and two black bands on the caudal fin, one at the base and the second at the posterior margin of the fin

Serrasalmus rhombeus

5-a Length of the adipose fin > interdorsal distance; maximum SL < 15 cm

Metynnis altidorsalis

5-b Length of adipose fin < interdorsal distance; maximum SL > 15 cm

6

6-a Absence of serrae in the preventral region

7

6-b Presence of serrae in the preventral region

8

7-a Mouth oblique and upturned (superior); absence of a humeral spot; more than 6 teeth on each side of the lower jaw

Tometes lebaili

7-b Mouth inferior; presence of a small humeral spot; less than 6 teeth on each side of the lower jaw

Acnodon oligacanthus

8-a The two median teeth of anterior premaxillary row in an advanced position resulting in a V-shaped row of outer teeth that does not follow the curvature of the upper jaw; the inner row is more or less on a straight line, and the two rows of premaxillary teeth thus form the letter 'A'; the adipose fin is relatively long (its length > height) and the dorsal fin short (21-24 branched rays); 29-31 branched rays in the anal fin

Myloplus rhomboidalis

8-b Teeth of the anterior (outer) row on the premaxilla (upper jaw) follow the curvature of the premaxillary bone; premaxillary teeth of the inner row also in the form of an arch

9

9-a Absence of two small, conical teeth behind the main row on the dentary

Myloplus ternetzi

9-b Small, conical teeth behind the main row on the lower jaw always present

10

10-a <33 ventral serrae (spines); 17-20 branched rays in the dorsal fin (Suriname, Saramacca, Coppename, Nickerie and Corantijn rivers)

Myleus setiger

10-b 33-46 ventral serrae

11

11-a Serrae 40-46 in adults; 34-37 branched rays in the anal fin; 22-24 branched rays in the dorsal fin; eye with a black, vertical bar; body depth 1.4-1.55 in SL; head length 3.35-3.65 in SL

Myloplus rubripinnis

11-b Serrae 33-39; 32-38 branched rays in the anal fin; 32-34 branched rays in the anal fin, 20-23 branched rays in the dorsal fin; well-defined black band along the distal margin of the anal fin (Marowijne)

Myloplus planquettei

An undescribed *Myloplus* species, close to *M. ternetzi*, is present in the Coppename River.

ACNODON OLIGACANTHUS (Müller & Troschel, 1844)
Local name(s): – (sheep pacu)

Diagnostic characteristics: length of adipose fin < interdorsal distance; maximum SL > 15 cm; absence of spines in the preventral region; presence of a very long predorsal spine; mouth inferior with the profile of the snout strongly hooked; presence of a small humeral spot; less than 6 teeth on each side of the lower jaw; snout strongly curved (aquiline 'nose')
Type locality: Suriname
Distribution: northern Guiana Shield rivers: French Guiana, Guyana and Suriname; in Suriname, only known from the Suriname and Marowijne rivers
Size: 20 cm TL

Acnodon oligacanthus (live), scale bar = 1 cm, Lawa River (Marowijne River), ANSP187112 (© M. Sabaj Pérez)

Habitat: upper reaches of rivers (upstream of the first rapids) in moderate to strong currents (e.g. in rapids); the juveniles live in small tributary creeks with sandy bottom substrate
Position in the water column: mid-water
Diet: aquatic plants of the family Podostemaceae ('kumalu-nyang') and fruits, leaves and flowers
Reproduction: in the rainy season they migrate into small tributaries to reproduce
Remarks: Géry (1977) suggests that *A. oligacanthus* may lead a crepuscular life (as suggested by its large eyes).

Metynnis altidorsalis Ahl, 1923
Local name(s): – (silver dollar fish)

Diagnostic characteristics: a very long adipose fin (its base larger than the inter-dorsal distance); maximum SL < 15 cm
Type locality: Paramaribo, Suriname
Distribution: north and eastern Guiana Shield rivers: French Guiana, Guyana, Suriname and Venezuela; in Suriname, known from Suriname (Para River) and Commewijne (Rikanau Creek) rivers
Size: 12 cm SL
Habitat: black-water streams and other coastal streams
Position in the water column: mid-water
Diet: aquatic macrophytes such as *Cabomba*
Reproduction:
Remarks: the only *Metynnis* species in Suriname.

Myleus setiger Müller & Troschel, 1844
Local name(s): –

Diagnostic characteristics: low number (<33) of ventral serrae; 17-20 branched rays in the dorsal fin
Type locality: Guyana
Distribution: tributaries of lower and middle Amazon River basin, Orinoco River basin, Upper Orinoco River and tributaries, and northern and eastern Guiana Shield rivers: Brazil, Guyana, Suriname and Venezuela; in Suriname known from Suriname, Saramacca, Coppename, Nickerie and Corantijn rivers
Size: 27 cm SL
Habitat:
Position in the water column:
Diet:
Reproduction:
Remarks:

Metynnis altidorsalis (live), Para River (Suriname River System)

Myloplus planquettei Jégu, Keith & Le Bail, 2003
Local name(s): –

Diagnostic characteristics: 33-39 ventral serrae; 32-38 branched rays in the anal fin; 32-34 branched rays in the anal fin, 20-23 branched rays in the dorsal fin; well-defined black band along the distal margin of the anal fin
Type locality: Twenke, Marowijne (Maroni) River, French Guiana
Distribution: Mana and Maroni basins, French Guiana and Suriname, and Essequibo basin, Guyana; in Suriname, only known from the Marowijne River
Size: 56 cm SL (6 kg)
Habitat:
Position in the water column:
Diet:
Reproduction:
Remarks:

Myloplus rhomboidalis (Cuvier, 1818)
Local name(s): kumalu, kumaru

Diagnostic characteristics: the two series of premaxillary teeth form a triangle (the form of an 'A'); crown of teeth on the lower jaw horizontal-straight; anterior-distal margin of dorsal fin dark; 29-31 branched rays in the short anal fin, 21-23 branched rays in the dorsal fin; length adipose fin > height anal fin; silvery, but may turn black with rosy patches in the reproductive season
Type locality: Brazil
Distribution: Amazon River basin and northern and eastern Guiana Shield rivers: Brazil, French Guiana, Guyana, Suriname; in Suriname probably present in all rivers
Size: 37 cm SL (5 kg)
Habitat: reaches with weak currents in the rivers of the Interior
Position in the water column: mid-water
Diet: in the high-water season they feed on fruits of palm trees (e.g. *Euterpe oleracea, Mauritia flexuosa*) and *Inga* trees and, in the dry season they eat Podostemaceae ('kumalu-nyang') (Boujard *et al.*, 1990); in addition they may also sometimes eat snails and fishes
Reproduction: males >15 cm develop a falciform anal fin and their interdorsal distance becomes shorter than the interdorsal distance in females; in the reproductive season, the silvery colors may change into black and large patches of rosy color, and scales may be lost; juveniles prefer the moderate to strong currents near rapids; juveniles <2.5 cm are marbled brown in color, larger specimen turn silvery
Remarks: a popular food fish in the Interior; the growth of *M. rhomboidalis* in two rivers in French Guiana was studied by Lecomte *et al.* (1993).

Myloplus planquettei (live) (© INRA-Le Bail)

Myloplus rhomboidalis (live) (© INRA-Le Bail)

Myloplus rubripinnis (Müller & Troschel, 1844)
Local name(s): pakusi

Diagnostic characteristics: 40-46 serrae in adults; 34-37 branched rays in the anal fin; 22-24 branched rays in the dorsal fin; eye with a black, vertical bar; body depth 1.4-1.55 in SL; head length 3.35-3.65 in SL
Type locality: Essequibo River, Guyana
Distribution: South America: Amazon and Orinoco River basins; north and eastern Guiana Shield rivers. Countries: Brazil, Colombia, Ecuador, French Guiana, Guyana, Peru, Suriname, Venezuela; in Suriname known from all rivers
Size: 30 cm SL
Habitat:
Position in the water column:
Diet:
Reproduction:
Remarks:

Myloplus ternetzi (Norman, 1929)
Local name(s): pakusi

Diagnostic characteristics: a small *Myloplus* species characterized by the absence of two small, conical teeth behind the main row on the dentary; low number of branched rays in the anal fin (29-31), large number of lateral line scales (75-82)
Type locality: Maparu Rapids, Approuague, French Guiana
Distribution: French Guiana and Suriname; in Suriname, known from Marowijne and Suriname rivers
Size: 23 cm SL (<1 kg)
Habitat:
Position in the water column:
Diet: vegetarian (Boujard *et al.*, 1990)
Reproduction:
Remarks: an undescribed *Myloplus* species, close to *M. ternetzi*, is present in the Coppename River.

Pristobrycon eigenmanni (Norman, 1929)
Local name(s): piren

Diagnostic characteristics: eye transparent, large humeral spot present; juveniles (spotted) with a black band at the base of the caudal fin, but lacking a black band on the posterior margin of the caudal fin and a black margin of the anal fin (both present in juvenile *S. rhombeus*)
Type locality: Rockstone, Guyana

Myloplus rubripinnis (live), scale bar = 5 mm, Lawa River (Marowijne River), ANSP189267 (© M. Sabaj Pérez)

Myloplus ternetzi (live) (© INRA-Le Bail)

Pristobrycon eigenmanni (live), Lawa River (Marowijne River), ANSP188683 (© M. Sabaj Pérez)

Distribution: Amazon River basin and northern and eastern Guiana Shield rivers: Brazil, French Guiana, Guyana, Suriname; in Suriname, known to occur in all rivers except Commewijne River
Size: 20 cm SL
Habitat: upper and lower reaches of rivers and larger tributaries, but always in freshwater
Position in the water column: mid-water
Diet: seems to feed mainly on fins of other fishes
Reproduction: in the reproductive season the anal fin of the male is falciform
Remarks: this species was identified as *S. humeralis* in Planquette *et al.* (1996); *P. eigenmanni* from the Nickerie and Corantijn rivers have the dorsal part of their flanks spotted.

PRISTOBRYCON STRIOLATUS (Steindachner, 1908)
Local name(s): piren

Diagnostic characteristics: presence of a black spot on the posterior part of the caudal peduncle and a 'fluorescent' spot on the apex of the caudal peduncle; relatively long adipose fin; absence of serrae (spines) posterior of the anus
Type locality: Rio Pará, Brazil
Distribution: Amazon and Orinoco River basins and northern and eastern Guiana Shield rivers: Brazil, French Guiana, Guyana, Suriname and Venezuela; in Suriname, only known from the Upper Marowijne River
Size: up to 30 cm TL, but often much smaller
Habitat: upstream reaches of Marowijne (Maroni) River and large tributaries
Position in the water column: mid-water
Diet: mainly fruits (it can be captured with riparian fruit on the hook)
Reproduction:
Remarks: a piranha with weak cheek armature that is not dangerous to humans.

PYGOPRISTIS DENTICULATA (Cuvier, 1819)
Local name(s): piren

Diagnostic characteristics: a small piranha that can be distinguished from the other Surinamese piranha species by the absence of black bands/spot on the caudal peduncle or base of the caudal fin; adipose fin long (3.3 in dorsal fin base; almost like *Metynnis*); snout blunt; absence of ventral spines posterior of the anus, interdorsal distance (distance between the adipose and dorsal fins) >2 in the base of the dorsal fin (long with 16 branched rays); teeth usually with 5 cusps (pentacuspid); eye, and (distal part) fins bright-red
Type locality: unknown
Distribution: Orinoco River basin, northern and eastern Guiana Shield rivers; tributaries of Lower Amazon: Brazil, French Guiana, Guyana, Suriname and Venezuela; in Suriname known, from Para River (Suriname River System), Wane

Pristobrycon striolatus (live), scale bar = 2 cm, Lawa River (Marowijne River), ANSP188688 (© M. Sabaj Pérez)

Pygopristis denticulata (live) Commewijne River (© R. Covain)

Creek (Commewijne River System) and rice fields in northwestern Suriname (Nickerie District)
Size: 20 cm TL
Habitat: black-water streams of the Savanna Belt and swamps of Coastal Plain; however, the species is apparently quite rare
Position in the water column: mid-water
Diet:
Reproduction: juveniles resemble tetras with the body less deep
Remarks: this beautiful, small piranha is not dangerous to humans.

Serrasalmus rhombeus (Linnaeus, 1766)
Local name(s): piren (redeye piranha)

Diagnostic characteristics: eye bright-red; humeral spot absent; juveniles (spotted), with transparent eyes, a black band along the margin of the anal fin and two black bands on the caudal fin, one at the base and the second at the posterior margin of the fin
Type locality: Brokopondo, Suriname River, Suriname
Distribution: Amazon and Orinoco River basins, northern and eastern Guiana Shield, and northeastern Brazilian coastal rivers: Argentina, Bolivia, Brazil, Colombia, Ecuador, French Guiana, Guyana, Peru, Suriname and Venezuela; in Suriname present in all rivers
Size: a large piranha, up to 41 cm SL (3 kg) (Willink *et al.*, 2011)
Habitat: upper and lower reaches of rivers and larger tributaries, but always in freshwater
Position in the water column: mid-water
Diet: fish and other animals that happen to fall into the water
Reproduction: in the rainy season (Vari, 1982); In the Rupununi River, Lowe-McConnell (2000) has observed the eggs of *S. rhombeus* attached to submerged roots and guarded by the parents
Remarks: although *S. rhombeus* did not attack humans in French Guiana (Planquette *et al.*, 1996) this large piranha may be dangerous to humans under certain conditions and, in Suriname, attacks on humans have taken place in Amerindian villages on the Wayombo River and two recreation parks on the Suriname River (Mol, 2006); in Brokopondo Reservoir, *S. rhombeus* is very abundant (Mol *et al.*, 2007), but apparently not often attacking humans in the water; this species apparently has a small home range (Cohen *et al.*, 1999).

Tometes lebaili Jégu, Keith & Belmont-Jégu, 2002
Local name(s): –

Diagnostic characteristics: length of adipose fin < interdorsal distance; maximum SL > 15 cm; absence of spines in the preventral region; mouth oblique and

Serrasalmus rhombeus (live), Coppename River (© B. Chernoff)

Tometes lebaili (live), Lawa River (Marowijne River), ANSP187099 (© M. Sabaj Pérez)

upturned (superior); absence of a humeral spot; more than 6 teeth on each side of the lower jaw
Type locality: rapids upstream of the Litany River at Antecume Pata village, 3°18'06.4"N, 54°04'54.1"W, Marowijne (Maroni) River
Distribution: Guianas and Suriname; in Suriname, only known from the Marowijne and Commewijne (Mapane Creek) rivers
Size: 51 cm SL
Habitat: in rapids and large to medium-sized streams with moderate to strong currents
Position in the water column: mid-water
Diet: mainly Podostemaceae ('kumalu-nyang'); *T. lebaili* can be captured with hook-and-line with Podostemaceae as bait
Reproduction:
Remarks: a threatened species due to its restricted distribution (Jégu & Keith, 2005).

This page is intentionally left blank.

FAMILY CHARACIDAE – SUBFAMILY APHYOCHARACINAE (BLOODFIN TETRAS)

The subfamily Aphyocharacinae was proposed by Eigenmann (1909) to include fishes with a single series of well developed teeth on the premaxilla, mandible and maxilla. They were also characterized by having large parietal and frontal fontanels, gill-membranes free from isthmus and from each other, and adipose fin always present. It included six genera: *Coelurichthys*, *Odontostilbe*, *Holoshesthes*, *Cheirodon*, *Aphyocharax* and *Holoprion*. However, since that time, this subfamily has been successively considered as valid or included in the Cheirodontinae. Nowadays the Aphyocharacinae is apparently a monophyletic group that comprises only one genus, *Aphyocharax*, and is defined by the reduction of the second infraorbital and the development of its third and fourth infraorbital bones, which together cover the whole cheek. They are usually small and have an incomplete lateral line and a short anal fin (12–22 rays). The maxilla can be short, forming a small mouth with only a few teeth on the proximal portion of the bone; or the maxilla can be longer, reaching almost the vertical through the middle of the eye and the third infraorbital, and is completely toothed. The mouth is terminal or superior. The teeth are always tricuspid, the median cusp much bigger than the others.

APHYOCHARAX ERYTHRURUS Eigenmann, 1912
Local name(s): – (flametail tetra)

Diagnostic characteristics: incomplete lateral line, 9-11 scales with pores; no caudal spot; short anal fin (17-27 rays), large first postorbital bone (4th circumorbital) and reduced 2nd postorbital bone, caudal fin blood-red in life
Type locality: Rockstone sand-bank, Guyana
Distribution: Essequibo River basin, Guyana, and Corantijn River, Suriname
Size: 5.8 cm SL
Habitat: lives in large rivers and medium-sized streams over sandy bottom
Position in the water column: mid-water and near the bottom
Diet:
Reproduction: males exhibit sexual hooks on the first anal-fin rays
Remarks: nothing is known about its biology.

Aphyocharax erythrurus (live) (© P. Willink)

FAMILY CHARACIDAE – SUBFAMILY CHARACINAE (CHARACINS, TETRAS)

The subfamily Characinae includes a number of characid genera whose phylogenetic relationships have not been adequately studied. The body is relatively deep especially anteriorly where a predorsal gibbosity is typical of most genera. There are more than 20 conical teeth on the maxilla (except in *Roeboexodon* and *Roeboides*), a pseudotympanum is present in front of the first pleural rib (except in *Gnathocharax*, *Hoplocharax*, and *Lonchogenys*) and a larval pectoral fin is retained in specimens up to 41.0 mm SL (except in *Gnathocharax*, *Heterocharax*, *Hoplocharax*, and *Lonchogenys*). Most species include fishes and insects (both adults and larvae) in their diets, but species of *Roeboexodon* and *Roeboides* have typical scale-eating behavior. For this they have mammiliform specialized teeth outside the mouth, pointing forward, to remove scales from other fishes (see Sazima, 1983 for details).

1-a Mouth inferior; presence of external, conical teeth on the extremity of the prominent, shark-like snout

Roeboexodon guyanensis

1-b Mouth superior, terminal or slightly inferior

2

2-a Characteristic body shape with the top of the head horizontal and the dorsal profile humped

3

2-b Dorsal profile not notably humped; pre-pelvic region flattened, covered with two series of elongate, overlapping scales bent at the sides (chevron-shape); teeth numerous, especially in the upper jaw (*Phenacogaster*)

6

3-a Scale-eaters with teeth transformed to mammilla-like tubercles pointing forward and outward on both jaws; normal teeth in two series on the upper jaw

Roeboides affinis

3-b Predators armed with conical or canine-like teeth, never with mammilla-like tubercles on snout

4

4-a Ctenoid (rough) scales; body elongated, rather large size (20 cm)

Cynopotamus essequibensis

4-b Cycloid (smooth) scales; up to 15 cm in length (*Charax*)

5

5-a Humeral spot and spot on caudal peduncle both large, about the size of the eye; pre-dorsal line partly without scales; absence of ectopterygoid teeth; Marowijne River

Charax aff. *pauciradiatus*

5-b Humeral spot and spot on caudal peduncle smaller than eye; pre-dorsal line scaled; ectopterygoid teeth present

Charax gibbosus

6-a Lateral line not complete, with 8-13 perforated scales

Phenacogaster carteri

6-b Lateral line complete (34-41 perforated scales)

7

7-a Lateral spot large, subcircular; a minute scale in the angle of each pair of the ventral series of scales, except the pair between the pectorals; a caudal spot, not continued to the end of the middle rays

Phenacogaster wayana

7-b Lateral spot minute, inconspicuous; few if any (minute) scales in the angle of the pairs of scales along the ventral surface; no caudal spot

Phenacogaster aff. *microstictus* (*pectinatus* group)

Charax gibbosus (Linnaeus, 1758)
Local name(s): –

Diagnostic characteristics: humeral spot and spot on caudal peduncle smaller than eye; pre-dorsal line scaled; ectopterygoid teeth present
Type locality: Suriname
Distribution: Essequibo River basins and coastal rivers of Suriname: Guyana and Suriname
Size: 12.5 cm SL
Habitat:
Position in the water column:
Diet: aquatic insect larvae, small fishes and shrimps
Reproduction: in the aquarium, eggs are deposited on vegetation; they hatch in about 30 hours
Remarks:

Charax aff. *pauciradiatus* Günther, 1864
Local name(s): –

Diagnostic characteristics: humeral spot and spot on caudal peduncle both large, about the size of the eye; pre-dorsal line partly without scales; absence of ectopterygoid teeth
Type locality: Rio Capin (= Capim), Para State, Brazil
Distribution: Lower Amazon River basin, Brazil
Size: 10.1 cm SL
Habitat: relatively deep pools (>1 m) with slow water current
Position in the water column:
Diet:
Reproduction: in the rainy season; males are a little smaller than females.
Remarks:

Cynopotamus essequibensis Eigenmann, 1912
Local name(s): –

Diagnostic characteristics: ctenoid (rough) scales; body elongated, rather large size; mouth large with caniniform teeth, snout very short; humeral spot and spot on caudal peduncle about the size of the eye diameter
Type locality: Potaro landing, Guyana
Distribution: coastal drainages of Amapá State (Brazil), Guyana, Suriname and French Guiana: Brazil, French Guiana, Guyana and Suriname; in Suriname, probably present in all rivers
Size: 16 cm SL
Habitat: most common (but never abundant) in upstream reaches of the rivers of the Interior

Charax gibbosus (live), Suriname River (© R. Covain)

Charax aff. *pauciradiatus* (live), Lawa River (Marowijne River) (© P. Willink)

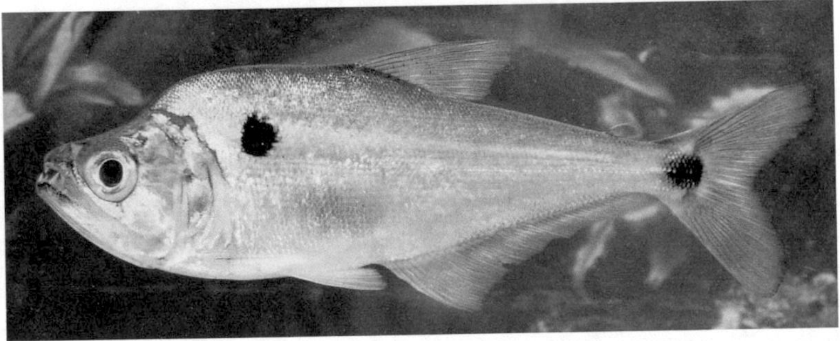

Cynopotamus essequibensis (live) (© INRA-Le Bail)

Position in the water column: mid-water
Diet: carnivorous
Reproduction:
Remarks:

PHENACOGASTER CARTERI (Norman, 1934)
Local name(s): – (glass tetra)

Diagnostic characteristics: dorsal profile not notably humped; easily distinguished from the two other Surinamese *Phenacogaster* species by its incomplete lateral line (only 8-13 perforated scales); absence of humeral and caudal spot
Type locality: Cuyuni River, Guyana
Distribution: Essequibo River, Caura River and Lower Orinoco River: Guyana and Venezuela; in Suriname only known from the (Upper) Corantijn River
Size: 3.6 cm SL
Habitat: forest creeks in the Interior
Position in the water column: mid-water
Diet:
Reproduction:
Remarks: *Phenacogaster* have a conspicuous triangular loss of muscles above the anterior part of the gas bladder (called a pseudotympanum or humeral hiatus) and a series of minute sensory pores on the head, and they are translucent in life (glass tetras).

PHENACOGASTER aff. *microstictus* Eigenmann, 1909 (*pectinatus* group)
Local name(s): – (glass tetra)

Diagnostic characteristics: dorsal profile not notably humped; pre-pelvic region flattened, covered with two series of elongate, overlapping scales bent at the sides (chevron-shape); lateral spot minute, inconspicuous; few if any (minute) scales in the angle of the pairs of scales along the ventral surface; no caudal spot
Type locality: Tumatumari, Lower Potaro River, Guyana
Distribution: Essequibo and Demerara River basins: Guyana and Suriname
Size: 4.2 cm SL
Habitat: forest creeks of the Interior
Position in the water column: mid-water
Diet:
Reproduction:
Remarks: the *P. pectinatus* group (under revision) includes *P. microstictus* Eigenmann 1909 (Lucena & Malabarba, 2010).

Phenacogaster carteri (alcohol), 26 mm SL, Sipaliwini River (Corantijn River) (© P. Willink)

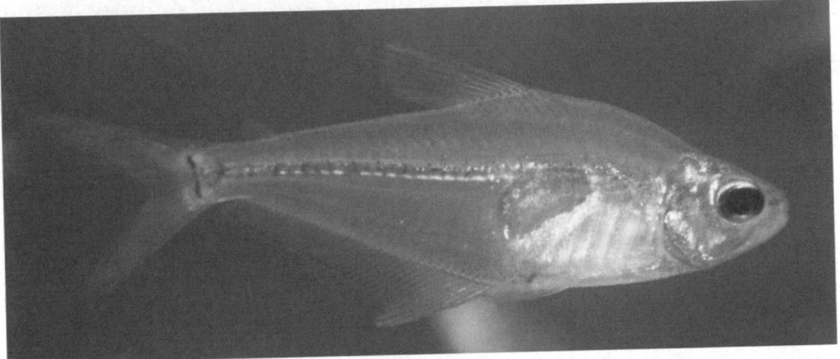

Phenacogaster carteri (live), Sipaliwini River (Corantijn River) (© K. Wan Tong You)

Phenacogaster aff. *microstictus* (live) (© P. Willink)

Phenacogaster wayana Le Bail & Lucena, 2010
Local name(s): – (glass tetra)

Diagnostic characteristics: dorsal profile not notably humped; pre-pelvic region flattened, covered with two series of elongate, overlapping scales bent at the sides (chevron-shape); lateral spot large, subcircular; a minute scale in the angle of each pair of the ventral series of scales, except the pair between the pectorals; an elongated caudal spot, not continued to the end of the middle rays
Type locality: Crique Japigny, Approuague River, French Guiana
Distribution: Approuague, Sinnamary, Mana, Marowijne (Maroni) and Corantijn rivers; French Guiana and Suriname
Size: 4.7 cm SL
Habitat: creeks with moderate to strong current and sandy or rocky bottom substrates
Position in the water column: mid-water
Diet: omnivorous
Reproduction:
Remarks:

Roeboexodon guyanensis (Géry, 1959)
Local name(s): –

Diagnostic characteristics: a very peculiar elongate species with a completely inferior mouth and a prominent, pointed snout ('nose') like that of a shark; presence of external, mammilla-like teeth on the extremity of the snout; presence of humeral and caudal spots
Type locality: Marowijne (Maroni) River, French Guiana
Distribution: Amazon, Tocantins, Xingu and Tapajós River basins, Brazil; in Suriname, known from the Corantijn, Suriname and Marowijne rivers
Size: 15 cm SL
Habitat: rivers and streams of the Interior, upstream of the first rapids
Position in the water column: mid-water
Diet: scale-eater (see Sazima, 1983 for details)
Reproduction:
Remarks: *Roeboexodon geryi* (Myers, 1960) is a junior synonym.

Roeboides affinis (Günther, 1868)
Local name(s): –

Diagnostic characteristics: scale-eaters with the body shape of *Charax* but teeth transformed to external, mammilla-like tubercles pointing forward and outward on both jaws; normal teeth in two series on the upper jaw; long anal fin; humeral spot present
Type locality: Río Huallaga, Amazon system, Peru

Phenacogaster wayana (alcohol), 49.5 mm SL, Lawa River (Marowijne River), ANSP189599 (© M. Sabaj Pérez)

Roeboexodon guyanensis (live), 45.6 mm SL, Lawa River (Marowijne River), ANSP189617 (© M. Sabaj Pérez)

Roeboides affinis (live), Corantijn River (© R. Smith)

Distribution: widespread in South America, including coastal rivers of Guyana and Suriname; in Suriname known from Corantijn, Saramacca and Suriname rivers
Size: 11 cm SL
Habitat:
Position in the water column: mid-water
Diet: scale-eaters
Reproduction:
Remarks: *R. thurni* is a junior synonym.

This page is intentionally left blank.

FAMILY CHARACIDAE – SUBFAMILY STETHAPRIONINAE (SILVER DOLLAR TETRAS)

Members of the subfamily Stethaprioninae can be readily distinguished from other characids by their deep, very compressed, sometimes disciform, body shape and by having an anteriorly directed bony spine preceding the first dorsal-fin ray (Reis, 1989). Mature male stethaprionines have very small, thin anal-fin hooks, varying from one to six per ray segment, that face in all directions. Fishes in this subfamily are very homogeneous in terms of body size (maximum 6.5-9.0 cm SL) and shape. Stethaprionine species are usually found in small to large rivers, where the current is not very strong.

1-a Predorsal spine straight, simple and sharp-pointed; first anal-fin element modified into a strong, forward directed spine
Brachychalcinus orbicularis
1-b Predorsal spine saddle-like, rounded distally in dorsal view, with two small posteroventrally directed processes; first anal-fin element simple or slightly laminar (*Poptella*)
2

2-a Seven or eight horizontal rows of scales between dorsal-fin origin and lateral line; first rays of dorsal and anal fins never more than twice the length of the following rays and not highly pigmented
Poptella brevispina
2-b Nine or ten horizontal rows of scales between dorsal-fin origin and lateral line; first rays of dorsal and anal fins usually much longer and darker than following rays in small specimens; Nickerie and Corantijn rivers
Poptella longipinnis

BRACHYCHALCINUS ORBICULARIS (Valenciennes, 1850)
Local name(s): –
Diagnostic characteristics: predorsal spine straight, simple and sharp-pointed; first anal-fin element modified into a strong, forward directed spine
Type locality: Essequibo River, Guyana
Distribution: Guyana and Suriname; in Suriname, known from the Corantijn, Coppename and Suriname rivers
Size: 9 cm SL
Habitat: forest creeks, small streams and rivers of the Interior
Position in the water column: mid-water
Diet: feeds on worms, insects, crustaceans and plant matter
Reproduction:
Remarks: *B. guianensis* Boeseman 1952 is a junior synonym (see Reis, 1989).

Brachychalcinus orbicularis (live), Kabalebo River (Corantijn River) (© R. Smith)

Poptella brevispina Reis, 1989
Local name(s): –

Diagnostic characteristics: predorsal spine short and saddle-like, rounded distally in dorsal view, with two small posteroventrally directed processes (it is movable); first anal-fin element simple or slightly laminar; seven or eight horizontal rows of scales between dorsal-fin origin and lateral line; first rays of dorsal and anal fins never more than twice length of following rays and not highly pigmented
Type locality: Igarape Apeu, Boa Vista, Castanhal, Para, Brazil
Distribution: Trombetas, Upper Branco, and Lower Tocantins River basins; coastal drainages of Guyana, Suriname and Pará State, Brazil: Brazil, Guyana and Suriname
Size: 7.6 cm SL
Habitat: small streams and forest creeks with clear water and sandy bottom substrate
Position in the water column: mid-water
Diet:
Reproduction: in the rainy season
Remarks: *P. brevispina* can be very abundant; a common species.

Poptella longipinnis (Popta, 1901)
Local name(s): –

Diagnostic characteristics: predorsal spine saddle-like, rounded distally in dorsal view, with two small posteroventrally directed processes; first anal-fin element simple or slightly laminar; nine or ten horizontal rows of scales between dorsal-fin origin and lateral line; first rays of dorsal and anal fins usually much longer and darker than following rays in small specimens
Type locality: Lower Nickerie River between Manilie-kreek and Arawara, Suriname
Distribution: Orinoco River basin and coastal drainages of Suriname and lower Tocantins R: Brazil, Colombia, Suriname and Venezuela; in Suriname, known from Nickerie and Corantijn rivers
Size: 6.9 cm SL
Habitat: small streams and forest creeks with clear water and sandy bottom substrate
Position in the water column: mid-water
Diet:
Reproduction:
Remarks: this species in not well known (see Reis, 1989).

Poptella brevispina (live), Paloemeu River (© P. Naskrecki)

Poptella longipinnis (live), Kabalebo River (Corantijn river) (© R. Smith)

FAMILY CHARACIDAE – SUBFAMILY TETRAGONOPTERINAE (CHARACINS, TETRAS)

Members of the subfamily Tetragonopterinae can be distinguished from other characids by their deep and lozenge-shaped body, flat preventral area, and long anal fin base. The phylogenetic relationship of this subfamily to other characids is unstudied. The Tetragonopterinae has been treated for many years (e.g. Géry, 1977) as a very large, all-encompassing subfamily, into which most genera as Incertae Sedis in Characidae by Reis *et al.* (2003) used to be allocated. Because of the lack of evidence that such a large Tetragonopterinae constitutes a monophyletic assemblage, Reis *et al.* (2003) preferred to emphasize that interrelationships among the Characidae are poorly known, and only recognized *Tetragonopterus* (with only three species currently known) as belonging to the Tetragonopterinae. They are usually found in small to large rivers, with slow to moderately strong water current. They are distributed in all major river drainages of cis-Andean South America.

1-a Absence of dark, longitudinal stripes between adjacent scale rows of the lateral surface of the body

Tetragonopterus chalceus

1-b Presence of dark, longitudinal stripes between adjacent scale rows of the lateral surface of the body

Tetragonopterus rarus

TETRAGONOPTERUS CHALCEUS Spix & Agassix, 1829
Local name(s): –

Diagnostic characteristics: beginning of lateral line descending steeply in stepwise fashion; two faint humeral spots; eye very large; first rays of dorsal fin black and elongated (reaching over the adipose fin); first rays of pelvic and anal fins whitish; absence of dark, longitudinal stripes between adjacent scale rows of the lateral surface of the body
Type locality: rivers of equatorial Brazil
Distribution: Amazon, São Francisco and Orinoco River basins and coastal drainages of the Guianas: Brazil, French Guiana, Guyana, Peru, Suriname and Venezuela; in Suriname known from Marowijne, Commewijne, Suriname, Saramacca and Coppename rivers
Size: 9.7 cm SL
Habitat: in rapids or in forest creeks with either moderate to fast currents on sandy bottom substrate or slow running water (e.g. in aquatic vegetation)

Tetragonopterus chalceus (alcohol), Coppename River (© M. Littmann)

Tetragonopterus chalceus (live) (© INRA-Le Bail)

Position in the water column:
Diet:
Reproduction:
Remarks:

TETRAGONOPTERUS RARUS (Zarske, Géry & Isbrücker, 2004)
Local name(s): –

Diagnostic characteristics: the species is distinguished from *T. chalceus* by the presence of approximately 12 dark, longitudinal zigzag stripes between adjacent scale rows of the lateral surface of the body and the number of scales between the tip of the supra-occipital spine and the base of the first dorsal-fin ray (8 in *T. rarus* versus 12-16 in *T. chalceus*); presence of a black spot on the caudal peduncle, but lacking a humeral spot; short head and large eye; large scales (32 in the lateral line row)
Type locality: at air strip, one day upstream of Oelemari (Ulemari), rapid in Oelemari (Ulemari) River, Upper Marowijne River, Suriname
Distribution: Marowijne and Corantijn River basins, Suriname
Size: 6.2 cm SL
Habitat: small rainforest streams with black water, a limited amount of emergent vegetation, moderate flow, and sandy bottom, but also in larger streams with swift currents
Position in the water column: mid-water
Diet:
Reproduction:
Remarks: *Tetragonopterus lemniscatus* Benine, Pelição & Vari, 2004 is a junior synonym.

Tetragonopterus rarus (live), Paloemeu River (Marowijne River)

FAMILY CHARACIDAE – SUBFAMILY STEVARDIINAE

The presence of glandular tissue associated with modified scales on the caudal-fin of sexually mature males has long been used to characterize the Glandulocaudinae, a group with a dorsal fin count of ii, 8 and four teeth on the inner row of the premaxilla that was initially thought to be monophyletic, but recently split into two subfamilies: Glandulocaudinae and Stevardiinae (Menezes & Weitzman, 2009). The Stevardiinae as proposed by Weitzman *et al.* (2005) can be separated from the Glandulocaudinae by three characters: (1) presence in the Stevardiinae of a hypertrophic extension of the body scales onto the rays of ventral caudal-fin lobe which is absent in all Glandulocaudinae; (2) caudal organs of the species of the Glandulocaudinae with a hypertrophic extension of the upper lobe body scales onto the rays of the dorsal caudal fin lobe rather than the lower caudal-fin lobe as in the Stevardiinae tribes; and (3) caudal-gland cells of the caudal organ of the Glandulocaudinae consisting of apparently specialized club cells, not the modified mucous cells reported for some Stevardiinae. In Suriname, the subfamily is represented by *Hemibrycon surinamensis*, three *Bryconamericus* species and *Creagrutus melanzonus*, all more or less elongated and small (with exception of *H. surinamensis*) species with a complete lateral line.

1-a Asymmetric spot on caudal peduncle, positioned on the lower portion of the caudal peduncle with extension on the caudal fin up to its posterior margin; body depth 2.8 times in standard length; relatively large species (up to 8 cm TL)

Hemibrycon surinamensis

1-b Spot on caudal peduncle absent or, when present, small and symmetrical; elongate species with body depth more than 3.5 times in standard length; small species (< 6 cm TL)

2

2-a Three rows of premaxillary teeth; anal fin short with less than 16 rays

Creagrutus melanzonus

2-b Two rows of premaxillary teeth; anal fin long with more than 16 rays (*Bryconamericus*)

3

3-a Absence of spot on caudal peduncle; snout pointed, mouth inferior; adipose fin colored red

Bryconamericus aff. *hyphesson*

3-b Presence of an elongated black spot on the caudal peduncle; snout rounded, mouth terminal; adipose fin hyaline

4

4-a Presence of a brilliant longitudinal band (black in preserved specimens); body depth 3.45-3.75 times in standard length, 40-41 scales in the lateral line; anal fin long iii, 19-22

Bryconamericus guyanensis

4-b Absence of brilliant longitudinal band along lateral line; body depth 3.8-4.0 times in standard length; 35-37 scales in lateral line; anal fin relatively short iii,16-19

Bryconamericus heteresthes

BRYCONAMERICUS aff. *HYPHESSON* Eigenmann, 1909
Local name(s): –

Diagnostic characteristics: a very small, elongate species with a long anal fin (iii14-16 rays) and a complete lateral line (33-36 scales); absence of a spot on caudal peduncle, humeral spot difficult to see in live specimens; snout pointed, mouth inferior; adipose fin and superior margin of the eye colored bright-red
Type locality (*B. hyphesson*): Tumatumari, Lower Potaro River, Guyana
Distribution: Potaro River basin, Guyana; in Suriname known from Marowijne (Planquette et al 1996) and Corantijn rivers
Size: 3.7 cm TL
Habitat: streams in the Interior (upstream of the first rapids) with moderate to strong current
Position in the water column:
Diet:
Reproduction:
Remarks: rare and not well-known.

BRYCONAMERICUS GUYANENSIS Zarske, Le Bail & Géry, 2010
Local name(s): –

Diagnostic characteristics: a small, elongate species, body depth 3.45-3.75 times in standard length; complete lateral line (40-42 scales); presence of an elongated black spot on the caudal peduncle; snout rounded, mouth terminal; adipose fin transparent; long anal fin, iii19-22
Type locality: Mana, crique Eau Claire at 7 kilometers of Saül, French Guyana
Distribution: river basins of French Guyana; in Suriname only known from the Marowijne River where it seems quite abundant
Size: 5.6 cm SL
Habitat: rapids and streams with moderate to strong currents
Position in the water column:
Diet:
Reproduction: females with ripe ovaries were collected in October (dry season)
Remarks: relatively abundant compared to other *Bryconamericus* species in the Marowijne River; identified as *Bryconamericus* aff. *stramineus* in Planquette *et al.* (1996).

Bryconamericus aff. *hyphesson* (alcohol), Lawa River (Marowijne River), ANSP189379 (© M. McIvor)

Bryconamericus aff. *hyphesson* (live) (© INRA-Le Bail)

Bryconamericus aff. *hyphesson* (live), Sipaliwini River (Corantijn River) (© K. Wan Tong You)

Bryconamericus heteresthes Eigenmann, 1908
Local name(s): –

Diagnostic characteristics: slender, elongated, very little compressed, body depth 3.8-4.0 times in standard length, body width about half of body depth; snout rounded, blunt; dorsal and ventral profiles equally arched, without depressions or humps; 35-37 scales in lateral line; anal fin relatively short iii16-19; presence of vertical humeral spot and faint elongated caudal spot; fins all hyaline
Type locality: Rio Tapajos, Brazil
Distribution: Brazil, Colombia, Ecuador, Guyana, and Venezuela; in Suriname known only from the Marowijne River (Planquette *et al.*, 1996)
Size: 5.1 cm TL
Habitat: streams in the Interior (upstream of the first rapids) with moderate to strong current
Position in the water column:
Diet:
Reproduction:
Remarks: rare and not well-known.

Creagrutus melanzonus Eigenmann, 1909
Local name(s): –

Diagnostic characteristics: small species with an elongated body (body depth more than 3.5 times in standard length) and the upper jaw at same level or lower than pupil of the eye; anal fin short (11 rays); lateral line complete (34-36 scales); eye large and with superior margin colored red; snout short and rounded; presence of a (faint) vertical, crescent-shaped humeral spot
Type locality: Crab Falls, Essequibo River, Guyana
Distribution: Cuyuni River of eastern Venezuela to Sinnamary River basin French Guiana: French Guiana, Guyana and Venezuela; in Suriname only known from the Corantijn and Marowijne rivers
Size: 3.6 cm SL
Habitat: upstream reaches and streams of the Interior with sandy or rocky bottom substrate
Position in the water column:
Diet:
Reproduction:
Remarks: apparently rare.

Bryconamericus guyanensis (alcohol), Lawa River (Marowijne River), ANSP189382 (© M. McIvor)

Bryconamericus guyanensis (live) (© INRA-Le Bail)

Bryconamericus heteresthes (live) (© INRA-Le Bail)

Creagrutus melanzonus (live), 33.8 mm SL, Lawa River (Marowijne River), ANSP189421 (© M. Sabaj Pérez)

Hemibrycon surinamensis Géry, 1962
Local name(s): –

Diagnostic characteristics: elongate species with two rows of teeth on the premaxilla, a complete lateral line, small eyes, and a short snout; presence of a characteristic asymmetric spot on the caudal peduncle, positioned on the lower portion of the caudal peduncle with extension on the caudal fin up to its posterior margin, and a vertically elongated humeral spot
Type locality: Suriname, Browns Creek, Saramacca River System
Distribution: coastal drainages of French Guiana and Suriname; in Suriname known from Saramacca, Suriname and Marowijne rivers
Size: 7.1 cm SL
Habitat: in small streams with moderate current and oxygen-rich water
Position in the water column: mid-water
Diet:
Reproduction:
Remarks: a rare species that is not well-known.

Hemibrycon surinamensis (live) (© INRA-Le Bail)

Hemibrycon surinamensis (live) (© INRA-Le Bail)

FAMILY CHARACIDAE – SUBFAMILY CHEIRODONTINAE

Members of the subfamily Cheirodontinae can be distinguished from all other Characiformes by the following characters: a characteristically structured pseudotympanum, represented by a gap or a reduction of the muscles covering the anterior portion of the swim bladder, anterolaterally on the body, between the first and second pleural ribs; lack of a humeral spot; jaw teeth with a basal peduncle or pedicle and a highly compressed and expanded distal tip, usually with several cusps; and a single tooth series on the premaxilla with teeth perfectly aligned and similar in shape and cusp number.

Odontostilbe gracilis (Géry, 1960)

Local name(s): –

Diagnostic characteristics: lateral line not complete, but rather long with 10-14 perforated scales, reaching to level of dorsal fin; pseudotympanum very evident (like a humeral spot); symmetrical black caudal spot present; black pigmentation at the base of the anal fin; short snout, small mouth and small eye
Type locality: Sable Creek, Upper Mana River, French Guiana
Distribution: Marowijne and Mana basins, French Guiana, and Corantijn River
Size: 2.4 cm SL
Habitat: in rivers and small streams of the Interior with moderate to strong water currents
Position in the water column: mid-water
Diet:
Reproduction:
Remarks: a small tetra that looks similar to *Hemigrammus guyanensis* and *H. boesemani* and that occurs syntopically with the latter two species.

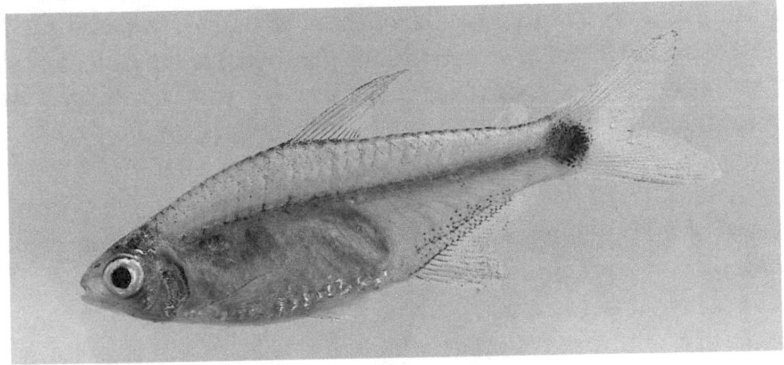

Odontostilbe gracilis (live) (© INRA-Le Bail)

FAMILY ACESTRORHYNCHIDAE (ACESTRORHYNCHIDS)

Fishes of this family are characterized by very elongate (pike-like) bodies covered with relatively small scales. All teeth are conical and strong canines are present on the premaxilla, anterior part of the maxilla and the dentary. Small conical teeth are present on the ectopterygoid and minute conical teeth have been detected on the mesopterygoid of some species. Other exclusive features of the group are: first infraorbital covering almost completely the maxilla when the mouth is closed; a branch of the laterosensory canal on the premaxilla; and rhinosphenoid bone in close contact with parasphenoid. The anal fin is falcate, never bearing hooks in sexually mature males and the origin of the dorsal fin is much nearer to caudal base than the tip of the snout. They are found in a variety of habitats, but primarily live in lakes, lagoons, areas near shore, and the smallest species are especially found in small streams (igarapés) of the Amazon basin. The peculiar dentition makes acestrorhynchid species very specialized predators among characiforms, most species feeding primarily on fishes.

1-a Presence of a large (larger than eye diameter) humeral spot; color of fins orange-red.
Acestrorhynchus falcatus
1-b Absence of a large humeral spot; color of fins transparent or yellowish
Acestrorhynchus microlepis

ACESTRORHYNCHUS FALCATUS (Bloch, 1794)
Local name(s): dagufisi

Diagnostic characteristics: elongated body, large teeth, large humeral spot, color of fins orange-red
Type locality: Suriname
Distribution: Amazon and Orinoco River basins and rivers of northern South America: Brazil, French Guiana, Guyana, Peru, Suriname and Venezuela; all rivers of Suriname
Size: 27 cm SL
Habitat: freshwater, Interior, mainly in small streams
Position in the water column: mid-water
Diet: fish
Reproduction: start of the wet season
Remarks: a crustacean parasite (Isopoda, Cymathoidae) can sometimes be found in the mouth of *A. falcatus*.

Acestrorhynchus falcatus (live), INHS48983 (© M. Sabaj Pérez)

Acestrorhynchus microlepis (Jardine, 1841)
Local name(s): dagufisi

Diagnostic characteristics: elongated body (more slender than *A. falcatus*), large teeth, very small humeral spot, color of fins transparent-yellowish
Type locality: Rio Negro, Rio Branco and Essequibo rivers
Distribution: Amazon and Orinoco River basins and rivers of northern South America: Brazil, French Guiana, Guyana, Peru, Suriname and Venezuela; all rivers of Suriname
Size: 26 cm SL
Habitat: freshwater, Interior, mainly larger streams and rivers
Position in the water column: mid-water
Diet: fish
Reproduction: start of the wet season
Remarks: *A. guianensis* Menezes 1969 is a junior synonym; abundant in Brokopondo Reservoir.

Acestrorhynchus microlepis (live), AUM36753 (© M. Sabaj Pérez)

FAMILY CYNODONTIDAE (CYNODONTIDS)

Fishes of the family Cynodontidae comprise a group of very distinctive Neotropical characiforms easily recognizable by their oblique mouth, well developed dentary canines, and relatively expanded pectoral fins. The group is not very diverse with 13 species currently recognized. They live in mid- and surface waters of rivers, lakes, and flooded forests in all water types, and are predatory fishes, mainly piscivorous, using their dentary canines to stab prey.

1-a Body depth 3.30-3.50 in standard length; 72-80 branched anal fin rays; humeral spot about the same size or larger than eye diameter
Cynodon gibbus
1-b Body depth 3.55-3.80 in standard length; 62-68 branched anal fin rays; humeral spot < eye diameter (Marowijne River)
Cynodon meionactis

CYNODON GIBBUS (Agassiz, 1829)
Local name(s): adyenu

Diagnostic characteristics: body depth 3.30-3.50 in standard length; 72-80 branched anal fin rays; humeral spot larger than eye diameter
Type locality: Lago Manacapuru, Amazonas State, Brazil
Distribution: Amazon and Orinoco River basins and rivers of Guyana: Bolivia, Brazil, Ecuador, Guyana, Peru and Venezuela; in Suriname known Corantijn River
Size: 28 cm SL
Habitat:
Position in the water column:
Diet:
Reproduction:
Remarks:

CYNODON MEIONACTIS Géry, Le Bail & Keith, 1999
Local name(s): adyenu

Diagnostic characteristics: body depth 3.55-3.80 in standard length; 62-68 branched anal fin rays; humeral spot < eye diameter
Type locality: Antecume Pata village at confluence Litani and Marouini, Upper Marowijne (Maroni) River, French Guiana and Suriname
Distribution: coastal rivers in French Guiana and Suriname; in Suriname known from Marowijne River (and possibly Suriname River)

Cynodon gibbus (live), Sipaliwini River (Corantijn River) (© P. Willink)

Cynodon gibbus (live), detail of head (© C. Covain)

Cynodon meionactis (alcohol), 190 mm SL, Litani River (Marowijne River), ANSP189129 (© M. Sabaj Pérez)

Size: 30 cm TL
Habitat: upper reaches of Marowijne River
Position in the water column: mid-water
Diet: probably carnivorous
Reproduction:
Remarks: only known from Upper Marowijne River and apparently not very common.

This page is intentionally left blank.

FAMILY ERYTHRINIDAE (TRAHIRAS)

Fishes of the family Erythrinidae are characterized by having a cylindrical body form, rounded caudal fin, dorsal fin with 8-15 rays, anterior to anal fin and usually above pelvic fins, anal fin short with 10-11 rays, no adipose fin, numerous teeth on the palate and lateral line with 34-47 scales. Three genera are currently recognized in the family, *Erythrinus*, *Hoplerythrinus* and *Hoplias*. The first two include medium size species, reaching at least 40 cm standard length. On the other hand, *Hoplias* is a genus with medium to large sized fishes, up to 100 cm of standard length in *H. aimara*. Erythrinid species are entirely restricted to South America, mainly in the Amazon basin, where the greatest species diversity occurs. They are found in a variety of habitats such as lakes, swamps, small and large rivers, and forest creeks. They are often important food fishes.

1-a 11-15 branched dorsal-fin rays; at least 37 lateral line scales; maxillary bone with 2 or 3 small canines plus a series of conical teeth (*Hoplias*)
3
1-b 8-9 branched dorsal-fin rays; 32-37 lateral line scales; maxillary bone without canines
2

2-a Maxillary bone not very elongate, not reaching posterior margin of eye in adults; eye very small, 22-25 times in standard length; caudal fin spotted; color pattern of flanks variable, usually marbled.
Erythrinus erythrinus
2-b Maxillary bone elongated, its tip reaching well beyond the posterior margin of the eye in adults; eye small, about 19 times in standard length; caudal fin plain; dark longitudinal band, caudal spot and opercular ocellus present
Hoplerythrinus unitaeniatus

3-a Medial margins of contralateral dentaries abruptly converging toward mandibular symphysis, forming a V-shaped margin in ventral view; eye moderate, about 16 times in standard length in adults; medium-sized species, up to 40 cm standard length
Hoplias malabaricus
3-b Medial margins of contralateral dentaries almost parallel to each other, resulting in a U-shaped margin in ventral view
4

4-a Dark oval spot on opercular membrane present; accessory ectopterygopid absent; large species >60 cm standard length
Hoplias aimara

4-b Dark oval spot on opercular membrane absent; accessory ectopterygoid present; not very large (up to 31 cm SL), dark color
Hoplias curupira

Erythrinus erythrinus (Bloch & Schneider, 1801)
Local name(s): matuli, stonwalapa

Diagnostic characteristics: 8-9 branched dorsal-fin rays; 32-37 lateral line scales; maxillary bone without canines; maxillary bone not very elongate, not reaching posterior margin of eye in adults; eye very small, 22-25 times in standard length; caudal fin spotted; color pattern of flanks variable, usually marbled
Type locality: Suriname
Distribution: Amazon and Orinoco River basins and coastal rivers of northern South America: Argentina, Brazil, French Guiana, Guyana, Suriname, Trinidad and Tobago and Venezuela
Size: 20 cm TL
Habitat: both forest creeks of the Interior, black-water streams of the Savanna Belt and freshwater swamps of the Coastal Plain
Position in the water column: mostly rests on the bottom
Diet: small fishes and aquatic insects
Reproduction: the male can be distinguished from the female by its elongated dorsal fin (adpressed against the body the posterior rays reach the vertical through the insertion of the anal fin)
Remarks: the posterior part of the swim bladder is modified to function in uptake of atmospheric oxygen and thus *E. erythrinus* can survive in (standing) water with low dissolved oxygen concentrations.

Hoplerythrinus unitaeniatus (Spix & Agassiz, 1829)
Local name(s): walapa

Diagnostic characteristics: 8-9 branched dorsal-fin rays; 32-37 lateral line scales; maxillary bone without canines; maxillary bone elongated, its tip reaching well beyond the posterior margin of the eye in adults; eye small, about 19 times in standard length; caudal fin plain; dark longitudinal band, caudal spot and characteristic opercular ocellus present
Type locality: Rio São Francisco, Brazil
Distribution: Amazon, Paraná, Orinoco, São Francisco, and Magdalena River basins and coastal rivers of northern South America and Panama: Argentina, Bolivia, Brazil, Ecuador, French Guiana, Guyana, Panama, Peru, Suriname, Trinidad and Tobago and Venezuela
Size: 31 cm TL (450 g)
Habitat: streams with slow water currents and swamps of the Coastal Plain
Position in the water column: mid-water or close to the bottom
Diet: omnivorous with predatory habits, feeding on aquatic insects and, to a lesser degree, small fishes
Reproduction: females mature at 16 cm TL; fecundity is approximately 2,000-6,000 eggs/female

Erythrinus erythrinus (alcohol), scale bar = 1 cm, ANSP175537 (© M. Sabaj Pérez)

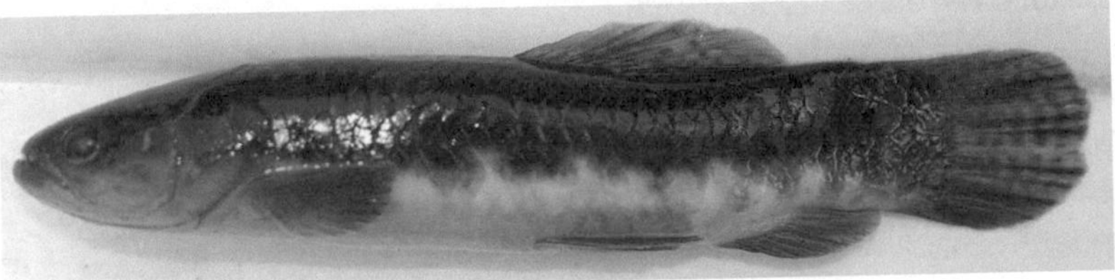

Erythrinus erythrinus (live), Nickerie River (© R. Smith)

Hoplerythrinus unitaeniatus (live) (© R. Covain)

Hoplerythrinus unitaeniatus (live), Nickerie River (© R. Smith)

Remarks: the posterior part of the swim bladder is modified to function in uptake of atmospheric oxygen (Kramer, 1978) and thus *H. unitaeniatus* can survive in (standing) water with low dissolved oxygen concentrations.

HOPLIAS AIMARA (Valenciennes, 1847)
Local name(s): anyumara

Diagnostic characteristics: a large species; medial margins of contralateral dentaries almost parallel to each other, resulting in a U-shaped margin in ventral view; dark oval spot on opercular membrane present; accessory ectopterygopid absent
Type locality: Cayenne, French Guiana
Distribution: widespread in South America
Size: 100 cm SL (and up to 40 kg)
Habitat: large rivers and smaller streams of the Interior
Position in the water column: sit-and-wait predator, resting on the bottom
Diet: piscivorous, a sit-and-wait predator feeding mainly on fishes, but attacks every animal that falls into the water
Reproduction: mainly at the start of the rainy season; dependent on the size, a female may contain 6,000-60,000 eggs.
Remarks: *Hoplias macrophthalmus* is a synonym; a very popular food fish in the Interior of Suriname; the species apparently has a small home range (Tito de Morais & Raffray, 1999). Maroons living in the Interior are very afraid of the bite of this fish (see chapter 8).

HOPLIAS CURUPIRA Oyakawa & Mattox, 2009
Local name(s): –

Diagnostic characteristics: medial margins of contralateral dentaries almost parallel to each other, resulting in a U-shaped margin in ventral view; dark oval spot on opercular membrane absent; accessory ectopterygoid present; color rather dark brown
Type locality: Rio Itacaiunas, Caldeirao, Serra dos Carajas, 05°45' S, 50°30' W, Tocantins Basin
Distribution: rivers of Amazon basin: Brazil, Guyana, Suriname and Venezuela; in Suriname known from Suriname, Saramacca, Coppename and Nickerie rivers, but probably present in all rivers
Size: 31 cm SL
Habitat: in Suriname possibly restricted to rivers and small streams of the Interior
Position in the water column:
Diet:
Reproduction:

Hoplias aimara (alcohol), ANSP176723 (© M. Sabaj Pérez)

Hoplias curupira (alcohol), Coppename River (© P. Willink)

Remarks: the distribution of *H. curupira* overlaps that of *H. aimara*, but *H. aimara* seems to prefer moderate to high current strength (e.g. rapids), while *H. curupira* may prefer slow currents; *H. curupira* does not grow as large as *H. aimara*.

Hoplias malabaricus (Bloch, 1794)
Local name(s): pataka

Diagnostic characteristics: medial margins of contralateral dentaries abruptly converging toward mandibular symphysis, forming a V-shaped margin in ventral view; eye moderate, about 16 times in standard length in adults
Type locality: South America, probably Suriname
Distribution: Costa Rica to Argentina in most river systems: Argentina, Bolivia, Brazil, Colombia, Costa Rica, Ecuador, French Guiana, Guyana, Paraguay, Peru, Suriname, Trinidad and Tobago, Uruguay and Venezuela
Size: 38 cm SL
Habitat: in Suriname possibly restricted to coastal swamps and black-water streams of the Savanna Belt
Position in the water column: benthopelagic
Diet: fish
Reproduction: in the coastal swamps of Suriname, *H. malabaricus* spawn after the first rains in the newly flooded swamps in very shallow (15-20 cm deep) water (Fig. 4.42; pers. observations; also see Prado *et al.* (2006) for reproduction of *H. malabaricus* in the Pantanal, Brazil); clutches of approximately 5000-10,000 yellowish eggs are deposited in a circular nest pit (10 cm diameter, 3 cm deep) and these are guarded by the male (that protects the eggs aggressively against other fish or observers that try to approach the nest); biparental care by male and female can also occur (Prado *et al.*, 2006)
Remarks: Oyakawa & Mattox (2009) described *Hoplias curupira* which is very similar to *H. malabaricus* and also present in Suriname (where it is possibly restricted to rainforest streams in the Interior).

Hoplias malabaricus (live) (© R. Covain)

Hoplias malabaricus (top), *Hoplias aimara* (middle) and *Hoplias curupira* (bottom) (live) (© M. Sabaj Pérez)

FAMILY LEBIASINIDAE (PENCILFISHES)

In size the lebiasinids range from the miniature *Nannostomus anduzei* (1.6 cm) to medium sizes in the Pyrrhulininae to 150 mm SL in the Lebiasininae. All species have a rather elongate, cylindrical body shape with fairly large scales, 17 to 33 in a longitudinal series. The laterosensory canal system on the body is reduced to seven scales, or absent. An adipose fin may be present or absent. The anal fin is short-based with up to 13 rays and the males of most species have well developed anal fins specialized for courtship and breeding. Most pencilfishes of the genus *Nannostomus* have different diurnal and nocturnal color patters, the diurnal pattern typically consisting of rather distinctive longitudinal black stripes and the very different nocturnal pattern usually composed of two or three comma-shaped blotches. Many of the species, especially the pencilfishes (*Nannostomus*), are important aquarium fishes.

1-a Mouth upturned; teeth conical on both jaws; lateral line completely lacking; upper caudal lobe longer than lower caudal lobe; black spot on dorsal fin (Pyrrhulinini)

2

1-b Mouth terminal; teeth multicuspidate (except the inner mandibular row, which is composed of a few conical teeth); lateral line incomplete; caudal-fin lobes equal or the lower longer; absence of black spot on dorsal fin (Nannostomini)

4

2-a Opercular membrane united to isthmus far forward; two complete rows of teeth on upper jaw; maxillary bone straight or slightly curved, never 'S'-shaped, usually toothed; nostrils close to each other (*Pyrrhulina*)

3

2-b Opercular membrane united to isthmus far backward; one row of teeth on upper jaw; maxillary bone forming a double curve in the shape of an 'S'; maxilla always toothed; nostrils separated from each other by a cutaneous bridge

Copella arnoldi

3-a Black stripe extending from the tip of snout through the eye up to the opercle

Pyrrhulina filamentosa

3-b Black stripe extending from the tip of snout through the eye and postorbital to the first two or three scales behind the head

Pyrrhulina stoli

4-a Long snout; adipose fin present; Para River (possibly an introduced/exotic species)

Nannostomus harrisoni

4-b Short snout; adipose fin absent

5

5-a Only one black, longitudinal stripe somewhat below the mid-axis of the body

Nannostomus beckfordi

5-b Two or more longitudinal stripes present

6

6-a Two longitudinal stripes, one similar to that of *N. beckfordi* and the other above mid-axis of body

Nannostomus bifasciatus

6-b Three stripes, the two first like in *N. bifasciatus* and the last one low on abdomen

Nannostomus marginatus

Copella arnoldi (Regan, 1912)
Local name(s): – (splashing tetra)

Diagnostic characteristics: a relatively large, upturned mouth with acute teeth (as contrasted with *Nannostomus* pencilfishes), a curved S-shaped maxillary bone, nostrils separated from each other by a cutaneous bridge, no adipose fin, no lateral line, a black stripe extending from the snout to the eye or opercle, dark spot on the dorsal fin
Type locality: Amazon River, Brazil
Distribution: South America: Lower Amazon, coastal Guianas to mouth of the Orinoco River; all rivers of Suriname
Size: 3.4 cm SL
Habitat: small forest creeks, both in the Interior (clear water) and Savanna Belt (black water)
Position in the water column: surface
Diet: small (allochthonous) invertebrates
Reproduction: the *splashing tetra* gets its name from its breeding behavior; during spawning, both the male and female will jump to an overhanging leaf that is out of the water; they use their fins to clamp on to the leaf, and the female will lay her eggs on the leaf with the male fertilizing them shortly after. After the eggs have been fertilized, the male will remain at the surface of the water where he keeps the eggs moist by splashing water on them. The eggs will then hatch in approximately 2 days at which time the fry will fall back into the water
Remarks: *C. carsevennensis* (Regan, 1912) is here considered a synonym of *C. arnoldi* (M. Marinho, pers. communication).

Pyrrhulina filamentosa Valenciennes, 1847
Local name(s): –

Diagnostic characteristics: resembles *Copella arnoldi* by its elongated shape, upturned mouth and coloration with black spot on dorsal fin and longitudinal stripe from snout to the opercle, but lacks the curved S-shaped maxillary and is also a larger species than *C. arnoldi*; nostrils close to each other
Type locality: Suriname
Distribution: Atlantic coastal rivers between mouth of Amazon and Orinoco rivers: Brazil, French Guiana, Guyana, Suriname and Venezuela;
Size: 8.5 cm SL
Habitat: small forest creeks in the Interior and Savanna Belt and herbaceous freshwater swamps of the Coastal Plain
Position in the water column: surface
Diet: small, allochthonous invertebrates (mainly ants that fall into the water)
Reproduction: it may spawn on submerged leafs
Remarks: *P. filamentosa* and *C. arnoldi* are commonly collected together in small streams and backwaters.

Copella arnoldi (live), 30.4 mm SL, Para River (Suriname River), ANSP189192 (© M. Sabaj Pérez)

Pyrrhulina filamentosa (live) (© INRA-Le Bail)

Pyrrhulina stoli Boeseman, 1953
Local name(s): –

Diagnostic characteristics: resembles *P. filamentosa*, but the horizontal black stripe extends 2 or three scales behind the head (opercle)
Type locality: Suriname, Marowijne System, Nassau Mountains
Distribution: Suriname and Guyana; in Suriname known from the Marowijne, Coppename and Corantijn rivers
Size: 6 cm SL
Habitat:
Position in the water column: surface
Diet:
Reproduction:
Remarks: nothing is known about its biology.

Nannostomus beckfordi Günther, 1872
Local name(s): – (golden pencilfish)

Diagnostic characteristics: diurnal color pattern with only one black, longitudinal stripe somewhat below the mid-axis of the body and extending anteriorly through the eye to the tip of the snout, base of the anal and caudal fins red, no adipose fin
Type locality: Goedverwagting, a plantation on the coast of Demerara, Guyana
Distribution: Rivers of Guiana south to Amazon River basin and up Amazon River to Negro River: Brazil, French Guiana, Guyana and Suriname; in Suriname known from Corantijn, Saramacca, Suriname and Commewijne rivers
Size: 6 cm SL
Habitat: mainly black-water creeks in the Savanna Belt and swamps of the Old Coastal Plain
Position in the water column: mid-water, often in aquatic vegetation (e.g. *Cabomba*)
Diet: small aquatic invertebrates
Reproduction: about 200 eggs are deposited among water plants; they hatch in 2-3 days
Remarks: all nannostomins have a so-called nocturnal color pattern that is very different from the distinctive diurnal horizontal-striped color pattern and usually composed of two or three vertical blotches.

Nannostomus bifasciatus Hoedeman, 1954
Local name(s): – (double-striped pencilfish)

Diagnostic characteristics: two longitudinal stripes, one similar to that of *N. beckfordi* and the second, shorter (only extending to the begin of the caudal peduncle) and above the mid-axis of the body, no adipose fin
Type locality: Berg en Dal, Suriname River, Suriname.
Distribution: coastal rivers of Suriname and French Guiana; in Suriname known from Marowijne, Commewijne and Suriname rivers

Pyrrhulina stoli (live) (© P. Willink)

Nannostomus beckfordi (live) (© INRA-Le Bail)

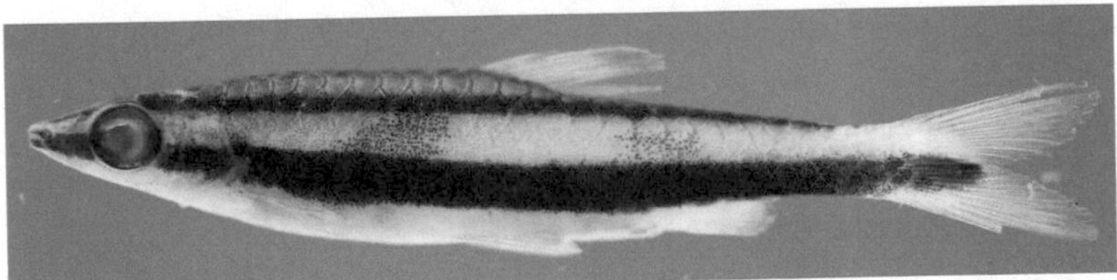

Nannostomus bifasciatus (alcohol), Marowijne River

Nannostomus bifasciatus (live) (© INRA-Le Bail)

Size: 3.5 cm SL
Habitat: small, clear-water streams of the Interior, apparently absent from black-water streams of the Savanna Belt
Position in the water column: mid-water
Diet: small invertebrates
Reproduction: eggs deposited on aquatic vegetation (when present)
Remarks:

NANNOSTOMUS HARRISONI (Eigenmann, 1909)
Local name(s): – (Harrison's pencilfish)

Diagnostic characteristics: resembles an elongate *N. beckfordi* with a long snout and an adipose fin present
Type locality: Canal at Christianburg, Guyana
Distribution: Demerara River basin, Guyana; in Suriname only known from the Para River, a black-water tributary of the Suriname River (probably an introduced species)
Size: 4.5 cm SL
Habitat: black-water tributaries of the Para River
Position in the water column: mid-water
Diet:
Reproduction:
Remarks: in Suriname, probably introduced by aquarium hobbyists.

NANNOSTOMUS MARGINATUS Eigenmann, 1909
Local name(s): – (dwarf pencilfish)

Diagnostic characteristics: three stripes, the two first like in *N. bifasciatus* and the last one low on the abdomen, adipose fin absent
Type locality: Maduni Creek, Guyana
Distribution: lower to middle Amazon River, Colombia east of the Andes, Guyana, Peru east of the Andes, Suriname and Venezuela east of the Andes: Brazil, Colombia, Guyana, Peru and Suriname; in Suriname only known from the Corantijn and Suriname rivers
Size: 3.5 cm TL
Habitat: apparently both black-water (Para River) and clear-water (Corantijn River) streams
Position in the water column: mid-water
Diet:
Reproduction:
Remarks: in the Para River, *N. marginatus* occurs together with both *N. beckfordi* and *N. harrisoni*, but it is not very common.

Nannostomus harrisoni (live), Para River (Suriname River) (© W. Kolvoort)

Nannostomus marginatus (live) (© P. Willink)

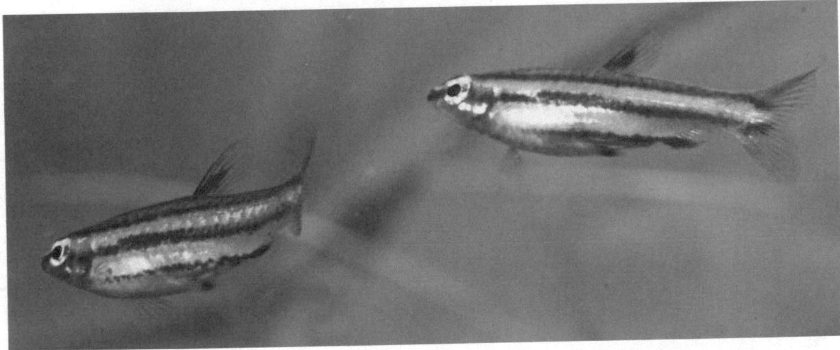

Nannostomus marginatus (live) Para River (Suriname River) (© W. Kolvoort)

FAMILY CETOPSIDAE (WHALE CATFISHES)

Cetopsidae are small to moderate sized fishes which share an anal fin with a long base, a lack of spines in the pectoral and dorsal fins other than in a few species of the subfamily Cetopsinae, and the lack of a nasal barbel, a free orbital margin and bony plates on the body (Vari *et al.*, 2005). Two subfamilies are recognized, the Cetopsinae and the Helogeninae. The Cetopsinae lack an adipose fin and have reduced or absent dorsal and pelvic fins and the eyes reduced or covered with an extensive integumentary layer; the dorsal fin when present is positioned far forward. The Helogeninae usually have an adipose fin and the dorsal fin positioned about halfway the body. Species of the genus *Cetopsis* are notorious for their feeding habits: attacking not only carrion, but also live fishes in gill nets (Burgess, 1989).

1-a More than 38 rays in the anal fin; adipose fin present (sometimes poorly visible); dorsal fin positioned approximately halfway the body
Helogenes marmoratus
1-b Less than 30 rays in the anal fin; no adipose fin; position of dorsal fin at about 1/3 of standard length in the anterior half of the body
2

2-a First pectoral-fin and dorsal-fin rays not spinous; robust and relatively large species (adults up to 9.5 cm TL)
Cetopsis sp.
2-b First pectoral-fin ray spinous for basal one-half of its length, first dorsal-fin ray spinous; small species (<6 cm TL)
3

3-a Tip of pelvic fin reaches posteriorly to anterior margin of vent; tip of pectoral fin falls distinctly short of vertical through pelvic-fin insertion, body depth 0.21-0.23 of SL
Cetopsidium orientale
3-b Tip of pelvic fin reaches beyond vent to anal fin origin; tip of pectoral fin reaches vertical through pelvic-fin insertion; body depth 0.17-0.19 of SL
Cetopsidium minutum

CETOPSIDIUM MINUTUM (Eigenmann, 1912)
Local name(s): –

Diagnostic characteristics: body laterally compressed and naked, lacking scales or bony plates; snout rounded; mouth terminal and large, with three pairs of barbels; dorsal fin with 1 spine and 6 soft rays; pectoral fins with 1 spine and 7-8 soft rays;

Cetopsidium minutum (alcohol), 13 mm SL, Koetari River (Corantijn River) (© P. Willink)

less than 30 rays in the anal fin; no adipose fin; position of dorsal fin at about 1/3 of standard length in the anterior half of the body; tip of pelvic fin reaches beyond the vent to anal fin origin; tip of pectoral fin reaches vertical through pelvic-fin insertion; body depth 0.17-0.19 of SL
Type locality: Amatuk waterfalls, Essequibo River, Guyana
Distribution: possibly restricted to Essequibo River System (Vari *et al.*, 2005); in Suriname observed in Upper Corantijn River (Willink *et al.*, 2011)
Size: 2.5 cm TL
Habitat:
Position in the water column: bottom; however, in Upper Corantijn a single specimen was collected at night on submersed leaves in flooded riparian forest together with *Otocinclus mariae* (K. Wan Tong You, pers. communication)
Diet:
Reproduction: Vari *et al.* (2005) observed no secondary sexual differences between males and females but only very few specimens were examined
Remarks: not common; like most catfishes active during the night, while hiding during the day.

Cetopsidium orientale (Vari, Ferraris & Keith, 2003)
Local name(s): –

Diagnostic characteristics: body laterally compressed and naked, lacking scales or bony plates; snout rounded; mouth terminal and large, with three pairs of barbels; dorsal fin with 1 spine and 6 soft rays; pectoral fins with 1 spine and 7-8 soft rays; less than 30 rays in the anal fin; no adipose fin; position of dorsal fin at about 1/3 of standard length in the anterior half of the body; tip of pelvic fin reaches posteriorly to anterior margin of vent; tip of pectoral fin falls distinctly short of vertical through pelvic-fin insertion, body depth 0.21-0.23 of SL
Type locality: Saramacca River basin, on Gros Rosebel Mining concession, about 5°07'08.8"N, 55°16'59.4"W, close to mouth of Maikaboe Creek, Mindrineti River, Brokopondo District, Suriname
Distribution: Atlantic coastal rivers of French Guiana and Suriname; in Suriname, known from Marowijne, Saramacca and Corantijn rivers
Size: 6 cm TL
Habitat: found hiding in large woody debris in small to medium sized streams of the Interior
Position in the water column: bottom
Diet: allochthonous terrestrial invertebrates (insects)
Reproduction: presumed males have filaments present on the dorsal and pectoral fins (absent in juveniles and females) (Vari *et al.*, 2005)
Remarks: not common; like most catfishes active during the night, while hiding during the day.

Cetopsidium orientale (live) (© INRA-Le Bail)

CETOPSIS sp.
Local name(s): –

Diagnostic characteristics: body rounded in cross-section and naked, lacking scales or bony plates; snout rounded, protruding anteriorly well beyond mouth; mouth subterminal and large, with three pairs of barbels; dorsal fin with 1 spine and 6 soft rays; pectoral fins without a spine and 7-8 soft rays; less than 30 rays in the anal fin; no adipose fin; position of dorsal fin at about 1/3 of standard length in the anterior half of the body; conspicuous blueish pigmentation on the back and blueish spots on the flanks; resembles *C. montana* in pigmentation pattern: the absence of a dark humeral spot, the presence of a posteriorly-rounded, variably-developed, bilobed patch of dark pigmentation at the base of the caudal fin, the presence of a spot of dark pigmentation on the base of the dorsal fin and absence of prominent dark pigmentation along the membrane behind the first ray of the dorsal fin, the presence of approximately eye-size, dark spots on the lateral surface of the body, the absence of finely scattered, dark pigmentation across the lateral and anterior surfaces of the snout
Type locality: –
Distribution: presently only known from two specimens collected in a small forest creek in Upper Commewijne River catchment
Size: 9.5 cm TL (largest specimen)
Habitat: at the collection site the stream was approximately 10 m wide, with water depth of 1 m and medium current strength; the bottom substrate was mainly sand, but there was also abundant woody debris and stands of the emergent aquatic macrophyte *Thurnia sphaerocephala*.
Position in the water column: it was collected in the water column during the day, but this diurnal activity may be related to an elevated turbidity in the stream caused by small-scale gold miners working upstream of the site of collection; in the aquarium *Cetopsis* sp. was most active during the night, while hiding during the day
Diet: in the aquarium the *Cetopsis* sp. accepted small tetras
Reproduction: presumed males have filaments present on the dorsal and pectoral fins (absent in juveniles and females) (Vari *et al.*, 2005)
Remarks: probably not very common; possibly an undescribed species.

HELOGENES MARMORATUS Günther, 1863
Local name(s): –

Diagnostic characteristics: body laterally compressed; snout rounded; mouth terminal and large, with three pairs of barbels; dorsal and pectoral fins both lacking spines; anal fin long (39-48 soft rays); adipose fin present (sometimes poorly visible); dorsal fin positioned approximately halfway the body; color light-brown, lateral line well marked
Type locality: Essequibo River, Guyana

Cetopsis sp. (live), Upper Commewijne River (© A. Gangadin)

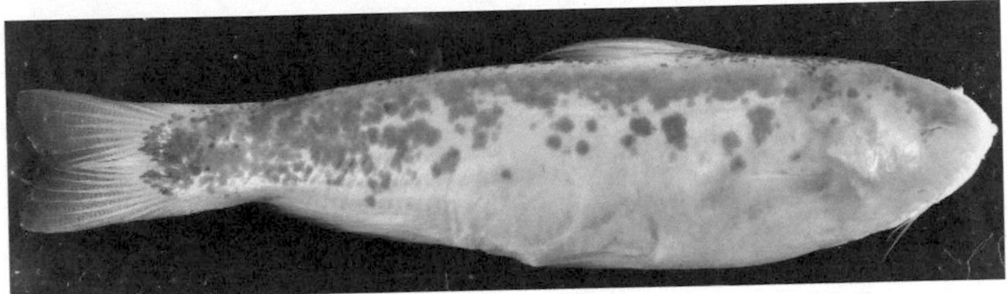

Cetopsis sp. (alcohol), 93 mm TL, Upper Commewijne River (© A. Gangadin)

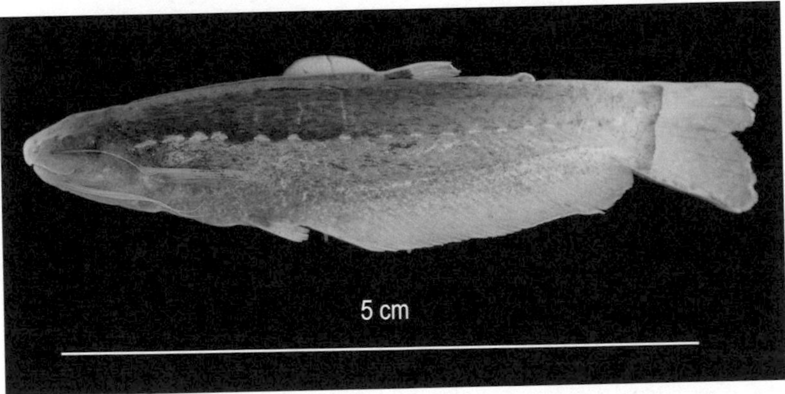

Helogenes marmoratus (alcohol), Suriname River, RMNH25668 (© Naturalis)

Distribution: Atlantic drainages of Guianas, Upper Orinoco and Negro systems and Upper Amazon River basin: Brazil, Ecuador, French Guiana, Guyana, Peru, Suriname and Venezuela
Size: 7.3 cm SL
Habitat: small clear-water forest creeks of the Interior and black-water streams of the Savanna Belt; during the day collected in leaf litter and submersed root masses
Position in the water column: bottom, in leaf litter or in submerged roots
Diet: allochthonous terrestrial invertebrates (insects)
Reproduction:
Remarks: at night *H. marmoratus* often swims while lying on its side; *H. marmoratus* can be locally abundant.

This page is intentionally left blank.

FAMILY ASPREDINIDAE (BANJO CATFISHES)

Aspredinidae have a characteristic body shape with a broad depressed head and body followed by a slender caudal peduncle, which is thought to resemble a banjo. They can further be distinguished by their rough skin covered by keratinized tubercles, opercular openings restricted to small slits, lack of a rigid dorsal spine, lack of an adipose fin, and 10 or fewer caudal-fin rays. In general, most species are cryptically pigmented, benthic and sluggish unless disturbed. Many species bury themselves shallowly in loose substrates. When handled some species produce stridulatory sounds while abducting and adducting their pectoral spines just like doradid catfishes. They are endemic to South America and widely distributed throughout most rivers in this area; they may be found in habitats ranging from shallow backwaters and small creeks to deep river channels and tidal estuaries.

1-a Body tapering, not very long; anal fin short with 6-8 rays; strictly freshwater
2

1-b Body very long and slender; anal fin long with 50-70 rays; in brackish water of river estuaries
4

2-a Head and body not greatly depressed; depth between the pectorals almost 0.6 times in head width; head externally turberculate (warty); coracoid process extending to beyond humeral process; Corantijn and Suriname rivers
Bunocephalus verrucosus

2-b Head and body very depressed; head only slightly warty; coracoids and coracoid process not prominent, humeral process extending past coracoids process
3

3-a Backward location of dorsal, anal and ventral fins (or elongation of abdominal region and decrease in caudal peduncle length), predorsal length 48.1% of SL, relatively short caudal peduncle (from insertion of last anal fin ray to end of hypural; 18.5% of SL), relatively long pectoral fin spine (23.4% of SL) with 16 serrae along the inner margin; Marowijne River
Bunocephalus aloikae

3-b Forward position of dorsal, anal, and ventral fins (short abdomen, long caudal peduncle), predorsal length 41.0% of SL, relatively long caudal peduncle (21.5% of SL), relatively short pectoral sine (20% of SL) with 18 serrae along the inner margin
Bunocephalus amaurus

4-a Ventral surface of the head and breast with 7 or more pairs of mental and postmental barbels; maxillary barbels free from head throughout their length

5

4-b Ventral surface of the head with only two pairs of mental barbels; the large maxillary barbels (one pair) adnate, joined to head by basal membrane, may have a second pair of barbels near their base

6

5-a Upper surface of snout with four broad, hooked spines; 9-10 pairs of mental and postmental barbels

Aspredinichthys tibicen

5-b Upper surface of snout smooth or with two hooked spines; 7 pairs of mental and postmental barbels

Aspredinichthys filamentosus

6-a Two pairs of maxillary barbels; no ridges on sides of body except for lateral line; caudal fin with 10 rays; coloration plain

Aspredo aspredo

6-b One pair of maxillary barbels; 3-4 longitudinal ridges along sides of the body; caudal fin with 9 rays; coloration mottled

Platystacus cotylephorus

Bunocephalus aloikae Hoedeman, 1961
Local name(s): braadpan meerval

Diagnostic characteristics: body tapering, not very long; anal fin short with 6-8 rays; head and body very depressed; head only slightly warty; coracoids and coracoid processes not prominent, humeral process extending past coracoids process 16 serrae along inner margin of long (23.4% of SL) pectoral spine.
Type locality: Litany River near the village Aloiké, Upper Marowijne River, French Guiana
Distribution: Marowijne (Maroni) River and rivers of French Guiana; French Guiana and Suriname; in Suriname, only known from Marowijne River
Size: 12 cm SL
Habitat: strictly freshwater; lives in leaf litter in small forest creeks of the Interior
Position in the water column: bottom
Diet:
Reproduction:
Remarks: they apparently swim both with fins and body undulation, and by forcefully ejecting water through their opercula.

Bunocephalus amaurus Eigenmann, 1912
Local name(s): braadpan meerval

Diagnostic characteristics: body tapering, not very long; anal fin short with 6-8 rays; head and body very depressed; head only slightly warty; coracoids and coracoid processes not prominent, humeral process extending past coracoids process 18 serrae along inner edge of short (20% of SL) pectoral spine.
Type locality: Konawaruk, Guyana
Distribution: coastal rivers between Orinoco and Amazon River mouths: Brazil, French Guiana, Guyana, Suriname and Venezuela; in Suriname, present in all rivers with possible exception of Marowijne River (where it is replaced by *B. aloikae*)
Size: 12 cm SL
Habitat: strictly freshwater; lives in leaf litter in small forest creeks of the Interior
Position in the water column: bottom (but in flooded forest of the Lower Kabalebo River a school of juveniles was observed swimming near the surface during day time)
Diet:
Reproduction: contrary to other aspredinids, female *B. amaurus* do not carry their eggs in cotylophores; 4,000-5,000 adhesive eggs are spawned on sandy bottom substrate
Remarks: a fourth *Bunocephalus* species, the Amazonian *B. coracoideus* (Cope, 1874) (with a long coracoid process, its length 0.7-1.0 in the distance between the extremities of the two coracoid processes) is reported from the Coppename River (A.R. Cardoso, pers. communication).

Bunocephalus aloikae (live), Paloemeu River (© P. Naskrecki)

Bunocephalus aloikae (live), detail, Paloemeu River (© P. Naskrecki)

Bunocephalus amaurus (alcohol), Nickerie River, RMNH29412 (© Naturalis)

Bunocephalus verrucosus (Walbaum, 1792)
Local name(s): –

Diagnostic characteristics: body tapering, not very long; anal fin short with 6-8 rays; head and body not greatly depressed; depth between the pectorals almost 0.6 times head width; head externally turberculate (warty); coracoids process extending to beyond humeral process
Type locality: no locality stated
Distribution: rivers of Guyana and Amazon River basin: Brazil, Ecuador, Guyana, Peru and Suriname; in Suriname only known from the Corantijn and Suriname rivers (in French Guiana it is known from Mana River; Le Bail *et al.*, 2012)
Size: 9.5 cm SL
Habitat: strictly freshwater
Position in the water column: bottom
Diet:
Reproduction:
Remarks:

Aspredinichthys filamentosus (Valenciennes, 1840)
Local name(s): banyaman, trompetvis

Diagnostic characteristics: body very long and slender; anal fin long with 50-70 rays; maxillary barbels free from head throughout their length; upper surface of snout smooth or with two hooked spines; 7 pairs of mental and postmental barbels
Type locality: Cayenne, French Guiana
Distribution: lower portions of coastal rivers and coastal waters of Venezuela and northern Brazil: Brazil, French Guiana, Guyana, Suriname, Trinidad and Tobago, and Venezuela; in Suriname known from Suriname and Marowijne estuaries
Size: 21.8 cm SL
Habitat: in brackish water of river estuaries, mainly in shallow water on muddy bottom substrates
Position in the water column: bottom
Diet:
Reproduction: the female carries the eggs with her in sacs or cotylophores attached to the ventral side of her body; this may provide oxygen and prevent sedimentation of the eggs in the muddy estuarine waters; in French Guiana the reproductive season is April-June.
Remarks:

Bunocephalus verrucosus (alcohol), ANSP180015 (© M. Sabaj Pérez)

Bunocephalus coracoideus (live) (© R. Covain)

Aspredinichthys filamentosus (live) (© INRA-Le Bail)

Aspredinichthys tibicen (Valenciennes, 1840)
Local name(s): banyaman, trompetvis

Diagnostic characteristics: body very long and slender; anal fin long with 50-70 rays; maxillary barbels free from head throughout their length; upper surface of snout with four broad, hooked spines; 9-10 pairs of mental and postmental barbels
Type locality: Suriname
Distribution: lower portions of coastal rivers and in coastal waters in Venezuela to northern Brazil: Brazil, French Guiana, Guyana, Suriname, Trinidad and Tobago, and Venezuela
Size: 21 cm SL
Habitat: in brackish water of river estuaries, mainly in shallow water on muddy bottom substrates
Position in the water column: bottom
Diet:
Reproduction: the female carries the eggs with her in sacs or cotylophores attached to the ventral side of her body; this may provide oxygen and prevent sedimentation of the eggs in the muddy estuarine waters; in French Guiana the reproductive season is March-June.
Remarks:

Aspredo aspredo (Linnaeus, 1758)
Local name(s): banyaman, trompetvis

Diagnostic characteristics: body very long and slender; anal fin long with 50-70 rays; two pairs of maxillary barbels; no ridges on sides of body except for lateral line; caudal fin with 10 rays; coloration plain; the pectoral fin spines are serrated and very strong
Type locality: South America
Distribution: lower portions of coastal rivers from Venezuela to northern Brazil: Brazil, French Guiana, Suriname, Trinidad and Tobago, and Venezuela.
Size: 38.3 cm SL
Habitat: in brackish water of river estuaries, mainly in shallow water on muddy bottom substrates
Position in the water column: bottom
Diet:
Reproduction: the female carries the eggs with her in sacs or cotylophores attached to the ventral side of her body; this may provide oxygen and prevent sedimentation of the eggs in the muddy estuarine waters; in French Guiana the reproductive season is in the first half of the year.
Remarks: the largest aspredinid species of Suriname.

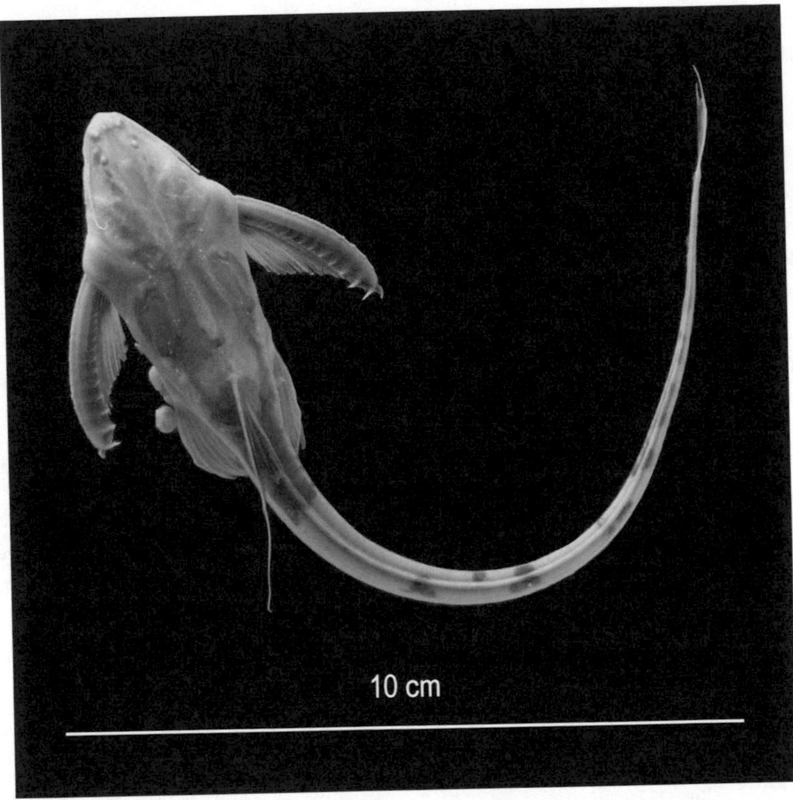

Aspredinichthys tibicen (alcohol), Saramacca/Coppename River Estuary, RMNH29293 (© Naturalis)

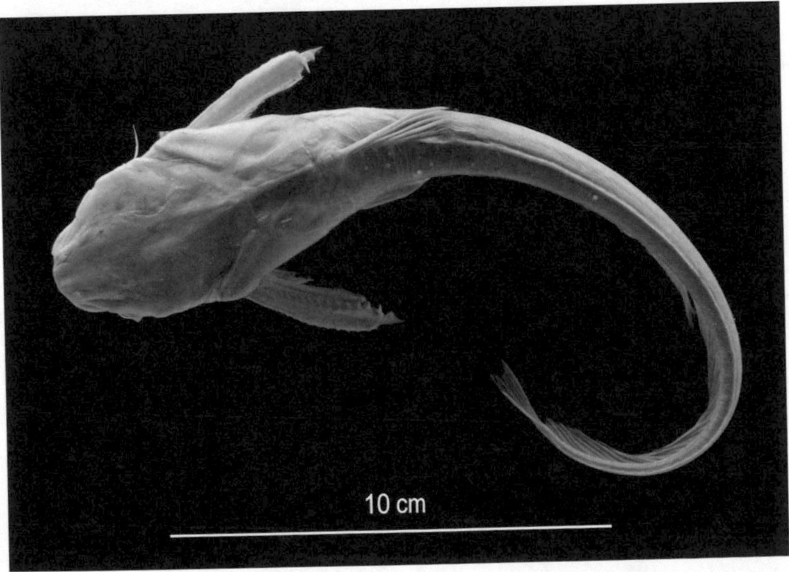

Aspredo aspredo (alcohol), RMNH10726 (© Naturalis)

Platystacus cotylephorus Bloch, 1794
Local name(s): banyaman, trompetvis

Diagnostic characteristics: body very long and slender; anal fin long with 50-70 rays; one pair of maxillary barbels; 3-4 longitudinal ridges along sides of the body; caudal fin with 9 rays; coloration mottled
Type locality: no locality; certainly from South America
Distribution: Venezuela to Brazil, including lower portions of coastal rivers: Brazil, French Guiana, Guyana, Suriname, Trinidad and Tobago, and Venezuela
Size: 31.8 cm SL
Habitat: in brackish water of river estuaries, mainly in shallow water on muddy bottom substrates
Position in the water column: bottom
Diet:
Reproduction: the female carries the eggs with her in sacs or cotylophores attached to the ventral side of her body; this may provide oxygen and prevent sedimentation of the eggs in the muddy estuarine waters; in French Guiana the reproductive season is in the first half of the year
Remarks:

Aspredo aspredo (live) (© INRA-Le Bail)

Platystacus cotylephorus (alcohol), Coppename River Estuary, RMNH29430 (© Naturalis)

FAMILY TRICHOMYCTERIDAE (PENCIL OR PARASITIC CATFISHES)

The most conspicuous external characteristic of the monophyletic family Trichomycteridae is the presence of a patch of odontodes on the interopercle, visible in ventral or ventrolateral aspect of the head. Other characteristics are the presence of a pair of barbels at the angle of the mouth, the lack of pectoral- and dorsal-fin spines and the posterior position of the dorsal fin, the presence of nasal barbels on the anterior nostrils (absent in many species, e.g. in *Ochmacanthus*), and the lack of an adipose fin. The distribution of the Trichomycteridae extends from Panama and Costa Rica in the north into Patagonia in the south. Several vandelliine and stegophiline species are parasitic; the hematophagous habits of vandelliines are widely known, although detailed information is still unavailable. Vandellines feed exclusively on blood (Spotte *et al.*, 2001), which they take by inserting their heads into the branchial chamber of the host (often a large pimelodid catfish) and lacerating a major vessel with their specialized teeth; after feeding they leave the larger fish immediately. The Stegophilinae (including *Ochmacanthus*) specialize in eating scales and mucus, and their feeding habits have been studied (Winemiller & Yan, 1989). With approximately half of the species in the family, the Trichomycterinae (including *Ituglanis* and the large, polyphyletic genus *Trichomycterus*) is the largest and most complex of trichomycterid subfamilies (De Pinna & Wosiacki in Reis *et al.*, 2003). The Trichomycterinae occur from near sea level to >1000 m elevation. The Surinamese species are not well known (probably with several undescribed species inhabiting mountain streams).

1-a Mouth terminal; 1 pair of nasal barbels, 2 pairs of equally-sized maxillary barbels

2

1-b Mouth inferior, very wide; no nasal barbels, 2 pairs of maxillary barbels, one of them miniscule (*Ochmacanthus*)

3

2-a Body covered with numerous well-defined, large rounded spots separated from each other by narrow unpigmented spaces; relatively large species (up to 7.5 cm TL).

Ituglanis amazonicus

2-b Body only obscurely spotted on the upper parts (more or less unpigmented, without large, rounded spots); relatively small species (< 5 cm TL).

Ituglanis sp. (Brownsberg Mountains)

3-a Depth of posterior part of caudal peduncle larger than depth of anterior part of caudal peduncle; interorbital distance equal to or slightly larger than the eye diameter

Ochmacanthus flabelliferus

3-b Depth of caudal peduncle not different between posterior and anterior part; interorbital distance 30% larger than eye diameter

Ochmacanthus reinhardtii

A trichomycterid, provisionally identified as *Trichomycterus* aff. *conradi*, is present in Nassau Mountains (Mol *et al.*, 2007b).

ITUGLANIS AMAZONICUS (Steindachner, 1882)
Local name(s): –

Diagnostic characteristics: body elongated, circular in cross-section; snout rounded, two pairs of maxillary barbels and one pair of nasal barbels; mouth subterminal; interoperculum with numerous small odontodes; dorsal positioned far backward on the body, no adipose fin, pelvic fins close to anal fin, pectoral fins with one spiny ray and 6-7 soft rays; body color with numerous rounded brown spots separated by narrow, unpigmented spaces
Type locality: Cudajas (=Codajas), South America
Distribution: Amazon River basin and rivers of French Guiana: Brazil and French Guiana
Size: 7.5 cm TL
Habitat: small forest creeks with sandy bottom
Position in the water column: bottom
Diet:
Reproduction:
Remarks: *Ituglanis gracilior* (Eigenmann, 1912), a species from Guyana, was identified by P. Willink from the Upper Corantijn River (Willink *et al.*, 2011); the specimens collected by P. Willink have the same (spotted) color pattern and size as *I. amazonicus*.

ITUGLANIS SP. (Brownsberg)
Local name(s): –

Diagnostic characteristics: a small species with the upper parts of the body only obscurely spotted
Type locality: –
Distribution: only known from Koemboe Creek, Brownsberg Mountains (about 400 m-asl)
Size: 4 cm TL
Habitat: a narrow (1 m width), shallow (50 cm) mountain stream with gravelly, rocky bottom, leaf litter and root masses
Position in the water column: bottom
Diet:
Reproduction:
Remarks: yet another undescribed *Ituglanis* species is known to occur at high elevation (>500 m) in Nassau Mountains.

OCHMACANTHUS AFF. *FLABELLIFERUS* Eigenmann, 1912
Local name(s): –

Diagnostic characteristics: width of head about equal to its length, depressed, snout semicircular in outline; mouth very large, inferior; upper jaw with three

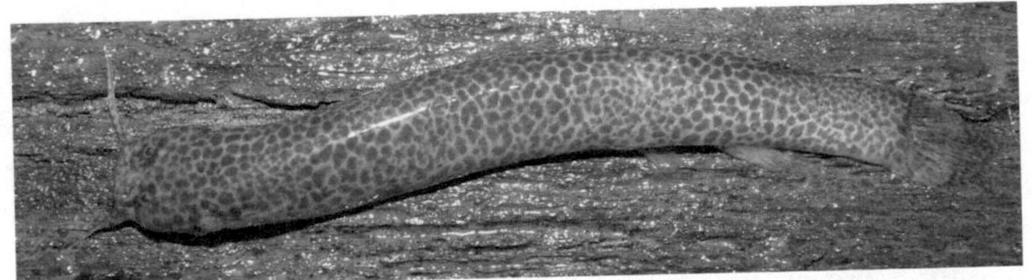

Ituglanis amazonicus (live) (© R. Covain)

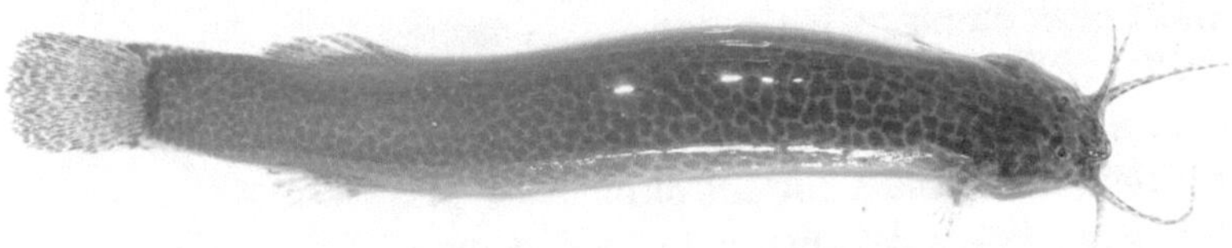

Ituglanis amazonicus (live), Nickerie River (© R. Smith)

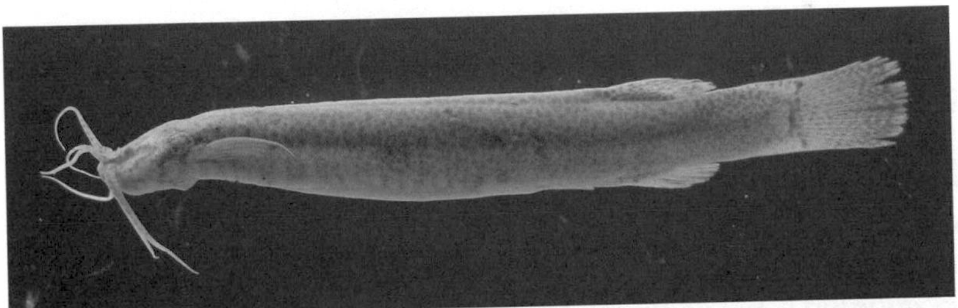

Ituglanis gracilior (alcohol), 43 mm SL, Koetari River (Corantijn River) (© P. Willink)

Ituglanis sp. Brownsberg (live), Koemboe Creek, Brownsberg (© K. Wan Tong You)

series of teeth, teeth in outer two series conical, those of the inner series broad, removed from the others; lower jaw with an outer series of long, curved, claw-like teeth in the lip, and four series in the jaw; interorbital distance equal to or slightly larger than the eye diameter; interopercle with nine claw-like erectile spines; gill-opening small, entirely above the level of the middle of the pectoral; ventrals small, reaching to anal fin; body color light, yellowish with numerous dark chromatophores, more or less aggregated in places, a black spot on the base of the caudal fin; depth of posterior part of caudal peduncle larger than depth of anterior part of caudal peduncle;
Type locality: Konawaruk, Guyana
Distribution: River drainages in Guyana and Venezuela
Size: 5 cm TL
Habitat: forest creeks
Position in the water column: bottom, often in leaf litter
Diet: *Ochmacanthus* species apparently feed on mucus and scales (Winemiller & Yan, 1989)
Reproduction:
Remarks:

Ochmacanthus reinhardtii (Steindachner, 1882)
Local name(s): –

Diagnostic characteristics: mouth large; interopercle with numerous hooks; interorbital distance 30% larger than eye diameter; depth of caudal peduncle not different between posterior and anterior part; body yellowish with small spots
Type locality: Rio Iça, Montalegre; Manacapuru Lake, Amazon River basin, Brazil
Distribution: Amazon River basin and drainages in French Guiana: Brazil and French Guiana; in Suriname, only known from the Marowijne (Maroni) River (Keith *et al.*, 2000)
Size: 4.5 cm TL
Habitat:
Position in the water column: bottom
Diet:
Reproduction:
Remarks:

Ochmacanthus reinhardtii (live) (© INRA-Le Bail)

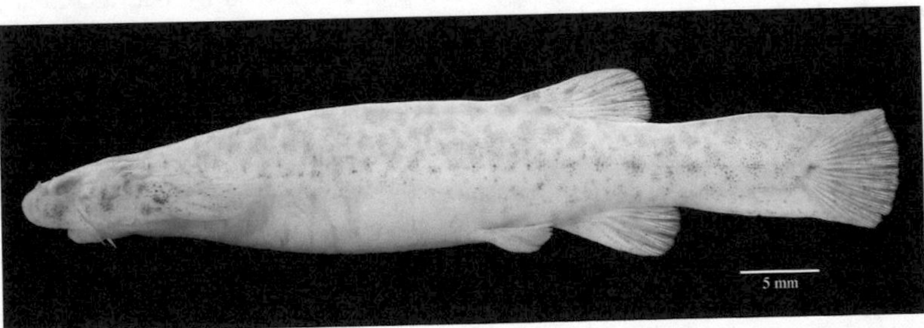

Ochmacanthus sp. (alcohol), 49.8 mm SL, Lawa River (Marowijne River), ANSP187117 (© M. Sabaj Pérez)

FAMILY CALLICHTHYIDAE (ARMORED CATFISHES)

Species of the family Callichthyidae can be readily distinguished from other fishes by their characteristic body armor consisting of a double series of bony plates ('scutes') on each side of the body. Two subfamilies are distinguished: the Callichthyinae, medium-sized, heavy bodied bottom-dwellers that build floating bubble nests, including *Callichthys*, *Megalechis* and *Hoplosternum*; and the Corydoradinae, small, substrate-spawning species, including *Corydoras*, the most species-rich genus of the family. Most species in the Corydoradinae are small and active during the day (exceptional in catfishes) and thus worldwide renowned as aquarium fishes. In Suriname, the Callichthyinae are known as kwikwi and very popular (expensive) food fishes. All three kwikwi species (*C. callichthys*, *H. littorale* and *M. thoracata*) are adapted to low-dissolved oxygen conditions in the standing water of coastal swamps by their air-breathing habit (intestinal uptake of atmospheric oxygen; Carter & Beadle, 1931; Gee & Graham, 1978) and the spawning of eggs in a floating bubble nest (i.e. the eggs embedded in oxygen-rich bubbles, above the oxygen-poor swamp water; Mol, 1993b; Fig. 4.41); in addition, *C. callichthys* is known to travel some distance over land in rainy weather and to survive unfavorable dry-season conditions in damp, terrestrial hiding places such as hollow tree trunks or burrows in the walls of irrigation canals. *Corydoras* species are only known from the Interior of Suriname. In *Corydoras* and *Hoplosternum*, fertilization of eggs involves sperm drinking; the female and male form the "T-position" with the female's mouth over the male's genital opening, and then the female drinks the sperms, which, after passing through the intestines, is released simultaneously with the eggs in a pouch formed by the females pelvic fins (Khoda *et al.*, 1995; Mazzoldi *et al.*, 2007).

1-a 2 pairs of long barbels that reach the pectoral spine; two pairs of bony scutes between the head and the dorsal fin; snout depressed, interorbital width greater than or equal to head depth at anterior margin of orbit (Callichthyinae)

2

1-b 2 or 3 pairs of short rictal barbels (rictus = junction of the lips at either side of the mouth) that rarely reach beyond the gill opening, one pair of short mental barbels at the lower lip; post-occipital process reaches to dorsal fin; snout compressed, interorbital width considerably less than head depth at anterior margin of orbit (Corydoradinae)

4

2-a Eyes small, with diameter equal to the distance between the anterior and posterior nostrils; bony coracoids not visible ventrally; head depressed

Callichthys callichthys

2-b Eye diameter larger than the distance between anterior and posterior nostrils; bony coracoids expanded on abdominal surface, visible ventrally; head compressed laterally

3

3-a Caudal fin forked; adult coloration uniform grey-black

Hoplosternum littorale

3-b Caudal fin emarginate; adult coloration with abdominal spots

Megalechis thoracata

4-a Pectoral spines strongly serrated at inner edge

5

4-b Pectoral spines not strongly serrated at inner edge

9

5-a Snout length (measured from tip of snout to anteriormost bony rim of orbit) 2.2 or more in head length (measured from tip of snout to highest point of gill opening)

Corydoras nanus

5-b Snout length 1.9 or less in head length

6

6-a 2 pairs of rictal barbels (but a pair of triangular skin notches may be present)

7

6-b 3 pairs of rictal barbels

8

7-a Lateral profile of snout slightly concave, with a sharply protruding tip; pectoral spine length (measured when pressed against the body, from articulation point to distal tip) 4.0-4.2 in standard length (measured from tip of snout to junction of posterior edges of last large body scutes); depth caudal peduncle (least depth measured) 2.3-2.5 in head length

Corydoras oxyrhynchus

7-b Lateral profile of snout fairly rectilinear, and a blunt tip of the snout; pectoral spine length 4.3-4.9 in standard length; depth caudal peduncle 2.1-2.3 in head length

Corydoras saramaccensis

8-a First soft dorsal fin ray strikingly elongated, extending far beyond the adipose fin

Corydoras filamentosus

8-b First dorsal fin ray not elongated

Corydoras geoffroy

9-a Black stripe along junctions of body scutes

10

9-b Black stripe absent
13

10-a Caudal fin without vertical bars or dark blotches arranged in vertical lines
Corydoras baderi
10-b Caudal fin with dark vertical bars or dark blotches arranged in vertical lines
11

11-a Caudal fin with 2 or 3 broad black vertical bars
Corydoras boesemani
11-b Caudal fin with 4 to 6 narrow vertical bars, sometimes fragmented into vertical series of blotches
12

12-a Circular dark spots on head and dorsal part of body, spotted region extending to ventral of adipose fin
Corydoras coppenamensis
12-b Irregular spots on head and dorsal part of body, and restricted to region anterior of dorsal fin
Corydoras sipaliwini

13-a A distinct mask across eyes and usually a dark blotch ventral to origin of dorsal fin or at basal part of dorsal fin
14
13-b Mask absent and no blotch under dorsal fin
17

14-a Lateral surfaces of body not densely mottled with dark spots, but uniform tan, with sometimes some scarce irregularly distributed brown pigment spots
Corydoras bicolor
14-b Dark spots all over lateral surfaces of body
15

15-a Dark spots arranged in 3 longitudinal rows
Corydoras surinamensis
15-b Dark spots scattered all over the body
16

16-a Caudal fin with spots arranged in irregular vertical bars
Corydoras brevirostris
16-b Caudal fin lacking the irregular vertical bars
Corydoras melanistius

17-a Large black blotch covering distal half of dorsal fin
Corydoras punctatus
17-b Large black blotch absent from distal part of dorsal fin
18

18-a Dorsal spine length (measured in vertical position from junction of the bases of the predorsal scute and dorsal spine to its distal tip) 3.9 or more in standard length
19
18-b Dorsal spine length 3.9 or less in standard length
21

19-a Pectoral spine length 2.9-3.4 in standard length; dorsal spine length 3.1-3.7 in standard length; bony orbit 2.7-3.4 in head length
Corydoras guianensis
19-b Pectoral spine length 3.5-4.3 in standard length; dorsal spine length 3.9-5.8 in standard length; bony orbit 3.0-4.1 in head length
20

20-a Interorbital width 2.0-2.4 in head length
Corydoras aeneus
20-b Interorbital width 2.8-3.3 in head length
Corydoras heteromorphus

21-a Interorbital width 2.1-2.4 in head length
Corydoras guianensis
21-b Interorbital width 2.4-2.9 in head length
22

22-a Dorsal spine and first soft dorsal ray greyish
Corydoras breei
22-b Dorsal spine and first soft dorsal ray black
Corydoras sanchesi

Callichthys callichthys (Linnaeus, 1758)
Local name(s): plata-ede-kwikwi, platkop kwikwi

Diagnostic characteristics: 2 pairs of long barbels that reach the pectoral spine; eyes small, with diameter equal to the distance between the anterior and posterior nostrils; bony coracoids not visible ventrally; head depressed
Type locality: in Americae rivulis (= South America)
Distribution: cosmopolitan in South America
Size: 25 cm TL
Habitat: in both costal swamp(-forests) and small rainforest creeks of the Interior (Mol, 1994); *C. callichthys* also occurs at high elevations (>500 m) in small mountain streams (e.g. Nassau Mountains; Mol *et al.*, 2007b)
Position in the water column: bottom
Diet: benthic invertebrates and detritus (Mol, 1995)
Reproduction: they reproduce in the rainy (wet) season, in the Coastal Plain of Suriname this is both in the short rainy season (December-January) and long rainy season (May-July); mature males have enlarged pectoral spines in the breeding season; in the morning females spawn about 1,600 eggs in the simple bubble nest consisting of a few leaves on top of a thin layer of foam (produced by the guarding male); a nest may contain up to 8,500 eggs (from successive spawnings) and spawning in the nest may continue for up to 17 days (all this time batches of eggs hatch and new batches are spawned in the nest); during this time the male guards the nest; eggs hatch in 2-3 days (Mol, 1993b, 1996a)
Remarks: *C. callichthys* is known for its semi-amphibious behavior; in the rainy season the platkop kwikwi are sometimes observed 'walking' (wriggling) with their pectoral fin spines over land; in dry-season pools of small forest creeks they may climb into hollow trees along the edge of the pool and wait in this damp, hiding place (above the water surface) for the return of more favorable (wet-season) conditions. In Suriname, *C. callichthys* is protected by law in its main reproductive season, 1 April – 15 July (Rondeel, 1965).

Corydoras aeneus (Gill, 1858)
Local name(s): seseiguse

Diagnostic characteristics: pectoral spines not strongly serrated at inner edge; absence of mask, blotch under dorsal fin, distal spot on dorsal fin, and black stripe along junctions of body scutes; dorsal spine length (measured in vertical position from junction of the bases of the predorsal scute and dorsal spine to its distal tip) 3.9 or more in standard length; pectoral spine length 3.5-4.3 in standard length; dorsal spine length 3.9-5.8 in standard length; bony orbit 3.0-4.1 in head length; interorbital width relatively large (2.0-2.4 in head length); the color of live fish is metallic-green
Type locality: Island of Trinidad

Callichthys callichthys (alcohol), ANSP179110 (© M. Sabaj Pérez)

Corydoras aeneus (live), Suriname River (© R. Covain)

Distribution: Colombia and Trinidad to La Plata River basin east of the Andes: Argentina, Bolivia, Brazil, Colombia, French Guiana, Guyana, Paraguay, Peru, Suriname, Trinidad and Tobago, and Venezuela
Size: 5.6 cm SL (up to 7.5 cm in the aquarium)
Habitat: in small forest creeks of the Interior
Position in the water column: bottom
Diet: omnivorous, feeding mostly on small, aquatic invertebrates during the night
Reproduction: the reproduction in the aquarium is described by Kohda *et al.* (1995) and includes the very unusual 'drinking' of sperm by the female (during the so-called 'T-position'); apparently the sperm pass the intestinal tract of the female before they are discharged together with the eggs into a 'pouch' formed by the female's pelvic fins. Thus the eggs are mixed with fresh, non-dispersed sperm in an enclosed space, ensuring effective insemination
Remarks: a beautiful species and very popular (like most *Corydoras*) in the aquarium hobby; *Corydoras* species are among the most exported South American aquarium fishes; under hypoxic conditions *Corydoras* species can resort to aerial respiration (Kramer & McClure 1980).

Corydoras baderi Geisler, 1969
Local name(s): seseiguse

Diagnostic characteristics: pectoral spines not strongly serrated at inner edge; presence of a black stripe along the junctions of body scutes; dorsal fin without a black spot and caudal fin without vertical bars or dark blotches arranged in vertical lines; the body is relatively deep (2.4-2.7 times in SL) and the snout is relatively long
Type locality: Rio Paru de Oeste und Bache bei der Missionsstation Tirio, Para, Brazil
Distribution: Pará State in Brazil, Marowijne River in Suriname: Brazil and Suriname; in Suriname, only known from the Upper Marowijne River
Size: 4.7 cm SL
Habitat:
Position in the water column: bottom
Diet:
Reproduction:
Remarks: *C. oelemariensis* is a junior synonym (but alternatively, *C. oelemariensis* may be a valid species, R. Covain, pers. communication).

Corydoras bicolor Nijssen & Isbrücker, 1967
Local name(s): seseiguse

Diagnostic characteristics: pectoral spines not strongly serrated at inner edge; mid-lateral, black stripe absent; presence of a distinct mask across the eyes and usually a dark blotch ventral to origin of dorsal fin or at basal part of dorsal fin;

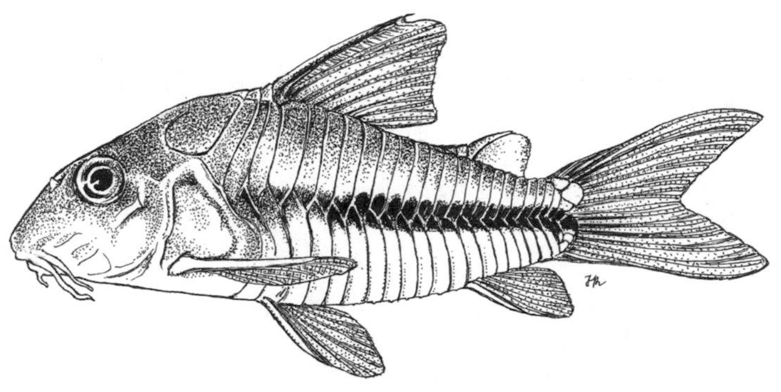

Corydoras baderi (*C. oelemariensis*) (© ZMA)

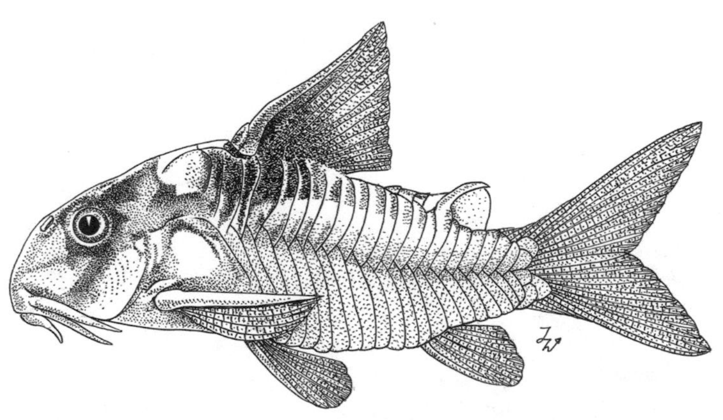

Corydoras bicolor (© ZMA)

lateral surfaces of body not densely mottled with dark spots, but uniform tan, with sometimes some scarce irregularly distributed brown pigment spots
Type locality: Sipaliwini River, Corantijn River System, Paru Savanna, Suriname
Distribution: Upper Corantijn River, Suriname
Size: 3.8 cm SL
Habitat:
Position in the water column: bottom
Diet:
Reproduction:
Remarks: *C. bicolor* is most closely related to *C. melanistius*, but differs from this species by the lack of pigment spots on the body.

CORYDORAS BOESEMANI Nijssen & Isbrücker, 1967
Local name(s): seseiguse

Diagnostic characteristics: pectoral spines not strongly serrated at inner edge; presence of a black stripe along the junctions of body scutes; dorsal fin without a black spot; caudal fin with 2 or 3 broad black vertical bars
Type locality: little tributaries of Gran Rio, between Logolio and Awaradam Falls, Suriname River, Brokopondo, Suriname
Distribution: Suriname and Corantijn rivers, Suriname
Size: 4.2 cm SL
Habitat: both in the main channel of rivers in the Interior (probably in shallow, near-shore water) and in small tributary forest creeks
Position in the water column: bottom
Diet:
Reproduction:
Remarks:

CORYDORAS BREEI Isbrücker & Nijssen, 1992
Local name(s): seseiguse

Diagnostic characteristics: pectoral spines not strongly serrated at inner edge; black stripe and mask absent, no blotch under the dorsal fin; large black blotch absent from distal part of the dorsal fin; dorsal spine length 3.9 or less in standard length; interorbital width 2.4-2.9 in head length; dorsal spine and first soft dorsal ray greyish
Type locality: Avanavero Falls, Kabalebo, River, Corantijn River system, Suriname
Distribution: Corantijn River basin, Suriname
Size: 3.9 cm SL
Habitat:
Position in the water column: bottom
Diet:
Reproduction:

Corydoras boesemani (live), Suriname River (© R. Covain)

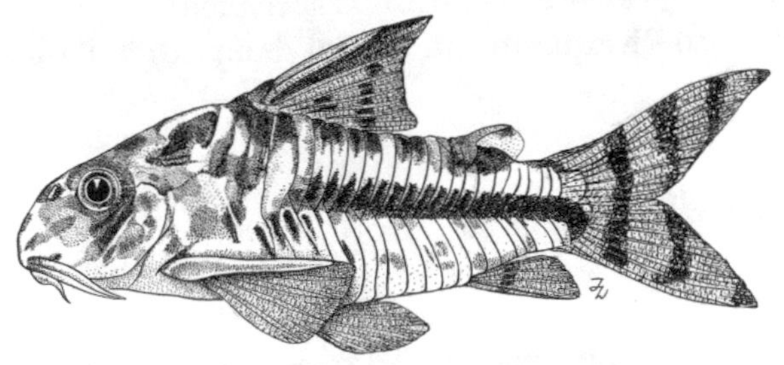

Corydoras boesemani (© ZMA)

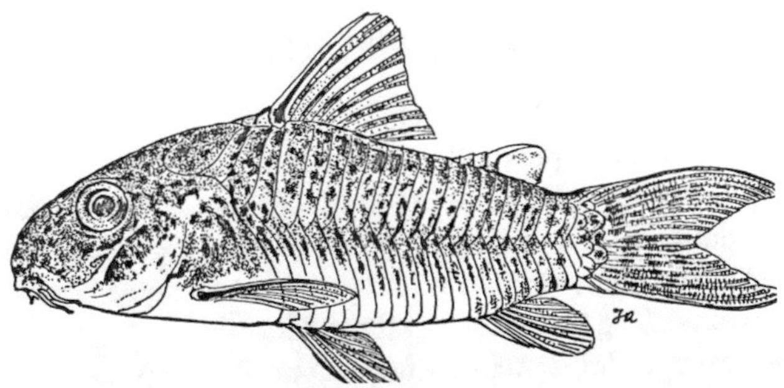

Corydoras breei (© ZMA)

Corydoras aff. *breei* (live), Tapanahony River (Marowijne River) (© R. Covain)

Remarks: *Corydoras osteocarus* is a junior synonym; a *Corydoras* species close to *C. breei* (*Corydoras* aff. *breei*) is present in the Marowijne River (Alexandrou *et al.*, 2011).

Corydoras brevirostris Fraser-Brunner, 1947
Local name(s): seseiguse

Diagnostic characteristics: pectoral spines not strongly serrated at inner edge; mid-lateral, black stripe absent; presence of a distinct mask across the eyes and usually a dark blotch ventral to origin of dorsal fin or at basal part of dorsal fin; dark spots scattered all over the body; four to six irregular vertical bars on the caudal fin
Type locality: Orinoco, Venezuela (aquarium specimen)
Distribution: Orinoco River basin and coastal drainages in Suriname: Suriname and Venezuela
Size: 5 cm SL
Habitat:
Position in the water column: bottom
Diet:
Reproduction:
Remarks: *C. wotroi* is a junior synonym; the closely related species *C. melanistius* differs from *C. brevirostris* in lacking the irregular vertical bars on the caudal fin.

Corydoras coppenamensis Nijssen, 1970
Local name(s): seseiguse

Diagnostic characteristics: pectoral spines not strongly serrated at inner edge; presence of a black stripe along the junctions of body scutes; dorsal fin without a black spot; caudal fin with 4 to 6 narrow vertical bars, sometimes fragmented into vertical series of blotches; circular dark spots on head and dorsal part of body, spotted region extending to ventral of adipose fin
Type locality: Creek at left bank of Coppename River system, 3°52'N, 56°55'W, District Saramacca, Suriname
Distribution: Coppename River basin, Suriname
Size: 3.1 cm SL
Habitat: both in tributary forest creeks and the main channel of the Upper Coppename River upstream of the first rapids
Position in the water column: bottom
Diet:
Reproduction:
Remarks:

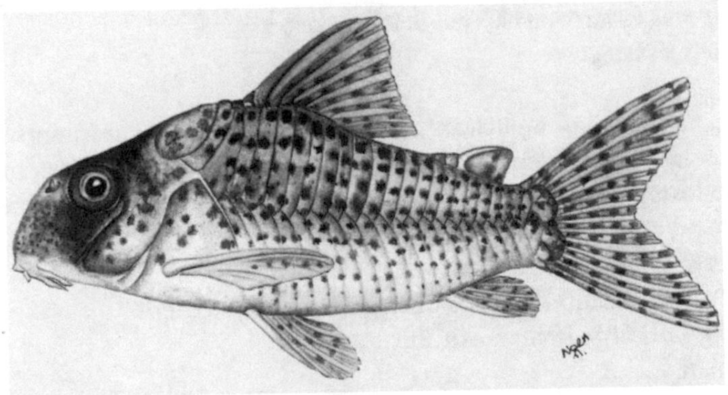

Corydoras brevirostris (© INRA-Le Bail)

Corydoras coppenamensis (alcohol), Coppename River (© M. Littmann)

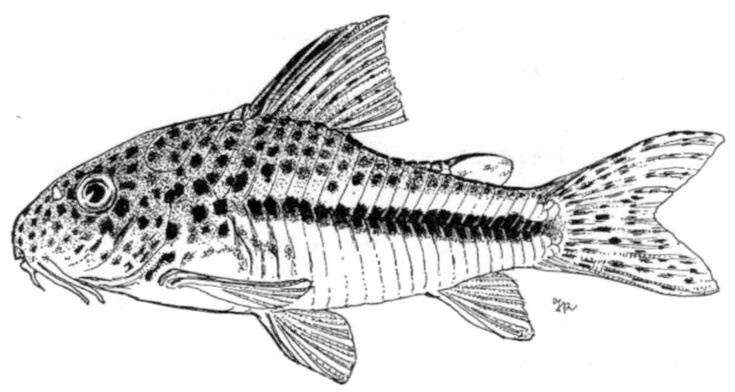

Corydoras coppenamensis (© ZMA)

Corydoras filamentosus Nijssen & Isbrücker, 1983
Local name(s): seseiguse

Diagnostic characteristics: pectoral spines strongly serrated at inner edge; snout length 1.9 or less in head length; 3 pairs of rictal barbels (and 1 pair of mental barbels at the lower lip); first soft dorsal fin ray strikingly elongated, extending far beyond the adipose fin
Type locality: Tributary of Sisa Creek, flowing northwards, Río Corantijn basin, 3°42'N, 57°42'W, Nickerie District, Suriname, depth 0-1 meters
Distribution: Corantijn River basin, Suriname
Size: 3.2 cm SL
Habitat:
Position in the water column: bottom
Diet:
Reproduction:
Remarks:

Corydoras geoffroy Lacépède, 1803
Local name(s): seseiguse

Diagnostic characteristics: pectoral spines strongly serrated at inner edge; snout length 1.9 or less in head length; 3 pairs of rictal barbels (and 1 pair of mental barbels at the lower lip); first dorsal fin ray not elongated; in adults there is a small, but very conspicuous black spot at the base of the 2-4 branched rays of the dorsal fin, and a grey band on the posterior margin of the lateral bony plates
Type locality: Marchall Creek, east of road Paranam-Afobaka, 1.5 kilometers north of Marchall Village, Suriname River System, Suriname
Distribution: coastal rivers of French Guiana and Suriname; in Suriname, known from Suriname and Marowijne rivers
Size: 7 cm SL
Habitat: most specimens were collected in tributary creeks with moderate current and sandy bottom substrate
Position in the water column: bottom
Diet:
Reproduction:
Remarks: *Corydoras octocirrus* is a junior synonym.

Corydoras filamentosus (live), Sipaliwini River (Corantijn River) (© R. Covain)

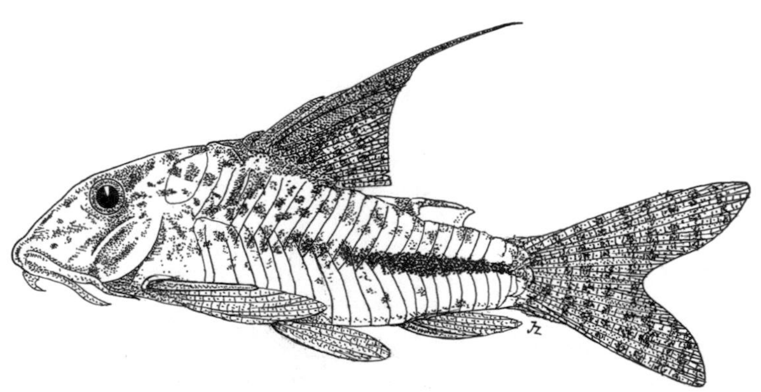

Corydoras filamentosus (© ZMA)

Corydoras geoffroy (live), Suriname River (© R. Covain)

Corydoras guianensis Nijssen, 1970
Local name(s): seseiguse

Diagnostic characteristics: pectoral spines not strongly serrated at inner edge; black stripe and mask absent, no blotch under the dorsal fin and no black spot on the distal part of the dorsal fin; dorsal spine length 3.9 or less in standard length; interorbital width 2.1-2.4 in head length; in adults, the dorsal series of bony plates have the posterior margin colored dark, and large specimens sometimes have a few brown spots on the gill cover
Type locality: creek at right bank of Nickerie River,12 km WSW of Stondansi Falls, Suriname
Distribution: coastal rivers of French Guiana and Suriname; in Suriname, known from Corantijn, Nickerie, Coppename, Saramacca and Marowijne rivers
Size: 4.2 cm SL
Habitat: lives in large schools in forest creeks with shallow water and sandy or muddy bottom
Position in the water column: bottom
Diet:
Reproduction:
Remarks:

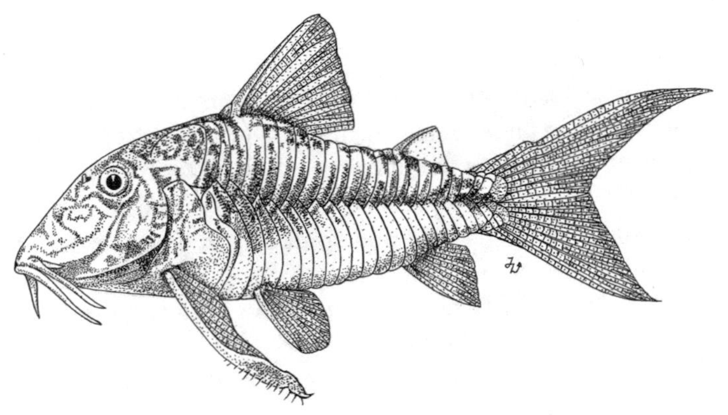

Corydoras geoffroy (© ZMA)

Corydoras guianensis (live) (© R. Covain)

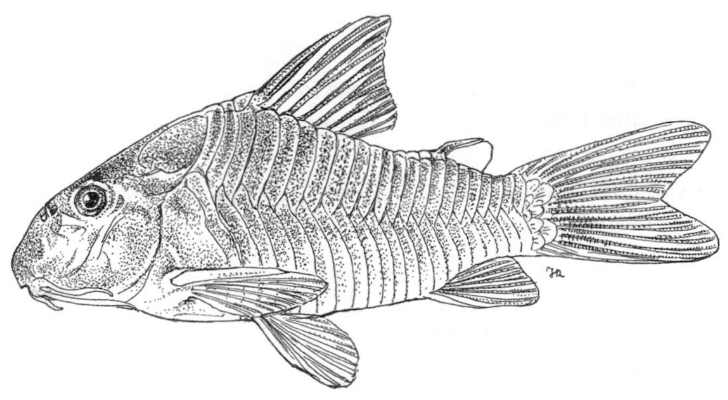

Corydoras guianensis (© ZMA)

Corydoras heteromorphus Nijssen, 1970
Local name(s): seseiguse

Diagnostic characteristics: pectoral spines not strongly serrated at inner edge; absence of mask, blotch under dorsal fin, distal spot on dorsal fin, and black stripe along junctions of body scutes; dorsal spine length (measured in vertical position from junction of the bases of the predorsal scute and dorsal spine to its distal tip) 3.9 or more in standard length; pectoral spine length 3.5-4.3 in standard length; dorsal spine length 3.9-5.8 in standard length; bony orbit 3.0-4.1 in head length; interorbital width relatively small (2.8-3.3 times in head length); the dorsal parts of head and body have very small spots
Type locality: Creek at right bank of Coppename River, Saramacca, 3°52'30"N, 56°53'W, Suriname
Distribution: Coppename and Nickerie River basins, Suriname
Size: 5.3 cm SL
Habitat: in small forest creeks (3-5 m width) of the Interior with shallow water and sandy or muddy bottom
Position in the water column: bottom
Diet:
Reproduction:
Remarks: the name '*heteromorphus*' was given to the species because it is intermediate between the blunt-snouted and the long-snouted groups in the genus *Corydoras*.

Corydoras melanistius Regan, 1912
Local name(s): seseiguse

Diagnostic characteristics: pectoral fin spines not strongly serrated at the inner edge; absence of a longitudinal black stripe along junctions of body scutes; a distinct mask across the eyes and usually a dark blotch ventral to the origin of the dorsal fin or at the basal part of the dorsal fin; dark spots scattered all over the body (not in distinct longitudinal rows); caudal fin lacking irregular vertical bars
Type locality: Essequibo River, Guyana
Distribution: coastal rivers of French Guiana, Guyana and Suriname; in Suriname, known from the Saramacca River
Size: 5.1 cm SL
Habitat:
Position in the water column: bottom
Diet:
Reproduction:
Remarks: *Corydoras wotroi* is a junior synonym.

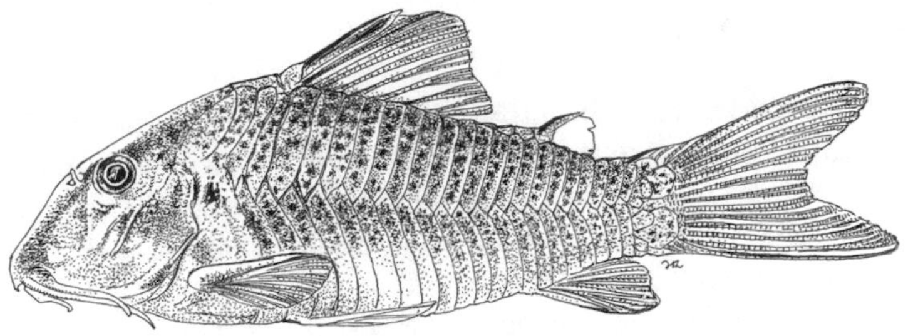

Corydoras heteromorphus (© ZMA)

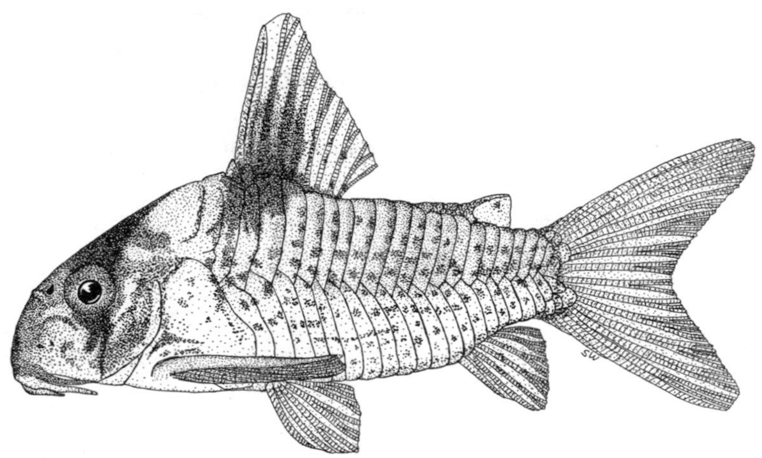

Corydoras melanistius (© ZMA)

Corydoras melanistius (live) (© R. Covain)

CORYDORAS NANUS Nijssen & Isbrücker, 1967
Local name(s): seseiguse

Diagnostic characteristics: a beautiful, small *Corydoras* species with pectoral spines strongly serrated at inner edge; snout length (measured from tip of snout to anteriormost bony rim of orbit) 2.2 or more in head length (measured from tip of snout to highest point of gill opening); two longitudinal, parallel brown stripes on the body, viz. a broad dorsal stripe extending from the nuchal plate to the caudal fin base, and a slender stripe extending across the scute junctions from dorsal part of cleithrum to caudal fin base; the dorsal fin may be reddish-brown or with irregular black spots
Type locality: Little tributaries of Gran-Rio between Ligoria (= Ligolio) and Awaradam Falls, Suriname River, Brokopondo, Suriname
Distribution: Suriname and Marowijne (Maroni) River basins in Suriname and Iracoubo River basin in French Guiana: French Guiana and Suriname
Size: 4.5 cm SL
Habitat: in small forest creeks of the Interior with shallow water, moderate current and sandy or muddy bottom
Position in the water column: bottom (and possibly also swimming in the water column)
Diet:
Reproduction:
Remarks: apparently a very rare species as only few specimens have been collected, both in Suriname and in French Guiana.

Corydoras nanus (live) (© INRA-Le Bail)

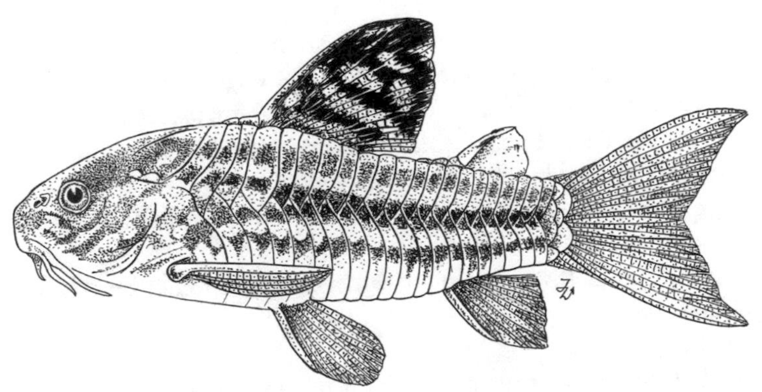

Corydoras nanus (© ZMA)

Corydoras oxyrhynchus Nijssen & Isbrücker, 1967
Local name(s): seseiguse

Diagnostic characteristics: pectoral spines strongly serrated at inner edge; snout length 1.9 or less in head length; 2 pairs of rictal barbels (but a pair of triangular skin notches may be present); lateral profile of snout slightly concave, with a sharply protruding tip; pectoral spine length (measured when pressed against the body, from articulation point to distal tip) 4.0-4.2 in standard length (measured from tip of snout to junction of posterior edges of last large body scutes); depth caudal peduncle (least depth measured) 2.3-2.5 in head length; caudal fin with about ten indistinct irregular vertical bars
Type locality: Gojo Creek, tributary of Saramacca River, 6 kilometers south of Poesoegroenoe, Brokopondo, Suriname
Distribution: Saramacca and Commewijne Rivers
Size: 5 cm SL
Habitat:
Position in the water column: bottom
Diet:
Reproduction:
Remarks:

Corydoras punctatus (Bloch, 1794)
Local name(s): seseiguse

Diagnostic characteristics: a beautiful, blunt-snouted *Corydoras* species with its pectoral spines not strongly serrated at the inner edge; mid-lateral, black stripe absent; mask absent and no blotch under the dorsal fin; presence of a large black blotch covering the distal half of dorsal fin; black markings on adipose fin variable; 6-8 irregular vertical black bars on the caudal fin
Type locality: Suriname (the species occurs for example in Compagnie Creek, Suriname River System)
Distribution: Suriname River basin in Suriname and Iracoubo River basin in French Guiana
Size: 6.6 cm SL
Habitat: in shallow (0.5-2 m water depth) forest creeks with sandy bottom and moderate current
Position in the water column: bottom
Diet:
Reproduction:
Remarks:

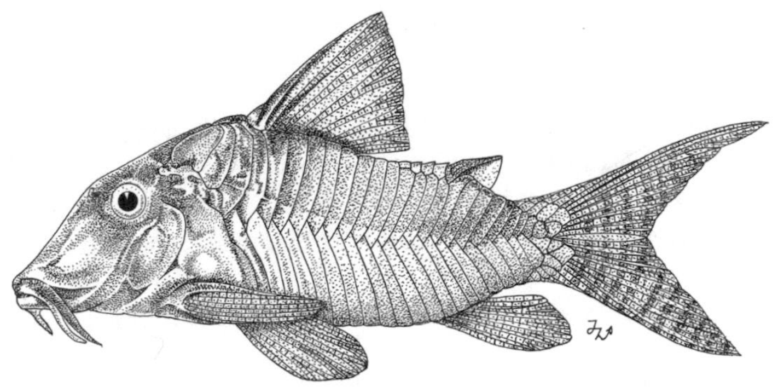

Corydoras oxyrhynchus (© ZMA)

Corydoras oxyrhynchus (live), Upper Commewijne River

Corydoras punctatus (live) (© INRA-Le Bail)

Corydoras sanchesi Nijssen & Isbrücker, 1967
Local name(s): seseiguse

Diagnostic characteristics: pectoral spines not strongly serrated at inner edge; black stripe and mask absent, no blotch under the dorsal fin; large black blotch absent from distal part of the dorsal fin; dorsal spine length 3.9 or less in standard length; interorbital width 2.4-2.9 in head length; dorsal spine and first soft dorsal ray black
Type locality: Gojo Creek above Poesoegroenoe, tributary of the Saramacca River, Brokopondo, Suriname
Distribution: Saramacca River basin, Suriname
Size: 4.1 cm SL
Habitat:
Position in the water column: bottom
Diet:
Reproduction:
Remarks: *C. sanchesi* is close to *C. breei* and may represent an eastern subspecies of *C. breei.*

Corydoras saramaccensis Nijssen, 1970
Local name(s): seseiguse

Diagnostic characteristics: pectoral spines strongly serrated at inner edge; snout length 1.9 or less in head length; 2 pairs of rictal barbels (but a pair of triangular skin notches may be present); a long-snouted species with the lateral profile of snout fairly rectilinear, and a blunt tip of the snout; pectoral spine length 4.3-4.9 in standard length; depth caudal peduncle 2.1-2.3 in head length; bony orbit length 3.5-3.9 in head length; dark grey blotch ventral to dorsal fin, extending from 2nd to 5th dorsolateral body scutes, and pigment extending onto dorsal fin spine and rays; interorbital area with darker pigmentation, faintly suggesting the presence of a mask
Type locality: Creek at right bank of Kleine Saramacca River, 11 kilometers east-southeast from junction with Saramacca River, Brokopondo, Suriname
Distribution: Saramacca River basin, Suriname
Size: 5.1 cm SL
Habitat: small forest creeks with sandy bottom substrate and moderate current (or even stagnant water)
Position in the water column: bottom
Diet:
Reproduction:
Remarks: close to *C. oxyrhynchus*; this latter species differs from *C. saramaccensis* in having a bonier head and a concave profile of the snout.

Corydoras punctatus (© ZMA) black/white photograph

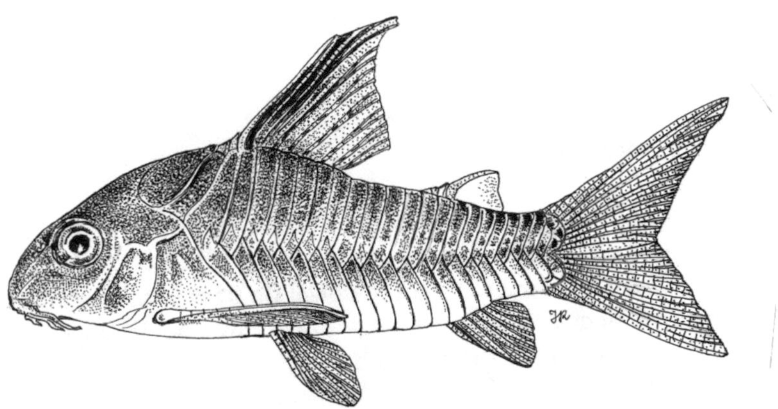

Corydoras sanchesi (© ZMA)

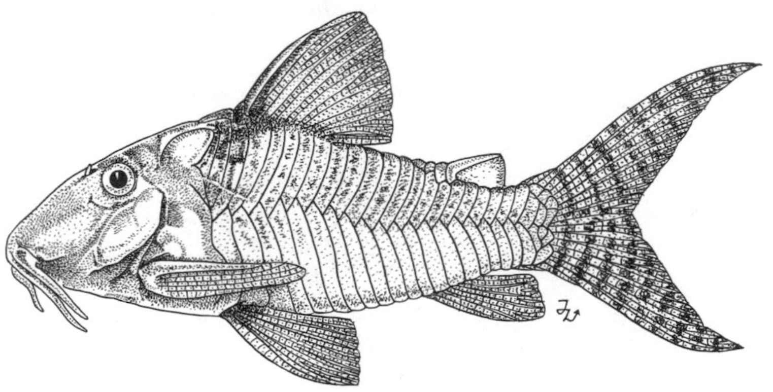

Corydoras saramaccensis (© ZMA)

Corydoras sipaliwini Hoedeman, 1965
Local name(s): seseiguse

Diagnostic characteristics: pectoral spines not strongly serrated at inner edge; presence of a black stripe along the junctions of body scutes; dorsal fin without a black spot; caudal fin with 4 to 6 narrow vertical bars, sometimes fragmented into vertical series of blotches; irregular spots on head and dorsal part of body, and restricted to region anterior of dorsal fin
Type locality: Sipaliwini River, Upper Corantijn River, at Paru Savanna, Suriname
Distribution: coastal river basins of Guyana and Suriname
Size: 4.6 cm SL
Habitat: in small, tributary forest creeks (4-5 m width, 0.3-1.5 m deep) of the Upper Corantijn River on sandy or rocky substrate
Position in the water column: bottom
Diet:
Reproduction:
Remarks:

Corydoras surinamensis Nijssen, 1970
Local name(s): seseiguse

Diagnostic characteristics: pectoral spines not strongly serrated at inner edge; mid-lateral, black stripe absent; presence of a distinct mask across the eyes and usually a dark blotch ventral to origin of dorsal fin or at basal part of dorsal fin; dark spots on the body arranged in 3 longitudinal rows; caudal fin with 4-5 vertical black bars, the first across the base of the fin
Type locality: unnamed creek at right bank of Coppename River, 3°52'30"N, 56°53'00"W, Suriname
Distribution: Coppename River system, Suriname
Size: 5.1 cm SL
Habitat: forest creeks with muddy-sand bottom substrate and moderate current
Position in the water column: bottom
Diet:
Reproduction:
Remarks:

Corydoras sipaliwini (live), Sipaliwini River (Corantijn River) (© R. Covain)

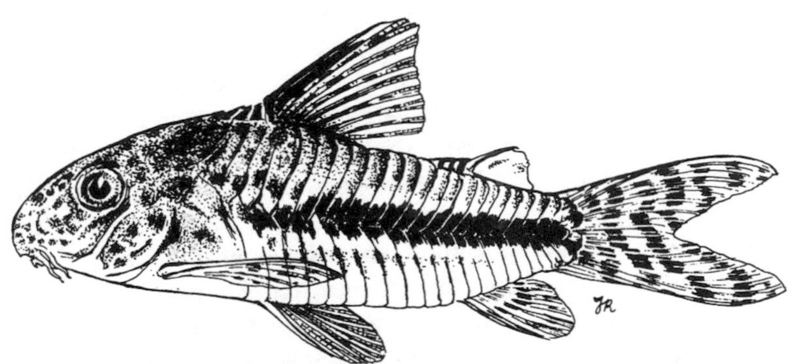

Corydoras sipaliwini (© ZMA)

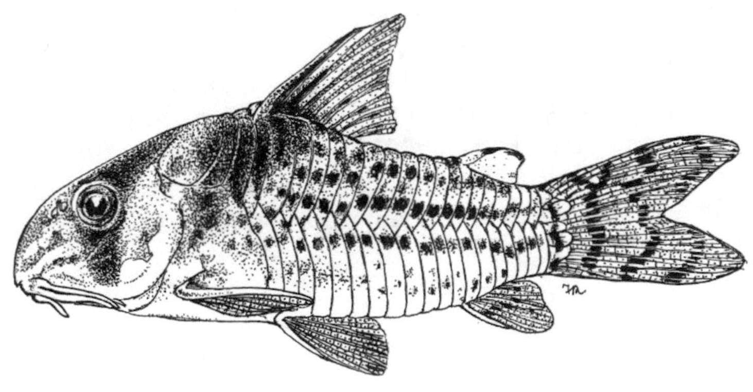

Corydoras surinamensis (© ZMA)

Hoplosternum littorale (Hancock, 1828)
Local name(s): soké-kwikwi, hé(i)-ède-kwikwi

Diagnostic characteristics: 2 pairs of long barbels that reach the pectoral spine; eye diameter larger than the distance between anterior and posterior nostrils; bony coracoids visible ventrally; head compressed laterally; caudal fin forked; adult coloration uniform grey-black (grey in muddy water of rice fields, black in 'clear' brown swamp water)
Type locality: Demerara (Guyana)
Distribution: Amazon, Orinoco and Paraguay River basins, and coastal rivers of the Guianas and northern Brazil: Bolivia, Brazil, Colombia, Ecuador, French Guiana, Guyana, Peru, Suriname, Trinidad and Tobago, and Venezuela
Size: 22 cm SL
Habitat: coastal swamps and rice fields; absent from the Interior (Mol, 1994)
Position in the water column: bottom
Diet: benthic invertebrates (and detritus) (Mol, 1995)
Reproduction: they reproduce in the rainy (wet) season, in the Coastal Plain of Suriname this is both in the short rainy season (December-January) and long rainy season (May-July); in the reproductive season, the male develops an enlarged pectoral spine with an upturned, ski-like curvature in its distal end (Winemiller, 1987); the soké or hé-ède kwikwi is well known for its spectacular, large bubble nests (up to 10 cm above the water surface and 42 cm in diameter; Fig. 4.43) which are easily spotted in shallow swamps and rice fields; the nests are build and aggressively guarded by the male; several females spawn their eggs simultaneously in the nest around noon (the nest can be observed shaking); after approximately three days, depending on nest temperature, the ~20,000 eggs (maximum 52,000 eggs/nest) hatch (Mol, 1993b, 1996a; Pascal *et al.*, 1994). After two days the newborn larvae start feeding on zooplankton (Mol, 1995). The reproductive behavior of male and female *H. littorale* was studied in the aquarium by Gautier *et al.* (1988)
Remarks: *H. littorale* is not only one of the best-studied Neotropical fish species, but, in Suriname, it is also an extremely popular and relatively expensive food fish; it is protected by law in its main reproductive season 1 April – 15 July (Rondeel, 1965) and, in the dry season, it is sold (alive) at the markets of Paramaribo and Nieuw Nickerie and along the road Paramaribo-Nickerie. Deep-frozen *H. littorale* kwikwi are imported from Brazil and Venezuela (where this species is mainly eaten by poor people); it is also exported from Brazil and Venezuela to the Netherlands, where it is eaten by Surinamese people living in that country.

Hoplosternum littorale (live), swamps of the Young Coastal Plain, Commewijne District

Hoplosternum littorale (live) (© INRA-Le Bail)

Megalechis thoracata (Valenciennes, 1840)
Local name(s): kat(a)rina-kwikwi

Diagnostic characteristics: 2 pairs of long barbels that reach the pectoral spine; eye diameter larger than the distance between anterior and posterior nostrils; bony coracoids visible ventrally; head compressed laterally; caudal fin emarginated; adult coloration with abdominal spots (juveniles with vertical, black and white bands)
Type locality: Mana River, French Guiana
Distribution: Amazon, Orinoco and Paraguay River basins, and coastal rivers of the Guianas and northern Brazil: Bolivia, Brazil, Colombia, Ecuador, French Guiana, Guyana, Peru, Suriname, Trinidad and Tobago, and Venezuela
Size: 16 cm SL
Habitat: in both costal swamp(-forests) and small rainforest creeks of the Interior (Mol, 1994)
Position in the water column: bottom
Diet: benthic invertebrates (and detritus) (Mol, 1995)
Reproduction: they reproduce in the rainy (wet) season, in the Coastal Plain of Suriname this is both in the short rainy season (December-January) and long rainy season (May-July); males can be distinguished from females by their enlarged, reddish pectoral spines and by the coracoids plates meeting ventrally (versus not meeting ventrally in females); eggs are spawned in simple bubble nests (much like the nests of *C. callichthys*; Fig. 4.28) that are build and guarded by the male; nests can receive multiple, successive spawning (eggs in different stages of developments can be observed in the same nest) and at any point in time they can contain up to 7,000 eggs; the eggs hatch in about three days (Mol, 1993b, 1996a). The impact of aquatic predators on eggs, larvae and juveniles of *M. thoracata* was studied in Surinamese swamps (Mol, 1996b)
Remarks: *Megalechis personata* is a synonym of *M. thoracata* (Reis *et al.*, 2005). A closely related species, *Megalechis picta* (Müller and Troschel, 1848), distinguished from *M. thoracata* by a vertical black band on the caudal fin, is present in the Amazon and the Upper Essequibo River in Guyana. The intestine of *M. thoracata* has both a respiratory and hydrostatic function (Gee & Graham, 1978). Kramer and Graham (1976) found group-induced synchronous air-breathing in *M. thoracata*, potentially an anti-predator behavior analogous to schooling. In Suriname, *M. thoracata* is protected by law in its main reproductive season 1 April – 15 July (Rondeel, 1965).

Megalechis thoracata (alcohol), scale bar = 1 cm, ANSP179795 (© M. Sabaj Pérez)

FAMILY LORICARIIDAE (ARMORED CATFISHES) – KEY TO THE SUBFAMILIES

The armored catfish family Loricariidae is the largest family of catfishes in the Neotropics and, indeed, in the world. At present, more than 700 species are recognized and more are described every year. The body armor consists of 3-5 rows of bony plates ('scutes'); the belly may be exposed. The mouth is ventral with or without noticeable barbels, but with a papillose ventral lip. The intestines are relatively long, reflecting a vegetarian diet (mainly periphyton). Six subfamilies are recognized in Reis *et al.* (2003), four of which occur in Suriname: Hypoptopomatinae, Loricariinae, Hypostominae and Ancistrinae. Schaefer (1986) and Montoya-Burgos *et al.* (1997) showed that the Ancistrinae are not monophyletic and thus the ancistrine catfishes are now included in the subfamily Hypostominae (although most publications still keep a tribal rank Ancistrini this is now meaningless as it is not diagnosed by any unique characters; S. Fisch-Muller, pers. communication).

1-a Caudal peduncle compressed laterally
2

1-b Caudal peduncle depressed dorso-ventrally
Loricariinae

2-a Caudal fin with 14 branched rays; interopercle with (Ancistrini) or without evertible odontodes; abdomen at least partly without bony plates
Hypostominae

2-b Caudal fin with less than 14 branched rays; abdomen entirely covered with bony plates
Hypoptopomatinae

This page is intentionally left blank.

FAMILY LORICARIIDAE – SUBFAMILY HYPOPTOPOMATINAE

Relatively small (mostly 2-3.5 cm SL) loricariids that occur in small to moderate-sized streams and rivers. They are distinguished from all other loricariids by the morphology of the pectoral fin skeleton, wherein the bones bear laminar extensions on their ventral surface that largely or entirely cover the fossae for the arrector muscles. The ventral surface of the pectoral fin skeleton is covered by thin skin and usually bears numerous odontodes, such that the bone appears to be exposed on the ventral surface. Body shape is quite variable among genera, from short and deep in *Otocinclus* to elongate with a depressed head in *Hypoptopoma*. The first unbranched ray of the dorsal, anal, pectoral, and pelvic fins are thickened and robust, bearing enlarged odontodes. An adipose fin may be present or absent. The dorsal fin has 7 branched rays (and sometimes a small spinelet), the pectoral fin has 6 rays, the pelvic fin has 4 rays, the anal fin 4-5 rays, and the caudal fin 12 or 14 branched rays. Hypoptopomatinae are mostly herbivorous and diurnal in habits; they are usually found at or near the water surface, typically in close association with marginal vegetation or subsurface structure (e.g. rocks, woody debris). The fishes can be observed clinging to the vegetation by clasping stems and leaves between the pelvic fins. Males of *Otocinclus* have a patch of odontodes near the caudal fin base which may function in adhesion during spawning and courtship (Aquino, 1994). The abundant odontodes covering the ventral surface may offer increased frictional surfaces relevant to ventral adhesion and attachment in strong currents (Alexander, 1965). Males are typically smaller than females and possess several sexual dimorphisms (Isbrücker & Nijssen, 1992).

1-a Dorsal fin without locking mechanism; 1st dorsal spinelet absent; head and snout strongly depressed; orbital rim reaching both the dorsal and the ventral profile of the head; Nickerie River

Hypoptopoma guianense

1-b Dorsal fin with a spine-locking mechanism; 1st dorsal spinelet present; head and snout not distinctly depressed; eyes not visible from ventral view

2

2-a Adipose present

Parotocinclus britskii

2-b Adipose absent

3

3-a Eye in lateral position; preopercle not visibly externally

Otocinclus mariae

3-b Eye in dorsolateral position; preopercle visibly externally; Marowijne River

Gen. nov. aff. *Parotocinclus* n. sp.

This page is intentionally left blank.

Hypoptopoma guianense Boeseman, 1974
Local name(s): –

Diagnostic characteristics: dorsal fin without locking mechanism; 1st dorsal spinelet absent; head and snout strongly depressed; orbital rim reaching both the dorsal and the ventral profile of the head
Type locality: left tributary of Nickerie River, a few km upstream Stondansi Falls, Suriname
Distribution: Essequibo (Guyana) and Nickerie (Suriname) rivers; in Suriname only known from the Nickerie River
Size: 6.2 cm SL
Habitat: apparently present in both forest creeks and the main river channel
Position in the water column: bottom
Diet:
Reproduction: male urogenital papilla well developed, pointed, joined at base to anterior flaplike anus. Males with patch of tightly arranged small odontodes oriented as a swirl, variably covering first to fourth plates of ventral series, lateral to urogenital papilla
Remarks:

Otocinclus mariae Fowler, 1940
Local name(s): –

Diagnostic characteristics: small species; dorsal fin with a spine-locking mechanism; 1st dorsal spinelet present; head and snout not distinctly depressed; eyes in lateral position, not visible from ventral view; adipose fin absent; preopercle not visibly externally; presence of a conspicuous, broad longitudinal black band from the tip of the snout, through the large eye, up to the base of the caudal fin
Type locality: Boca Chapare, Rio Chimore [tributary to Ichilo River] [Cochabauba, Rio Marmore River system] Bolivia
Distribution: Upper Madeira and Lower Amazon River basins, Bolivia; in Suriname known from Upper Marowijne (Maroni), Suriname and Corantijn rivers
Size: 3.3 cm SL
Habitat: well-oxygenated, clear-water creeks of the Interior, often associated with submersed aquatic vegetation
Position in the water column: bottom and other substrate (woody debris, submersed aquatic vegetation)
Diet: feeds on periphyton ('aufwuchs')
Reproduction: females 10-20% larger than males; males with urogenital papillae and a contact organ consisting of a patch of modified odontodes near the caudal fin base; females deposit about 50 eggs on submersed leaves
Remarks: an Amazonian species that is also known from French Guiana (Maroni; Le Bail *et al.*, 2000).

Hypoptopoma guianense (live), dorsal view (© P. Willink)

Hypoptopoma guianense (live), lateral view (© P. Willink)

Otocinclus mariae (live), Sipaliwini River (Corantijn River) (© P. Willink)

Parotocinclus britskii Boeseman, 1974
Local name(s): –

Diagnostic characteristics: a small species; dorsal fin with a spine-locking mechanism; 1st dorsal spinelet present; head and snout not distinctly depressed; eyes not visible from ventral view; adipose fin present; body colored light yellowish-brown with three or four dark vertical bands
Type locality: left tributary of Coppename River, Suriname, 3°51'N, 56°55'W
Distribution: Orinoco River basin and Atlantic coastal drainages of the Guianas: Guyana, Suriname and Venezuela; in Suriname known from the Coppename and Nickerie rivers
Size: 2.5 cm SL
Habitat: small forest creeks (4-5 m width, 30-100 cm depth) with a sandy or muddy bottom, shallow upper reaches and sandy river beaches of the main channel
Position in the water column: bottom
Diet:
Reproduction:
Remarks: in the low-water season *P. britskii* was collected in large numbers (school?) in a shallow beach of the Upper Coppename River and also observed in large numbers (dispersed) on the bottom and large woody debris in the Upper Nickerie River.

GEN. NOV. AFF. *Parotocinclus* n. sp. (see Le Bail *et al.*, 2000; p. 262-263)
Local name(s): –

Diagnostic characteristics: a very small species; dorsal fin with a spine-locking mechanism; 1st dorsal spinelet present; head and snout not distinctly depressed; eyes not visible from ventral view; adipose fin absent; eye in dorsolateral position; preopercle visibly externally; eyes in dorso-lateral position; ventral part of the body white, presence of a thin longitudinal dark band from the tip of the snout to the base of the caudal fin and 4-5 transversal brown bands from the back to the mid-lateral dark band
Type locality:
Distribution: Upper Marowijne River
Size: 2 cm SL
Habitat: small, shallow forest creeks with sandy or rocky bottom and moderate to strong current
Position in the water column: bottom
Diet:
Reproduction:
Remarks: may live in schools (Le Bail *et al.*, 2000).

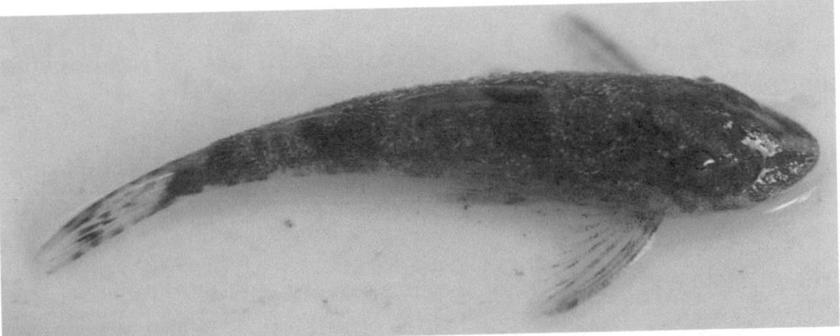

Parotocinclus britskii (live), Nickerie River (© R. Smith)

Gen. nov. aff. *Parotocinclus* sp. (live) (© INRA-Le Bail)

FAMILY LORICARIIDAE – SUBFAMILY LORICARIINAE

The subfamily Loricariinae can readily be distinguished from other loricariids by having a depressed caudal peduncle with no adipose fin and, often, a depressed snout. Most species are small to medium-sized fishes (the smallest are *Harttiella* species that grow to about 5 cm SL). Loricariines exhibit a wide array of sexually dimorphic features including the size and placement of odontodes on the head, body, and pectoral fin and the form of the lower lip. Male *Loricariichthys* with enlarged lower lips carry fertilized eggs attached to their lip until the larvae hatch.

1-a Presence of rostrum; dorsal fin with five branched rays, the last one split to its base; dorsal fin about opposite to anal fin (*Farlowella*)

2

1-b Absence of rostrum; dorsal fin with 6 branched rays, the last one split to its base; dorsal fin about opposite to pelvic fins

3

2-a Presence of 7-8 predorsal bony plates; 2 rows of abdominal scutes, except for a single median scute at base of pelvic fins; preorbital ridge and eye itself slightly elevated over head

Farlowella rugosa

2-b Presence of 8-9 predorsal bony plates; 3 rows of abdominal scutes; preorbital ridges and eyes not elevated

Farlowella reticulata

3-a Caudal fin with 12 branched rays (Harttiini)

4

3-b Caudal fin with 10 branched rays (Loricariini)

10

4-a Each jaw segment with 27 or less teeth; upper and lower lips, together with the rectal barbels form a horseshoe-shape, with three buccal papillae on the surface of the lip, lateral papillae trilobate (*Metaloricaria*)

5

4-b Each jaw segment with 40 or many more teeth; upper and lower lip outline and barbels other as described in couplet 4a

6

5-a Dorsal and lateral sides of body and head with numerous vague brown spots or without spots; often up to 5 broad transverse brown or blackish bars posterior to base of last dorsal fin ray; juveniles however with color pattern strongly reminiscent of that of juvenile *paucidens*, namely about 3 narrow transverse stripes on dorsal and lateral sides of caudal peduncle

Metaloricaria nijsseni

5-b Dorsal and lateral sides of body and head with numerous conspicuous dark drown spots; no broad transverse bars posterior to base of last dorsal fin ray; juveniles with about 3 narrow transverse stripes on dorsal and lateral sides of caudal peduncle; Marowijne River

Metaloricaria paucidens

6-a Caudal peduncle weakly depressed, more or less circular in cross-section; scutes on body and tail without evident lateral keels; scutes covered with odontodes; small species, standard length up to 5 cm; Nassau Mountains

Harttiella crassicauda

6-b Caudal peduncle strongly depressed; scutes on body and tail with distinct lateral keels; scutes at most slightly roughened; standard length at least 17 cm

7

7-a Abdomen completely covered with small scutes; caudal fin deeply forked, with a large median black crescent; head and body relatively slender

Cteniloricaria platystoma

7-b Abdomen naked or cover restricted to lateral abdominal plates and numerous minute scutelets; caudal fin not deeply forked, often with black basicaudal blotch; head and body relatively broad (*Harttia*)

8

8-a Abdominal cover constituted of small granular platelets; Suriname River

Harttia surinamensis

8-b Abdominal cover restricted to preanal and lateral plates; a row of platelets may join these two series

9

9-a Presence of a row of platelets joining preanal to lateral abdominal plates; Coppename River

Harttia fluminensis

9-b Absence of a row of platelets joining preanal to lateral abdominal plates; Marowijne River

Harttia guianensis

10-a Abdomen completely covered with relatively large scutes; lower lip bilobate with median furrow, surface of this lip more or less smooth or weakly papillose

Loricariichthys maculatus

10-b Abdomen naked or covered with small scutelets in different patterns; lower lip strongly papillose or filamentous

11

11-a Margin of lower lip with long filamentous barbels (*Loricaria*)

12

11-b Margin of lower lip smooth, or fringed with papillae or short inconspicuous filamentous barblets

13

12-a Standard length 10.4-12.0 cm; thoracic scutes elongated and extending further on the sides of the abdomen, abdominal region narrower; in adult males, pectoral spines thickened near the distal tip, which is acute; dorsal side of head and body with irregular brownish spots

Loricaria nickeriensis

12-b Standard length up to 29 cm; thoracic scutes less elongated onto the abdomen, and abdominal region relatively wider; in adult males, pectoral spines club-shaped; dorsal side of head and body yellowish tan without irregular brownish spots

Loricaria cataphracta

13-a A very distinct, circular, dark brown blotch just in front of the base of the dorsal spine; interorbital width about 5.3 times in head length; Corantijn River

Rineloricaria fallax

13-b Predorsal region without a single, highly distinct dark brown blotch; interorbital width 2.8-4.5 times in head length

Rineloricaria stewarti, R. aff. stewarti and Rineloricaria sp.

Cteniloricaria platystoma (Günther, 1868)
Local name(s): –

Diagnostic characteristics: caudal peduncle strongly depressed; scutes on body and tail with distinct lateral keels; scutes at most slightly roughened; abdomen completely covered with small scutes; caudal fin deeply forked, with a large median black crescent; head and body relatively slender
Type locality: Suriname
Distribution: widespread in Atlantic coastal drainages from Essequibo River (Guyana) to Sinnamary River (French Guiana); Guyana, French Guiana and Suriname; in Suriname probably present in all rivers, upstream of the first rapids
Size: 18 cm SL
Habitat: main channel of rivers on rocky or sandy substrates in fast flowing water; sometimes observed in forest creeks
Position in the water column: bottom
Diet:
Reproduction: no sexual dimorphism is known in this species
Remarks: *C. maculata* (Boeseman, 1971) is a junior synonym.

Farlowella reticulata Boeseman, 1971
Local name(s): –

Diagnostic characteristics: presence of rostrum; dorsal fin with five branched rays, the last one split to its base; dorsal fin about opposite to anal fin; presence of 8-9 predorsal bony plates; 3 rows of abdominal scutes; preorbital ridges and eyes not elevated

Ctenilaricaria platystoma (alcohol), scale bar = 1 cm, 149.5 mm SL, Lawa River (Marowijne River), ANSP187330 (© M. Sabaj Pérez)

Ctenilaricaria platystoma (live), Raleighvallen, Coppename River (© W. Kolvoort)

Farlowella reticulata (live), Marowijne River (© R. Covain)

Type locality: Maka Creek, left tributary of the Lawa River, 10 km south of Stoelmanseiland, Marowijne River system, Suriname
Distribution: Essequibo, Marowijne and Oyapock River basins: French Guiana, Guyana and Suriname
Size: 15 cm SL
Habitat: small (1-5 m width), shallow (<1 m) forest creeks with moderate to strong current and rocky or sandy bottom
Position in the water column: bottom and between woody debris and aquatic macrophytes
Diet: periphyton
Reproduction: approximately 50 eggs are spawned in a crevice
Remarks: *Farlowella* species show very little activity (especially during the day) and, while remaining more or less motionless among woody debris and submersed aquatic vegetation, are well camouflaged; this may partly explain their apparent rarity.

Farlowella rugosa Boeseman, 1971
Local name(s): –

Diagnostic characteristics: presence of rostrum; dorsal fin with five branched rays, the last one split to its base; dorsal fin about opposite to anal fin; presence of 7-8 predorsal bony plates; 2 rows of abdominal scutes, except for a single median scute at base of pelvic fins; preorbital ridge and eye itself slightly elevated over head
Type locality: Kamaloea (or Saloea) Creek, right tributary of Marowijne River, 9 kilometers southeast of the outlet of Gran Creek, French Guiana
Distribution: Essequibo, Marowijne and Corantijn River basins: French Guiana, Guyana and Suriname; in Suriname, only known from the Marowijne, Nickerie and Corantijn rivers
Size: 21.4 cm SL
Habitat: forest creeks of the Interior
Position in the water column:
Diet:
Reproduction: in the reproductive season, males develop short odontodes on the bony plates of the snout
Remarks: apparently a very rare species; *Farlowella parvicarinata* (type locality Nickerie River) is a junior synonym.

Harttia fluminensis Covain & Fisch-Müller, 2012
Local name(s): –

Diagnostic characteristics: abdomen naked or cover restricted to lateral abdominal plates and numerous minute scutelets; caudal fin not deeply forked, often with black basicaudal blotch; head and body relatively broad (*Harttia*); presence

Farlowella reticulata (live), detail, Marowijne River (© R. Covain)

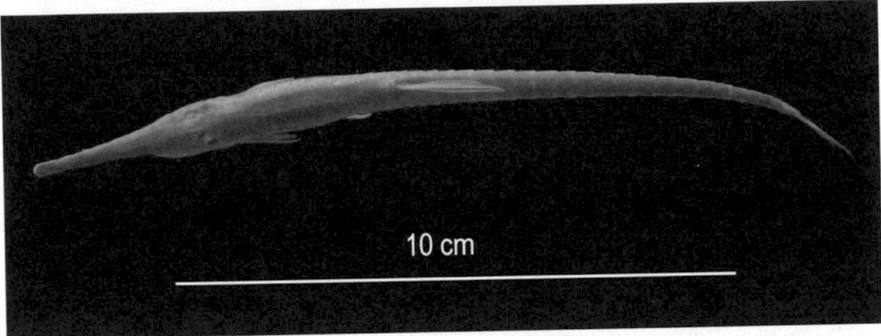

Farlowella rugosa (alcohol), Nickerie River, RMNH27517 (© Naturalis)

Farlowella rugosa (© INRA-Le Bail)

of a row of platelets joining preanal to lateral abdominal plates; background color of dorsal surface of head and body dark brown, with 5-6 indistinct postdorsal dark bands and dark marbling; ventral surface lighter, yellowish; Coppename River
Type locality: Coppename River at Raleighvallen rapids
Distribution: Coppename River
Size: 15 cm SL
Habitat: main channel of the river on rocky bottom substrate and large boulders in fast flowing water
Position in the water column: bottom
Diet:
Reproduction: males have a larger head and thickened pectoral spines bearing hypertrophied odontodes
Remarks:

Harttia guianensis Rapp Py-Daniel & Olivieira, 2001
Local name(s): –

Diagnostic characteristics: absence of a row of platelets joining preanal to lateral abdominal plates; background color of dorsal surface yellowish tan to beige, dark, almost black marbling covers the dorsal surface, presence of 5 black postdorsal bands; caudal fin deeply forked, with large, black basal blotch and medial part bright yellow; a back blotch present on the tip of the dorsal fin; Marowijne River
Type locality: Saut Athanase, Approuague River, French Guiana
Distribution: coastal drainages of French Guiana and Suriname, from Approuague River to Marowijne (Maroni) River; Brazil, French Guiana and Suriname; in Suriname, only known from Marowijne River
Size: 16.7 cm SL
Habitat: main channel of rivers of the Interior on rocky and sandy substrate in fast flowing water
Position in the water column: bottom
Diet:
Reproduction: in the breeding season males develop hypertrophied odontodes on the upper surface of the thickened pectoral spines, the snout margin, and on the keels of the lateral plates
Remarks: locally very abundant; often occurs syntopically with *Cteniloricaria platystoma*.

Harttia surinamensis Boeseman, 1971
Local name(s): –

Diagnostic characteristics: abdominal cover constituted of small granular platelets (complete in large specimens only, i.e. >15 cm SL); head large, with large elliptic mouth and papillose lips and numerous teeth (about 80 on each jaw); Suriname River
Type locality: Grandam, Gran Rio, Upper Suriname River, Suriname

Harttia guianensis (live), 133 mm SL, Lawa River (Marowijne River),ANSP189126 (© M. Sabaj Pérez)

Harttia surinamensis (live), 133.5 mm SL, ANSP187328 (© M. Sabaj Pérez)

Harttia surinamensis (live), Suriname River (© R. Covain)

Distribution: restricted to the Suriname River
Size: 19 cm SL
Habitat: main channel on rocky and sandy bottom substrate, in fast flowing water
Position in the water column: bottom
Diet:
Reproduction: males develop hypertrophied odontodes on the upper surface of the thickened pectoral spines
Remarks: locally very abundant; often occurs syntopically with *Cteniloricaria platystoma.*

Harttiella crassicauda (Boeseman, 1953)
Local name(s): –

Diagnostic characteristics: caudal peduncle weakly depressed, more or less circular in cross-section; scutes on body and tail without evident lateral keels; scutes covered with odontodes; small species, standard length up to 5 cm
Type locality: unnamed creek (Paramaka Creek) in Nassau Mountains, Suriname
Distribution: restricted to Paramaka Creek, Nassau Mountains
Size: 4.7 cm SL
Habitat: small, shallow (<50 cm water depth) mountain stream (250-500 m altitude) with moderate to strong current, clear, well-oxygenated water and rocky to sandy bottom substrate (Mol *et al.*, 2007b)
Position in the water column: bottom
Diet: periphyton (diatoms and filamentous red algae) and detritus
Reproduction: males develop hypertrophied odontodes on the entire body, and particularly on the S-shaped pectoral spines and around the snout; females have an extremely low fecundity of only 3-7 eggs/female (Mol *et al.*, 2007b)
Remarks: the extremely restricted *distribution* in a single creek, coupled with small population size and low fecundity, make *H. crassicauda* highly vulnerable. Urgent protection measures should be taken to protect this species and its immediate environment that is endangered by a proposed bauxite mine in Nassau Mountains.

Loricaria cataphracta Linnaeus, 1758
Local name(s): basyafisi, basyakakaku

Diagnostic characteristics: large species (standard length up to 29 cm) with the margin of lower lip with long filamentous barbels (*Loricaria*); thoracic scutes less elongated onto the abdomen, and abdominal region relatively wider; presence of a double keel on the supra-occipital and the first two dorsal plates; in adult males, pectoral spines club-shaped; dorsal side of head and body yellowish tan without irregular brownish spots; caudal fin dark, at least the distal part of the lower lobe
Type locality: mouth of the Marowijne River, near Galibi, 5°45'N, 54°00'W, Suriname

Harttiella crassicauda (live), Paramaka Creek, Nassau Mountains (© T. Larsen)

Loricaria cataphracta (live), Suriname River (© R. Covain)

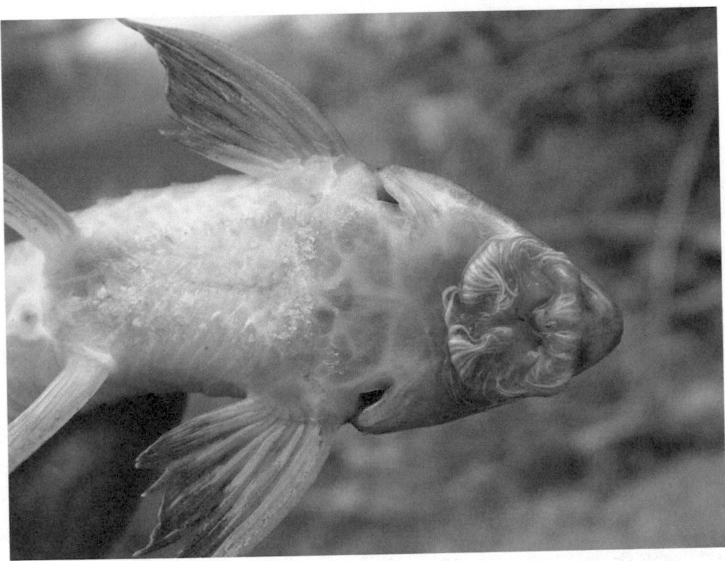

Loricaria cataphracta (live), detail head (© R. Covain)

Distribution: Amazon River basin and coastal rivers in the Guianas: Brazil, French Guiana, Guyana and Suriname
Size: 29.5 cm SL
Habitat: mainly in estuaries, canals in the Young Coastal Plain, and lower reaches of rivers (and tributaries), with muddy bottom and turbid, slow-flowing freshwater
Position in the water column: bottom
Diet:
Reproduction: males develop a club-shaped, thickened distal end of their pectoral spines at about 18 cm; males are abdomino-lip brooders (carry fertilized eggs attached to the enlarged lower lip until the larvae hatch)
Remarks:

LORICARIA NICKERIENSIS Isbrücker, 1979
Local name(s): –

Diagnostic characteristics: margin of lower lip with long filamentous barbels (*Loricaria*); standard length 10.4-12.0 cm; thoracic scutes elongated and extending further on the sides of the abdomen, abdominal region narrower; in adult males, pectoral spines thickened near the distal tip, which is acute; dorsal side of head and body with irregular brownish spots
Type locality: Fallawatra River, a tributary of Nickerie River, 5 kilometers south-southwest of Stondansi falls, Nickerie District, Suriname
Distribution: Corantijn, Nickerie and Marowijne rivers
Size: 12.5 cm SL
Habitat: rivers and small streams of the Interior with sandy bottom and clear water
Position in the water column: bottom
Diet:
Reproduction: males develop a club-shaped, thickened distal end of their pectoral spines at about 10 cm; males are abdomino-lip brooders
Remarks:

LORICARIICHTHYS MACULATUS (Bloch, 1794)
Local name(s): basyafisi, basyakakaku

Diagnostic characteristics: abdomen completely covered with relatively large scutes; lower lip bilobate with median furrow, surface of this lip more or less smooth or weakly papillose; ground color brownish grey dorsally, yellowish ventrally, with three faint transverse black bands on the dorsal side and a few dark spots irregularly distributed on the body; spines and rays with a regular series of dark grey spots, except in the anal fin; a large faint brownish grey blotch in the terminal third of the lower caudal fin lobe
Type locality: South America (probably Suriname)

Loricaria nickeriensis (live), Corantijn River (© R. Covain)

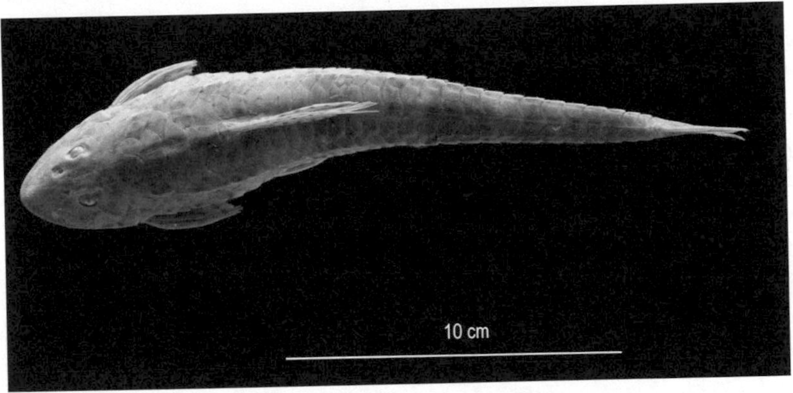

Loricariichthys maculatus (alcohol), Suriname River, RMNH17253 (© Naturalis)

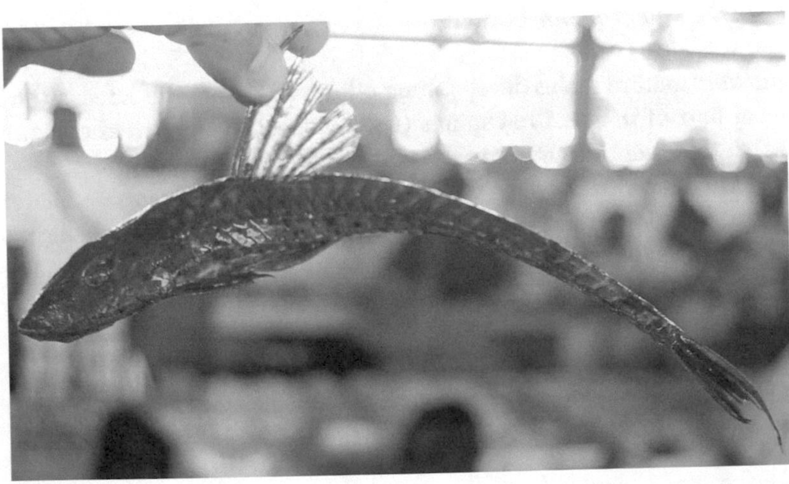

Loricariichthys maculatus (live), swamps of the Young Coastal Plain, Saramacca District

Distribution: rivers of Suriname; known from coastal freshwater swamps and canals from Paramaribo to Nickerie
Size: 21.5 cm SL
Habitat: coastal swamps and canals
Position in the water column: bottom
Diet:
Reproduction: sexual dimorphism in the lips; male *Loricariichthys* with enlarged lower lips carry fertilized eggs attached to their lip until the larvae hatch
Remarks: can be locally abundant; sold at the markets of Paramaribo.

METALORICARIA NIJSSENI (Boeseman, 1976)
Local name(s): –

Diagnostic characteristics: each jaw segment with 27 or less teeth; upper and lower lips, together with the rectal barbels form a horseshoe-shape, with three large buccal papillae on the surface of the lip, the lateral ones trilobite; dorsal and lateral sides of body and head with numerous vague brown spots or without spots; often up to 5 broad transverse brown or blackish bars posterior to base of last dorsal fin ray; juveniles however with color pattern strongly reminiscent of that of juvenile *paucidens*, namely about 3 narrow transverse stripes on dorsal and lateral sides of caudal peduncle
Type locality: Sipaliwini River, Upper Corantijn River system, Suriname
Distribution: Suriname, Saramacca, Nickerie and Corantijn River basins, Suriname
Size: 29.5 cm SL
Habitat: main channel, especially near rapids in sunlit water with moderate to strong current
Position in the water column: bottom
Diet:
Reproduction: mature males develop long odontodes along the sides of the head, the dorsal part of the pectoral spines (these are much thickened compared to females) and the ventral surface of the pelvic spines
Remarks: *Metaloricaria* is a curious genus because of its geographic isolation and unique combination of morphological characteristics, such as the length of the maxillary barbels (longer than in all other Harttiini), low number of small teeth, reduction of the number of caudal-fin rays (i–11–i), and sexual dimorphism reminiscent of that seen in *Rineloricaria*.

METALORICARIA PAUCIDENS Isbrücker, 1975
Local name(s): –

Diagnostic characteristics: each jaw segment with 27 or less teeth; upper and lower lips, together with the rectal barbels form a horseshoe-shape, with three large buccal papillae on the surface of the lip, lateral ones trilobite; dorsal and lateral sides of body and head with numerous conspicuous dark drown spots; no broad

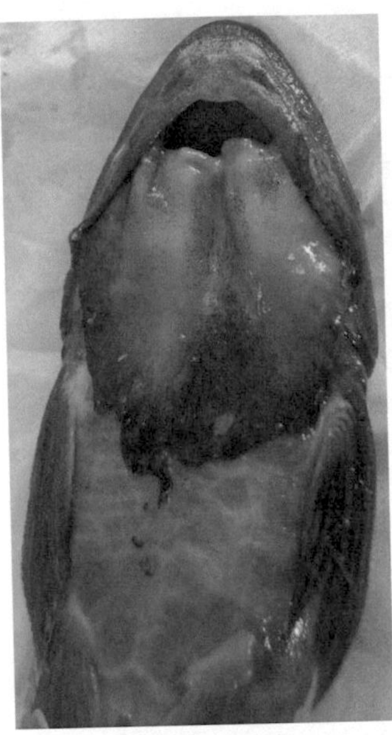

Loricariichthys maculatus (live), male, ventral view of head showing enlarged lower lips, swamps of the Young Coastal Plain, Saramacca District

Metaloricaria nijsseni (live) Nickerie River (© R. Smith)

Metaloricaria nijsseni (live), detail of head, Nickerie River (© R. Smith)

transverse bars posterior to base of last dorsal fin ray; juveniles with about 3 narrow transverse stripes on dorsal and lateral sides of caudal peduncle
Type locality: creek at right bank of Ouaqui River, upstream of Sant Bali, Marowijne (Maroni) River System, French Guiana
Distribution: Oyapock, Sinnamary and Marowijne River basins: French Guiana and Suriname; in Suriname only known from Marowijne River
Size: 27 cm SL
Habitat: main channel and large tributaries in the Interior (upstream of the first rapids), especially near rapids
Position in the water column: bottom
Diet:
Reproduction:
Remarks:

Rineloricaria fallax (Steindachner, 1915)
Local name(s): –

Diagnostic characteristics: margin of lower lip smooth, or fringed with papillae or short; a very distinct, circular, dark brown blotch just in front of the base of the dorsal spine; interorbital width about 5.3 times in head length; inconspicuous filamentous barblets
Type locality: Iguarape de Carauna (=Sa. Grande) near Boa Vista, Rio Branco drainage, Rupununi River
Distribution: Upper Rupununi and Branco River basins: Brazil and Guyana; in Suriname known from Corantijn River
Size: 15.7 cm SL
Habitat: forest creeks with moderate current and sandy bottom
Position in the water column: bottom
Diet:
Reproduction: sexual dimorphism includes hypertrophied development of the odontodes along the sides of the head, on the pectoral spines and rays, and predorsal area of mature males. *Rineloricaria* are cavity brooders (R. Covain, pers. comm.): numerous eggs (often more than 100) are laid attached to one another in single layer masses on the cavity floor, and are brooded by males
Remarks:

Rineloricaria stewarti (Eigenmann, 1909)
Local name(s): –

Diagnostic characteristics:
Type locality: Chipoo Creek, Ireng River, Guyana
Distribution: coastal rivers of the Guianas and Upper Branco River basin: French Guiana, Guyana and Suriname
Size: 10 cm SL

Metaloricaria paucidens (live), Lawa River (Marowijne River), ANSP187325 (© M. Sabaj Pérez)

Rineloricaria fallax (live) (© W. Kolvoort)

Rineloricaria aff. *stewarti* (live) (© R. Covain)

Habitat: shallow (<60 cm) forest creeks of the Interior with moderate current, sandy bottom, and clear water
Position in the water column: bottom
Diet:
Reproduction: see *R. fallax*
Remarks: two undescribed *Rineloricaria* species occur in Suriname, one species is known from the Corantijn en Nickerie rivers and the second species is known from Marowijne River (R. Covain, pers. communication)

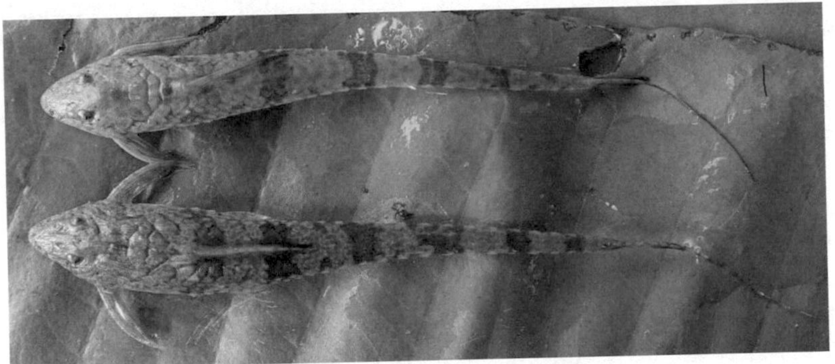

Rineloricaria fallax (live, above) and *R. stewarti* (live, below) (© R. Covain)

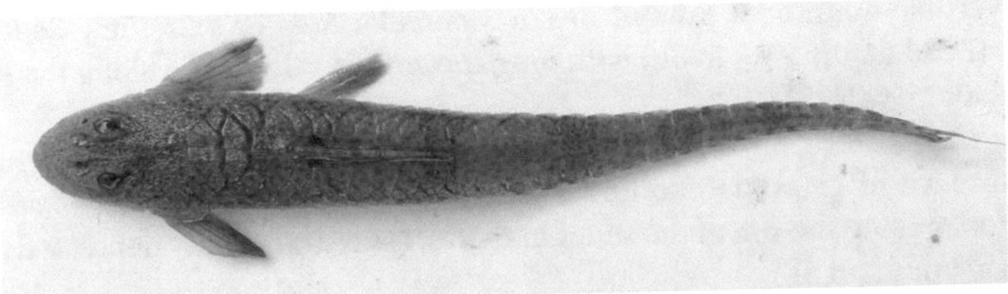

Rineloricaria sp. (live), Sipaliwini River (Corantijn River) (© R. Covain)

FAMILY LORICARIIDAE – SUBFAMILY HYPOSTOMINAE

Armored catfishes in the large (>100 species) genus *Hypostomus* are robust, medium-sized loricariid catfishes lacking evertible, bristle-like odontodes in the interopercular area, having the pectoral girdle not exposed and having a compressed or cylindrical caudal peduncle. Except for *Hypostomus watwata*, which lives in brackish water of estuaries, they seem restricted to freshwaters, mostly living on the bottom of sandy or rocky rivers. During the day they stay under rocks or woody debris (Fig. 4.8b) and most of them become active after sunset. Sexual dimorphism is poorly developed. *H. macrophthalmus* and *H. pseudohemiurus*, with only juveniles as type specimens, are considered doubtful species (Weber *et al.*, 2012). Ancistrine armored catfishes (formerly placed in the subfamily Ancistrinae) are characterized by cheek odontodes that can be everted, conferring them a defensive as well as offensive advantage. Most species are small to medium sized (except *Pseudacanthicus*). Secondary sexual dimorphism often includes more developed odontodes on pectoral fins or on body scutes (in males) as in many other Loricariidae, but also spectacularly developed odontodes along the snout in *Pseudancistrus* and fleshy branched tentacles on the snout in male *Ancistrus*. They are active during the night while staying in their woody or rock refuges during the day.

1-a Evertible odontodes present in interopercular area or extremely depressed fishes (head depth > 5.5 in SL) with long inevertible odontodes along the rim of the snout in males

14 ('Ancistrini')

1-b Absence of evertible odontodes in the opercular area or long inevertible odontodes along the rim of the snout in males; fishes not grealty depressed (head depth < 5 times in SL)

2 (*Hypostomus*)

2-a Tooth number reduced; teeth with short, solid, thick stem, the crown spoon-shaped; teeth not curved; Corantijn and Nickerie rivers

Hypostomus taphorni

2-b Teeth numerous, filiform, with bifid crown that curves abruptly toward Interior of buccal cavity

3

3-a Depth caudal peduncle (least depth of caudal peduncle measured) 1.35-1.7 in interdorsal length (measured from dorsum under posterior lower rim of dorsal fin to distal tip of scale at the anterior basis of adipose fin)

4

3-b Depth of caudal peduncle 1.8-2.7 in interdorsal length

10

4-a Dorsal and caudal fins completely or almost completely dusky, or finely marked

5

4-b Dorsal and caudal fins mostly spotted or with bands

6

5-a Dark spots on body and fins very small and usually elongate, present on posterior part of caudal peduncle, several on each scute; interorbital width 2.6-3.3 times in head length (measured from tip of snout to posterior tip of occipital process); diameter orbit 1.6-3.85 in snout length (measured from tip of snout to anteriormost bony rim of orbit); Suriname River

Hypostomus micromaculatus

5-b Dark spots on body relatively large, ovate, lacking on the posterior part of caudal peduncle, where at most traces of a few longitudinal stripes seem present; interorbital width 3.1-3.35 times in head length; diameter orbit 2.45-2.9 in snout length; Corantijn River

Hypostomus pseudohemiurus

6-a Venter completely covered with rather large and distinct dark spots

Hypostomus plecostomus

6-b Venter with at most a few indistinct spots

7

7-a When dorsal fin adpressed, dorsal fin rays hardly or almost reaching the adipose fin spine; mandibular ramus 1.8-2.1 in interorbital width

8

7-b When dorsal fin adpressed, some dorsal fin rays overlapping the adipose fin; mandibular ramus 2.5-3.3 in interorbital width

9

8-a Dorsal spine length (measured in vertical position from junction of the bases of the predorsal scute and dorsal spine to its distal tip) slightly surpassing predorsal length (distance measured from tip of snout to posterior rim of last predorsal scute before articulation of dorsal spine); dorsal spine length 2.25-2.5 in standard length; head depth (measured vertically from distal tip of occipital process dorsally to coracoidal band ventrally) 5.21-5.6 in standard length; head depth 1.75-1.85 in head length; spots dense on complete dorsal fin; dark spots on body less small, ovate with a diameter about equaling interspaces; dark spots on pectoral spines small, numerous and dense; Corantijn River

Hypostomus crassicauda

8-b Dorsal spine length shorter than predorsal length; dorsal spine length 2.45-3.2 in standard length; head depth 5.4-6.3 in standard length; head depth 1.85-2.25 in head length; spots on distal dorsal fin usually vague or lacking; dark spots on

body small and circular with a diameter distinctly less than interspaces; dark spots on pectoral spines large, round and widely interspaced; Saramacca River
Hypostomus saramaccensis

9-a Depth head 5.0-5.7 in standard length; spots on body and fins intense, of moderate size, very widely interspaced, even smaller spots on snout not dense; spots on caudal fin irregularly arranged; Suriname River
Hypostomus paucimaculatus

9-b Depth head 4.45-5.1 in standard length; spots on body and fins vague, large, on snout small and densely distributed; dark spots on caudal fin forming transverse bands
Hypostomus plecostomus

10-a Depth caudal peduncle about 2.7 in interdorsal length
Squaliforma tenuis

10-b Depth caudal peduncle 1.8-2.35 in interdorsal length
11

11-a Occipital with well-marked median ridge (crest on the head); post-occipital scute multiple; venter completely covered with moderate to large spots and small bony plates (rough when touched); cleithral width considerable, 3.05-3.45 in standard length; mandibular ramus 2.6-3.5 in interorbital width; relatively large species from estuaries with brackish-water
Hypostomus watwata

11-b Absence of occipital median ridge; post-occipital scute single (except sometimes in juveniles of <5 cm standard length); venter without spots and without bony plates (smooth); cleithral width moderate, 3.25-3.28 in standard length; mandibular ramus 2.0-3.0 in interorbital width; freshwater species
12

12-a Dark spots moderate, variably intense and usually less well-defined; in adults, mandibular teeth usually 20-40 on each side; length of dorsal spine 2.55-3.65 in standard length
Hypostomus gymnorhynchus

12-b Dark spots rather small, very intense and numerous, well-defined; in adults, mandibular teeth usually about 30-60 on each side; length dorsal spine 2.4-3.1, rarely more, in standard length
13

13-a Depth caudal peduncle 2.4-2.5 in interdorsal length; ground color rather light, beige-brownish, and completely covered with relatively few small to moderate dark round spots, the interspaces between them much larger than spot diameters; length of dorsal spine 2.85-2.95 in standard length; Coppename River
Hypostomus coppenamensis

13-b Depth caudal peduncle 2.0-2.3 in interdorsal length; ground color usually darker, and completely covered with relatively numerous intensely dark round or oblong spots, the interspaces between them about as large as spot diameters;

length of dorsal spine 2.4-3.1, rarely more, in standard length; Corantijn and Nickerie rivers
Hypostomus corantijni

14-a Head without inevertible odontodes; head not greatly depressed (head depth < 5.5 times in standard length); absence in adults, a row of odontodes along the rim of the snout, strikingly elongated in males
15

14-b Head and/or body with very prominent, inevertible odontodes; head depressed, head depth at least 5.5 times in standard length; in adults, a row of odontodes along the rim of the snout, strikingly elongated in males (*Pseudancistrus*)
31

15-a Mouth small, width included more than 2 times in width of sucking disk; sucking disk with a posterior suture; small species, standard length up to 7 cm (*Lithoxus*)
16

15-b Mouth large, width included less than 2 times in sucking disk width; sucking disk without a posterior suture; standard length variable
21

16-a Adipose fin absent
17

16-b Adipose fin present
18

17-a Dorsal and lateral sides with light, circular spots; mandibular ramus length 2.8-3.0 times in interorbital width; interorbital width 4.0-4.8 times in head length; interopercle with 18-23 enlarged odontodes
Lithoxus pallidimaculatus

17-b Dorsum dusky, with or without suggestions of darker cross-bands; mandibular ramus length 2.3-2.4 times in interorbital width; interorbital width 3.0-4.1 times in head length; interopercle with 8-12 enlarged odontodes
Lithoxus surinamensis

18-a 3 branched anal fin rays (last ray split to its base); adipose with a posterior membranous extension; lateral body scutes 23 or, rarely, 24; body depth 8.65-10.5 times in standard length; Nickerie River
Lithoxus sp.

18-b 4 branched anal fin rays (last ray split to its base); adipose fin without posterior extension; lateral scutes usually 25; body depth 5.85-9.9 times in standard length
19

19-a Caudal fin bright with brownish horizontal bands or small blotches; Marowijne River
Lithoxus planquettei

19-b Caudal fin dark with a white distal margin
20

20-a Acute teeth; absence of numerous well-developed odontodes on anterior half of pectoral fin spine; length pectoral fin spine 4.0-4.6 in SL; orbit diameter 6.0-6.7 times in head length; caudal peduncle depth 11.2-12.0 in SL; Nickerie and Corantijn rivers
Lithoxus group *bovallii*
20-b Teeth with truncate tips (bilobed); presence of numerous well-developed odontodes on anterior half of pectoral fin spine; length pectoral fin spine 2.8-3.6 in SL; orbit diameter 5.7-6.1 times in head length; caudal peduncle depth 9.0-11.2 in SL; Marowijne River
Lithoxus stocki

21-a Considerably less than 20 teeth in each half-jaw; teeth strong and with a spoon- or cup-shaped crown
22
21-b More than 20 teeth in each half-jaw; teeth weak and slender or filiform
24

22-a Body color black or dark-brown; presence of at least one single strong spine on each body scute; standard length up to 36 cm; premaxillae fused (*Pseudacanthicus*)
23
22-b Body color mottled brown; spines on the body scutes small; premaxillae separated from each other; standard length <10 cm; Marowijne River
Panaqolus koko

23-a Ground color of body brown with small bluish white dots on the scutes, on fins, and more distinct ones on venter; orbit diameter 8 times in head length; each mandible with 8 teeth
Pseudacanthicus fordii
23-b Ground color of body brownish with distinct dark blotches or only indistinct lighter dots on body, fins and venter; orbit diameter 4.3-6.5 times in head length; each mandible with 10-15 teeth
Pseudacanthicus serratus

24-a Snout with a wide naked margin; evertible interopercular odontodes hook-like; presence of long, forked tentacles on the snout of males (*Ancistrus*)
25
24-b Snout granular to its margin; evertible odontodes needle-like; absence of forked tentacles on the snout of males
28

25-a Absence of spots on the body and fins
26
25-b Presence of whitish spots on the fins and body
27

26-a Head color brown; Commewijne and Marowijne rivers
Ancistrus aff. *hoplogenys*
26-b Head with reticulated pattern; Saramacca River
Ancistrus sp. 'reticulate'

27-a Numerous small, well-defined spots with bright orange color
Ancistrus temminckii
27-b Spots less numerous, but larger
Ancistrus cf. *leucostictus*

28-a Head and body depressed; margin of head and snout usually with short, bristle-like odontodes (*Guyanancistrus*)
29
28-b Head and body robust; margin of head and snout without bristle-like odontodes
30

29-a Medium-sized species (15 cm)
Guyanancistrus brevispinis group
29-b Small species (< 8 cm TL); only known from Nassau Mountains
Guyanancistrus sp. 'bigmouth'

30-a Ground color of body and fins brownish with numerous large clearly delineated, circular, dark blotches; maximum orbital diameter 4-4.5 times in head length; body depth 4-4.5 times in standard length; dorsal fin reaching beyond adipose fin, almost extending to caudal fin base; large species (39 cm TL); Marowijne River
Hemiancistrus medians
30-b Ground color yellowish, body with obscure dark-brownish transverse bandings (most outspoken on caudal fin); abdominal surface partly covered with patches of scutelets; maximum orbital diameter 4.4-5.5 times in head length; body depth 4.5-5.2 times in standard length; dorsal fin not reaching beyond adipose fin; small species (8 cm SL); Marowijne River
Peckoltia otali

31-a Coloration dark with small, white spots
32
31-b Coloration mottled or with oblique bars; Coppename River
Pseudancistrus kwinti

32-a Hypertrophied snout odontodes brown-reddish, increasing gradually in size from snout tip to cheeks; very distinct whitish spots covering the body, increasing gradually in size from snout to caudal peduncle
33
32-b Hypertrophied odontodes whitish, of homogeneous size except shorter in the snout tip and a few larger in the cheeks; tiny whitish spots on the anterior part of the head enlarging abruptly in size posterior to the eyes; Corantijn River
Pseudancistrus corantijniensis

33-a Minimum interorbital distance <10.8 times in standard length (valid for specimens larger than 8.5 cm), French Guiana and Marowijne River
Pseudancistrus barbatus
33-b Minimum interorbital distance >10.8 times in standard length, Suriname and Coppename rivers
Pseudancistrus depressus

The type locality of *Squaliforma tenuis* (Paramaribo) is doubtful since the species is only known from its holotype and no other representative of *Squaliforma* is known from Suriname (Weber *et al.*, 2012).

Recently (February 2012) an *Hypostomus*-like armored catfish was offered for sale at the central market of Paramaribo. This catfish was identified as *Pterygoplichthys multiradiatus* (Hancock, 1828) (type locality [Probably near Santa Catalina, Orinoco system], Demerara; distribution Orinoco River System, Guyana (?); introduced elsewhere) (Weber, 1992). *Pterygoplichthys* can be distinguished from *Hypostomus* by the number of rays in the dorsal fin: I, 10-13 in *Pterygoplichthys* versus I, 7 in *Hypostomus*.

Ancistrus aff. *hoplogenys* (Günther, 1864)
Local name(s): – (bristlenose catfish)

Diagnostic characteristics: more than 20 teeth in each half-jaw; teeth weak and slender or filiform; snout with a wide naked margin; evertible interopercular odontodes hook-like; presence of long, forked tentacles on the snout of males; absence of spots on the body and fins; body color brown
Type locality: Rio Capim, tributary of Guama River, Para State, Brazil
Distribution: Amazon, Essequibo, and Paraguay River basins: Argentina, Brazil, Guyana, Paraguay and Peru; in Suriname, known from tributaries of Corantijn, Coppename, Commewijne and Marowijne rivers
Size: 11.5 cm SL
Habitat: forest creeks of the Interior with moderate current and sandy bottom; during the day they hide in holes in large woody debris
Position in the water column: bottom
Diet:
Reproduction: *Ancistrus* is a cavity-nester; mature males develop branched tentacles on the snout and they typically select a dark cavity in wood or rock as a nest site; the nesting male is territorial and defends the site aggressively against all intruders, especially conspecific males; courtship consists largely of displays of raised dorsal and caudal fins, and leading behaviors whereby the male attempts to escort a female back to his nest; the female deposits a clump of 20-200 eggs usually on the ceiling, but also on the sides and floor of the cavity; after oviposition, the female either leaves the nest or is forcibly evicted, and takes no more interest in eggs or fry; male *Ancistrus* care for the eggs and fry in the nest cavity, using the

Ancistrus aff. *hoplogenys* (live) (© INRA-Le Bail)

fins and mouth to clean the eggs and clear the cavity of detritus (Burgess, 1989). Sabaj *et al.* (1999) hypothesize that female *Ancistrus* prefer to spawn with males guarding larvae, and that the male's snout tentacles stimulate this bias by mimicking the presence of larvae in an otherwise empty nest
Remarks:

ANCISTRUS GROUP *LEUCOSTICTUS* (Günther, 1864)
Local name(s): – (bristlenose catfish)

Diagnostic characteristics: more than 20 teeth in each half-jaw; teeth weak and slender or filiform; snout with a wide naked margin; evertible interopercular odontodes hook-like; presence of long, forked tentacles on the snout of males; body brown with large, whitish spots
Type locality: Essequibo River, Guyana
Distribution: Essequibo River basin and main Guianan rivers to Oyapock River in the east; French Guiana (?), Guyana, and Suriname (?)
Size: 10 cm SL (?)
Habitat: small forest creeks of the Interior with sandy or rocky bottom, sometimes in small rapids; apparently not as closely associated with large woody debris as other *Ancistrus* species (Le Bail *et al.*, 2000)
Position in the water column: bottom
Diet:
Reproduction: mature males develop branched tentacles on the snout
Remarks:

ANCISTRUS SP. 'RETICULATE'
Local name(s): – (bristlenose catfish)

Diagnostic characteristics: more than 20 teeth in each half-jaw; teeth weak and slender or filiform; snout with a wide naked margin; evertible interopercular odontodes hook-like; presence of long, forked tentacles on the snout of males; absence of spots on the body and fins; head with a conspicuous reticulated pigmentation pattern
Type locality:
Distribution: only known from Mindrineti River, Saramacca River System, and the Upper Commewijne River, Suriname
Size: 10 cm SL
Habitat: medium-sized forest creeks, often in/on large woody debris
Position in the water column: bottom
Diet:
Reproduction: mature males develop branched tentacles on the snout
Remarks:

Ancistrus gr. *leucostictus* (live), Sipaliwini River (Corantijn River) (© R. Covain)

Ancistrus sp. 'reticulate' live, Mindrineti River, Saramacca River System

Ancistrus temminckii (Valenciennes, 1840)
Local name(s): – (bristlenose catfish)

Diagnostic characteristics: more than 20 teeth in each half-jaw; teeth weak and slender or filiform; snout with a wide naked margin; evertible interopercular odontodes hook-like; presence of long, forked tentacles on the snout of males; numerous small, well-defined spots with bright orange color
Type locality: Suriname
Distribution: Saramacca, Suriname and Maroni River basins, Suriname
Size: 9.8 cm SL
Habitat: small and medium-sized forest creeks in the Interior with moderate current and sandy or rocky bottom
Position in the water column: bottom
Diet: periphyton
Reproduction: mature males develop branched tentacles on the snout, females are smaller and do not develop branched tentacles on the snout; the female deposits 50-100 eggs in a hole in large woody debris; the male guards the eggs and the newborn larvae
Remarks:

Guyanancistrus brevispinis (Heitmans, Nijssen & Isbrücker, 1983)
Local name(s):

Diagnostic characteristics: snout covered with bony plates up to its margin; more than 20 teeth in each half-jaw; teeth weak and slender or filiform; head and body depressed; margin of head and snout usually with short, bristle-like odontodes; body color variable, usually brownish, marbled with yellowish spots (juveniles have a more banded color pattern); medium-sized species (15 cm)
Type locality: Fallawatra River, Nickerie River System, Suriname
Distribution: Atlantic coastal drainages of French Guiana and Suriname (possibly restricted to Nickerie River, Suriname (J. Montoya-Burgos pers. comm.; also see Cardoso & Montoya-Burgos, 2009)
Size: 14.2 cm SL
Habitat: relatively abundant in clear-water rivers and large to medium-sized tributaries of the Interior in running water over rocky bottom
Position in the water column: bottom
Diet:
Reproduction:
Remarks: a very variable species (-group) that may actually consist of several closely related species, with *G. brevispinis* restricted to Nickerie River, its type locality (see Cardoso & Montoya-Burgos, 2009). See Covain & Fisch-Muller (2012) for the validity of the genus name *Guyanancistrus*.

Ancistrus temminckii (live) (© INRA-Le Bail)

Ancistrus temminckii (live), Suriname River (© R. Covain)

Guyanancistrus brevispinis (live) (© INRA-Le Bail)

Guyanancistrus sp. 'bigmouth'
Local name(s): –

Diagnostic characteristics: more than 20 teeth in each half-jaw; teeth weak and slender or filiform; head and body depressed; margin of head and snout usually with short, bristle-like odontodes; small species (<6 cm SL), only known from Nassau Mountains
Type locality: Paramaka Creek, Nassau Mountains
Distribution: only known from Paramaka Creek, Nassau Mountains
Size: 6 cm SL
Habitat: small mountain creeks in Nassau Mountains
Position in the water column: bottom
Diet:
Reproduction: males develop large odontodes on the pectoral fin spines
Remarks: very restricted distribution; threatened with extinction by a proposed bauxite mine in Nassau Mountains.

Hemiancistrus medians (Kner, 1854)
Local name(s): -

Diagnostic characteristics: a large, robust species (body depth 4-4.5 times in SL) with a short (1.7-1.8 times in head length), rounded snout; the rim of the snout is covered with bony plates and without odontodes; the eyes are large (4-4.5 times in head length); interopercular area with long, evertible odontodes; teeth filliform, bifid and numerous; dorsal fin large, reaching to the adipose fin; caudal fin obliquely emarginated; the bony plates, 23-24 in lateral series, are carinate (with keels) and strongly spinulose; color of the body brown with large (approximately the size of the pupil) black spots, both on the body and all the fins
Type locality: no locality (= Suriname [see Isbrücker, 1992b])
Distribution: only known from Marowijne (Maroni) River, French Guiana and Suriname
Size: 39 cm TL
Habitat: mainly in the main river channel and large tributary streams, especially in and near rapids with strong current between rock/boulders
Position in the water column: bottom
Diet:
Reproduction: no secondary sexual dimorphism
Remarks: a beautiful species that is endemic to the Marowijne River System.

Hypostomus coppenamensis Boeseman, 1969
Local name(s): warawara

Diagnostic characteristics: depth caudal peduncle 1.4-2.5 in interdorsal length; when dorsal fin adpressed, dorsal fin rays usually falling distinctly short of adipose fin spine; post-occipital scute single (except sometimes in juveniles of <5 cm

Guyanancistrus brevispinis (live), Suriname River (© R. Covain)

Guyanancistrus brevispinis (live), Nickerie River (type locality) (© R. Smith)

Guyanancistrus sp. ('bigmouth') (live), Paramaka Creek, Nassau Mountains (© A. Flynn)

Hemiancistrus medians (live), Lawa River (Marowijne River), ANSP187122 (© M. Sabaj-Pérez)

standard length); venter without spots; cleithral width moderate, 3.25-3.28 in standard length; mandibular ramus 2.0-3.0 in interorbital width; dark spots rather small, very intense and numerous, well-defined; in adults, mandibular teeth usually about 30-60 on each side; length dorsal spine 2.4-3.1, rarely more, in standard length; ground color rather light, beige-brownish, and completely covered with relatively few small to moderate dark round spots, the interspaces between them much larger than spot diameters; length of dorsal spine 2.85-2.95 in standard length
Type locality: Left Coppename River, Suriname, 3°54'N, 56°46'W
Distribution: Upper Coppename River basin, Suriname
Size: 12.5 cm SL
Habitat: main channel and tributaries of Upper Coppename River, upstream of the first rapids, clear water, moderate current, rocky (Fig. 4.8b) or sandy bottom and among large woody debris (Fig. 4.33b)
Position in the water column: bottom
Diet:
Reproduction:
Remarks:

Hypostomus corantijni Boeseman, 1968
Local name(s): warawara

Diagnostic characteristics: depth caudal peduncle 2.0-2.3 in interdorsal length; when dorsal fin adpressed, dorsal fin rays usually falling distinctly short of adipose fin spine; post-occipital scute single (except sometimes in juveniles of <5 cm standard length); venter without spots; cleithral width moderate, 3.25-3.28 in standard length; mandibular ramus 2.0-3.0 in interorbital width; dark spots rather small, very intense and numerous, well-defined; in adults, mandibular teeth usually about 30-60 on each side; length dorsal spine 2.4-3.1, rarely more, in standard length; ground color usually darker, and completely covered with relatively numerous intensely dark round or oblong spots, the interspaces between them about as large as spot diameters; length of dorsal spine 2.4-3.1, rarely more, in standard length; 3 or 4 scutes between adipose fin and caudal fin; 13 postanal scutes, not including 1 or 2 small additional scutes which cover the origins of these fins; lateral body scutes usually 27 but sometimes 28; cleithral width 3.5-3.85 in standard length
Type locality: Sipaliwini River, Corantijn River system, Suriname
Distribution: Corantijn and Nickerie rivers, upstream of the first rapids, Suriname
Size: 18.8 cm SL
Habitat:
Position in the water column: bottom
Diet:
Reproduction:
Remarks: *H. nickeriensis* and *H. sipaliwinii* are junior synonyms (Weber *et al.*, 2012).

Hypostomus coppenamensis (live), (© R. Covain)

Hypostomus corantijni (live), Nickerie River (© R. Smith)

Hypostomus corantijni (live), Sipaliwini River (Corantijn River) (© R. Covain)

HYPOSTOMUS CRASSICAUDA Boeseman, 1968
Local name(s): warawara

Diagnostic characteristics: depth caudal peduncle 1.35-1.7 in interdorsal length; dorsal and caudal fins mostly spotted or with bands; dark spots present on posterior part of caudal peduncle; lower caudal lobe with dark spots or banded; venter with at most a few indistinct spots; when dorsal fin adpressed, dorsal fin rays hardly or almost reaching adipose fin spine; mandibular ramus 1.8-2.1 in interorbital width; dorsal spine length (measured in vertical position from junction of the bases of the predorsal scute and dorsal spine to its distal tip) slightly surpassing predorsal length (distance measured from tip of snout to posterior rim of last predorsal scute before articulation of dorsal spine); dorsal spine length 2.25-2.5 in standard length; head depth (measured vertically from distal tip of occipital process dorsally to coracoidal band ventrally) 5.21-5.6 in standard length; head depth 1.75-1.85 in head length; spots dense on complete dorsal fin; dark spots on body less small, ovate with a diameter about equaling interspaces; dark spots on pectoral spines small, numerous and dense
Type locality: Sipaliwini River, Corantijn River system, Suriname
Distribution: Upper Corantijn River, Suriname
Size: 14.3 cm SL
Habitat:
Position in the water column: bottom
Diet:
Reproduction:
Remarks:

HYPOSTOMUS GYMNORHYNCHUS (Norman, 1926)
Local name(s): warawara

Diagnostic characteristics: depth caudal peduncle 2.25-2.3 in interdorsal length; when dorsal fin adpressed, dorsal fin rays usually falling distinctly short of adipose fin spine; post-occipital scute single (except sometimes in juveniles of <5 cm standard length); venter without spots; cleithral width moderate, 3.25-3.28 in standard length; mandibular ramus 2.0-3.0 in interorbital width; dark spots moderate, variably intense and usually less well-defined; in adults mandibular teeth usually 20-40 on each side; length of dorsal spine 2.55-3.65 in standard length; when dorsal fin adpressed, dorsal fin rays falling far short of reaching adipose fin spine; the vague spots on body and fins less intense and slightly smaller; diameter orbit 3.3 in snout length; spots on body and caudal peduncle less densely distributed, but more intense in coloration
Type locality: Iponcin Creek, into Approuague River, French Guiana
Distribution: Guianan coastal drainages, from Oyapock westward to Maroni basin: French Guiana and Suriname
Size: 17 cm SL

Hypostomus crassicauda (live), Corantijn River (© R. Covain)

Hypostomus gymnorhynchus (live), 107.8 mm SL, Lawa River (Marowijne River), ANSP189118 (© M. Sabaj Pérez)

Habitat: main channel and large tributaries of rivers in the Interior, strong currents and rocky bottom
Position in the water column: bottom
Diet:
Reproduction:
Remarks: *H. surinamensis*, *H. occidentalis* and *H. tapanahoniensis* are junior synonyms (Weber *et al.*, 2012).

Hypostomus macrophthalmus Boeseman, 1968
Local name(s): warawara

Diagnostic characteristics: depth caudal peduncle 1.35-1.7 in interdorsal length; dorsal and caudal fins mostly spotted or with bands; dark spots relatively large, ovate, lacking on posterior part of caudal peduncle; lower caudal lobe dusky without spots
Type locality: Sipaliwini River, near air strip, Suriname
Distribution: Sipaliwini River, Upper Corantijn River basin, Suriname
Size: 7.9 cm SL
Habitat:
Position in the water column: bottom
Diet:
Reproduction:
Remarks: a questionable species, the type specimen is juvenile (Weber *et al.*, 2012).

Hypostomus micromaculatus Boeseman, 1968
Local name(s): warawara

Diagnostic characteristics: depth caudal peduncle 1.35-1.7 in interdorsal length; dorsal and caudal fins completely or almost completely dusky, or finely marked; dark spots on body and fins very small and usually elongate, present on posterior part of caudal peduncle, several on each scute; interorbital width 2.6-3.3 times in head length (measured from tip of snout to posterior tip of occipital process); diameter orbit 1.6-3.85 in snout length (measured from tip of snout to anteriormost bony rim of orbit)
Type locality: Mamadam falls, Suriname River, Suriname
Distribution: Upper and middle Suriname River basin, Suriname, i.e. upstream of the first rapids
Size: 18.5 cm SL
Habitat:
Position in the water column: bottom
Diet:
Reproduction:
Remarks:

Hypostomus micromaculatus (live), Suriname River (© R. Covain)

Hypostomus paucimaculatus Boeseman, 1968
Local name(s): warawara

Diagnostic characteristics: depth caudal peduncle 1.35-1.7 in interdorsal length; dorsal and caudal fins mostly spotted or with bands; dark spots present on posterior part of caudal peduncle; lower caudal lobe with dark spots or banded; venter with at most a few indistinct spots; when dorsal fin adpressed, some dorsal fin rays overlapping adipose fin; mandibular ramus 2.5-3.3 in interorbital width; depth head 5.0-5.7 in standard length; spots on body and fins intense, of moderate size, very widely interspaced, even smaller spots on snout not dense; spots on caudal fin irregularly arranged
Type locality: Suriname River near Brokopondo, Suriname
Distribution: Upper and middle Suriname River basin, Suriname, upstream of the first rapids
Size: 12 cm SL
Habitat:
Position in the water column: bottom
Diet:
Reproduction:
Remarks:

Hypostomus plecostomus (Linnaeus, 1758)
Local name(s): warawara

Diagnostic characteristics: depth caudal peduncle 1.35-1.7 in interdorsal length; dorsal and caudal fins mostly spotted or with bands; dark spots present on posterior part of caudal peduncle; lower caudal lobe with dark spots or banded; venter with at most a few indistinct spots; when dorsal fin adpressed, some dorsal fin rays overlapping adipose fin; mandibular ramus 2.5-3.3 in interorbital width; depth head 4.45-5.1 in standard length; spots on body and fins vague, large, on snout small and densely distributed; dark spots on caudal fin forming transverse bands
Type locality: Suriname River, Suriname
Distribution: lower freshwater reaches of coastal drainages in French Guiana, Guyana and Suriname;
Size: 50 cm SL
Habitat:
Position in the water column: bottom
Diet:
Reproduction:
Remarks: *H. ventromaculatus* is a junior synonym (Weber *et al.*, 2012).

Hypostomus paucimaculatus (live), Suriname River (© R. Covain)

Hypostomus plecostomus (live), Suriname River (© R. Covain)

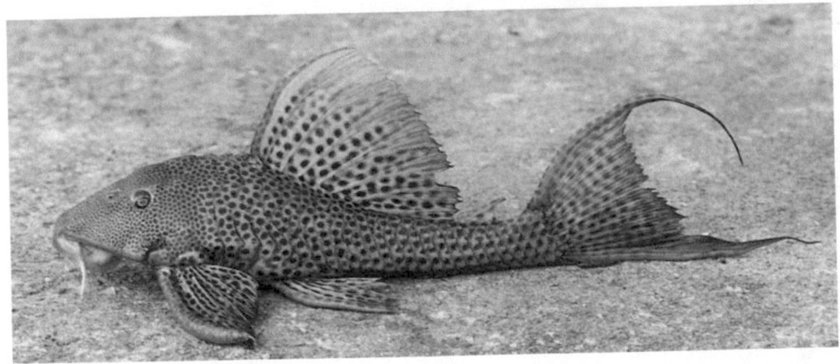

Hypostomus plecostomus (live) (© INRA-Le Bail)

Hypostomus pseudohemiurus Boeseman, 1968
Local name(s): warawara

Diagnostic characteristics: depth caudal peduncle 1.35-1.7 in interdorsal length; dorsal and caudal fins completely or almost completely dusky, or finely marked; dark spots on body relatively large, ovate, lacking on the posterior part of caudal peduncle, where at most traces of a few longitudinal stripes seem present; interorbital width 3.1-3.35 times in head length; diameter orbit 2.45-2.9 in snout length
Type locality: Kabalebo River, Corantijn River system, Suriname
Distribution: Corantijn River basin, Suriname
Size: 6.2 cm SL
Habitat:
Position in the water column: bottom
Diet:
Reproduction:
Remarks: a questionable species, only known from (juvenile) type specimen (Weber *et al.*, 2012).

Hypostomus saramaccensis Boeseman, 1968
Local name(s): warawara

Diagnostic characteristics: depth caudal peduncle 1.35-1.7 in interdorsal length; dorsal and caudal fins mostly spotted or with bands; dark spots present on posterior part of caudal peduncle; lower caudal lobe with dark spots or banded; venter with at most a few indistinct spots; dorsal spine length shorter than predorsal length; when dorsal fin adpressed, dorsal fin rays hardly or almost reaching adipose fin spine; mandibular ramus 1.8-2.1 in interorbital width ; dorsal spine length 2.45-3.2 in standard length; head depth 5.4-6.3 in standard length; head depth 1.85-2.25 in head length; spots on distal dorsal fin usually vague or lacking; dark spots on body small and circular with a diameter distinctly less than interspaces; dark spots on pectoral spines large, round and widely interspaced
Type locality: Feddiprati (rapids), middle Saramacca River, Suriname
Distribution: endemic for the Saramacca River System, Suriname
Size: 11.5 cm SL
Habitat:
Position in the water column: bottom
Diet:
Reproduction:
Remarks:

This page is intentionally left blank.

Hypostomus taphorni (Lilyestrom, 1984)
Local name(s): warawara

Diagnostic characteristics: a robust species with a reduced number of teeth; teeth very characteristic with short, solid, thick stem and spoon-shaped crown; teeth not curved
Type locality: Rio Botanamo, Rio Cuyuni drainage near the bridge on road to Bochinche, Edo, Bolivar State, Venezuela
Distribution: Cuyuni River basin, Essequibo drainage, Venezuela; Rupununi River and Takutu River basins, Guyana; in Suriname, only known from Corantijn and Nickerie rivers
Size: 18.5 cm SL
Habitat:
Position in the water column: bottom
Diet: the spoon-shaped, *Panaque*-like teeth may indicate wood-eating in *H. taphorni* (Nelson *et al.*, 1999)
Reproduction:
Remarks:

Hypostomus watwata Hancock, 1828
Local name(s): libakwikwi, sekwikwi (warawara)

Diagnostic characteristics: depth caudal peduncle 1.8-2.35 in interdorsal length; when dorsal fin adpressed, dorsal fin rays usually falling distinctly short of adipose fin spine; occipital with a well-marked median ridge; post-occipital scute multiple; venter completely covered with moderate to large spots; cleithral width considerable, 3.05-3.45 in standard length; mandibular ramus 2.6-3.5 in interorbital width
Type locality: off Berbice River, Guyana
Distribution: lower freshwater and estuarine reaches of Guiana Shield river drainages from Oyapock River to Demerara River: French Guiana, Guyana, Suriname and Venezuela
Size: 35 cm SL
Habitat: lives in estuaries with muddy bottom substrate and turbid, brackish water; also present in swamps of the Young Coastal Plain
Position in the water column: bottom
Diet:
Reproduction: mature males develop large odontodes on the pectoral fin spines and the rays of the caudal fin
Remarks: sold at the markets of Paramaribo.

Hypostomus taphorni (live) (© P. Willink)

Hypostomus taphorni (live), detail of the head, Corantijn River (© R. Covain)

Hypostomus taphorni (live), detail of mouth with teeth, Corantijn River (© R. Covain)

LITHOXUS GROUP *BOVALLII* (Regan, 1906)
Local name(s):

Diagnostic characteristics: mouth small, width included more than 2 times in sucking disk width; sucking disk with a posterior suture; 3-7 acute teeth on the lower jaw; standard length up to 6 cm; adipose fin present; 4 branched anal fin rays (last ray split to its base); adipose fin without posterior extension; lateral scutes usually 25; body depth 5.85-9.9 times in standard length; caudal fin dark with a white distal margin; absence of numerous well-developed odontodes on anterior half of pectoral fin spine; length pectoral fin spine 4.0-4.6 in SL; orbit diameter 6.0-6.7 times in head length; caudal peduncle depth 11.2-12.0 in SL
Type locality: Kaat River, tributary of Treng River [= Ireng], Upper Potaro, Guyana
Distribution: Ireng River, Guyana, and Corantijn and Nickerie rivers, Suriname
Size: 5.9 cm SL
Habitat:
Position in the water column: bottom
Diet:
Reproduction: mature males develop large odontodes on the pectoral fin spines and the rays of the caudal fin
Remarks:

LITHOXUS PALLIDIMACULATUS Boeseman, 1982
Local name(s): -

Diagnostic characteristics: mouth small, width included more than 2 times in sucking disk width; sucking disk with a posterior suture; standard length up to 5 cm ; adipose fin absent ; dorsal and lateral sides with light, circular spots; mandibular ramus length 2.8-3.0 times in interorbital width; interorbital width 4.0-4.8 times in head length; interopercle with 18-23 enlarged odontodes
Type locality: Kwambaolo Creek, right tributary of Sara Creek above dam, Suriname River system, Suriname
Distribution: Upper Suriname River, Suriname; also present in Nassau Mountains (Marowijne River System)
Size: 4.7 cm SL
Habitat:
Position in the water column: bottom
Diet:
Reproduction: mature males develop large odontodes on the pectoral fin spines and the rays of the caudal fin
Remarks:

Hypostomus watwata (live), Suriname River

Hypostomus watwata (live) (© INRA-Le Bail)

Lithoxus gr. *bovallii* (alcohol), Nickerie River

LITHOXUS PLANQUETTEI Boeseman, 1982
Local name(s): -

Diagnostic characteristics: mouth small, width included more than 2 times in sucking disk width; sucking disk with a posterior suture; standard length up to 7 cm; adipose fin present; 4 branched anal fin rays (last ray split to its base); adipose fin without posterior extension; lateral scutes usually 25; body depth 2.1-2.5 times in head length; caudal fin bright with brownish horizontal bands or small blotches
Type locality: Crique Boulenger, Comte River System, French Guiana
Distribution: Atlantic coastal drainages from Marowijne (Maroni) to Kaw river basins, French Guiana and Suriname
Size: 7 cm SL
Habitat:
Position in the water column: bottom
Diet:
Reproduction: mature males develop large odontodes on the pectoral fin spines and the rays of the caudal fin
Remarks:

LITHOXUS STOCKI Nijssen & Isbrücker, 1990
Local name(s):

Diagnostic characteristics: mouth small, width included more than 2 times in sucking disk width; sucking disk with a posterior suture; 8-10 teeth with truncate tips (bilobed) on the lower jaw; standard length up to 7 cm ; adipose fin present; 4 branched anal fin rays (last ray split to its base); adipose fin without posterior extension; lateral scutes usually 25; an extremely depressed species with body depth 2.6-3.3 times in head length; caudal fin dark with a white distal margin; presence of numerous well-developed odontodes on anterior half of pectoral fin spine; length pectoral fin spine 2.8-3.6 in SL; orbit diameter 5.7-6.1 times in head length; caudal peduncle depth 9.0-11.2 in SL
Type locality: Marouini River, downstream of village Epoia, Upper Marowijne (Maroni) River, French Guiana
Distribution: Marowijne (Maroni) and Mana River basins, French Guiana and Suriname
Size: 6.6 cm SL
Habitat: small, shallow forest creeks of the Interior with fast flowing water and rocky bottom and rapids in the main channel
Position in the water column: bottom
Diet:
Reproduction: mature males develop strong odontodes on the (enlarged) pectoral fin spines
Remarks:

Lithoxus gr. *bovallii* (live), Sipaliwini River (Corantijn River) (© R. Covain)

Lithoxus pallidimaculatus (live), Commewijne River (© R. Covain)

Lithoxus planquettei (live) (© INRA-Le Bail)

Lithoxus planquettei (live), scale bar = 1 cm, Lawa River (Marowijne River), ANSP189131 (© M. Sabaj Pérez)

Lithoxus surinamensis Boeseman, 1982
Local name(s):

Diagnostic characteristics: mouth small, width included more than 2 times in sucking disk width; sucking disk with a posterior suture; standard length up to 4 cm; adipose fin absent; dorsum dusky, with or without suggestions of darker cross-bands; mandibular ramus length 2.3-2.4 times in interorbital width; interorbital width 3.0-4.1 times in head length; interopercle with 8-12 enlarged odontodes
Type locality: near Awaradam, Gran Rio, Suriname River, Suriname
Distribution: Gran Rio basin in Upper Suriname River drainage and Coppename River, Suriname
Size: 4.1 cm SL
Habitat:
Position in the water column: bottom
Diet:
Reproduction: mature males develop large odontodes on the pectoral fin spines and the rays of the caudal fin
Remarks:

Panaqolus koko Fisch-Muller & Covain, 2012
Local name(s): –

Diagnostic characteristics: considerably less than 20 teeth in each half-jaw; teeth strong and with a spoon- or cup-shaped crown; body color mottled brown; spines on the body scutes small; premaxillae separated from each other; standard length <10 cm
Type locality: Marowijne River
Distribution: in Suriname only known from rapids of the Upper Marowijne River
Size: 8 cm SL
Habitat: rapids with fast-flowing, clear water and rocky bottom
Position in the water column: bottom
Diet: *Panaque* /*Panaqolus* are wood-eating catfishes with spoon-shaped teeth (Nelson *et al.*, 1999) that are active during the night
Reproduction:
Remarks: apparently quite rare as only few specimens have been collected; a related, undescribed *Panaqolus* species is known from Mapane Creek, Upper Commewijne River ((S. Fisch-Muller, pers. communication). Le Bail *et al.* (2000) listed *P. koko* as *Panaque* cf. *dentex*.

Lithoxus surinamensis (alcohol), Coppename River (© P. Willink)

Panaqolus koko (live), Marowijne River (© INRA-Le Bail)

PECKOLTIA OTALI Fisch-Muller & Covain, 2012
Local name(s): -

Diagnostic characteristics: a robust species with a small head (Head Length 2.6-2.8 times in SL) and small eyes (orbit diameter 4.4-5.5 times in HL); body depth 4.5-5.2 times in standard length; ground color yellowish, body with obscure brownish transverse bandings; abdominal surface partly covered with patches of scutelets; adpressed dorsal fin not reaching beyond adipose fin
Type locality: Marowijne River
Distribution: in Suriname only known from rapids of the Upper Marowijne River
Size: 7.7 cm SL
Habitat: rapids with fast-flowing, clear water and rocky bottom
Position in the water column: bottom
Diet:
Reproduction:
Remarks: body coloration closely resembles that of juvenile *Guyanancistrus brevispinis*; apparently quite rare as only few specimens have been collected; a related, undescribed *Peckoltia* species is present in Mapane Creek, Upper Commewijne River (S. Fisch-Muller, pers. communication). Le Bail *et al.* (2000) listed *P. otali* as *Hemiancistrus* aff. *braueri.*

PSEUDACANTHICUS SERRATUS (Valenciennes, 1840)
Local name(s):

Diagnostic characteristics: body color black or dark-brown with whitish spots on the ventral surface; presence of at least one (1-5) strong spine on each body scute; standard length up to 36 cm; premaxillae fused, 11-17 teeth on the lower jaw; numerous strong odontodes on the pectoral spines and the head; dorsal fin very large reaching to the adipose fin; adipose positioned very close to the caudal fin
Type locality: region around Paramaribo, Suriname
Distribution: coastal drainages of French Guiana and Suriname; in Suriname, known from the Marowijne, Suriname and, possibly, Corantijn rivers
Size: 36 cm SL
Habitat: known to occur in both lower and upstream reaches of large rivers, but probably present in low numbers
Position in the water column: bottom
Diet:
Reproduction:
Remarks: the related *Pseudacanthicus fordii* (Günther, 1868) is only known from its type that was collected in 'Suriname'; only very few specimens of *P. serratus* have been collected from the Suriname and Marowijne rivers; a *Pseudacanthicus*

Peckoltia otali (live), Marowijne River (© INRA-Le Bail)

Pseudacanthicus serratus (live), Marowijne River (© INRA-Le Bail)

Pseudacanthicus sp. (alcohol), dorsal view, Sipaliwini River (Corantijn River) (© P. Willink)

species has recently been collected from Upper Corantijn River (Willink *et al.*, 2011).

PSEUDANCISTRUS BARBATUS (Valenciennes, 1840)
Local name(s): -

Diagnostic characteristics: head and/or body with very prominent, inevertible odontodes; head depressed, head depth at least 5.5 times in standard length; in adults, a row of odontodes along the rim of the snout, strikingly elongated in males; more than 20 teeth in each half-jaw; teeth weak and slender or filiform; coloration dark with white spots; hypertrophied snout odontodes brown-reddish, increasing gradually in size from snout tip to cheeks; very distinct whitish spots covering the body, increasing gradually in size from snout to caudal peduncle; minimum interorbital distance <10.8 times in standard length (valid for specimens larger than 8.5 cm), French Guiana and Marowijne River
Type locality: Mana River, French Guiana
Distribution: Oyapock, Mana, Marowijne (Maroni) rivers; French Guiana and Suriname
Size: 20 cm SL
Habitat: in rapids and other reaches of the main channel of rivers in the Interior with strong currents and rocky bottom
Position in the water column: bottom
Diet:
Reproduction: mature males may develop spectacular large odontodes along the snout up to the cheeks ('beard')
Remarks:

PSEUDANCISTRUS CORANTIJNIENSIS De Chambrier & Montoya-Burgos, 2008
Local name(s):

Diagnostic characteristics: head and/or body with very prominent, inevertible odontodes; head depressed, head depth at least 5.5 times in standard length; in adults, a row of odontodes along the rim of the snout, strikingly elongated in males; more than 20 teeth in each half-jaw; teeth weak and slender or filiform; coloration dark with white spots; tiny whitish spots on the anterior part of the head enlarging abruptly in size posterior to the eyes; hypertrophied odontodes whitish, of homogeneous size except shorter in the snout tip and a few larger in the cheeks; Corantijn River
Type locality: Corantijn River at Cow Falls, 4°59'48.3"N, 57°37'49.5"W, Corantijn River drainage, Sipaliwini District, Suriname
Distribution: Corantijn River, Suriname
Size: 18 cm SL

Pseudacanthicus sp. (alcohol), lateral view, Sipaliwini River (Corantijn River) (© P. Willink)

Pseudancistrus barbatus (live), Marowijne River (© P. Willink)

Pseudancistrus barbatus (live) (© INRA-Le Bail)

Habitat:
Position in the water column: bottom
Diet:
Reproduction: mature males may develop large odontodes along the rim of the snout up to the cheeks
Remarks:

PSEUDANCISTRUS DEPRESSUS (Günther, 1868)
Local name(s): -

Diagnostic characteristics: head and/or body with very prominent, inevertible odontodes; head depressed, head depth at least 5.5 times in standard length; in adults, a row of odontodes along the rim of the snout, strikingly elongated in males; more than 20 teeth in each half-jaw; teeth weak and slender or filiform; coloration dark with white spots; hypertrophied snout odontodes brown-reddish, increasing gradually in size from snout tip to cheeks; very distinct whitish spots covering the body, increasing gradually in size from snout to caudal peduncle; minimum interorbital distance >10.8 times in standard length, Suriname and Coppename rivers
Type locality: probably Suriname
Distribution: rivers of central Suriname: Suriname and Coppename rivers
Size: 13 cm SL
Habitat:
Position in the water column: bottom
Diet:
Reproduction: mature males may develop large odontodes along the snout up to the cheeks
Remarks:

Pseudancistrus corantijniensis (live), Sipaliwini River (Corantijn River)

Pseudancistrus depressus (live), Raleighvallen rapids, Coppename River (© W. Kolvoort)

Pseudancistrus depressus (live), Suriname River (© R. Covain)

Pseudancistrus kwinti Willink, Mol & Chernoff, 2010
Local name(s):

Diagnostic characteristics: head and/or body with very prominent, inevertible odontodes; head depressed, head depth at least 5.5 times in standard length; in adults, a row of odontodes along the rim of the snout, strikingly elongated in males; more than 20 teeth in each half-jaw; teeth weak and slender or filiform; coloration mottled or with oblique bars (juveniles; Fig. 4.8a); absence of two dentary papillae; Coppename River
Type locality: Coppename River at base of Sidonkroetoe-val rapids, Central Suriname Nature Reserve, 4°31'51"N, 56°30'56"W, Sipaliwini State, Suriname
Distribution: Coppename River, Suriname
Size: 9.6 cm SL
Habitat: rapids with swift current, clear water, and rocky bottom substrate
Position in the water column: bottom
Diet:
Reproduction: mature males develop large hypertrophied odontodes along the snout; juveniles have a more pronounced body coloration with alternating (oblique) dark and pale bars; the four dark bars about twice as wide as pale bars (Fig. 4.8a)
Remarks:

Pseudancistrus kwinti (alcohol), adult, Coppename River (© P. Willink)

Pseudancistrus kwinti (alcohol), juvenile, Coppename River (© P. Willink)

SUPERFAMILY PSEUDOPIMELODOIDEA – KEY TO THE SPECIES IN THE FAMILIES HEPTAPTERIDAE, PIMELODIDAE AND PSEUDOPIMELODIDAE

The families Heptapteridae, Pimelodidae and Pseudopimelodidae were, until recently, united in an expanded family Pimelodidae (Nelson, 2006). Although the three families are not necessarily closely related (De Pinna, 1998), Diogo *et al.* (2004) concluded that they constitute a monophyletic assemblage (and thus should be recognized as subfamilies). The three families are endemic to the Neotropics and can easily be distinguished based on osteological and anatomical features. However, it is not always possible to assign species to one of the three families based on external features.

1-a Origin of anal fin anterior to the dorsal fin; eyes small and ventrolaterally placed at the same horizontal plane of (or below) the mouth; head broad and depressed anteriorly, but higher posteriorly; body laterally compressed
Hypophthalmus marginatus
1-b Origin of anal fin posteriorly to dorsal fin; eyes placed well above the mouth
2

2-a Eye freely movable in orbit (orbital rim free)
3
2-b No free orbital rim (eye continuous with epidermis of head)
23

3-a Large patches of teeth on vomer and palatines; large species (>30 cm)
4
3-b No teeth on vomer and palatines; small to medium-sized species (<30 cm)
10

4-a Body coloration with spots or black cross bands; head strongly depressed
5
4-b Body coloration lacking spots and black cross bands; head not strongly depressed
7

5-a Lower jaw strongly protruding over upper jaw; eyes large with diameter less than 4 times in snout length; body coloration with a few (about 7) well-developed black spots more or less in a mid-lateral series
Hemisorubim platyrhynchos
5-b Lower jaw distinctly shorter than upper jaw; eyes small with diameter more than 4 times in snout length; body coloration with striking pattern of black spots and (reticulated) cross bands, caudal fin spotted
6

6-a Posterior end of fontanel closer to horizontal plane through the eyes than to the base of the post-occipital process; <45 spots on the caudal fin

Pseudoplatystoma fasciatum

6-b Posterior end of fontanel closer to the base of the post-occipital process than to the horizontal plane through the eyes; >50 spots on the caudal fin, no black spots on the lateral region of the body; Marowijne River

Pseudoplatystoma tigrinum

7-a Body coloration with a dark upper half and a white lower flank and partly or wholly (caudal fin) bright-orange fins; adipose fin partly rayed; Corantijn River

Phractocephalus hemioliopterus

7-b Body coloration uniform brown, grey or silvery

8

8-a Barbels short, not reaching the origin of the dorsal fin; body with uniform silvery color

Brachyplatystoma rousseauxii

8-b Barbels long, reaching well over the origin of the dorsal fin; body coloration brown to grey

9

9-a Adipose fin short, equal in size to the anal fin; upper jaw distinctly protruding over lower jaw

Brachyplatystoma filamentosum

9-b Adipose fin longer than anal fin (anal fin base about 2 times in adipose fin base); upper and lower jaws of approximately the same length

Brachyplatystoma vaillantii

10-a Post-occipital process distinctly connected with the pre-dorsal plate, and with a broad base and tapering towards the posterior end where it meets the heart-shaped pre-dorsal plate; body width generally less than 5 times in standard length (except *Pimelabditus moli*)

11

10-b Post-occipital process generally not reaching the pre-dorsal plate, but if it reaches the pre-dorsal plate then narrow and with an uniform width from its base to posterior end; slender species with body width generally more than 5 times in standard length

14

11-a Body elongated with body depth more than 5 times in standard length; adipose fin long (base of anal fin >2.5 times in base of adipose fin); eyes very large, horizontal diameter less than 3 times in head length and 1.2-1.4 times in snout length; small species (14 cm standard length); Upper Marowijne River

Pimelabditus moli

11-b Body robust with body depth less than 5 times in standard length; short adipose fin (base anal fin <2 times in base adipose fin); medium-sized species (>15 cm standard length)
12

12-a Well-defined black blotch on the dorsal fin and broad longitudinal black bands on both lobes of the caudal fin
Pimelodus ornatus
12-b Absence of black blotch in dorsal fin and black markings on caudal fin
13

13-a Color plain, light liver-brown; eye small, 1.0-1.7 in interorbital
Pimelodus blochii
13-b Color dark grayish with a conspicuous broad white longitudinal band along the sides; eye larger 0.7-1.0 in interorbital
Pimelodus albofasciatus

14-a Dorsal fin reaches adipose fin when adpressed
15
14-b Adpressed dorsal fin falling distinctly short of adipose fin
19

15-a Absence of a well-defined longitudinal band on the flanks
16
15-b Presence of a well-defined longitudinal band on the flanks
18

16-a Very small species (<4 cm standard length); body mostly with light pigmentation; dorsal spine and first dorsal ray colored blackish; Corantijn River
Brachyrhamdia heteropleura
16-b Medium-sized species (>15 cm standard length); dorsal spine and first dorsal ray not blackish
17

17-a Eyes small, eye diameter >2 times in interorbital distance; fontanel does not continue posterior of the eyes
Rhamdia quelen
17-b Eyes large, eye diameter greater than the interorbital distance; fontanel prolonged backward to the base of the post-occipital process
Rhamdia foina

18-a Longitudinal black band thin and not continuing on the head; upper lobe of the caudal fin shorter or equal to the lower lobe of the caudal fin
Pimelodella cristata
18-b Longitudinal black band thick and continuing on the head up to the tip of the snout; upper lobe of the caudal fin longer than the lower lobe of the caudal fin
Pimelodella geryi

19-a Presence of both a black blotch at the tip of the dorsal fin; longitudinal band on the flanks well defined
20

19-b Absence of both a black blotch on the dorsal fin and a well-defined longitudinal band on the flanks
21

20-a Eye diameter distinctly smaller than snout length; maximum depth of adipose fin situated halfway its base
Pimelodella procera

20-b Eye diameter about equal to snout length; maximum depth of adipose fin in the posterior half of the fin base
Pimelodella macturki

21-a Maxillary barbels not reaching the pelvic fins; tips of the first rays of the pectoral and pelvic fins soft; up to 10 cm SL
Imparfinis pijpersi

21-b Maxillary barbels reaching over the pelvic fins; tip of first rays of the pectoral and pelvic fins soft; small species <5 cm SL
Imparfinis aff. *stictonotus*

21-c Maxillary barbels reaching over the pelvic fins; tips of the first rays of the pectoral and pelvic fins hard
22

22-a Lobes of the caudal fin of equal size; posterior end of adipose fin at the same vertical plane as the posterior end of the anal fin; post-occipital process reaches to pre-dorsal plate
Pimelodella megalops

22-b Upper lobe of the caudal fin distinctly longer than the lower lobe; posterior end of adipose fin reaches over vertical plane at the posterior end of the anal fin; post-occipital process does not reach the pre-dorsal plate
17-b

23-a First ray of the pectoral fin very elongated reaching over the pelvic fin
Mastiglanis cf. *asopos*

23-b First ray of the pectoral fin not elongated never reaching the pelvic fin
24

24-a Long and slender species, length of head more than 4 times in standard length; dorsal and pectoral fins without spines; dorsal fin positioned posterior or equal to the pelvic fins
25

24-b Robust species with a blunt, heavy, more or less depressed head, head length <4 times in standard length; dorsal and pectoral fins with strong spines; dorsal fin positioned in front of the pelvic fins
29

25-a Adipose fin broadly connected with caudal fin; caudal fin lanceolate with upper lobe much better developed than lower lobe; anal fin with 20-22 rays
Heptapterus bleekeri
25-b Adipose fin entirely free from caudal fin or separated from it by a deep notch; caudal fin with two lobes; anal fin with less than 20 rays
26

26-a Anal fin with 17-19 rays; adipose fin separated from caudal by a deep notch
Chasmocranus surinamensis
26-b Anal fin with 10-16 rays; adipose fin entirely free from caudal fin
27

27-a Anal fin with 16 rays
Phenacorhamdia tenuis
27-b Anal fin with 13 (rarely), 14 or 15 rays
Heptapterus tapanahoniensis
27-c Anal fin with 10-12 rays
28

28-a Tail forked, but the lobes posteriorly rounded; length of adipose fin equal to or longer than distance between dorsal and adipose fins
Chasmocranus longior
28-b Tail forked with pointed lobes; length of adipose fin shorter than distance between dorsal and adipose fins
Chasmocranus brevior

29-a Lateral line complete; large to medium-sized species, standard length >4 cm; maxillary teeth laterally with a backward projection
30
29-b Lateral line not complete; small species, standard length <4 cm; maxillary teeth without backward projection
33

30-a Caudal fin forked; adpressed dorsal fin reaching no more than about half-way the distance between dorsal and adipose fins; pectoral fins with 7 rays
Pseudopimelodus bufonius
30-b Caudal fin truncated or rounded, never forked; adpressed dorsal fin usually reaching well beyond half way to the adipose fin; pectoral fin with 6 rays
31

31-a Caudal fin rounded; barbels short, only reaching to the pectoral fins
Cephalosilurus nigricaudus
31-b Caudal fin emarginate to truncated; barbels long, reaching to the dorsal fin
32

32-a Head and body dark above, with a fairly narrow, but conspicuous pale band across the nape, reaching from pectoral to pectoral; caudal fin with a conspicuous vertical pale band

Batrochoglanis raninus

32-b Body coloration dark grey-brown, more or less mottled, without a nape band or other clear markings

Batrochoglanis villosus

33-a Presence of a vertical black band on the caudal fin; pectoral with I,5 rays; tip of pectoral spine pointed

Microglanis poecilus

33-b Absence of vertical black band on caudal fin; pectoral fin with I,6 rays; tip of pectoral spine divided in two teeth, one pointing outwards and one pointing backwards

Microglanis secundus

IMPARFINIS HASEMANI (Steindachner, 1915) (an *Imparfinis* species with a dark longitudinal band) was identified from Upper Marowijne River by M. Sabaj-Pérez.

FAMILY PSEUDOPIMELODIDAE (BUMBLEBEE AND DWARF MARBLED CATFISHES)

Members of the family Pseudopimelodidae are robust catfishes with a blunt, heavy, more or less depressed head, strong spines in the dorsal and pectoral fins, and the dorsal fin positioned in front of the pelvic fins; they can further be distinguished by their wide mouth, small eyes without free orbital rim, and short barbels; the size is variable ranging from 5 to 40 cm. The biology of the species of the family is poorly known, but some of the larger species may be sedentary sit-and-wait predators. The small *Microglanis* species are often found in dead leaf litter where they probably feed on macro-invertebrates. Pseudopimelodid species occur in both small forest creeks and in the main river channel, in calm water and in rapids with strong currents. Some species are popular aquarium fishes, noted for their body coloration of dark brown blotches.

BATROCHOGLANIS RANINUS (Valenciennes, 1840)
Local name(s): –

Diagnostic characteristics: lateral line complete; medium-sized species, standard length >4 cm; maxillary teeth laterally with a backward projection; caudal fin emarginate to truncated; barbels long, reaching to the dorsal fin; head and body dark above, with a fairly narrow, but conspicuous pale band across the nape, reaching from pectoral to pectoral; caudal fin with a conspicuous vertical pale band
Type locality: La Mana, near Rio de Janeiro. Brazil
Distribution: Amazon River basin: Bolivia, Brazil, French Guiana, Guyana, Peru and Suriname; in Suriname, known from the Marowijne and Suriname rivers
Size: 9 cm SL
Habitat: main channel of rivers of the Interior and large tributaries with slow current; during the day it rests under stones or large woody debris
Position in the water column: bottom
Diet: feeds on animal prey that can pass through its mouth, micro-crustaceans and aquatic insects when young, and fish later in its life cycle
Reproduction:
Remarks:

BATROCHOGLANIS VILLOSUS (Eigenmann, 1912)
Local name(s): –

Diagnostic characteristics: lateral line complete; large to medium-sized species, standard length >4 cm; maxillary teeth laterally with a backward projection;

Batrochoglanis raninus (live), Marowijne River

Batrochoglanis raninus (live) (© INRA-Le Bail)

Batrochoglanis villosus (live) (© R. Covain)

caudal fin emarginate to truncated; barbels long, reaching to the dorsal fin; body coloration dark grey-brown, more or less mottled, without a nape band or other clear markings
Type locality: Potaro Landing, Potaro River, Guyana
Distribution: Demerara, Essequibo, Orinoco and Amazon River basins: Brazil, Guyana, Suriname and Venezuela; in Suriname, known only from Corantijn River
Size: 17 cm SL
Habitat:
Position in the water column: bottom
Diet:
Reproduction:
Remarks: morphologically very similar to *B. raninus*, but appears to attain a larger size and differs also by its much duller coloration (especially the pale, spotted pelvic and caudal fins).

Cephalosilurus nigricaudus (Mees, 1974)
Local name(s): –

Diagnostic characteristics: lateral line complete; large to medium-sized species with a very broad and depressed head, angular at the gape, above brownish with a complex pattern of black markings; fins generally black with a white margin (especially the dorsal and caudal fins of juveniles); maxillary teeth laterally with a backward projection; caudal fin long, lanceolate, truncated or rounded, never forked, with a white margin; adpressed dorsal fin usually reaching well beyond half way to the adipose fin; pectoral fin with 6 rays; barbels short, only reaching to the pectoral fins
Type locality: Sipaliwini River, Upper Corantijn River, Suriname
Distribution: Upper Corantijn (Sipaliwini), Suriname and Marowijne rivers, Suriname
Size: 27.1 cm TL
Habitat: rivers of the Interior (i.e. upstream of the first rapids), in pools and counter-current habitats of rapids with sandy or rocky bottom and large woody debris
Position in the water column: bottom
Diet:
Reproduction:
Remarks:

Microglanis poecilus Eigenmann, 1912
Local name(s): –

Diagnostic characteristics: lateral line not complete; small species, standard length <4 cm; maxillary teeth without backward projection; presence of a vertical black band on the caudal fin; pectoral with I,5 rays; tip of pectoral spine pointed
Type locality: below Packeoo Falls, Guyana

Cephalosilurus nigricaudus (alcohol), 310 mm SL, Lawa River (Marowijne River), ANSP189087 (© M. Sabaj Pérez)

Microglanis poecilus (alcohol), Coppename River (© M. Littmann)

Distribution: Essequibo River basin and rivers of French Guiana: Guyana and French Guiana
Size: 3.2 cm SL
Habitat: often found in leaf litter in small or medium-sized forest streams in the Interior
Position in the water column: bottom
Diet:
Reproduction:
Remarks:

MICROGLANIS SECUNDUS Mees, 1974
Local name(s): –

Diagnostic characteristics: lateral line not complete; small species, standard length <4 cm; maxillary teeth without backward projection; absence of vertical black band on caudal fin; pectoral fin with I,6 rays; tip of pectoral spine divided in two teeth, one pointing outwards and one pointing backwards
Type locality: Sipaliwini, Suriname
Distribution: Catatumbo River basin, Guyana; Sipaliwini River basin. Suriname: Venezuela and Colombia: Guyana, Suriname, Colombia and Venezuela
Size: 3.8 cm SL
Habitat: occurs in the same leaf litter habitat and the same rainforest streams as *M. poecilus*
Position in the water column: bottom
Diet:
Reproduction:
Remarks: the disruptive coloration of both *M. poecilus* and *M. secundus* provides for good camouflage in their natural environment. *M. secundus* is a slightly larger species than *M. poecilus.*

PSEUDOPIMELODUS BUFONIUS (Valenciennes, 1840)
Local name(s): –

Diagnostic characteristics: lateral line complete; large to medium-sized species, head less broad and compressed as in other pseudopimelodids, rounded at the gape; its color whitish to pale brownish and dark, the head fairly dark, mottled, with a broad pale nuchal band; maxillary teeth laterally with a backward projection; caudal fin forked, pale with a blackish base and one blackish vertical bar; adpressed dorsal fin reaching no more than about halfway the distance between dorsal and adipose fins; pectoral fins with 7 rays
Type locality: probably Cayenne, French Guiana (see Mees, 1974)
Distribution: rivers of northeastern South America from Lake Maracaibo basin to eastern Brazil: Brazil, Colombia, French Guiana, Suriname, Venezuela and Guyana; in Suriname, known from Corantijn, Nickerie, Suriname and Marowijne rivers

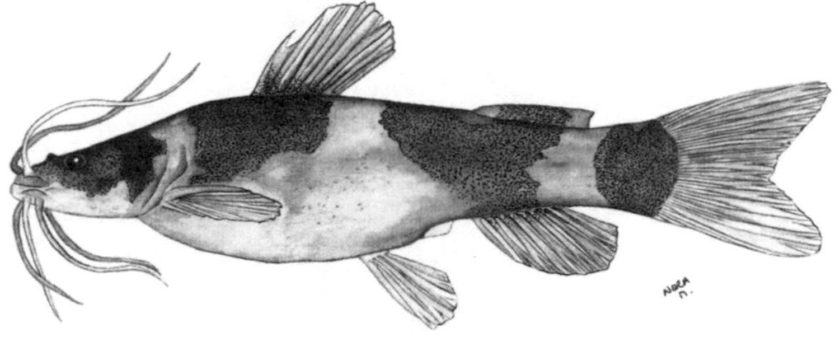

Microglanis secundus (© INRA-Le Bail)

Pseudopimelodus bufonius (live), scale bar = 1 cm, Litani River (Marowijne River), ANSP189098 (© M. Sabaj Pérez)

Size: 25.7 cm SL
Habitat: rapids in rivers of the Interior and large forest streams with fast flowing water and rocky bottom
Position in the water column: bottom
Diet:
Reproduction:
Remarks:

This page is intentionally left blank.

FAMILY HEPTAPTERIDAE (HEPTAPTERIDS)

The family Heptapteridae includes many catfishes long classified together in the Pimelodidae, but now placed in a more restricted version of Pimelodidae. The family was first recognized by Regan in 1911 based on osteological characters, but heptapterids lack unique externally visible characters, making difficult their distinction from some members of the families Pimelodidae and Pseudopimelodidae. Most species are small-sized (<20 cm; species of *Rhamdia* and *Pimelodella* can exceed) and slender (body width generally more than 5 times in standard length) and have three pairs of barbels (maxillary, inner and outer mentals).

BRACHYRHAMDIA HETEROPLEURA (Eigenmann, 1912)
Local name(s): –

Diagnostic characteristics: no free orbital rim (eye continuous with epidermis of head); the occipital process does not taper towards the dorsal base but is of almost equal width over its whole length; in this character it agrees with *Pimelodella* from which genus it differs, however, in having the fontanel reaching backwards only to level with the posterior border of the eyes, whereas *Pimelodella* has the fontanel continued as a narrow slit right to the base of the occipital process; dorsal fin reaches adipose fin when adpressed; absence of a well-defined longitudinal band on the flanks; very small species (<5 cm standard length); body mostly with light pigmentation; dorsal spine and first dorsal ray colored blackish; caudal fin deeply forked; teeth very small and poorly developed, in bands in each jaw; Corantijn River
Type locality: Rupununi Pan (Essequibo River basin, Guyana)
Distribution: Corantijn, Essequibo and Negro River basins; Guyana, Suriname and northern Brazil; in Suriname, only known from 3 specimens from Baruba Creek, left tributary of Dalbana Creek, Kabalebo River, Corantijn River System (Mees, 1985)
Size: 4.6 cm SL
Habitat:
Position in the water column: bottom
Diet:
Reproduction:
Remarks:

CHASMOCRANUS BREVIOR Eigenmann, 1912
Local name(s): –

Diagnostic characteristics: no free orbital rim (eye continuous with epidermis of head); first ray of pectoral fin not elongated, never reaching the pelvic fin; long

Brachyrhamdia heteropleura (alcohol), ANSP179740 (© M. Sabaj Pérez)

Chasmocranus brevior (alcohol), ANSP179713 (© M. Sabaj Pérez)

and slender species, length of head more than 4 times in standard length; dorsal and pectoral fins without spines; dorsal fin positioned posterior or equal to the pelvic fins; anal fin with 10-12 rays; adipose fin entirely free from caudal fin; caudal fin forked with pointed lobes; length of adipose fin shorter than distance between dorsal and adipose fins
Type locality: Warratuk, Potaro River, Guyana
Distribution: Mana and Marowijne/Maroni River basins and Potaro River: Guyana, Suriname and French Guiana
Size: 7.9 cm SL
Habitat:
Position in the water column: bottom
Diet:
Reproduction:
Remarks: a very rare species, only known from its *type locality* in Guyana, two localities in Marowijne River (Nassau Mountains and Antécume pata) and one locality in Mana River (French Guiana)

CHASMOCRANUS LONGIOR Eigenmann, 1912
Local name(s): –

Diagnostic characteristics: no free orbital rim (eye continuous with epidermis of head); first ray of pectoral fin not elongated, never reaching the pelvic fin; long and slender species, length of head more than 4 times in standard length; dorsal and pectoral fins without spines; dorsal fin positioned posterior or equal to the pelvic fins; anal fin with 10-12 rays; adipose fin entirely free from caudal fin; caudal fin forked, but the lobes posteriorly rounded; length of adipose fin equal to or longer than distance between dorsal and adipose fins; color dark grey-brown with a whitish vertical band just behind the head, the dorsal and caudal fins have a transparent/whitish margin
Type locality: Amatuk, Essequibo River, Guyana
Distribution: widespread in northern South America: Brazil, Guyana, Suriname and Venezuela; in Suriname present in all rivers
Size: 12 cm SL
Habitat: between pebbles and rocky bottom substrate in shallow water with moderate to strong current, in rapids (between Podostemaceae) or riffles in forest creeks of the Interior
Position in the water column: bottom
Diet:
Reproduction:
Remarks: a relatively common species.

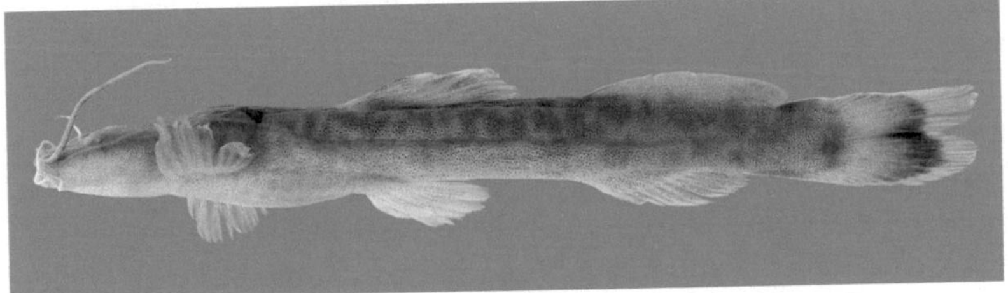

Chasmocranus longior (alcohol), ANSP179708 (© M. Sabaj Pérez)

Chasmocranus surinamensis (Bleeker, 1862)
Local name(s): –

Diagnostic characteristics: no free orbital rim (eye continuous with epidermis of head); long and slender species, length of head more than 4 times in standard length; dorsal and pectoral fins without spines; dorsal fin positioned posterior or equal to the pelvic fins; anal fin with 17-19 rays; adipose fin separated from caudal by a deep notch; color dull brown without markings, the ventral surface paler, along the side-line slightly darker (the species is somewhat lighter in color than *C. longior* and *Heptapterus tapanahoniensis*)
Type locality: Suriname [Suriname River]
Distribution: only known from Suriname River basin, Suriname
Size: 13 cm SL
Habitat: main river channel
Position in the water column: bottom
Diet:
Reproduction:
Remarks:

Heptapterus bleekeri Boeseman, 1953
Local name(s): –

Diagnostic characteristics: no free orbital rim (eye continuous with epidermis of head); long and slender species, length of head more than 4 times in standard length; dorsal and pectoral fins without spines; dorsal fin positioned posterior or equal to the pelvic fins; adipose fin broadly connected with caudal fin; caudal fin either lanceolate or with unequally developed lobes; anal fin with 20-22 rays
Type locality: Marowijne basin [*type locality* was corrected to "creek on eastern side of Nassau Mountains, belonging to the Marowijne/Maroni River basin, Suriname" by Mees (1983:55)]
Distribution: Marowijne (Maroni) River basin (Nassau Mountains and Upper Marowijne River), Oyapock River, and state of Amapá: Brazil, French Guiana and Suriname
Size: 15.5 cm SL
Habitat: upper headwater reaches of rivers in the Interior and small forest creeks
Position in the water column: bottom
Diet:
Reproduction:
Remarks: apparently a very rare species.

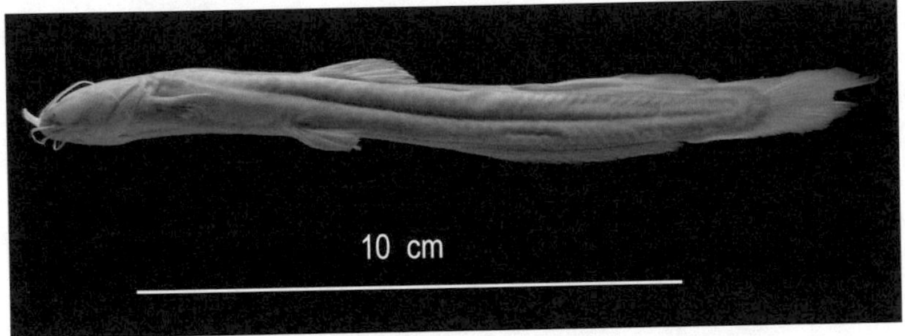

Heptapterus bleekeri (alcohol), RMNH28594 (© Naturalis)

Heptapterus bleekeri (live) (© INRA-Le Bail)

Heptapterus tapanahoniensis Mees, 1967
Local name(s): –

Diagnostic characteristics: no free orbital rim (eye continuous with epidermis of head); long and slender species, length of head more than 4 times in standard length; dorsal and pectoral fins without spines; dorsal fin positioned posterior or equal to the pelvic fins; anal fin with 13 (rarely), 14 or 15 rays; adipose fin entirely free from caudal fin
Type locality: Tapanahony River, about two km downstream from its confluence with the Paloemeu River, Suriname
Distribution: Marowijne (Maroni) and Sinnamary River basins: French Guiana and Suriname; apparently also present in Guyana and Venezuela
Size: 13 cm SL
Habitat: in shallow, clear water of rapids and forest creeks in the Interior (i.e. upstream of the first rapids) with moderate to strong currents and rocky bottom
Position in the water column: bottom
Diet:
Reproduction:
Remarks: can be locally abundant.

Imparfinis aff. *stictonotus* (Fowler, 1940)
Local name(s): –

Diagnostic characteristics: a small *Imparfinis* species (<5 cm SL) without distinctive pigmentation (e.g. absence of both a black blotch on the dorsal fin and a well-defined longitudinal band on the flanks); adpressed dorsal fin falling distinctly short of adipose fin; maxillary barbels reaching over the pelvic fins; tip of first rays of the pectoral and pelvic fins soft
Type locality: Todos Santos, Rio Chapare, Bolivia
Distribution: Mamoré/Madeira, Paraguay and Ucayali River basins: Bolivia, Brazil and Ecuador; the Surinamese species is known from Upper Corantijn River
Size: 4.8 cm SL
Habitat: small forest creeks
Position in the water column: bottom
Diet:
Reproduction:
Remarks:

Imparfinis pijpersi (Hoedeman, 1961)
Local name(s): –

Diagnostic characteristics: no free orbital rim (eye continuous with epidermis of head); post-occipital process generally not reaching the predorsal plate, but if it

Heptapterus tapanahoniensis (alcohol), ANSP180024 (© M. Sabaj Pérez)

Imparfinis aff. *stictonotus* (alcohol), Sipaliwini River (Corantijn River) (© P. Willink)

Imparfinis pijpersi (live), Nickerie River (© R. Smith)

reaches the predorsal plate then narrow and with an uniform width from its base to posterior end; body width generally more than 5 times in standard length; adpressed dorsal fin falling distinctly short of adipose fin; absence of a black blotch at the tip of the dorsal fin and a well-defined longitudinal band on the flanks; maxillary barbels not reaching the pelvic fins; tips of the first rays of the pectoral and pelvic fins soft

Type locality: Sipaliwini River, Upper Corantijn River, Suriname, 20 kilometers from frontier with Brazil

Distribution: Corantijn, Nickerie, Suriname and Marowijne rivers, Suriname; also present in French Guiana (Le Bail *et al.*, 2000, 2012)

Size: 9.5 cm SL

Habitat: present in both rapids with strong currents and small streams with moderate current and sandy bottom

Position in the water column: bottom

Diet:

Reproduction:

Remarks: Le Bail *et al.* (2000) identified this species as *Imparfinis minutus*.

Mastiglanis cf. *asopos* Bockmann, 1994

Local name(s): –

Diagnostic characteristics: no free orbital rim (eye continuous with epidermis of head); first ray of the pectoral fin very elongated reaching over the pelvic fin, translucent body

Type locality: Igarape Saracazinho, tributary of Rio Trombetas, near Porto Trombetas, Para, Brazil

Distribution: Amazon, Capim and Orinoco River basins: Brazil and Venezuela; in Suriname, known from the Upper Marowijne River

Size: 7 cm SL

Habitat: in side channels of rapids, both at sites with strong current and rocky substrate and at sites with slow current and sandy bottom and leaf litter; hides in the sand during the day (Zuanon *et al.*, 2006)

Position in the water column: bottom

Diet: a sit-and-wait predator that feeds during the night on aquatic insect larvae by spreading its long barbels and filamentous pectoral-fin rays in a drift trap-like device (Zuanon *et al.*, 2006)

Reproduction: probably in the rainy season (Zuanon *et al.*, 2006)

Remarks: Le Bail *et al.* (2000) identified this species as *Megalonema* cf. *platycephalus*. *M. asopos* is a translucent sand-dwelling species that hunts at night and hides in the sand during the day, much like the gymnotiform knifefish *Gymnorhamphichthys rondoni*; adaptations for this psammophilous way of life are discussed by Zuanon *et al.* (2006).

Mastiglanis cf. *asopos* (live), 43.2 mm SL, Lawa River (Marowijne River), ANSP189106 (© M. Sabaj Pérez)

Phenacorhamdia tenuis (Mees, 1986)
Local name(s): –

Diagnostic characteristics: no free orbital rim (eye continuous with epidermis of head); an extremely slender species with a free adipose fin, an anal fin with 16 rays, and a rather small, distinctly forked caudal fin; coloration much like *Heptapterus tapanahoniensis*, earth brown on the upper parts, laterally with fine spotting, unpigmented ventral surface, fins hyaline, except for the base of the caudal fin which is blackish; there is a narrow blackish band along the lateral line
Type locality: Crique Cascade, tributary to Lower Marowijne (Maroni) River
Distribution: Marowijne, Mana and Approuague rivers, French Guiana (and probably Suriname)
Size: 7.3 cm SL
Habitat: only known from small forest creeks with moderate to strong current and rocky bottom
Position in the water column: bottom
Diet:
Reproduction:
Remarks: a very rare species, until now only collected in French Guiana (Le Bail *et al.*, 2000), but probably also present in Surinamese tributaries of the Marowijne River.

Pimelodella cristata (Müller & Troschel, 1848)
Local name(s): liba dyaki

Diagnostic characteristics: no free orbital rim (eye continuous with epidermis of head); post-occipital process narrow and with an uniform width from its base to posterior end, reaching the predorsal plate; body width generally more than 5 times in standard length; dorsal fin reaches adipose fin when adpressed; presence of a well-defined longitudinal band on the flanks; longitudinal black band thin and not continuing on the head; upper lobe of the caudal fin shorter or equal to the lower lobe of the caudal fin
Type locality: Takutu and Mahu [tributaries of Upper Rio Branco, Guyana]
Distribution: Argentina, French Guiana, Suriname and Guyana
Size: 34 cm SL
Habitat: *P. cristata* and *P. geryi* live in the same habitat, i.e. in the upper and middle reaches of rivers of the Interior and large to medium-sized tributaries, in water with moderate to slow current
Position in the water column: bottom
Diet:
Reproduction: mature males have larger eyes (<4 times in head length) than females (>4 times in head length)

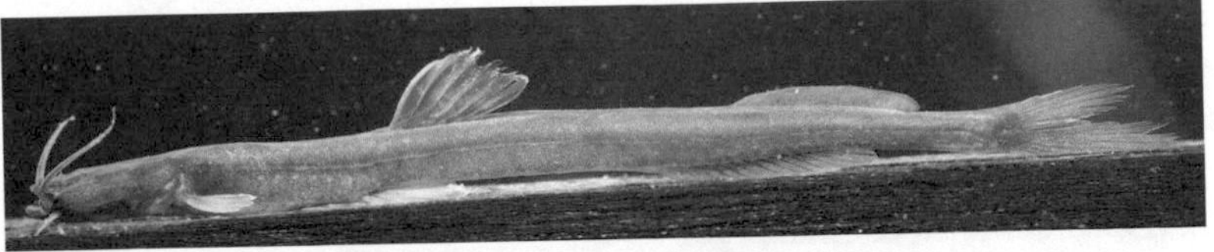

Phenacorhamdia tenuis (© INRA-Le Bail)

Pimelodella cristata (live) (© INRA-Le Bail)

Pimelodella cristata (live), Corantijn River (© R. Covain)

Remarks: very common and often abundant; the very sharp pectoral and dorsal spines of *Pimelodella* species may inflict painful wounds due to the presence of poisonous mucus.

PIMELODELLA GERYI Hoedeman, 1961
Local name(s): liba dyaki

Diagnostic characteristics: no free orbital rim (eye continuous with epidermis of head); post-occipital process narrow and with an uniform width from its base to posterior end, reaching the predorsal plate; body width generally more than 5 times in standard length; dorsal fin reaches adipose fin when adpressed; presence of a well-defined longitudinal band on the flanks; longitudinal black band thick and continuing on the head up to the tip of the snout; upper lobe of the caudal fin longer than the lower lobe of the caudal fin
Type locality: Litani River, Upper Marowijne (Maroni) River, village Aloiké, French Guiana
Distribution: Marowijne (Maroni) River basin, Suriname
Size: 5.8 cm SL
Habitat: *P. cristata* and *P. geryi* live in the same habitat, i.e. in the upper and middle reaches of rivers of the Interior and large to medium-sized tributaries, in water with moderate to slow current
Position in the water column: bottom
Diet: ontogenetic diet shifts are mentioned for this species by Le Bail *et al.* (2000) with early stages feeding on micro-crustaceans and aquatic insects and adults also feeding on fish
Reproduction:
Remarks: *P. geryi* is a much smaller species than *P. cristata*.

PIMELODELLA MACTURKI Eigenmann, 1912
Local name(s): liba dyaki

Diagnostic characteristics: no free orbital rim (eye continuous with epidermis of head); post-occipital process generally not reaching the predorsal plate, but if it reaches the predorsal plate then narrow and with an uniform width from its base to posterior end; body width generally more than 5 times in standard length; adpressed dorsal fin falling distinctly short of adipose fin; presence of a black blotch at the tip of the dorsal fin; longitudinal band on the flanks well defined, starting below the dorsal fin; presence of a humeral blotch of the size of the eye-pupil; eye diameter about equal to snout length; maximum depth of adipose fin in the posterior half of the fin base; pectoral spine with 12 hooks along its posterior margin; dorsal spine with serrae along its anterior and posterior margin
Type locality: creek in Mora Passage, coastal Guyana

Pimelodella geryi (live), scale bar = 1 cm, Lawa River (Marowijne River), ANSP189109 (© M. Sabaj Pérez)

Pimelodella macturki (live) (© INRA-Le Bail)

Distribution: Approuague, Corantijn and Nickerie River basins and eastern coastal drainages of Suriname (Marowijne River, Le Bail *et al.*, 2012): French Guiana, Guyana and Suriname
Size: 11 cm SL
Habitat: in small tributaries of lower reaches of rivers
Position in the water column: bottom
Diet:
Reproduction:
Remarks:

PIMELODELLA MEGALOPS Eigenmann, 1912
Local name(s): –

Diagnostic characteristics: no free orbital rim (eye continuous with epidermis of head); post-occipital process generally not reaching the predorsal plate, but if it reaches the predorsal plate then narrow and with an uniform width from its base to posterior end; body width generally more than 5 times in standard length; adpressed dorsal fin falling distinctly short of adipose fin; absence of both a black blotch at the tip of the dorsal fin and a well-defined longitudinal band on the flanks; maxillary barbels reaching over the pelvic fins; tips of the first rays of the pectoral and pelvic fins hard; lobes of the caudal fin of equal size; posterior end of adipose fin at the same vertical plane as the posterior end of the anal fin; post-occipital process reaches to predorsal plate
Type locality: Tumatumari (Guyana)
Distribution: Essequibo River, Guyana, and Approuague River, French Guiana; in Suriname, only known from Upper Marowijne River
Size: 7.9 cm SL
Habitat: sandy bottom substrate in main river channel
Position in the water column: bottom
Diet:
Reproduction:
Remarks: apparently a rare species.

PIMELODELLA PROCERA Mees, 1983
Local name(s): liba dyaki

Diagnostic characteristics: no free orbital rim (eye continuous with epidermis of head); post-occipital process generally not reaching the predorsal plate, but if it reaches the predorsal plate then narrow and with an uniform width from its base to posterior end; body width generally more than 5 times in standard length; adpressed dorsal fin falling distinctly short of adipose fin; presence of a black blotch at the tip of the dorsal fin; longitudinal band on the flanks well defined; eye diameter distinctly smaller than snout length; maximum depth of adipose fin situated halfway its base

Pimelodella megalops (alcohol), ANSP179751 (© M. Sabaj Pérez)

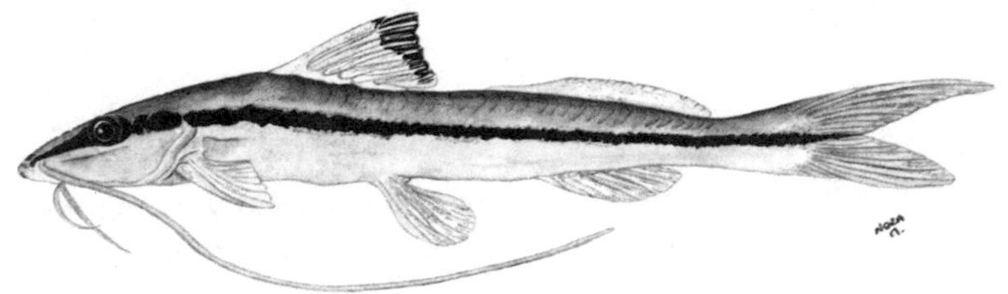

Pimelodella procera (© INRA-Le Bail)

Pimelodella procera (live), Marowijne River (© INRA-Le Bail)

Type locality: Crique Balaté, Marowijne River system, French Guiana
Distribution: only known from its type locality in a tributary of the Lower Marowijne (Maroni) River basin, French Guiana
Size: 10 cm SL
Habitat:
Position in the water column: bottom
Diet:
Reproduction:
Remarks:

RHAMDIA FOINA (Müller & Troschel, 1849)
Local name(s): dyaki

Diagnostic characteristics: free orbital rim (eye freely movable in orbit); eyes directed upward and outward; eyes large, eye diameter 4 times in head length, two-thirds of interorbital distance; post-occipital process very short, extending about one fourth of the distance to the pre-dorsal plate; occipital fontanel long and narrow, prolonged backward to the base of the post-occipital process; body width generally more than 5 times in standard length; dorsal fin reaches adipose fin when adpressed; absence of a well-defined longitudinal band on the flanks; medium-sized species (>15 cm standard length); dorsal spine and first dorsal ray not blackish
Type locality: Takutu, Guyana
Distribution: Essequibo, Branco, Negro, Tocantins and Trombetas River basins: Brazil and Guyana; in Suriname, known only from the (Upper) Marowijne River
Size: 16.5 cm SL
Habitat: clear, running water
Position in the water column: bottom
Diet:
Reproduction:
Remarks: Le Bail *et al.* (2000) identify this species as *Rhamdella* cf. *leptosoma* (now *Pimelodella leptosoma*; see Le Bail *et al.*, 2012), but *P. leptosoma* is a much smaller species (8 cm TL) with a black longitudinal band; the occurrence of *P. leptosoma* in Suriname needs confirmation.

RHAMDIA QUELEN (Quoy & Gaimard, 1824)
Local name(s): dyaki

Diagnostic characteristics: no free orbital rim (eye continuous with epidermis of head); post-occipital process not reaching the pre-dorsal plate; body width generally more than 5 times in standard length; dorsal fin reaches adipose fin when adpressed; absence of a well-defined longitudinal band on the flanks; medium-sized species (>15 cm standard length); dorsal spine and first dorsal ray not blackish; fins hyaline, but the dorsal fin has a clear band at its base; eyes small, eye

Rhamdia foina (live), scale bar = 1 cm, Lawa River (Marowijne River), ANSP189115 (© M. Sabaj Pérez)

Rhamdia quelen (live), probably Suriname River

Rhamdia quelen (live) (© INRA-Le Bail)

diameter >2 times in interorbital distance; fontanel does not continue posterior of the eyes
Type locality: between Caño Pastos and Hamburgo, tributary to Río Samiria, Depto. Loreto, Peru
Distribution: widespread in South America, from Mexico to Argentina; in Suriname, present in all rivers
Size: 34.8 cm SL
Habitat: prefers standing water or slow current and muddy bottoms with leaf litter in both coastal swamps and rivers of the Interior, but also caught in the swamps, canals and streams of the Coastal Plain; it is active during the night
Position in the water column: bottom
Diet: ontogenetic diet shifts are described by Le Bail *et al.* (2000), with newborn larvae feeding on micro-crustaceans (zooplankton), juveniles feeding on aquatic insects and adults feeding on shrimps and fish
Reproduction: fecundity is 1-5 x 10^5 eggs per kg; the eggs (diameter 1.1-2.8 mm) are non adhesive and hatch in about 48 hours at 22°C. After ten days the larvae weigh 100 mg, but growth is slow (0.5-1.15 g/day)
Remarks: poison glands are present at the base of the pectoral and dorsal fin spines.

This page is intentionally left blank.

FAMILY PIMELODIDAE (LONG-WHISKERED CATFISHES)

As delimited by Lundberg & Littmann (2003; p. 432-446 in Reis *et al.*, 2003) the Pimelodidae includes *Hypophthalmus* and excludes many catfishes long classified in Pimelodidae, but now placed in the families Heptapteridae and Pseudopimelodidae. Its members may be identified by a combination of features, including naked skin, nares well separated and lacking barbels, 3 pairs of barbels (maxillary, inner and outer mentals), adipose fin well developed, caudal fin emarginated, lobed or forked, gill membranes free, branchial openings not restricted, orbital rim free, and dorsal- and pectoral-fin spines pungent/stiff. The adults mostly measure between 20-80 cm SL (extremes include the giant *Brachyplatystoma filamentosum* > 3 m). In body shape most pimelodids have moderately depressed snouts and laterally compressed tails. They are mainly macrophagous carnivores or omnivores that consume large numbers of small fishes and invertebrates. Most are benthic, but *Hypophthalmus* is distinctly pelagic. Sex dimorphism is scarcely developed; pimelodids are externally fertilizing and are not known to practice parental care. Some species of *Brachyplatystoma* undertake long distance upriver migrations, presumably for spawning (Barthem *et al.*, 1991; Barthem & Goulding, 1997); the larvae and juveniles float downstream high in the water column of the main channel. They are most common in base level and lowland rivers with strong currents, and the migrating species are able to negotiate rapids. However, pimelodids do not reach high-gradient upland or mountain streams, and they are rare or absent from small forest streams and stagnant swamps. Many large-size pimelodids are important food resources throughout much of tropical South America. Several species are highly-priced ornamental species (e.g. the large *Phractocephalus hemioliopterus*).

BRACHYPLATYSTOMA FILAMENTOSUM (Lichtenstein, 1819)
Local name(s): lalaw

Diagnostic characteristics: eye small (about 5 times in snout length) and freely movable in orbit (orbital rim free); post-occipital process long, ending in a fork at the pre-dorsal plate; large patches of teeth on vomer and palatines; a very large species (>300 cm); body coloration uniform brown to grey, whitish ventrally; barbels long, maxillary barbels reaching well over the origin of the dorsal fin, up to the caudal fin; adipose fin short, equal in size to the anal fin; snout short and flattened, upper jaw distinctly protruding over lower jaw; a very large species
Type locality: Brazil
Distribution: Amazon and Orinoco River basins and major rivers of Guianas and northeastern Brazil: Argentina, Bolivia, Brazil, Colombia, Ecuador, French Guiana, Peru, Suriname and Venezuela

Brachyplatystoma filamentosum (live), scale bar = 5 cm, Suriname River, ANSP187105 (© M. Sabaj Pérez)

Size: 360 cm TL (>200 kg) (a photograph of a large specimen of about 170 cm TL from the middle Suriname River near the village Brokopondo, km 185, is shown in Mees, 1974, plate 15 [wrongly identified as *B. vaillantii*])
Habitat: estuaries and the main channel of large rivers in the Interior
Position in the water column: bottom
Diet: they feed on fishes about 1/3 of their own size
Reproduction: in the Amazon, some species of *Brachyplatystoma* undertake long distance upriver migrations, presumably for spawning (Barthem & Goulding, 1997); the larvae and juveniles then float downstream high in the water column of the main channel to the estuary which is the main feeding ground for the juveniles. In Suriname, large *B. filamentosum* are also caught both in the estuaries and far upriver, e.g. in the Suriname River near the village Brokopondo, km 185 (Mees, 1974) and in the Corantijn River near the Wonotobo rapids, km 350 (R. Covain, pers. communication)
Remarks: the largest catfish and also the largest freshwater fish of Suriname.

BRACHYPLATYSTOMA ROUSSEAUXII (Castelnau, 1855)
Local name(s): –

Diagnostic characteristics: eye small and freely movable in orbit (orbital rim free); large patches of teeth on vomer and palatines; large species (>130 cm; 50 kg); body coloration uniform silvery; barbels short, not reaching the origin of the dorsal fin; snout short and flattened, mouth terminal (with upper and lower jaw of the same length); post-occipital process short and triangular, not reaching the pre-dorsal plate
Type locality: Amazon River, Brazil
Distribution: South America: Bolivia, Colombia, Ecuador, French Guiana, Suriname, Peru, Venezuela; in Suriname, known from the estuaries and lower freshwater reaches of the Suriname and Corantijn rivers
Size: 192 cm TL
Habitat: in estuaries of large rivers, but in the Amazon this species possibly migrates far upriver (Rio Negro, Rio Madeira) to spawn
Position in the water column: bottom
Diet: a nocturnal piscivore
Reproduction: in the Amazon, upriver spawning migrations are known for *B. rousseauxii*, but these have not been observed in Suriname or French Guiana; Le Bail *et al.* (2000) identified this species as *B. flavicans* and speculate that *B. rousseauxii* caught in French Guiana are of Amazon origin
Remarks: Le Bail *et al.* (2000) identify this species as *B. flavicans*.

BRACHYPLATYSTOMA VAILLANTII (Valenciennes, 1840)
Local name(s): pasisi

Diagnostic characteristics: eye small and freely movable in orbit (orbital rim free); large patches of teeth on vomer and palatines; large species (>100 cm); body color-

Brachyplatystoma rousseauxii (live), scale bar = 5 cm, Suriname River, ANSP187109 (© M. Sabaj Pérez)

Brachyplatystoma vaillantii (live), scale bar = 5 cm, Suriname River, ANSP187107 (© M. Sabaj Pérez)

ation uniform brown or grey; barbels long, reaching well over the origin of the dorsal fin; adipose fin longer than anal fin (anal fin base about 2 times in adipose fin base); upper and lower jaws of approximately the same length
Type locality: Cayenne (French Guiana), Suriname
Distribution: Amazon and Orinoco River basins and major rivers of Guianas and northeastern Brazil: Bolivia, Brazil, Colombia, Ecuador, French Guiana, Peru, Suriname, and Trinidad; in Suriname, known from the lower freshwater reaches of the Suriname, Coppename and Corantijn rivers
Size: 150 cm TL
Habitat: estuaries of large rivers and, in the Amazon, far upriver in the main channel
Position in the water column: bottom
Diet: piscivore
Reproduction: in the Amazon, upriver spawning migrations are known
Remarks:

HEMISORUBIM PLATYRHYNCHOS (Valenciennes, 1840)
Local name(s): –

Diagnostic characteristics: eye freely movable in orbit (orbital rim free); large patches of teeth on vomer and palatines; large species (>50 cm); body coloration with a few (about 7) well-developed black spots; head strongly depressed (resembling a duck-bill); lower jaw strongly protruding over upper jaw; eyes large with diameter about 3 times in snout length; maxillary barbels reach up to the adipose fin; post-occipital process well developed, reaches pre-dorsal plate
Type locality: no locality [Brazil]
Distribution: Amazon, Maroni, Orinoco and Paraná River basins: Argentina, Bolivia, Brazil, Colombia, Ecuador, French Guiana, Guyana, Paraguay, Peru, Suriname and Uruguay; in Suriname, known from the Corantijn, Nickerie, Suriname and Marowijne rivers
Size: 52.5 cm SL
Habitat: main channel of rivers upstream of the first rapids
Position in the water column: bottom
Diet: fish
Reproduction:
Remarks: apparently not very abundant, but it was observed on the central market of Paramaribo.

HYPOPHTHALMUS MARGINATUS Valenciennes, 1840
Local name(s): kwasimama

Diagnostic characteristics: origin of anal fin anterior to the dorsal fin; eyes small and ventrolaterally placed at the same horizontal plane or below the mouth; head broad and depressed anteriorly, but higher posteriorly; body laterally compressed;

Hemisorubim platyrhynchos (live), Nickerie River (© A. Gangadin)

Hemisorubim platyrhynchos (live), lateral view, Suriname River

Hemisorubim platyrhynchos (live), dorso-lateral view, Suriname River

Hypophthalmus marginatus (live), scale bar = 5 cm, Suriname River, ANSP187103 (© M. Sabaj Pérez)

caudal fin deeply forked; interorbital distance 25-50 % of head length; pectoral fin length 13.2-18.2 % of SL
Type locality: Cayenne, French Guiana; Suriname
Distribution: Amazon and Orinoco River basins and major rivers of French Guiana and Suriname: Brazil, French Guiana, Peru and Suriname
Size: 45 cm SL
Habitat: lower freshwater reaches of rivers (i.e. downstream of the first rapids)
Position in the water column: mid-water
Diet: pelagic zooplanktivores or detrital filter feeders (Carvalho, 1980) feeding on cladocerans, copepods, ostracods, aquatic insect (larvae) and algae
Reproduction: fecundity 50,000-100,000 eggs per female (depending on the size)
Remarks: previously classified in the family Hypophthalmidae (lookdown catfishes), but molecular evidence places it in the Pimelodidae (J. Lundberg, pers. communication). This species is sold at the markets of Paramaribo.

PHRACTOCEPHALUS HEMIOLIOPTERUS (Bloch & Schneider, 1801)
Local name(s): mototyar (motorcar), switwatra jarabaka

Diagnostic characteristics: eye freely movable in orbit (orbital rim free); body coloration with a dark upper half and a white lower flank and partly or wholly (caudal fin) bright-orange fins; adipose fin partly rayed; head large and wide-mouthed, with a coarsely rugose bony structure covering the upper head and the large predorsal shield; Corantijn River
Type locality: Rio Maranham, Brazil
Distribution: Amazon and Orinoco River basins: Brazil, Colombia, Ecuador, Suriname, Guyana, Peru and Venezuela; in Suriname, only known from Corantijn River
Size: 132 cm TL (50 kg)
Habitat: main channel of Corantijn River and large tributaries (e.g. Kabalebo River), upstream of the first rapids
Position in the water column: bottom
Diet: fish
Reproduction:
Remarks: a popular sport fish and food fish; one of the most colorful catfishes, thus popular in large public aquariums.

PIMELABDITUS MOLI Parisi & Lundberg, 2009
Local name(s): –

Diagnostic characteristics: eye freely movable in orbit (orbital rim free); no teeth on vomer and palatines; small to medium-sized species (<15 cm SL); post-occipital process distinctly connected with the pre-dorsal plate, and with a broad base and tapering towards the posterior end where it meets the heart-shaped pre-dorsal plate; body elongated with body depth more than 5 times in standard length;

Phractocephalus hemioliopterus (live), Corantijn River (© P. Willink)

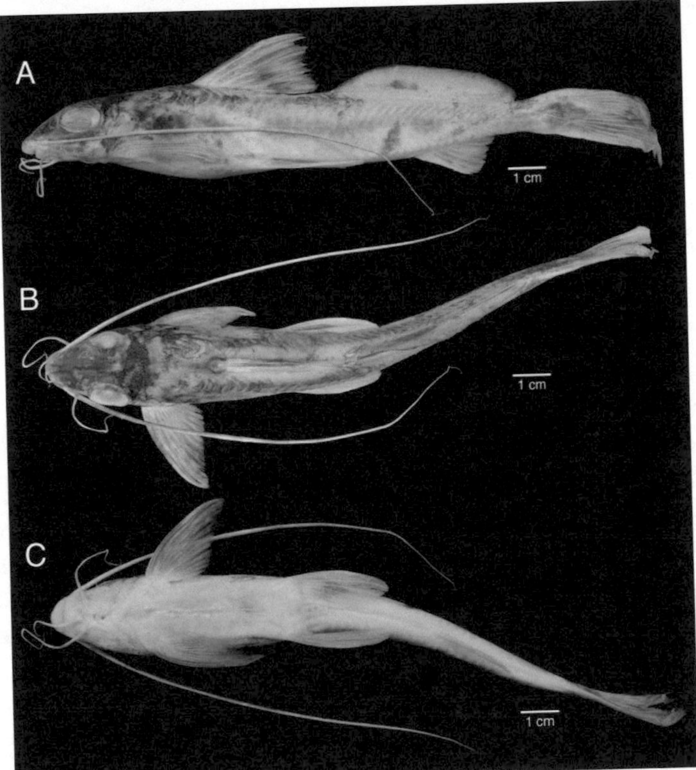

Pimelabditus moli (alcohol), holotype, Tapanahoni River (Marowijne River) (© J. Lundberg)

Pimelabditus moli (live), Tapanahoni River (Marowijne River)

adipose fin long (base of anal fin >2.5 times in base of adipose fin); eyes very large, horizontal diameter less than 3 times in head length and 1.2-1.4 times in snout length; mouth small, ventral, with enlarged, heavily toothed upper jaw; snout (2.2-2.3 in head length), with steep (45°) profile; Upper Marowijne River
Type locality: Tapanahony River, Marowijne River, Kumaru Konde Sula, 3°21.960'N, 55°25.926'W, Sipaliwini District, Suriname
Distribution: Upper Marowijne (Maroni) River and large tributaries: French Guiana and Suriname
Size: 14.1 cm SL
Habitat: in the immediate environment of rapids, but not in strong currents in the rapids themselves
Position in the water column: bottom
Diet:
Reproduction:
Remarks: the recent (2009) discovery of this new species and genus shows that the fish faunas of remote upper reaches of the Surinamese rivers are still not well known; molecular evidence shows that *P. moli* is most closely related to *Pimelodus ornatus* (Lundberg *et al.*, 2012).

PIMELODUS ALBOFASCIATUS Mees, 1974
Local name(s): –

Diagnostic characteristics: eye freely movable in orbit (orbital rim free); no teeth on vomer and palatines; small to medium-sized species (<30 cm); post-occipital process distinctly connected with the pre-dorsal plate, and with a broad base and tapering towards the posterior end where it meets the heart-shaped pre-dorsal plate; body width generally less than 5 times in standard length; absence of black blotch in dorsal fin and black marking on caudal fin; color dark grayish with a conspicuous broad white longitudinal band along the sides; diameter of eye larger than interorbital width (0.7-1.0 in interorbital)
Type locality: Sipaliwini River, Upper Corantijn River, Suriname
Distribution: Amazon, Orinoco, Upper Corantijn and Sipaliwini River basins: Brazil, Suriname and Venezuela; in Suriname only known from Corantijn and Nickerie rivers
Size: 15 cm SL
Habitat: upper reaches of Corantijn and Nickerie rivers
Position in the water column: bottom
Diet:
Reproduction:
Remarks: resembles *P. blochii*, but has larger eyes and a conspicuous white longitudinal band along the sides; in Suriname, *P. albofasciatus* is present in upper reaches, whereas *P. blochii* occurs in the lower reaches of rivers.

Pimelodus albofasciatus (live) (© R. Covain)

Pimelodus blochii Valenciennes, 1840
Local name(s): kaweri

Diagnostic characteristics: eye freely movable in orbit (orbital rim free); no teeth on vomer and palatines; small to medium-sized species (<30 cm); post-occipital process distinctly connected with the pre-dorsal plate, and with a broad base and tapering towards the posterior end where it meets the heart-shaped pre-dorsal plate; body width generally less than 5 times in standard length; absence of black blotch in dorsal fin and black marking on caudal fin; color plain, light liver-brown to silvery-green; eye small (1.0-1.7 in interorbital width); adipose fin triangular
Type locality: Suriname
Distribution: Gulf of Paria, Amazon, Corantijn, Essequibo and Orinoco River basins: Bolivia, Brazil, Colombia, Ecuador, French Guiana, Guyana, Peru, Suriname and Venezuela; in Suriname probably present in all rivers
Size: 22 cm SL
Habitat: estuaries and lower freshwater reaches of large rivers, downstream of the first rapids
Position in the water column: bottom
Diet: omnivore, feeding on fruits, insects and little fishes, but also on detritus
Reproduction: may undertake upriver spawning migrations; fecundity is about 50,000 eggs per female, fertilization is extern
Remarks: Mees (1974) mentions the presence of a crustacean parasite (Cymathoidae) in the mouth of a *P. blochii* specimen. Le Bail *et al.* (2000; p. 98) described a spotted form of *P. blochii* from French Guiana and this form is also present in Suriname in Para River (Le Bail *et al.*, 2000) and Nickerie River (personal observations).

Pimelodus ornatus Kner, 1858
Local name(s): kaweri, katfisi

Diagnostic characteristics: eye freely movable in orbit (orbital rim free); no teeth on vomer and palatines; small to medium-sized species (<30 cm); post-occipital process distinctly connected with the pre-dorsal plate, and with a broad base and tapering towards the posterior end where it meets the heart-shaped pre-dorsal plate; body width generally less than 5 times in standard length; presence of a well-defined black blotch on the dorsal fin, a yellow band above the lateral line, a broad longitudinal blackish-grey band along the lateral line and blackish bands on both lobes of the caudal fin; paired and adipose fins yellowish
Type locality: Suriname, Rio Negro and Cujaba [Brazil]
Distribution: widespread in South America; in Suriname, possibly present in all rivers
Size: 38.5 cm SL
Habitat: inhabits greater and smaller rivers of the Interior, not often found in small forest creeks

Pimelodus blochii (live), Mindrineti River (Saramacca River)

Pimelodus blochii spotted form (live), Nickerie River

Pimelodus ornatus (live), 169 mm SL, Lawa River (Marowijne River), ANSP187113 (© M. Sabaj Pérez)

Position in the water column: bottom
Diet: fish
Reproduction: Le Bail *et al.* (2000) mention the presence of spermatozoids in the genital tract of the female, which suggest internal fertilization in this species
Remarks: easily distinguished, especially by the black blotch on the dorsal fin. *Pimelodus ornatus* is related to *Pimelabditus moli* (Lundberg *et al.*, in press) and may be transferred to the genus *Pimelabditus* in the future.

PSEUDOPLATYSTOMA FASCIATUM (Linnaeus, 1766)
Local name(s): spigrikati

Diagnostic characteristics: eye freely movable in orbit (orbital rim free); large patches of teeth on vomer and palatines; large species (>60 cm SL) with big, moderately depressed head; lower jaw distinctly shorter than upper jaw; eyes small with diameter more than 4 times in snout length; body coloration with striking pattern of black spots and cross bands; <45 spots on the caudal fin; posterior end of fontanel closer to horizontal plane through the eyes than to the base of the post-occipital process
Type locality: Brazil and Suriname; Mees (1974) restricted the type locality to Suriname (also see Buitrago-Suárez & Burr, 2007)
Distribution: Guyana region: Guyana, Suriname and (French) Guyana; in Suriname, widely distributed, possibly in all rivers
Size: 90 cm Fork Length (12 kg)
Habitat: in the main channel of large rivers of the Interior, but also in deeper parts of smaller tributaries
Position in the water column: bottom
Diet: nocturnal piscivore, but also feeding opportunistically on shrimps and crabs
Reproduction: mature males are >45 cm, mature females >56 cm; fecundity may be as high as 8 million eggs per kg
Remarks: growth and maximum age of *P. fasciatum* in Bolivia was studied by Loubens and Panfili (2000).

PSEUDOPLATYSTOMA TIGRINUM (Valenciennes, 1862)
Local name(s): spigrikati

Diagnostic characteristics: eye freely movable in orbit (orbital rim free); large patches of teeth on vomer and palatines; large species (>100 cm); lower jaw distinctly shorter than upper jaw; eyes small with diameter more than 4 times in snout length; body coloration with striking pattern of >50 black spots in the caudal fin, reticulated loops and cross bands along the side of the body, no spots on the lateral region of the body; posterior end of fontanel closer to the base of the post-occipital process than to the horizontal plane through the eyes; Marowijne River
Type locality: Brazil

Pseudoplatystoma fasciatum (live), scale bar = 5 cm, Suriname River, ANSP187106 (© M. Sabaj Pérez)

Pseudoplatystoma tigrinum (live), Paloemeu River

Distribution: Amazon River: Bolivia, Brazil, Colombia, Ecuador, French Guiana, Suriname, Peru and Venezuela; in Suriname, only known from Marowijne River (Le Bail *et al.*, 2000), but its occurrence in this river needs confirmation
Size: 130 cm TL
Habitat: main channel of large rivers
Position in the water column: bottom
Diet: nocturnal piscivore, but also feeding opportunistically on shrimps and crabs
Reproduction:
Remarks: Le Bail *et al.* (2000) mention the presence of *Pseudoplatystoma tigrinum* in Marowijne (Maroni) River. Growth and maximum age of *P. tigrinum* in Bolivia was studied by Loubens and Panfili (2000).

This page is intentionally left blank.

FAMILY ARIIDAE (SEA CATFISHES)

The family Ariidae is a large group of medium to large-sized (20-120 cm TL) catfishes with a worldwide distribution associated with their occurrence in estuaries, brackish-water lagoons, tidal rivers and shallow marine habitats of warm-temperate and tropical regions; only few species are confined to freshwater or to deep, marine habitats. The species are all rather similar in general appearance; shape and arrangement of the vomer and associated tooth plates are used to distinguish species. The head has a conspicuous bony shield covered with thin skin or thick skin and muscles. Maxillary and mandibular (mental) barbels are usually present; anterior and posterior nostrils are positioned close together; eyes usually have a free orbital margin; caudal fin is deeply forked, the lateral line complete, branching at the caudal fin onto the dorsal and or ventral lobes. A well developed adipose fin is present in all species that occur in Suriname. In most, if not all species, the male carries the relatively large eggs in its mouth until hatching (oral incubation; Fig. 4.21). Names of the genera follow Marceniuk & Menezes (2007). Leopold (2004) presents information on the ariid catfishes from French Guiana (most of which also occur in Suriname).

1-a Two pairs of barbels, one pair of maxillary barbels and one pair of mental (mandibular) barbels; maxillary barbels and filaments of dorsal and pectoral fins appearing as long, flattened ribbons (*Bagre*)
2

1-b Three pairs of barbels, one pair of maxillary barbels and two pairs of mental barbels, all round in cross-section
3

2-a Anal fin long, 29-37 rays
Bagre bagre

2-b Anal fin short, 22-28 rays
Bagre marinus

3-a Presence of a furrow, partly covered by a flap of skin, extending across the snout between the posterior nostrils
4

3-b No fleshly furrow between the posterior nostrils
6

4-a Snout short and square, mouth terminal or nearly so
Sciades passany

4-b Snout long and rounded; mouth inferior
5

5-a Supra-occipital slightly keeled; usually 21-23 gill rakers on second arch; palatine teeth patch always with an U-form; medium-sized species (40 cm)

Sciades herzbergii

5-b Supra-occipital process without a median keel, rounded from above; usually 19 or 20 gill rakers on second arch; palatine teeth patches forming bands in juveniles, with U-form in adults; large species (up to 80 cm)

Sciades couma

6-a Presence of a longitudinal fleshly groove in the median depression of the head; palate with one patch of molariform teeth on each side (*Cathorops*)

12

6-b Absence of a fleshy groove in the median depression of the head; palatine teeth filiform

7

7-a Predorsal plate enlarged, shield-shaped, its length more than ½ of supra-occipital process

8

7-b Predorsal plate small, chevron- or crescent-shaped, much shorter than supra-occipital process

10

8-a Predorsal plate notched anteriorly, enclosing the tip of the narrow supra-occipital process; predorsal plate uniformly rugose; maxillary barbels extending only to pectoral fins

Sciades proops

8-b Predorsal plate without anterior notch or with shallow notch only (*Aspistor quadriscutis*)

9

9-a Predorsal plate with saddle (butterfly) shape and a shallow anterior notch; maxillary barbels extending to pectorals only; medium-sized species (50 cm)

Aspistor quadriscutis

9-b Predorsal plate shield shaped without anterior notch; maxillary barbels extending to anal fin in juveniles, becoming shorter with age; large species (up to 150 cm)

Sciades parkeri

10-a Supra-occipital process usually narrower at the base than distally, with near parallel sides or the sides variously extended distally into a large rounded plate

Notarius grandicassis

10-b Supra-occipital process much broader at the base than distally, the sides converging posteriorly to meet the predorsal plate

11

11-a Total anterior gill rakers on first arch 12-15; total anterior gill rakers on second arch 13-16

Amphiarius phrygiatus

11-b Total anterior gill rakers on first arch 14-17; total anterior gill rakers on second arch 16-20

Amphiarius rugispinis

12-a Live fishes colored yellowish (yellow color absent in fixed specimens); eyes small, eye diameter more than 6 times in head length

Cathorops arenatus

12-b Color of fishes (both live and in alcohol) silvery; eyes large, with eye diameter less than 6 times in head length

Cathorops spixii

AMPHIARIUS PHRYGIATUS (Valenciennes, 1840)
Local name(s): – (kukwari sea catfish)

Diagnostic characteristics: no fleshly furrow between the posterior nostrils; head broad, flattened above, snout rounded transversely, mouth inferior; predorsal plate small, chevron- or crescent-shaped, much shorter than supra-occipital process; supra-occipital process much broader at the base than distally, the sides converging posteriorly to meet the predorsal plate; total anterior gill rakers on first arch 12-15; total anterior gill rakers on second arch 13-16; color grey-brown dorsally, lighter ventrally
Type locality: Cayenne (interpreted by Boeseman (1972) as having come from Suriname)
Distribution: Atlantic coastal rivers and estuaries from Guyana to mouth of Amazon River: Brazil, French Guiana, Guyana, Suriname and Venezuela
Size: 30 cm TL
Habitat: mainly in estuaries
Position in the water column: bottom
Diet:
Reproduction: the female deposits the eggs in a depression on the bottom; the male practices oral incubation and guards the eggs until they hatch
Remarks:

AMPHIARIUS RUGISPINIS (Valenciennes, 1940)
Local name(s): – (softhead sea catfish)

Diagnostic characteristics: no fleshly furrow between the posterior nostrils; head flattened above, exposed head shield well visible, rugose, short, not extending forward to the eyes; predorsal plate small, chevron- or crescent-shaped, much shorter than the long supra-occipital process; supra-occipital process much broader at the base than distally, the sides converging posteriorly to meet the predorsal plate; total anterior gill rakers on first arch 14-17; total anterior gill rakers on second arch 16-2; color grey to reddish-brown dorsally, lighter ventrally
Type locality: Cayenne, French Guiana

Amphiarius rugispinis (live), ANSP178749 (© M. Sabaj Pérez)

Distribution: Atlantic coastal rivers from Guyana to mouth of Amazon River: Brazil, French Guiana, Guyana, Suriname, Trinidad and Tobago, and Venezuela
Size: 45 cm TL (common to 30 cm)
Habitat: turbid waters of estuaries and shallow coastal areas
Position in the water column: bottom
Diet: crabs, amphipods and shrimps
Reproduction: gonads develop at size >12 cm; 54 eggs (each 10 mm diameter) in a 27 cm female; oral incubation of the eggs by the male
Remarks: looks much like *Aspistor quadriscutis*, but is smaller and occurs in deeper water (20-30 m). *A. phrygiatus* has shorter barbels (not reaching the pectoral fins) and is even smaller (26 cm) than *A. rugispinis* (Leopold, 2004).

ASPISTOR QUADRISCUTIS (Valenciennes, 1840)
Local name(s): katfisi
Diagnostic characteristics: no fleshly furrow between the posterior nostrils; snout rounded, mouth inferior, dorsal head shield well visible and posteriorly very rough; absence of a fleshy groove in the median depression of the head; palatine teeth filiform; predorsal plate enlarged, shield-shaped, its length more than ½ of supra-occipital process (which is very short); predorsal plate with shallow notch only; predorsal plate with saddle (butterfly) shape and a shallow anterior notch; maxillary barbels extending to pectorals only; medium-sized species (50 cm) with a yellow-green-brown back and ventrally whitish
Type locality: Suriname
Distribution: Atlantic coastal rivers from Guyana to northeastern Brazil: Brazil, French Guiana, Guyana and Suriname
Size: up to 50 cm TL (1 kg)
Habitat: mainly marine, in shallow turbid water, but also in estuaries
Position in the water column: bottom
Diet: bottom-living invertebrates
Reproduction: in French Guiana spawning takes place in the dry season between September and November; eggs 9-11 mm in diameter, incubated by the male in his mouth
Remarks:

BAGRE BAGRE (Linnaeus, 1766)
Local name(s): barbaman, koko (coco sea catfish)
Diagnostic characteristics: two pairs of barbels, one pair of maxillary barbels and one pair of mental (mandibular) barbels; maxillary barbels and filaments of dorsal and pectoral fins appearing as long, flattened ribbons; anal fin long, 29-37 rays; color silvery grey to bluish grey above, lighter below; a large black spot usually present on anterior anal-fin rays
Type locality: America meridionali (Central America, probably erroneous); Brazil
Distribution: coastal rivers of Caribbean and Atlantic: Colombia to Amazon River mouth: Brazil, Colombia, French Guiana, Guyana and Venezuela

Aspistor quadriscutis (live), ANSP178740 (© M. Sabaj Pérez)

Bagre bagre (live), ANSP178751 (© M. Sabaj Pérez)

Size: 55 cm TL
Habitat: estuaries and shallow coastal marine areas
Position in the water column: bottom
Diet: small fishes and macroinvertebrates such as shrimps and polychaetes
Reproduction: reproductive season in French Guiana is from May to November; male incubates the eggs in his mouth until they hatch; for a few days the newborn (30-40 mm) may hide into the mouth of the male in the presence of danger
Remarks:

BAGRE MARINUS (Mitchill, 1815)
Local name(s): barbaman (gafftopsail sea catfish)

Diagnostic characteristics: two pairs of barbels, one pair of maxillary barbels and one pair of mental (mandibular) barbels; maxillary barbels and filaments of dorsal and pectoral fins appearing as long, flattened ribbons; anal fin short, 22-28 rays; color bluish grey to dark brown above, lighter below
Type locality: New York, U.S.A.
Distribution: Western Atlantic from northern U.S.A. to southern Brazil: Belize, Brazil, Colombia, Costa Rica, Cuba, French Guyana, Guyana, Honduras, Mexico, Nicaragua, Panama, Surinam, U.S.A., and Venezuela
Size: up to 100 cm TL (4.5 kg) (Leopold, 2004)
Habitat: common in estuaries and shallow coastal marine areas
Position in the water column: bottom
Diet: feeds mainly on small fishes and macroinvertebrates
Reproduction:
Remarks: when compared to *B. bagre*, *B. marinus* is less common in estuarine waters (more marine) and grows to a larger size.

CATHOROPS ARENATUS (Valenciennes, 1840)
Local name(s): katfisi

Diagnostic characteristics: no fleshly furrow between the posterior nostrils; presence of a longitudinal fleshly groove in the median depression of the head; palate with one patch of molariform teeth on each side; live fishes colored yellowish (yellow color absent in fixed specimens); eyes small, eye diameter more than 6 times in head length
Type locality: Suriname
Distribution: Atlantic coastal rivers from Guyana to northeastern Brazil: Brazil, French Guiana, Guyana and Suriname
Size: 25 cm SL
Habitat: estuaries and shallow coastal areas
Position in the water column: bottom
Diet: small fishes and macroinvertebrates
Reproduction:
Remarks: *C. fissus* is a junior synonym.

Cathorops arenatus (live) (© INRA-Le Bail)

Cathorops spixii (Spix & Agassiz, 1829)
Local name(s): katfisi (madamango sea catfish)

Diagnostic characteristics: no fleshly furrow between the posterior nostrils; presence of a longitudinal fleshly groove in the median depression of the head; palate with one patch of molariform teeth on each side; color of fishes (both live and in alcohol) silvery; eyes large, with eye diameter less than 6 times in head length
Type locality: equatorial Brazil
Distribution: Atlantic and Caribbean rivers and estuaries from Mexico to Brazil: Brazil, Colombia, French Guiana, Guyana, Mexico, Suriname and Venezuela
Size: 30 cm TL
Habitat: estuaries and shallow coastal areas
Position in the water column: bottom
Diet: small fishes and macroinvertebrates
Reproduction:
Remarks:

Notarius grandicassis (Valenciennes, 1840)
Local name(s): kodoku (Thomas sea catfish)

Diagnostic characteristics: no fleshly furrow between the posterior nostrils; mouth inferior; predorsal plate small, chevron- or crescent-shaped, much shorter than supra-occipital process; supra-occipital process usually narrower at the base than distally, with near parallel sides or the sides variously extended distally into a large rounded plate; color yellowish-grey dorsally, lighter ventrally
Type locality: Guyana
Distribution: rivers and estuaries from Gulf of Venezuela to mouth of Amazon River: Brazil, French Guiana, Guyana, Suriname and Venezuela
Size: 63 cm TL (mostly 40-50 cm)
Habitat: turbid waters of estuaries and shallow coastal areas
Position in the water column: bottom
Diet: detritus, mud and small shrimps and fishes
Reproduction: apparently in the period October-November, during which males incubating large eggs (10-12 mm) have been observed (Leopold, 2004)
Remarks:

Sciades couma (Valenciennes, 1840)
Local name(s): kumakuma (couma sea catfish)

Diagnostic characteristics: presence of a furrow, partly covered by a flap of skin, extending across the snout between the posterior nostrils; snout long and rounded, mouth slightly inferior, head only slightly flattened above; exposed head shield well visible and extending forward to opposite the eyes; supra-occipital process without a median keel, rounded from above; usually 19 or 20 gill rakers on

Cathorops spixii (live) (© INRA-Le Bail)

Notarius grandicassis (live) (© INRA-Le Bail)

Sciades couma (live), ANSP178747 (© M. Sabaj Pérez)

second arch; palatine teeth patches forming bands in juveniles, with U-form in adults; large species (up to 80 cm); yellowish grey to dark greyish brown above, whitish below
Type locality: Cayenne, French Guiana
Distribution: Gulf of Paria to mouth of Amazon River: Brazil, Colombia, French Guiana, Guyana, Suriname and Venezuela
Size: 97 cm TL
Habitat: estuaries and shallow coastal areas, but may migrate upriver to lower freshwater reaches of coastal rivers
Position in the water column: bottom
Diet: fish, but mainly crustaceans (especially crabs) (Rojas-Beltran, 1989)
Reproduction: females deposit about 100-165 eggs (2 cm diameter) in a depression in the bottom and, after fertilization, the male incubates the eggs in its mouth
Remarks: in French Guiana, *S. couma* has two periods of growth, corresponding with the two dry seasons of March and August-November (Lecomte *et al.*, 1985).

SCIADES HERZBERGII (Bloch, 1794)
Local name(s): wetkati (pemecou sea catfish)

Diagnostic characteristics: presence of a furrow, partly covered by a flap of skin, extending across the snout between the posterior nostrils; snout long and rounded, mouth slightly inferior; supra-occipital process slightly keeled; usually 21-23 gill rakers on second arch; palatine teeth patch always with an U-form; medium-sized species; soft rays in pectoral fins usually 10 or 11; color grey to dark brown above, whitish below
Type locality: Suriname
Distribution: Caribbean and Atlantic draining rivers and estuaries from Colombia to Brazil: Brazil, Colombia, French Guiana, Guyana, Suriname and Venezuela
Size: 54 cm TL (common to 40 cm)
Habitat: a euryhaline species, very tolerant of changes in salinity, known to occur in turbid estuaries, mangrove-lined coastal lagoons (e.g. Bigi Pan Lagoon) and lower reaches of rivers
Position in the water column: bottom
Diet: fishes, worms, pelagic micro-crustaceans, benthic shrimps, detritus
Reproduction: takes place in the period September-December; about 20-30 large (10-12 mm) eggs per female, which are incubated by the male in its mouth (Fig. 4.21)
Remarks: spines are poisonous and may inflict painful wounds that sometimes result in long-lasting effects, e.g. a fisherman that stepped on a *S. herzbergii* catfish when walking bare-footed through a shallow lagoon developed symptoms of the 'Guillain-Barré' syndrome, i.e. 'weak muscles in both arms and legs' that were slow to heal (see chapter 8).

This page is intentionally left blank.

SCIADES PARKERI (Trail, 1832)
Local name(s): yarabaka, geribaka, geelbagger (gillbacker sea catfish)

Diagnostic characteristics: no fleshly furrow between the posterior nostrils; head flattened above, exposed head shield well visible, rugose posteriorly, but smoother anteriorly, extending forward to opposite eyes; absence of a fleshy groove in the median depression of the head; mouth moderately inferior; palatine teeth filiform; predorsal plate enlarged, shield-shaped, its length more than ½ of supraoccipital process; predorsal plate without anterior notch; maxillary barbels extending to anal fin in juveniles, becoming shorter with age; large species with yellow color (up to 190 cm and 40 kg according to Leopold, 2004)
Type locality: Guiana [as British Guiana]
Distribution: Atlantic draining Coastal rivers from Guyana to northern Brazil: Brazil, French Guiana, Guyana, Suriname, and Venezuela
Size: 150 cm TL (Leopold 2004, states a maximum size of 190 cm)
Habitat: occurs in shallow (<20 m) turbid waters over muddy bottoms in coastal areas and estuaries; sometimes in the lower reaches of rivers
Position in the water column: bottom
Diet: fishes and crustaceans (Rojas-Beltran, 1989)
Reproduction: approximately 20 large (18-21 mm) eggs per spawn; males practice oral incubation of the eggs until hatching; newborns measure 60-65 mm; sexual maturity is reached at 50-60 cm
Remarks: a very popular (and expensive) food fish in Paramaribo; the large yellowish eggs are also sold on the market. The maximum age of *Sciades parkeri*, *S. proops* and *S. couma* is only 3-5 years (Leopold, 2004).

SCIADES PASSANY (Valenciennes, 1840)
Local name(s): pani(pani), pasani (passany sea catfish)

Diagnostic characteristics: presence of a furrow, partly covered by a flap of skin, extending across the snout between the posterior nostrils; head broad, flattened above, snout short and square, mouth terminal or nearly so, the lower jaw equal to, or slightly longer than, upper jaw; exposed bony head shield well visible, very rugose and extending forward to opposite the eyes; supraoccipital slightly keeled; color grey to dark brown or blackish above, lighter below
Type locality: Cayenne, French Guiana
Distribution: Northern South America from Guyana to mouth of the Amazon River: Brazil, French Guiana, Guyana, Suriname, and Venezuela
Size: up to 100 cm TL (15 kg)
Habitat: shallow, estuarine, coastal waters and estuaries
Position in the water column: bottom
Diet: fish, crustaceans, detritus and mud (Leopold, 2004)

Sciades parkeri (live), ANSP178741 (© M. Sabaj Pérez)

Sciades passany (live), ANSP179579 (© M. Sabaj Pérez)

Reproduction: in French Guiana this species apparently reproduces twice in a year, in April and in September-December; females spawn 20-25 large eggs (12-15 mm diameter), which are incubated by the male in its mouth
Remarks:

SCIADES PROOPS (Valenciennes, 1840)
Local name(s): kupila (crucifix sea catfish)

Diagnostic characteristics: no fleshly furrow between the posterior nostrils; head more or less flattened above; exposed head shield very rugose, extending forward approximately to the eyes; absence of a fleshy groove in the median depression of the head; palatine teeth filiform; predorsal plate enlarged, shield-shaped, its length more than half of the (very short, spine-like) supra-occipital process; predorsal plate notched anteriorly, enclosing the tip of the narrow supra-occipital process; predorsal plate uniformly rugose; maxillary barbels extending only to pectoral fins; color medium grey, bluish grey, or dark brown to dark blue above, lighter whitish below
Type locality: Guianas; Puerto Rico
Distribution: Caribbean Sea and northern coast of South America, Colombia to Brazil: Brazil, Colombia, French Guiana, Guyana, Puerto Rico, Suriname and Venezuela
Size: 100 cm TL (common to 55 cm)
Habitat: predominantly found in shallow (<20 m) brackish water estuaries and lagoons, but may also occur in lower freshwater reaches of rivers
Position in the water column: bottom, sometimes observed swimming at the surface
Diet: fish and shrimps (Rojas-Beltran, 1989)
Reproduction: in French Guiana, reproduction is in the period November to April, but mainly in March-May (Leopold, 2004); sexual maturity is attained at an age 1-2 years; the species may live for 3 or 4 years; females are batch spawners and may spawn about 50 large (17 mm diameter) eggs/kg
Remarks: two periods of growth per year, i.e. in the dry seasons of March and August-November (Lecomte *et al.*, 1989); bones in the skull of this species have some resemblance with a crucifix.

Sciades proops (live), ANSP178742 (© M. Sabaj Pérez)

FAMILY DORADIDAE (THORNY CATFISHES)

Freshwater catfishes of the family Doradidae can be distinguished most easily by the ossified tubules of the lateral-line, variously expanded to form a conspicuous row of midlateral scutes each with a median backward-directed spine or thorn. Most species also have a well-developed head shield, subterminal mouth, large exposed humeral processes, sturdy serrated pectoral and dorsal-fin spines, and ventrally flattened bodies. Most doradids attain lengths between 10 and 20 cm SL, but a few large river forms (e.g. *Pterodoras*) can exceed 50 cm. Doradids are often separated into two easily-recognized groups, one with simple, and the other with fringed, maxillary barbels. The fringed-barbel taxa (like *Doras*) are further characterized by long, conical snouts, and relatively narrow heads with deep bodies. Doradids are also called 'talking catfishes' and, when caught, most doradids produce considerable noise either by movements of their pectoral fin spines or by vibrating the swim bladder.

1-a Maxillary barbels fringed; head convex and laterally compressed (head width < head length); eyes positioned in posterior half of the head (*Doras*)

2

1-b Maxillary barbels simple; head depressed (width equal to or larger than head length); eyes in anterior half of the head

3

2-a More than 30 lateral bony plates; lateral bony plates present from the end of the caudal peduncle to the dorsal fin

Doras carinatus

2-b Less than 25 lateral bony plates; lateral bony plates present from the end of caudal peduncle to approximately halfway the distance between the pelvic and anal fins

Doras micropoeus

3-a Caudal fin rounded to more or less truncate; small species (<15 cm SL)

4

3-b Caudal fin distinctly forked; adipose fin extending forward as a long keel; medium to large-sized species (>20 cm SL)

5

4-a Dorsal fin-spine serrated in front but not behind; caudal peduncle lacking plates above and below; caudal fin rounded

Acanthodoras cataphractus

4-b Dorsal fin-spine grooved, without spines on sides, front, or back; post-cleithral process serrated (with spines); nasal bone serrated and forming anterior margin of orbit; caudal fin truncate; Corantijn and Nickerie rivers
Amblydoras affinis

5-a Caudal peduncle provided with plates above and below
Platydoras costatus

5-b Caudal peduncle naked above and below; Corantijn and Nickerie rivers
Pterodoras aff. *granulosus*

Acanthodoras cataphractus (Linnaeus, 1758)
Local name(s): merkikwikwi

Diagnostic characteristics: maxillary barbels simple; head depressed (width equal to or larger than head length); small eyes (>3.5 times in interorbital distance) in anterior half of the head; caudal fin rounded; anal fin with 10-11 branched rays; small species (<15 cm SL) with brown color and a light yellow-brown longitudinal band along the lateral line
Type locality: America
Distribution: Amazon River basin, coastal drainages of French Guiana and Suriname: Bolivia, Brazil, Colombia, French Guiana, Guyana, Peru and Suriname; in Suriname, known from Para River, Wane Creek, and tributaries of the Lower Corantijn, Coppename and Marowijne rivers
Size: 11.5 cm SL
Habitat: black-water streams of the Savanna Belt
Position in the water column: bottom
Diet: a nocturnal omnivore
Reproduction:
Remarks: the local name of *A. cataphractus* 'merkikwikwi' (milk kwikwi) reflects the emission of a milk-white fluid from glands at the base of its pectoral spines when disturbed; this substance is apparently toxic to fishes and in the aquarium hobby *Acanthodoras* are transported in separate containers so that they do not kill other fishes by 'fouling' the water (M. Sabaj Pérez, pers. communication); like most doradids, *A. cataphractus* produce noise by moving the pectoral fin spines in their sockets when caught.

Amblydoras affinis (Kner, 1855)
Local name(s): –

Diagnostic characteristics: mouth large, almost terminal, with three pairs of long barbels (reaching the origin of the pectorals); head wide with large eyes (<2.5 times in interorbital distance); the body scutes narrow, but high; postcleithral process serrated (with spines); nasal serrated and forming the anterior margin of the orbit; caudal fin emarginated; spine of dorsal fin not serrated; adipose fin short; body color marbled brown, with a vertical black marking at the base of the caudal fin and a black line running from the caudal peduncle along body just below the (white) lateral line; irregular spots and blotches on the flanks and head area
Type locality: Rio Branco, Rio Guaporé
Distribution: Guaporé, Branco and Essequibo River basins: Bolivia, Brazil and Guyana; in Suriname, this species is only known from the Nickerie and Corantijn rivers
Size: 10 cm SL

Acanthadoras cataphractus (alcohol), Para River (Suriname River), NZCS-F1618 (© M. Sabaj Pérez)

Amblydoras affinis (alcohol), Nickerie River, ZMA102372 (© Naturalis)

Habitat: lives in small rainforest creeks; in Matapi Creek (Corantijn River), it was collected together with *Acanthodoras cataphractus*; it apparently lives in schools (e.g. it was collected in considerable numbers in a tributary of Marataka River)
Position in the water column: bottom; during day time they hide under woody debris or leaves
Diet: algae and detritus
Reproduction: the belly of females is uniform yellow-brown, while that of males has small brown spots; it is reported that they construct a nest of leaves in the substrate and that the eggs are laid during the rainy season (high-water season), and that once laid the eggs are covered with leaves; both parents guard the eggs and emerging fry
Remarks: like most doradids it can create a sound by grating its fin bones in each socket and amplifying the noise via the swim bladder; *A. hancockii* is a junior synonym.

Doras carinatus (Linnaeus, 1766)
Local name(s): agunoso

Diagnostic characteristics: maxillary barbels fringed, mouth inferior, snout 17% of SL; head convex and laterally compressed (head width < head length); large eyes positioned dorsally in the posterior half of the head; more than 30 lateral bony plates; lateral bony plates present from the end of the caudal peduncle to the dorsal fin; caudal fin forked; color of body yellowish-grey
Type locality: Lawa River (Marowijne (Maroni) drainage) about 8 kilometers south-southwest of Anapaike/Kawemhakan, 3°19'31"N, 54°03'48"W, Sipaliwini, Surinam
Distribution: Essequibo River basin and other coastal drainages to mouth of Amazon River: Brazil, French Guiana, Guyana, Suriname and Venezuela
Size: 30 cm SL
Habitat: rivers of the Interior
Position in the water column: bottom
Diet:
Reproduction:
Remarks: produces sounds (duration 40-70 ms, 60-90 Hertz) by moving its pectoral fin spines when caught.

Doras micropoeus (Eigenmann, 1912)
Local name(s): agunoso

Diagnostic characteristics: maxillary barbels fringed, mouth inferior, snout 20% of SL; head convex and laterally compressed (head width < head length); large eyes positioned dorsally in the posterior half of the head; less than 25 lateral bony plates; lateral bony plates present from the end of caudal peduncle to approxi-

Amblydoras affinis (live) (© P. Willink)

Doras carinatus (live), Lawa River (Marowijne River), ANSP187114 (© M. Sabaj Pérez)

Doras micropoeus (live), scale bar = 1 cm, Lawa River (Marowijne River), ANSP187110 (© M. Sabaj Pérez)

mately halfway the distance between the pelvic and anal fins; caudal fin forked with rounded lobes; color of body yellowish, beige or greyish
Type locality: Upper Demerara River at Wismar, Guyana
Distribution: Essequibo, Demerara and Corantijn River basins, and possibly other drainages to mouth of Amazon River: French Guiana, Guyana and Suriname
Size: 35 cm SL
Habitat: rivers of the Interior
Position in the water column: bottom
Diet:
Reproduction:
Remarks:

PLATYDORAS COSTATUS (Linnaeus, 1758)
Local name(s): soké

Diagnostic characteristics: maxillary barbels simple, mouth almost terminal; large head depressed (width equal to or larger than head length); large eyes (<2 times in interorbital distance) in the anterior half of the head; caudal fin emarginated to deeply forked; adipose fin extending forward as a long keel; medium to large-sized species (>20 cm SL); caudal peduncle provided with plates above and below; color dark brown dorsally, grey-white ventrally, sometimes with a yellow-white longitudinal band along the lateral line (this yellow-white lateral band is especially pronounced in juveniles)
Type locality: Indiis [South America]
Distribution: Amazon, Tocantins, Parnaíba, Orinoco and Essequibo River basins and coastal drainages in French Guiana and Suriname: Argentina, Bolivia, Brazil, Colombia, French Guiana, Guyana, Peru, Suriname and Venezuela. But perhaps only coastal drainages of Suriname and French Guiana
Size: 24 cm SL
Habitat: rivers of the Interior, mainly on sandy bottom substrates
Position in the water column: bottom, often found on sandy bottom substrates
Diet: mollusks, crustaceans and detritus
Reproduction:
Remarks: a slightly different *Platydoras* sp. with distinctly shallower scutes than those of typical *P. costatus* occurs in Marowijne River; preliminary molecular data based on one specimen of each of the two Surinamese *Platydoras* (*P. costatus* and *Platydoras* 'shallow scute') found that they do not group together (M. Sabaj-Pérez, pers. comm.); in the Brazilian Amazon, the conspicuously colored juvenile *Platydoras costatus* were observed picking parasites ('cleaning') from the piscivorous pataka *Hoplias* cf. *malabaricus* during the day (Carvalho *et al.*, 2003).

Platydoras costatus (alcohol), 177.1 mm SL, Suriname River, RMHN31843 (© M. Sabaj Pérez)

Platydoras sp. ('shallow scute') (live), Lawa River (Marowijne River) (© M. Sabaj Pérez)

Pterodoras aff. *granulosus* (Valenciennes, 1821)
Local name(s): –

Diagnostic characteristics: maxillary barbels simple; head depressed (width equal to or larger than head length); very small eyes in anterior half of the head; caudal fin emarginated to deeply forked; adipose fin extending forward as a long keel; medium to large-sized species (>20 cm SL); caudal peduncle naked above and below; Corantijn and Nickerie rivers
Type locality: South America
Distribution: Amazon and Paraná River basins and coastal drainages in Guyana and Suriname: Argentina, Bolivia, Brazil, Colombia, Guyana, Paraguay, Peru, Suriname and Uruguay; in Suriname, only known from the Corantijn and Nickerie rivers
Size: 70 cm TL
Habitat: rivers of the Interior, but was also caught in the lower reaches of Corantijn River (i.e. downstream of the first rapids)
Position in the water column: bottom
Diet: *Pterodoras granulosus* is nocturnal predator feeding opportunistically on a variety of foods including fish, mussels and the fruits of *Astrocaryum*
Reproduction:
Remarks:

Pterodoras aff. *granulosus* (alcohol), Sipaliwini River (Corantijn River) (© P. Willink)

FAMILY AUCHENIPTERIDAE (DRIFTWOOD CATFISHES)

The Auchenipteridae, a family of small to medium sized catfishes, can be readily recognized by the following features: body without bony plates, dorsal region of body, between head and dorsal fin origin, covered with bony plates that are sutured together and readily visible beneath a thin skin, usually three pairs of barbels (nasal barbels absent), adipose fin small, eye covered with adipose tissue and without free orbital rim. Auchenipterids are unique among catfishes in their reproductive biology: all species are thought to undergo internal insemination (Mazzoldi *et al.*, 2007) and the female does not necessarily expel her mature ova immediately following spawning. Instead, females may carry mature, unfertilized eggs and packets of sperm inside her reproductive organs for an extended period of time before triggering fertilization and deposition of the eggs. There is a pronounced sexual dimorphism of the anal fin and, in some species, the dorsal fin, maxillary barbels, and other parts of the body. Some of the anterior rays of the anal fin in males become enlarged and modified in shape, forming an intromittent organ for the deposition of sperm packets inside the female. Elongated dorsal-fin spines and stiffened maxillary barbels in nuptial males (Fig. 12.4) appear to act as clasping organs, which hold the female near to the male during spawning. Except for the modified anal-fin rays, the secondary sexually dimorphic characters found in male auchenipterids disappear at the end of the reproductive season. The Auchenipteridae now includes the previously recognized family Ageneiosidae. Auchenipterids are typically nocturnal, although some species of *Auchenipterus*, *Ageneiosus* and *Centromochlus* appear to feed during the day. Most species feed on insects, especially those that fall onto the surface of the water. At night auchenipterids can be observed swimming just below the surface of the water, picking at insects that struggle on the water's surface. *Auchenipterus* species are planktivorous and *Ageneiosus* species appear to be primarily piscivorous. During the day, many of the nocturnal species hide in deep water and crevices in submerged logs, giving rise to the common name for the family: driftwood catfishes.

1-a Mandibular barbels absent; maxillary barbels short (*Ageneiosus*)
2

1-b Mandibular barbels present
4

2-a Caudal fin deeply forked, with sharply pointed lobes and 8+9 principal rays; anal fin long with 41-50 fin rays; Corantijn River
Ageneiosus ucayalensis

Fig. 12.4. Nuptial male *Ageneiosus inermis* with short, ossified, fringed maxillary barbel (© B. Chernoff)

2-b Caudal fin obliquely truncate or weakly emarginated, with 8+10 principal rays; 34-40 anal fin rays;

3

3-a Body and head heavily marbled, pigment consisting of prominent, large, dark black blotches and spots; coloration on body not uniformly countershaded; paired fins nearly uniform black; blotches on sides extending onto anal and caudal fins

Ageneiosus marmoratus

3-b Body and head not heavily marbled; coloration on body countershaded, upper half yellowish-brown to gray, diminishing below midline; paired fins often striped or mottled, never uniformly black; caudal fin without dark black blotches; anal fin pale

Ageneiosus inermis

4-a Caudal fin obliquely truncate or obliquely rounded, never forked

Trachelyopterus galeatus

4-b Caudal fin forked, lobes pointed or rounded

5

5-a More than 16 anal fin rays

6

5-b Less than 14 anal fin rays

8

6-a Less than 30 anal fin rays; ventral fin with 8 fin rays; maxillary barbels not ossified in males; lateral line clearly visible with zigzag

Pseudauchenipterus nodosus

6-b More than 35 anal fin rays; ventral fin with 12-15 rays; maxillary barbels ossified in males; presence of grooves in the ventral surface of the head that accommodate adducted mental barbels (*Auchenipterus*)

7

7-a First branchial arch with 26-32 gill rakers; Corantijn, Saramacca and Suriname rivers

Auchenipterus dentatus

7-b First branchial arch with 36-46 gill rakers; Marowijne River

Auchenipterus nuchalis

8-a Pectoral fin rays I,7-12 (*Centromochlus*)

9

8-b Pectoral fin rays I,4-6

10

9-a Body color plain, dark grey above, white below, with some pigment spots on the chin; standard length 29-33 mm; Coppename River

Centromochlus concolor

9-b Body color densely mottled and dotted with dark grey on a pale (unpigmented) background; standard length 34-45 mm; Suriname and Marowijne rivers
Centromochlus punctatus

10-a Dorsal and pectoral fin spines smooth; upper surface of head covered with thick skin; pectoral fin rays I,5-6; anal fin rays 10-13
Glanidium leopardum

10-b Dorsal and pectoral fin spines with well developed teeth and hooks; upper surface of head and nuchal region covered with bony plates; pectoral fin rays I,4; anal fin rays 7-11 (*Tatia*)
11

11-a Relatively large size up to 12 cm standard length; body color dark brown with many large roundish to elliptical white/yellowish dots
Tatia intermedia

11-b Smaller size (up to 6 cm standard length), body coloration lacking large white dots
12

12-a Body coloration dark brown, with only vague pale areas; pectorals, dorsal and adipose spotted with brown
Tatia brunnea

12-b Body coloration with upper surface of heads and back dull brown and sides longitudinally marbled brown and white, a broad blackish longitudinal band on the sides
Tatia gyrina

Ageneiosus inermis (Linnaeus, 1766)
Local name(s): plarplari, prarprari, paypay

Diagnostic characteristics: head dorsoventrally flattened, mouth large, eyes positioned laterally at the same level as the mouth; mandibular barbels absent; maxillary barbels short; caudal fin emarginated; pectoral reaching up to or beyond insertion of pelvic; 34-40 anal fin rays; body color yellowish-brown with sometimes a longitudinal black band on the flanks; a large species compared to *A. marmoratus*
Type locality: Suriname
Distribution: widespread in South America: Argentina, Bolivia, Brazil, Colombia, Ecuador, French Guiana, Guyana, Paraguay, Suriname and Venezuela
Size: 47 cm SL
Habitat: main channel of rivers and large tributaries, both in the Interior and in lower reaches (i.e. downstream of the first rapids)
Position in the water column: mid-water (a pelagic catfish)
Diet: fish
Reproduction: sexual dimorphism with mature males showing an elongated dorsal-fin spine crowded with curved hooks, (short) ossified, fringed maxillary barbels (Fig. 12.4), and enlarged, modified anal-fin rays; fertilization is internally and the female may store sperm packets and ripe eggs for some time; small juveniles with a very distinct spotted coloration and black pectoral fins (Fig. 9.2a)
Remarks: *Ageneiosus brevifilis* is a synonym; a very popular food fish in Suriname.

Ageneiosus marmoratus Eigenmann, 1912
Local name(s): plarplari, prarprari, paypay

Diagnostic characteristics: a handsome catfish distinguished from all other species by its striking coloration pattern, consisting of large, irregular, and sharply contrasting black blotches on the head, body, and fins; further characterized by a combination of its large, robust head, a broad, parabolic snout and wide gape, an emarginate tail with 8+10 principal caudal rays, and the first dorsal and pectoral lepidotrichia flexible, segmented, and not serrated; paired fins black; dark black blotches of sides extending onto anal and caudal fins
Type locality: creek below Potaro landing, Guyana
Distribution: known at present only from isolated localities in the Essequibo and Corantijn Rivers, Guyana and Suriname, the Upper Amazon basin in Peru, and the middle Parang River in Argentina (these widely disjunct records are presumably indicative of a widespread *distribution* throughout central and northeastern South America); in Suriname, only known from the Corantijn and Nickerie rivers
Size: 30 cm SL
Habitat: in Suriname collected in large to medium-sized tributaries of Corantijn (Walsh, 1990) and Nickerie (pers. observations) rivers
Position in the water column: mid-water

Ageneiosus inermis (live), 200 mm SL, Litani River (Marowijne River), FMNH-SUR07-05 (© M. Sabaj Pérez)

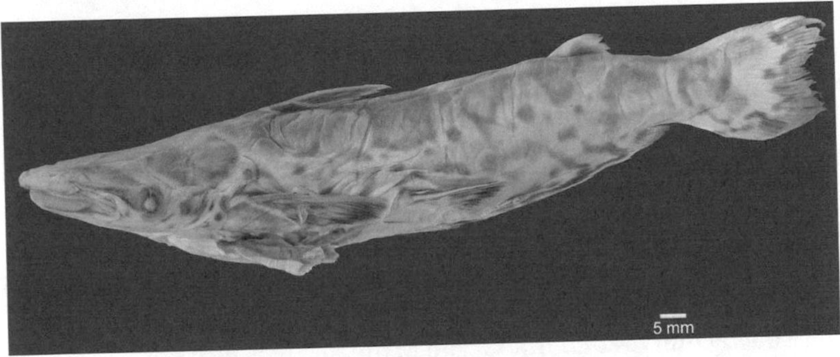

Ageneiosus marmoratus (alcohol, juvenile), type, FMNH53245 (© M. Littmann)

Ageneiosus marmoratus (live, adult, approx. 30 cm SL), Kofimaka Creek, Lower Nickerie River (© A. Gangadin)

Diet: fish
Reproduction:
Remarks: nothing is known about its biology; the coloration of juveniles and adults may differ.

AGENEIOSUS UCAYALENSIS Castelnau, 1855
Local name(s): plarplari, prarprari, paypay

Diagnostic characteristics: readily distinguished from other species of the genus by a combination of the deeply forked tail, a very elongate, compressed body, a greatly flattened head with a distinctly inferior mouth, and a very long anal fin; pigmentation variable, the typical pattern, consisting of dark stripes along the cranial fontanel, an hourglass-shaped patch on the nuchal plate, and a thin mid-dorsal stripe along the back from behind the rayed dorsal fin and extending onto the base of the upper caudal lobe, further serves to distinguish this species from congeners; the high number of anal fin rays (41-50), and numerous (18-25), long, crenulated gill rakers on the first arch separates it from all other *Ageneiosus* species
Type locality: Rio Ucayali, Peru
Distribution: it is found throughout most of the major lowland drainages of South America, including the Orinoco and Amazon basins and various river systems draining directly into the Atlantic Ocean in the northeastern portion of the continent; in Suriname, only known from Corantijn River
Size: 28.5 cm SL
Habitat: main channel of rivers and large tributaries
Position in the water column: mid-water
Diet: fish
Reproduction:
Remarks: junior synonyms are *Ageneiosus dentatus* Kner 1858 (type locality 'Suriname') and *A. guianensis* Eigenmann, 1912 (see Walsh, 1990)

AUCHENIPTERUS DENTATUS Valenciennes, 1840
Local name(s): –

Diagnostic characteristics: body elongated and laterally compressed (depth> width); eyes large, positioned laterally; presence of grooves in the ventral surface of the head that accommodate adducted mental barbels; dorsal- and pectoral-fin spines thin and sharply pointed; more than 35 anal fin rays, anal-fin origin anterior to the middle of the body; ventral fin with 12-15 rays; adipose fin rudimentary; maxillary barbels ossified in males; first branchial arch with 26-32 gill rakers ; body dorsally yellowish-grey, flanks silvery-white with a longitudinal black band along the (complete) lateral line
Type locality: near Paramaribo, Suriname (Boeseman, 1972; Ferraris & Vari, 1999)

Auchenipterus dentatus (live), Litani River (Marowijne River), ANSP189102 (© M. Sabaj Pérez)

Distribution: Northern coastal rivers of the Guianas; Corantijn, Suriname, Arataye, Sinnamary and Oyapock rivers: French Guiana and Suriname; in Suriname, known from Corantijn, Saramacca and Suriname rivers, and probably Lower Marowijne River (Mees, 1974; Ferraris & Vari (1999) point out that the specimens identified as *A. nuchalis* in Mees (1974) are, in fact, specimens of *A. dentatus*)
Size: 12.1 cm SL
Habitat: lower courses of the great rivers and large to medium-sized tributaries; also known to occur in black-water streams (e.g. Para River)
Position in the water column: mid-water (a free swimming pelagic species)
Diet: insects and microcrustaceans (zooplankton)
Reproduction: sexual dimorphism with females achieving a slightly greater maximum size than do males and mature males showing an elongated dorsal-fin spine covered with rounded papillae on the anterior margin but otherwise smooth, presence of papillae on the snout and the dorsal and median surface of the (short) ossified maxillary barbels, and enlarged, modified anterior anal-fin rays forming an intromittent organ (Ferraris & Vari, 1999); fertilization is internally and the female may store sperm packets and ripe eggs for some time
Remarks: *Euanemus colymbetes* Müller & Troschel (1842) is a junior synonym.

AUCHENIPTERUS NUCHALIS (Spix & Agassiz, 1829)
Local name(s): –

Diagnostic characteristics: body elongated and laterally compressed (depth> width); eyes large, positioned laterally; presence of grooves in the ventral surface of the head that accommodate adducted mental barbels; more than 35 anal fin rays, anal-fin origin anterior to the middle of the body; ventral fin with 12-15 rays; maxillary barbels ossified in males; first branchial arch with 36-46 gill rakers
Type locality: Rio Capim, near São Domingos do Capim, at Igarapé Pirajauara, about 1°41'S, 47°47'W, Pará, Brazil
Distribution: Lower Amazon and Tocantins rivers northward to Marowijne River; possibly also Rupununi River, Guyana and Negro River, Brazil: Brazil and French Guiana ; in Suriname, only known from the lower reaches of Marowijne River and large tributaries such as Gran Creek (Ferraris & Vari, 1999)
Size: 15.4 cm SL
Habitat: restricted to the lower course of Marowijne River
Position in the water column: mid-water
Diet: insects and microcrustaceans (zooplankton)
Reproduction: sexual dimorphism with females achieving a slightly greater maximum size than do males and mature males showing an elongated dorsal-fin spine covered with rounded papillae on the anterior margin but otherwise smooth, presence of papillae on the snout and the dorsal and median surface of the (short) ossified maxillary barbels, and enlarged, modified anterior anal-fin rays forming an intromittent organ (Ferraris & Vari, 1999); fertilization is internally and the female may store sperm packets and ripe eggs for some time
Remarks:

This page is intentionally left blank.

CENTROMOCHLUS CONCOLOR (Mees, 1974)
Local name(s): –

Diagnostic characteristics: dorsal shield with horns which are more or less diamond-shaped at the tip; anal fin with 8 rays; posterior border of anal fin of males rounded, reaching to behind posterior border of adipose fin; pectoral fin rays I,7-12; body color plain, dark grey above, white below, with some pigment spots on the chin; small size, standard length 29-33 mm; Coppename River
Type locality: headwaters of Coppename River, 3°49'N, 56°57'W, Suriname
Distribution: only known from Upper Coppename River basin, Suriname
Size: 3.3 cm SL
Habitat: headwaters of Coppename River, in the main channel
Position in the water column: bottom
Diet:
Reproduction:
Remarks: nothing is known about its biology.

CENTROMOCHLUS PUNCTATUS (Mees, 1974)
Local name(s): –

Diagnostic characteristics: nuchal plate with blunt horns, moderately bent outwards; usually 10 anal fin rays; posterior border of adpressed anal fin falling behind hind-border of adipose fin or almost opposite; pectoral fin rays I,7-12; body color densely mottled and dotted with dark grey on a pale (unpigmented) background, ventrally white, unpigmented; small size, standard length 34-45 mm; Suriname and Marowijne rivers
Type locality: creeks between Kabel and Lombé, Suriname River basin, Suriname
Distribution: Atlantic coastal rivers above Amazon River mouth, Suriname; in Suriname, known from Suriname and Marowijne rivers
Size: 4.5 cm SL
Habitat: rivers of the Interior, upstream of the first rapids
Position in the water column: bottom
Diet:
Reproduction:
Remarks: nothing is known about its biology.

GLANIDIUM LEOPARDUM (Hoedeman, 1961)
Local name(s): –

Diagnostic characteristics: very close to *Tatia* and *Centromochlus*, but often larger in size; snout rounded, eye diameter larger than snout length; anterior borders of dorsal and pectoral spines smooth or with small teeth only; horns of nuchal plate slender, almost stalked; 12-14 anal fin rays (9-10 branched); pectoral fin rays I,5-6; upper surface of head covered with thick skin; caudal shallowly forked, its lobes

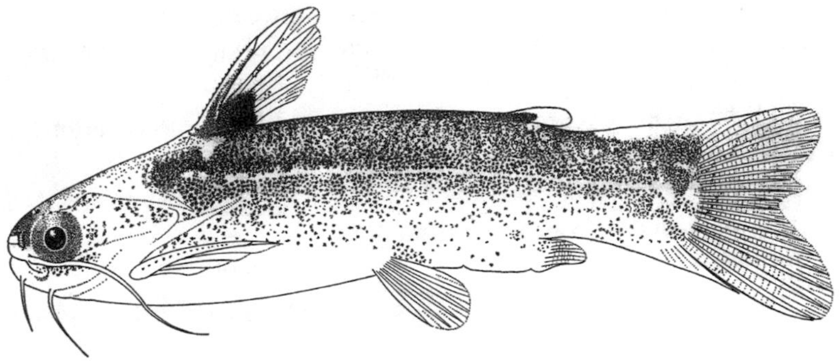

Centromochlus concolor (© Brill)

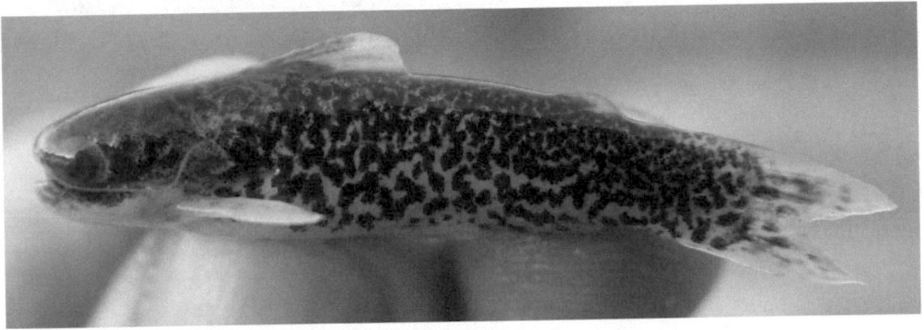

Centromochlus punctatus (live) (© R. Covain)

Glanidium leopardum (live), 48.5 mm SL, Litani River (Marowijne River), ANSP189104 (© M. Sabaj Pérez)

rounded; the conspicuous color pattern with very large often confluent dark brown blotches on a brownish-white to white background is diagnostic for the species
Type locality: Litani River (Upper Marowijne River), village Aloiké, French Guiana [Alowike, Suriname – Mees 1974:144]
Distribution: coastal rivers of the Guianas: French Guiana, Guyana and Suriname; in Suriname, only known from Marowijne River
Size: 10.9 cm SL
Habitat: shallow forest creeks of the Interior with sandy-muddy bottom and moderate current
Position in the water column: bottom
Diet:
Reproduction: sexual dimorphism shown in the enlarged anal fin of males (transformed into a copulatory organ) and the position of the urogenital papillae of the males
Remarks: *G. leopardum* it is not collected often and usually in low numbers, suggesting a largely solitary life for this species.

PSEUDAUCHENIPTERUS NODOSUS (Bloch, 1794)
Local name(s): botromanki

Diagnostic characteristics: rather heavily-bodied fishes with head and anterior part of the body roundish in circumference, the posterior part compressed; eyes of moderate size, lateral in position; head with thick skin covering the remarkable honeycomb-like frontal bones; caudal fin forked, lobes pointed or rounded; more than 16, but less than 30 anal fin rays; ventral fin with 8 fin rays; spines strong and pungent, smooth along their anterior border, serrated behind; barbels well developed, longer than in most other auchenipterids, maxillary barbels not ossified in males; lateral line complete, whitish, with clearly visible zigzags in its anterior part
Type locality: Tranquebar [South America]
Distribution: lower reaches of rivers and in estuaries from Venezuela to Brazil: Brazil, French Guiana, Guyana, Suriname, Trinidad and Tobago, and Venezuela; in Suriname, known from the Corantijn, Saramacca, Suriname and Marowijne rivers, but probably present in all rivers (lower reaches)
Size: 22 cm SL
Habitat: lower reaches of rivers and estuaries with fresh, brackish and even salt water, i.e. downstream the first rapids
Position in the water column: bottom
Diet: omnivore with tendency to detritivory
Reproduction: sexual dimorphism in the position of the urogenital pore and the shape of the anal fin with elongated anterior rays; in December 2006, an upriver spawning migration was observed in Mindrineti River, a tributary of Saramacca

Pseudauchenipterus nodosus (live) (© INRA-Le Bail)

River; in December, at the start of the reproductive season, the gonads of mature females may represent 10-25% of the total weight of the fish
Remarks:

TATIA BRUNNEA Mees, 1974
Local name(s): –

Diagnostic characteristics: a relatively large *Tatia* species with 8-9 anal fin rays; pectoral fin rays I,4; dorsal and pectoral fin spines with well developed teeth and hooks; upper surface of head and nuchal region covered with bony plates; body coloration dark earth brown, with only vague pale areas; pectorals, dorsal and adipose spotted with brown; caudal fin more deeply forked than in other species, hyaline with large irregular blackish-brown dots; dorsal shield usually distinctively paler than the body
Type locality: Compagnie Creek, Suriname River system, Suriname
Distribution: river basins in Brazil, French Guiana and Suriname; in Suriname, only known from the Suriname and Marowijne rivers
Size: 7 cm SL
Habitat: in forest creeks, in reaches with slow current
Position in the water column: bottom
Diet:
Reproduction: anal fin of the male is transformed into a copulatory organ (first apparent in specimens of 35-40 mm SL)
Remarks: the species seems quite rare; it can be confused with young *G. leopardum*, but differs from the latter by its number of anal fin rays and the serrated dorsal and pectoral spines (versus smooth in *G. leopardum*).

TATIA GYRINA (Eigenmann & Allen, 1942)
Local name(s): –

Diagnostic characteristics: small species with a relatively heavy body and 8-11 anal fin rays; pectoral fin rays I,4; dorsal and pectoral fin spines with well developed teeth and hooks; upper surface of head and nuchal region covered with bony plates; caudal fin shallowly forked; body coloration diagnostic with upper surface of heads and back dull brown and sides longitudinally marbled brown and white, a conspicuous broad blackish longitudinal band on the sides
Type locality: Río Itaya, Iquitos, Peru (*type locality* of *T. creutzbergi* is Djaikreek, Suriname River system, Suriname)
Distribution: upper and central Amazon River basin: Brazil, Colombia, Peru; and in northern Suriname; in Suriname, known from Corantijn, Nickerie, Saramacca, Suriname and Commewijne rivers
Size: 4 cm SL
Habitat: forest creeks in the northern lowlands
Position in the water column: bottom

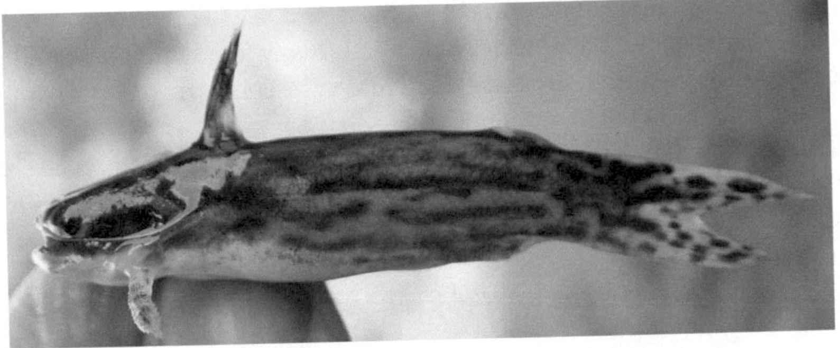

Tatia brunnea (live), Sipaliwini River (Corantijn River) (© R. Covain)

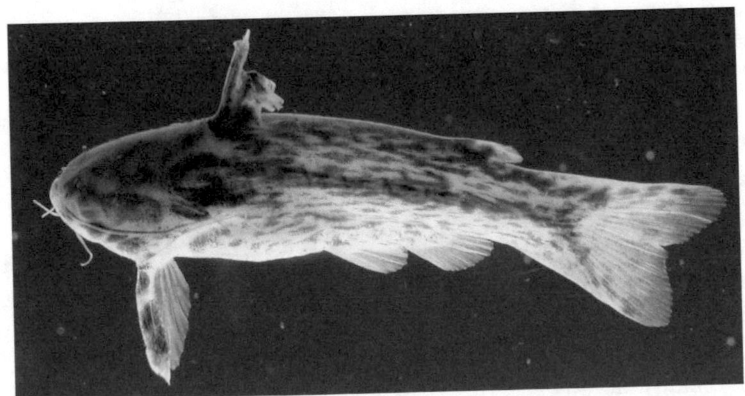

Tatia gyrina (live) (© P. Willink)

Diet:
Reproduction:
Remarks: *Tatia creutzbergi* (Boeseman, 1953) is considered junior synonym (Sarmento-Soares and Martins-Pinheiro, 2008: 519), but may be a valid species.

TATIA INTERMEDIA (Steindachner, 1877)
Local name(s): –

Diagnostic characteristics: a relatively large species (up to 12 cm SL) with 10 (iii.7) anal fin rays; pectoral fin rays I,4-6; dorsal and pectoral fin spines with well developed teeth and hooks; upper surface of head and nuchal region covered with bony plates; body color dark earth brown with many large roundish to elliptical white dots (in large specimens these dots tend to be smaller)
Type locality: Marabitanos, Para, Brazil
Distribution: Amazon River basin and rivers of the Guianas: Brazil, French Guiana, Guyana and Suriname; in Suriname, known from Corantijn, Nickerie and Suriname rivers
Size: 12 cm SL
Habitat: small shallow forest creeks of the Interior, mainly in upper basins of the great rivers
Position in the water column: bottom
Diet: omnivorous with preference for insects
Reproduction: sexual dimorphism in males (enlarged anal fin) apparent in specimens of 35-40 mm SL
Remarks:

TRACHELYOPTERUS GALEATUS (Linnaeus, 1766)
Local name(s): noya

Diagnostic characteristics: short, heavy, round-bodied fish; the head not much depressed, only a little wider than deep; maxillary barbels present; 6 rays in the pelvic fins; 20-25 rays in the anal fin; caudal fin obliquely truncate or obliquely rounded, never forked; small specimens mottled brown or blackish on a pale brown or cream colored background, or with longitudinally-elongated black dots, large specimens very dark, blackish brown, irregularly mottled with black; caudal fin usually with a black vertical bar
Type locality: South America
Distribution: widespread in northern South America: Argentina (?), Brazil, French Guiana, Peru, Suriname and Trinidad and Tobago; in Suriname, present in all rivers
Size: 22 cm SL
Habitat: rivers and small tributary forest creeks in the Interior and Savanna Belt, but also in coastal swamps; often very abundant (Fig. 5.5)
Position in the water column: bottom

Tatia intermedia (live), Sipaliwini River (Corantijn River) (© R. Covain)

Trachelyopterus galeatus (live) (© R. Covain)

Trachelyopterus galeatus (live) (© INRA-Le Bail)

Diet: small fishes, arthropods and sometimes fruits
Reproduction: internal insemination; the female may carry mature, unfertilized eggs and packets of sperm inside her reproductive organs for several months before triggering fertilization and deposition of the eggs; mature females can have ovaries weighing 20% of their body weight; the eggs (3 mm diameter) hatch after a few days and the newborn larvae (15 mm) start feeding on zooplankton approximately 9 hours after hatching; after 11 days the young catfish show negative phototropism and start hiding under stones and woody debris during the day
Remarks: *Parauchenipterus galeatus* is a synonym; the noya can survive extended periods (days, weeks) under hypoxic conditions, such as may occur in dry-season pools in small forest creeks (Fig. 5.5).

Trachelyopterus galeatus (live), Saramacca River (© R. Covain)

ORDER GYMNOTIFORMES

The ostariophysan order of Neotropical electric knifefishes (Gymnotiformes) is endemic to the South American continent and currently considered a sister group of the catfishes (Albert & Campos-da-Paz, 1998; Albert & Fink, 2007). Lissmann (1958) hypothesized that the electric organ discharges were derived from the muscle action potentials. Electric fish can be classified with respect to their electric organ discharges (EOD) as 'wave species' (e.g. Sternopygidae), in which the electric organ fires in a nearly sinusoidal manner, or as 'pulse species' (e.g. Gymnotidae), in which the EOD impulses are much shorter than the interval between discharges and repeated less regularly. In weakly electric fish, the electric field produced by the EOD of special large electrocyte cells (Stoddard & Markham, 2010) is used in electrolocation, i.e. to 'see' their world within half a body length both at night and in turbid water by analyzing distortions in these electric fields caused by nearby objects with different impedances than the surrounding water. The electrogenic ability is also used in social electric communication, with males singing courtship songs to females and engaging in energetically expensive contests of electric one-upmanship with their rivals. The electric eel is the only Neotropical electric freshwater fish capable of producing not only weak signals for navigation and communication, but also a strong discharge (hundreds of volts) to stun prey and deter would-be predators. The presence of knifefishes and their exact location in a stream (and often their specific identity) can be detected with a electric-fish-finder apparatus, an audio-frequency amplifier that transforms the electric organ discharges (EODs) of the fish into sounds, and with practice most species can be identified acoustically (e.g. Hopkins & Heiligenberg, 1978; Westby, 1988; Fig. 12.5).

FAMILY GYMNOTIDAE (NAKED-BACK KNIFEFISHES)

The family Gymnotidae (i.e., the *Gymnotus* plus *Electrophorus* group) is viewed as the sister-group to all remaining gymnotiforms (Albert & Campos-da-Paz, 1998). Compared to other gymnotiforms, *Gymnotus* and *Electrophorus* have a more cylindrical body (sometimes referred to as "sub-cylindrical"; versus compressed body), depressed head, and considerably shorter tails (i.e., ca. 4-10% of body length to the end of anal fin in *Gymnotus* and juveniles of *Electrophorus* [adults in this latter genus have no tail at all] vs. 20-80% of LEA). *Gymnotus* species have small, cycloid scales, a superior mouth, the lower jaw conspicuously longer than the upper jaw, and are only weakly electric. *Electrophorus*, on the other hand, has no scales, and a terminal mouth. The single species currently assigned to that genus, *E. electricus* (Linnaeus, 1766) (i.e., the notorious electric eel) is strongly electric, with larger adults capable of shocks approaching 500-600 V (1-2 amperes).

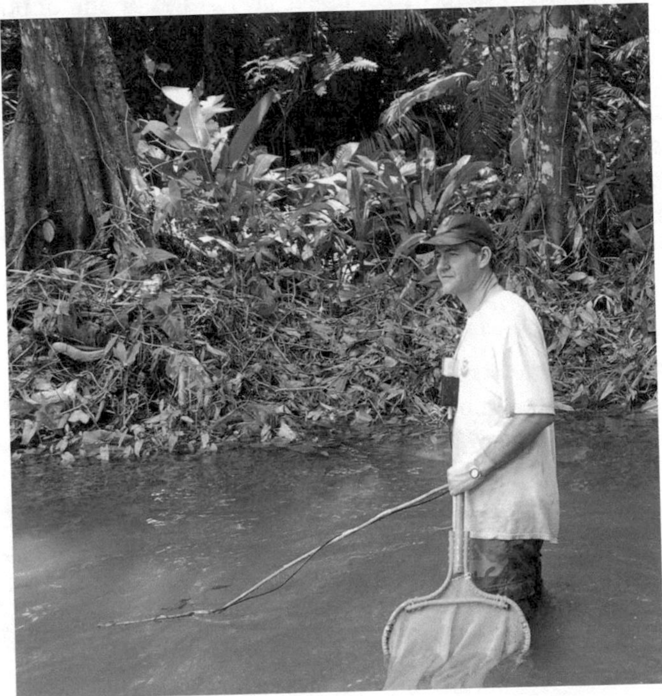

Fig. 12.5. Dr. Will Crampton, specialist in the order Gymnotiformes, collecting electric fishes in a clear-water forest creek in northeastern Suriname with an electric fish finder apparatus and a dip net.

Gymnotids are usually aggressive nocturnal predators, with a number of species exhibiting territorial behavior and building nests. They also can breathe atmospheric air: individuals of *Gymnotus* species are able to use part of their swimbladder system for doing that in certain occasions, while the obligatory air-breather *Electrophorus* has a vascularized oral respiratory organ.

1-a Upper and lower jaw equal in length; posterior end of body rounded; anal fin up to the posterior end of the body
Electrophorus electricus
1-b Lower jaw distinctly longer than upper jaw; posterior end of body pointed; anal fin falling short to the posterior end of the body
2

2-a Oblique transversal white bands all over the body; medium-sized species, up to 40 cm Total Length
Gymnotus carapo
2-b Transversal white bands restricted to posterior half of the body, in the anterior part of the body they are absent or not reaching the back
3

3-a Small, slender species (maximum body size <17 cm Total Length)
Gymnotus coropinae
3-b Large species (maximum body size >30 cm TL)
*Gymnotus anguillaris**

* *Gymnotus anguillaris* Hoedeman, 1962 is described from Coropina Creek, Para River, Suriname River system. Extensive recent collecting effort (including use of an electric fish finder apparatus) failed to collect the species from its type locality or other tributaries of the Para River (W. Crampton, pers. communication).

Electrophorus electricus (Linnaeus, 1766)
Local name(s): praké, stroomfisi, maisi, sidderaal (electric eel)

Diagnostic characteristics: upper and lower jaw equal in length; posterior end of body rounded; anal fin up to the posterior end of the body
Type locality: Suriname; South American rivers
Distribution: Amazon and Orinoco River basins and other areas in northern Brazil: Brazil, French Guiana, Guyana, Peru, Suriname and Venezuela; in Suriname, present in all rivers
Size: 250 cm SL
Habitat: in Suriname, the electric eel occurs in both large rivers (often near rapids) and small forest streams in the Interior, upstream of the first rapids; however, in French Guiana and Brazil, *E. electricus* also occurs in coastal freshwater swamps where it can survive low dissolved oxygen concentrations by breathing air

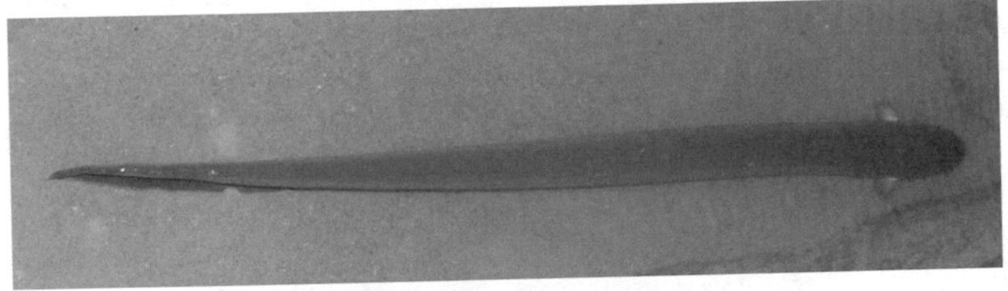

Electrophorus electricus (live), Corantijn River (© R. Covain)

Electrophorus electricus (live), detail of the head, Corantijn River (© R. Covain)

Position in the water column: near the bottom; however, the electric eel is an obligate air-breather that has to surface regularly to breathe air; at night the electric eel was observed cruising at the water surface (presumably hunting)
Diet: *Electrophorus* is a formidable predator of smaller, weakly electric fishes; it begins hunting in the late afternoon when other Gymnotiformes are still inactive (Westby, 1988). After locating weakly electric fish by their electric signals, *Electrophorus* stuns them with its high-voltage discharge and gulps them down before they can recover
Reproduction: in coastal swamps of Marajó Island (Pará, Brazil) the electric eel is known to construct floating bubble nests (Assunção & Schwassmann, 1995); however, such bubble nests are not known from electric eels in the Interior of Suriname
Remarks: the strong electric discharges (500-600 V; 1-2 A) of the electric eel can be dangerous to humans because they may knock out an adult (i.e. render one unconscious) for several minutes (J. Montoya-Burgos, pers. communication) and in this period one may drown after falling in the water leaving the mouth and nose under the water surface (see chapter 8). Although the powerful shocks of the electric eel are a formidable weapon to deter potential predators, about 20% of the *E. electricus* specimens in three collections in Venezuela had their tails mutilated with bite marks by piranhas (T. Roberts, pers. communication), indicating that at least piranhas do attack electric eels. Two distinct forms of the electric eel are recognized in Suriname (and other localities in Amazonia; T. Roberts, pers. communication), a dark-brown, slender morph, and a light-brown, fat morph (Fig. 12.6); the two morphs produced distinct Electric Organ Discharge signals (W. Crampton, pers. communication), but the EODs of only few specimens have been analyzed and additional EOD data are needed.

GYMNOTUS CARAPO Linnaeus, 1758
Local name(s): logologo

Diagnostic characteristics: lower jaw distinctly longer than upper jaw; posterior end of body pointed; anal fin falling short to the posterior end of the body; oblique transversal white bands all over the body; medium-sized species, up to 40 cm TL
Type locality: America [South America]; all syntypes were collected in the 18th Century near Paramaribo, Suriname (Albert & Crampton, 2003)
Distribution: Southern Mexico to Paraguay, including Trinidad: Argentina, Bolivia, Brazil, French Guiana, Guatemala, Mexico, Paraguay, Trinidad and Tobago, Uruguay and Venezuela; in Suriname, it is present in all rivers and in coastal freshwater swamps
Size: 38 cm TL
Habitat: in coastal freshwater swamps and black-water streams of the Savanna Belt, it is often associated with aquatic vegetation in standing water (*G. carapo* is an air-breather that can survive low dissolved oxygen concentrations in these environments), but it is also present in running water habitats of large rivers and

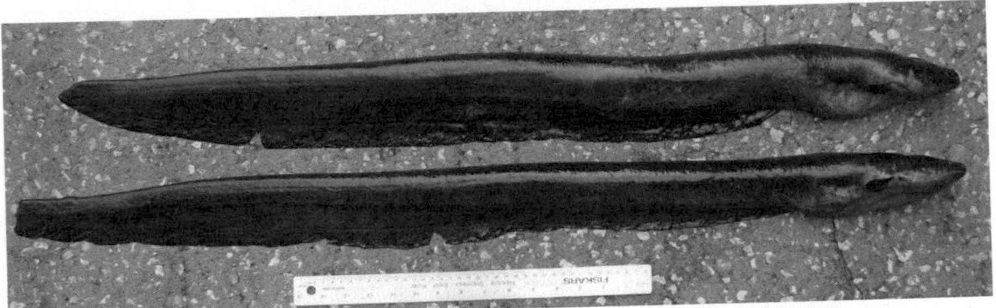

Fig. 12.6. Lateral view of the slender and fat morphs of the electric eel (*Electrophorus electricus*), Raleighvallen Rapids, Upper Coppename River (© W. Crampton).

Gymnotus carapo (alcohol), scale bar = 1 cm, Suriname River (© W. Crampton)

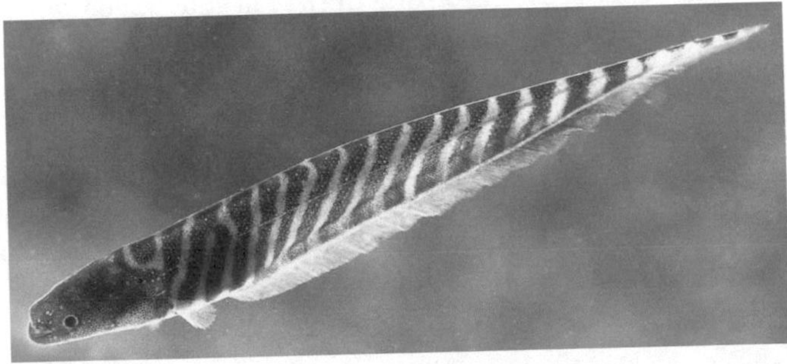

Gymnotus carapo (live) (© INRA-Le Bail)

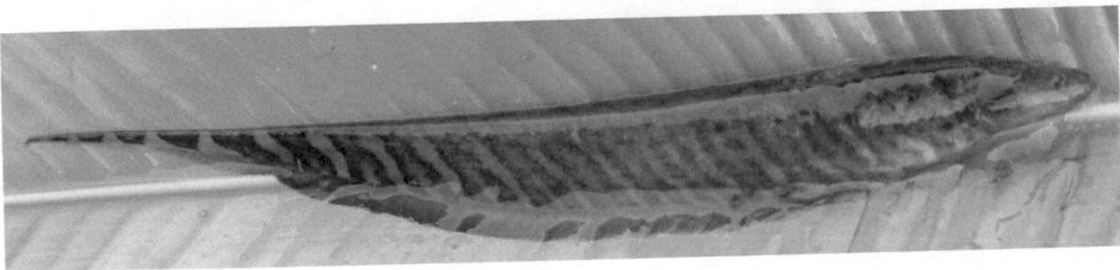

Gymnotus carapo (live), Suriname River (© R. Covain)

forest creeks of the Interior where it may hide during the day in leaf litter or submerged root masses
Position in the water column: mid-water
Diet: a nocturnal predator; juveniles feed on aquatic macro-invertebrates, adults are piscivorous
Reproduction: a nesting species (Crampton & Hopkins, 2005)
Remarks: abundant in coastal swamps and irrigation canals.

Gymnotus coropinae Hoedeman, 1962
Local name(s): –

Diagnostic characteristics: lower jaw distinctly longer than upper jaw; posterior end of body pointed; anal fin falling short to the posterior end of the body; transversal white bands restricted to posterior half of the body, in the anterior part of the body they are absent or not reaching the back
Type locality: Coropina Creek, Para River, Lower Suriname River, Suriname
Distribution: widely distributed throughout the Amazon and Orinoco basins and the Guyana Shield, South America; in Suriname, known to occur in Corantijn, Coppename, Saramacca, Suriname, Commewijne and Marowijne rivers, but probably present in all rivers
Size: 8.4-16.2 cm TL
Habitat: in small forest creeks with moderate to slow current, both black-water streams in the Savanna Belt and clear-water streams in the Interior
Position in the water column: hiding during the day in submerged root masses, leaf litter, woody debris and overhanging vegetation
Diet: both allochthonous and aquatic invertebrates
Reproduction:
Remarks: often confused with *G. anguillaris* (e.g. in Planquette *et al.*, 1996), a much larger species (maximum length 30 cm TL) (see Crampton & Albert, 2003); the presence of *G. anguillaris* in Suriname needs confirmation (it is possible that the type locality of *G. anguillaris* was erroneously denoted Coropina Creek, Suriname, because Hoedeman, who described *G. anguillaris*, did not collect the type specimen himself).

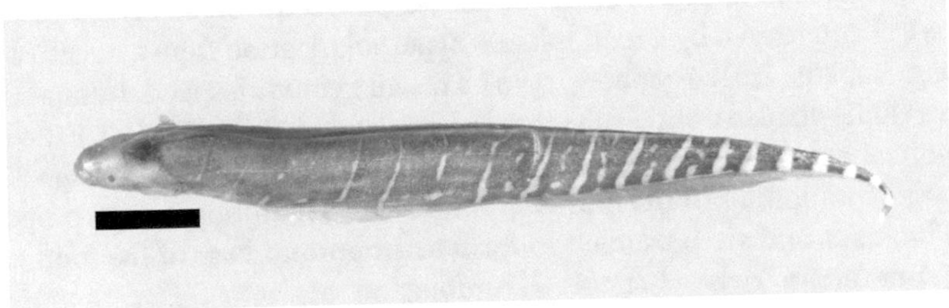

Gymnotus coropinae (alcohol), scale bar = 1 cm, Para River (Suriname River) (© W. Crampton)

Gymnotus coropinae (live) (© INRA-Le Bail)

FAMILY STERNOPYGIDAE (GLASS KNIFEFISHES, RATTAIL KNIFEFISHES)

Sternopygid species possess the following unique combination of characters among gymnotiforms: multiple rows of small, filiform (brush-like) teeth in both jaws; large eye (diameter equal to or greater than distance between nares); infraorbital bones large and bag-like, with expanded bony arches; anterior nares located outside gape; anal-fin origin at isthmus; no urogenital papilla, caudal fin or dorsal organ. Sternopygids possess a tone-type electric organ discharge ('wave' species), characterized by a monophasic hyperpolarization from a negative baseline (e.g. Hopkins & Heiligenberg, 1978). Like all gymnotiforms the shape is cultriform (knife-shaped), with an elongate body and anal fin. There is little direct commercial exploitation of sternopygid species. Some *Eigenmannia* species are ecologically important in riverine systems, often constituting a large proportion of the biomass, and are presumed to form an important base of the food web in main Amazonian river channels (Lundberg *et al.*, 1987). *Eigenmannia* and *Sternopygus* also seem to thrive in muddy streams disturbed by small-scale gold miners (Mol & Ouboter, 2004). Two species, *Sternopygus macrurus* and *Eigenmannia* cf. *virescens* are common in the aquarium trade. Sternopygids are not usually eaten by humans. The genus *Eigenmannia* is in need of revision.

1-a Eye free in orbit
Sternopygus macrurus
1-b Orbital margin continuous with the skin of the head
2

2-a Body with 7 broad transversal bands, alternatively dark and light
Japigny kirschbaum
2-b Absence of transversal bands on the body
3

3-a Scales missing from the anterior part of the back and upper sides; branchial openings large; ossified gill rakers present on the branchial arches
Rhabdolichops jegui
3-b Postcranial body scalation complete; gill openings small; gill rakers fleshy
4

4-a Body pigmentation darker dorsally and fading ventrally; a dark fringe on the distal margin of the anal fin (otherwise dusky); absence of three longitudinal dark lines on body
Eigenmannia sp. 1
4-b Anal fin clear; three longitudinal dark lines on body, first adjacent to anal fin, second line dorsal to first, third middle of body
Eigenmannia sp. 2 (*E.* aff. *virescens*)

This page is intentionally left blank.

EIGENMANNIA SP. 1
Local name(s): –

Diagnostic characteristics: orbital margin continuous with the skin of the head; body pigmentation darker dorsally and fading ventrally; a dark fringe on the distal margin of the anal fin; lacks three longitudinal dark lines on body
Type locality:
Distribution: in Suriname present in Commewijne, Coppename and Corantijn rivers
Size:
Habitat:
Position in the water column:
Diet:
Reproduction:
Remarks: see Willink & Sidlauskas (2004) for comments on this species and *Eigenmannia* sp. 2.

EIGENMANNIA SP. 2 (*E.* aff. *virescens*)
Local name(s): –

Diagnostic characteristics: orbital margin continuous with the skin of the head; anal fin clear; three longitudinal dark lines on body, first adjacent to anal fin, second line dorsal to first, third middle of body
Type locality: South America; no types known (*E. virescens*)
Distribution: *Eigenmannia virescens* is widely distributed east of Andes from Orinoco to La Plata River basins: Argentina, Bolivia, Brazil, Colombia, Ecuador, French Guiana, Guyana, Paraguay, Peru, Suriname, Uruguay and Venezuela
Size: up to 35 cm TL
Habitat: in small forest creeks of the Interior
Position in the water column:
Diet: nocturnal predators of small invertebrates
Reproduction: males up to 35 cm TL, females 20 cm TL; females produce 100-200 adhesive eggs
Remarks: P. Willink (Willink & Sidlauskas, 2004) notes the presence of a urogenital papilla in *Eigenmannia* sp. 2 whereas Sternopygidae do not have a urogenital papilla according to Albert (2001), so this species may be *E. virescens*, *E. lineatus* (Muller & Troschel 1845), a completely different *Eigenmannia* species or it may not even be in the genus *Eigenmannia*. The genus *Eigenmannia* is in need of revision (Willink & Sidlauskas, 2004).

JAPIGNY KIRSCHBAUM Meunier, Jégu & Keith, 2011
Local name(s): –

Diagnostic characteristics: a short snout with a subterminal mouth; body with 7 broad transversal bands, alternatively dark and light

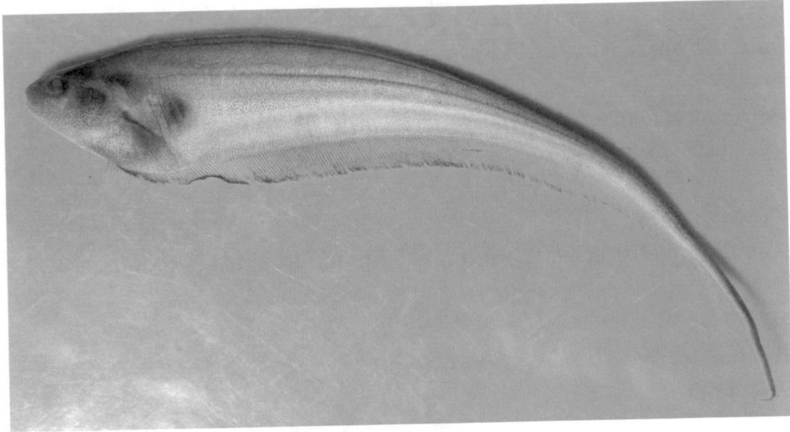

Eigenmannia sp. 1 (alcohol), Coppename River (© P. Willink)

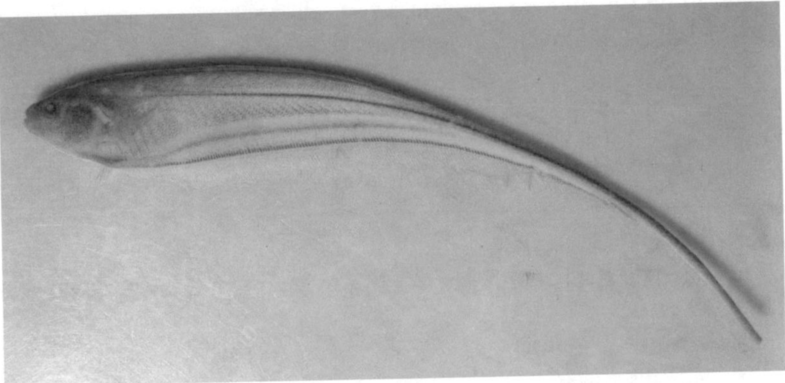

Eigenmannia sp. 2 (alcohol), Coppename River (© P. Willink)

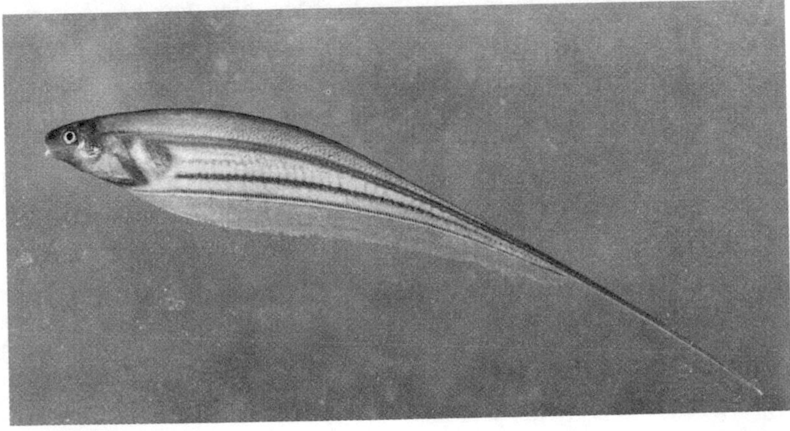

Eigenmannia sp. 2 (live) (© INRA-Le Bail)

Type locality: Saut Fracas, Mana River, French Guiana
Distribution: French Guiana; in Suriname known from Corantijn River
Size: 20 cm TL
Habitat: forest streams of the Interior, in parts with slow flow
Position in the water column:
Diet:
Reproduction:
Remarks: Planquette *et al.* (1996, p.386) identified this species as *Eigenmannia* n.sp.

RHABDOLICHOPS JEGUI Keith & Meunier, 2000
Local name(s): –

Diagnostic characteristics: mouth large and sub-terminal; scales missing from the anterior part of the back and upper sides; branchial openings large; ossified gill rakers present on the branchial arches; body yellow-brown to rose-yellow on the flanks, head darker, fins hyaline, insertion of the anal fin marked by small, parallel grey bars
Type locality: Antecume Pata, Litany River (Marowijne/Maroni Basin), French Guiana, 3°17'45"N, 54°04'13"W
Distribution: Mana and Maroni (Marowijne) River basins, French Guiana and Suriname
Size: 31.8 cm TL
Habitat: in the main river channel of the Upper Marowijne River, downstream of rapids in quiet water
Position in the water column:
Diet: *Rhabdolichops* species are planktivorous, consuming insect and crustacean larvae, but also small fishes
Reproduction:
Remarks: apparently not very common (Keith & Meunier, 2000).

STERNOPYGUS MACRURUS (Bloch & Schneider, 1801)
Local name(s): saprapi

Diagnostic characteristics: eye free in orbit; nostrils well separated; all anal rays simple; presence of a characteristic longitudinal white band on the posterior 2/3 part of the body
Type locality: Brazil
Distribution: widespread in South America: Argentina, Bolivia, Brazil, Colombia, Ecuador, French Guiana, Guyana, Paraguay, Peru, Suriname, and Venezuela; in Suriname, known from Corantijn, Saramacca, Suriname, Commewijne and Marowijne rivers, but probably present in all rivers
Size: 140 cm TL

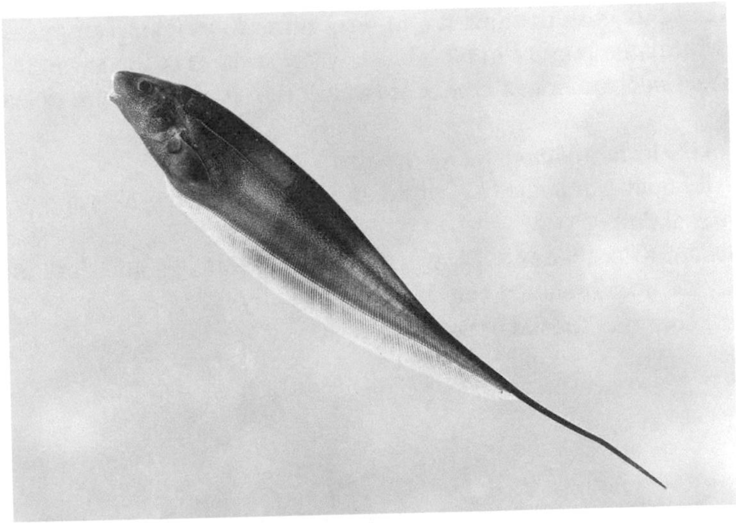

Japigny kirschbaum (live) (© INRA-Le Bail)

Japigny kirschbaum (live), Corantijn River (© R. Covain)

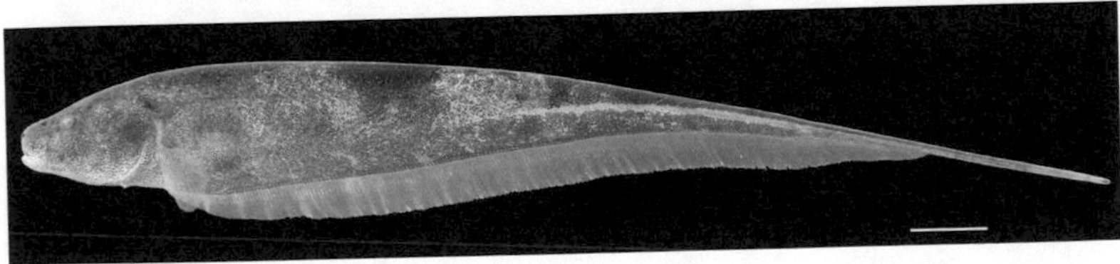

Sternopygus macrurus (live), scale bar = 1 cm, Lawa River (Marowijne River), ANSP189018 (© M. Sabaj Pérez)

Habitat: inhabits both floodplain and terra firme (non-floodplain) streams and rivers; in Suriname present in both black-water streams of the Savanna Belt and clear-water rivers and forest creeks of the Interior; it seems to prefer slow water currents
Position in the water column:
Diet: small aquatic animals (e.g. dipteran larvae) from the benthos or roots of aquatic vegetation
Reproduction: Kirschbaum & Schugardt (2002) report egg guarding in male *S. macrurus*, but no subsequent care of the hatched fry
Remarks: a common species in Suriname.

This page is intentionally left blank.

FAMILY RHAMPHICHTHYIDAE (SAND KNIFEFISHES)

The Rhamphichthyidae are a small family of electric knifefishes that can be easily recognized by following combination of characters: body highly compressed; snout elongated and tubular; mouth small, devoid of jaw teeth; dorsal and caudal fins absent. Rhamphichthyids live in tributary streams, marginal lagoons of large rivers and deeper portions of main river channels (Mago-Leccia, 1994). Species of *Gymnorhamphichthys* remain partially buried in sand or mud during the day; little else is known of their biology.

1-a Body translucent, naked, except the tail where there are small, embedded scales; tail long; pectoral rays 11-13, anal rays 140-180; relatively small (<22 cm TL), sand-dwelling species of small forest creeks
Gymnorhamphichthys rondoni

1-b Body marbled, brownish, covered by small, cycloid scales; tail (posterior to anal fin) short, pectoral rays 18-20, anal rays 300-415; relatively large species (up to 90 cm TL), mostly in the main channel of large to medium-sized rivers
Rhamphichthys rostratus

GYMNORHAMPHICHTHYS RONDONI (Miranda Ribeiro, 1920)
Local name(s): –

Diagnostic characteristics: body translucent, naked, except the tail where there are small, embedded scales; snout produced, straight and tubular, mouth very small; tail long; anal opening positioned anteriorly, slightly posteriorly to the eye, or close to a vertical extending from the pectoral fin base; pectoral rays 11-13, anal rays 140-180; relatively small (<22 cm TL), sand-dwelling species of small forest creeks
Type locality: Rio 17 de Fevereiro, trib. of Alto Cautario, Amazonas, Brazil
Distribution: Amazon, Paraná, Orinoco River basins and coastal rivers of the Guianas: Brazil, Colombia, Guyana, Paraguay, Suriname and Venezuela; in Suriname, known from Corantijn, Nickerie, Coppename and Suriname rivers (Nijssen *et al.*, 1976)
Size: 22 cm TL
Habitat: small forest creeks with moderate to strong current and sandy bottom in the Interior
Position in the water column: bottom
Diet: hides in the sand during the day and actively searches for interstitial prey (aquatic insect larvae and microcrustaceans) during the night by probing into the sand with its long snout (Zuanon *et al.*, 2006)
Reproduction:

Gymnorhamphichthys rondoni (alcohol), scale bar = 1 cm, Coppename River (© W. Crampton)

Gymnorhamphichthys rondoni (live), detail of the head, Nickerie River (© R. Smith)

Remarks: *G. rondoni* spends the daylight hours buried in the sand and emerges after sunset to swim about actively until shortly before dawn, when it again enters the sand. With an electric fish finder apparatus (Fig. 12.33.1), the fish can easily be detected when buried in the sand (and scooped out of the sand with a dip net). EOD activity is low when the fish is resting in the sand (but it does not stop completely), and high when it is freely swimming. Adaptations for the psammophilous (sand-dwelling) way of life of *G. rondoni* are discussed by Zuanon *et al.* (2006).

RHAMPHICHTHYS ROSTRATUS (Linnaeus, 1766)
Local name(s): –

Diagnostic characteristics: body marbled, brownish, covered by small, cycloid scales; tail (posterior to anal fin) short, pectoral rays 18-20, anal rays 300-415; relatively large species (up to 90 cm TL), mostly in the main channel of large to medium-sized rivers
Type locality: America [South America]
Distribution: coastal rivers of northeastern South America: Argentina, Brazil, French Guiana, Guyana and Suriname; in Suriname, known from Corantijn, Saramacca and Marowijne rivers, but possibly present in all rivers
Size: 90 cm TL
Habitat: main channel of rivers and large tributaries in the Interior, upstream of the first rapids
Position in the water column:
Diet: aquatic, benthic invertebrates
Reproduction:
Remarks: much larger than *Gymnorhamphichthys rondoni*, *R. rostratus* probably does not hide in the sand during daylight hours because it was caught in surface water with a dip net during daylight hours. It is not often collected in Suriname (but may still be relatively abundant) because it lives in the main stem habitat that is not easily sampled (except at low water levels in upper reaches, and under such conditions it can be caught in some numbers).

Rhamphichthys rostratus (live), Lawa River (Marowijne River), FMNH-SUR07-07 (© M. Sabaj Pérez)-

FAMILY HYPOPOMIDAE (BLUNTNOSE KNIFEFISHES)

Hypopomid species possess the following unique combination of characters among gymnotiforms: teeth absent from both jaws; snout moderate to short length (preorbital region less than 38% of head length); eye small (diameter less than the distance between nares); nasal capsule near eye, anterior nares located outside gape. The long anal fin originates below or posterior to the pectoral fins. There is no caudal fin. Like all gymnotiforms the shape is typically culteriform (knife-shaped), with an elongate body and anal fin. The ecology and natural history of most hypopomid species is very poorly understood. *Brachyhypopomus* is perhaps the most well studied group; it is also the most widely distributed hypopomid genus, in terms of both geography and habitat. *Brachyhypopomus* is able to tolerate protracted periods of hypoxic water conditions, at least in part because they can hold air bubbles in the gill chamber for use in aerial respiration. Hypopomid species are not common in the aquarium trade and are not usually eaten. The electric organ discharge (EOD) of these fish is multiphasic (usually biphasic), and produced in distinct pulses. Certain predators, such as catfish and predatory knifefish (e.g. the electric eel; Westby, 1988), are able to detect these EODs and use this to their advantage in finding prey. However, it has been found that species in the genus *Brachyhypopomus* restrict the low-frequency spectrum of their electric field close to their body, allowing higher-frequencies to spread further; this makes it more difficult for predators to detect them (Stoddard & Markham, 2008).

1-a Long snout (head length / snout length = 2.6-2.8); posterior nostrils equidistant between tip of snout and anterior margin of eye

Hypopomus artedi

1-b Short snout (head length / snout length = 3.0-4.6); posterior nostril positioned much more close to the eye than to the tip of the snout

2

2-a Presence of an accessory electric organ at the base of the pectoral fin; long tail (posterior of anal fin), approximately 1/3 of Total Length

Hypopygus lepturus

2-b Absence of accessory electric organ at the base of the pectoral fin; tail short, 10-20% of TL (*Brachyhypopomus*)

3

3-a Presence of electric accessory organ over opercular region

Brachyhypopomus sp. 1

3-b Absence of electric accessory organ over opercular region

4

4-a Total anal-fin rays 250-293
Brachyhypopomus brevirostris
4-b Total anal-fin rays 161-230
5

5-a Scale rows above lateral line 8-12, anal-fin rays 194-230, back with alternatively dark and lighter bands
Brachyhypopomus beebei
5-b Scale rows above lateral line 5-7, anal-fin rays 161-196, back brownish from the head to the end of the tail
Brachyhypopomus pinnicaudatus

An undescribed *Brachyhypopomus* species with dark transverse bands on the flanks, resembling *B. beebei*, but with a smaller body size (matures at less than 95 mm TL) and an electric organ discharge distinct from that of *B. beebei* (*Brachyhypopomus* sp. 2), occurs in Suriname and is known from a single specimen collected in a small tributary of Cottica River, Commewijne River system (W. Crampton, pers. communication). This species resembles an undescribed species of *Brachyhypopomus* that is common around Belem (Brazil) (W. Crampton, pers. communication).

Brachyhypopomus beebei (Schultz, 1944)
Local name(s): –

Diagnostic characteristics: short snout, sub-terminal mouth, conical head; back with alternatively dark and lighter bands; dark transversal bands much more narrow than the light transversal bands; anal fin with 194-230 rays
Type locality: Rio Caripe near Caripito, Estado Monagas, Venezuela
Distribution: tropical areas east of Andes: Bolivia, Brazil, Colombia, Ecuador, Guyana, Paraguay, Peru, Suriname and Venezuela; in Suriname known to occur in Nickerie, Coppename, Saramacca, Suriname and Commewijne rivers
Size: 20.4 cm TL
Habitat: small, shallow forest creeks with moderate to strong current, both black-water streams in the Savanna Belt and clear-water streams in the Interior
Position in the water column: bentho-pelagic
Diet:
Reproduction:
Remarks:

Brachyhypopomus brevirostris (Steindachner, 1868)
Local name(s): –

Diagnostic characteristics: back with alternatively dark and lighter bands; dark transversal bands about equal in size as light bands; anal fin with 250-293 rays
Type locality: Rio Guapore, Brazil
Distribution: Eastern South America from Orinoco to La Plata River: Argentina, Bolivia, Brazil, Colombia, Ecuador, Guyana, Peru, Suriname and Venezuela; in Suriname, known from Corantijn, Suriname and Commewijne rivers
Size: 34.7 cm TL
Habitat: forest creeks with moderate to strong current, both black-water streams in the Savanna Belt and clear-water streams in the Interior
Position in the water column: bentho-pelagic
Diet:
Reproduction:
Remarks: one of the largest *Brachyhypopomus* species.

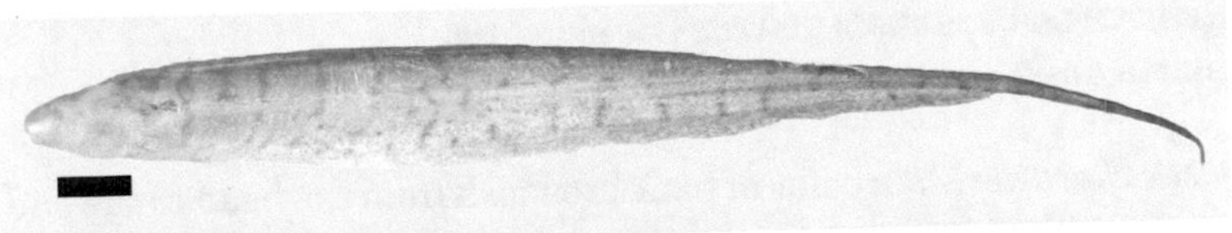

Brachyhypopomus beebei (alcohol), scale bar = 1 cm, Suriname River (© W. Crampton)

Brachyhypopomus beebei (live) (© INRA-Le Bail)

Brachyhypopomus brevirostris (alcohol), scale bar = 1 cm, Suriname River (© W. Crampton)

Brachyhypopomus brevirostris (live) (© INRA-Le Bail)

Brachyhypopomus pinnicaudatus (Hopkins, 1991)
Local name(s): –

Diagnostic characteristics: color of back brownish from the head to the end of the tail; anal fin with 161-196 rays
Type locality: coastal swamp called 'Grand Pripris', 3.5 kilometers northwest of the center of old Kourou, French Guiana, approx. 0.1 km N of old route Nationale #1, French Guiana
Distribution: South America: Bolivia, Brazil, Colombia, Ecuador, French Guiana, Guyana, Paraguay, Peru, Suriname, Uruguay and Venezuela; in Suriname, only known from the Commewijne River
Size: 18.6 cm TL
Habitat: seems to prefer slow currents
Position in the water column: bentho-pelagic
Diet:
Reproduction: sexual dimorphism with males having long wide tails, while females have shorter, more slender tails
Remarks:

Brachyhypopomus sp. 1 (Crampton & De Santana, in prep)
Local name(s): –

Diagnostic characteristics: presence of an accessory electric organ over the opercular region
Type locality:
Distribution: in Suriname, known from tributaries of the Suriname and Commewijne rivers (W. Crampton, pers. communication)
Size: about 20 cm TL
Habitat: in small forest creeks with leaf-litter on the streambed and slow water flow
Position in the water column:
Diet:
Reproduction:
Remarks:

Hypopomus artedi (Kaup, 1856)
Local name(s): saprapi

Diagnostic characteristics: a fairly large species (common to 30 cm TL) with a long snout (head length / snout length = 2.6-2.8), inferior mouth and a laterally compressed body (grey with small, irregular dark spots); posterior nostrils equidistant between tip of snout and anterior margin of eye
Type locality: Mana River, French Guiana

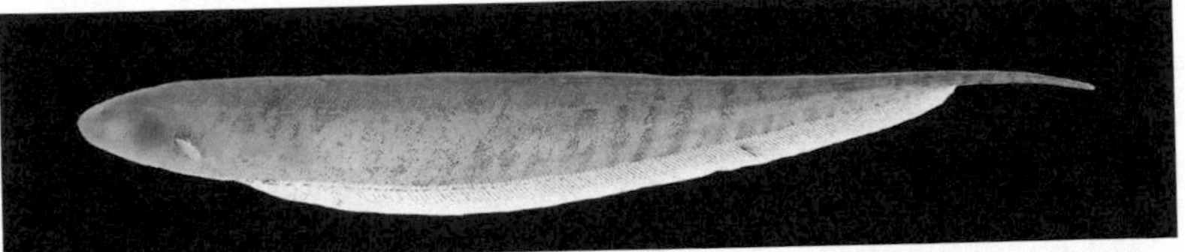

Brachyhypopomus pinnicaudatus (alcohol) (© P. Willink)

Brachyhypopomus pinnicaudatus (live) (© INRA-Le Bail)

Brachyhypopomus sp. 1 (alcohol), scale bar = 1 cm, Suriname River (© W. Crampton)

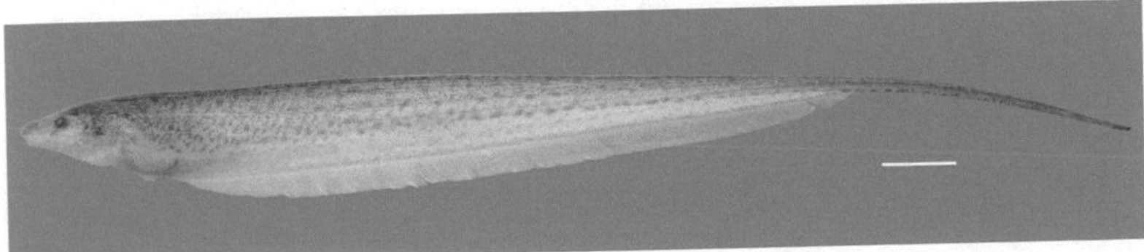

Hypopomus artedi (alcohol), scale bar = 1 cm, Lawa River (Marowijne River), ANSP189266 (© M. Sabaj Pérez)

Distribution: Guianas and Uruguay and Paraguay rivers: Argentina, Brazil, French Guiana, Guyana and Suriname; in Suriname, known from Corantijn, Coppename, Saramacca, Suriname, Commewijne and Marowijne rivers
Size: 50 cm TL
Habitat: forest streams with running water and sandy bottom
Position in the water column: bentho-pelagic
Diet:
Reproduction:
Remarks: synonym is *Parupygus savannensis* Hoedeman, 1962.

Hypopygus lepturus Hoedeman, 1962
Local name(s): –

Diagnostic characteristics: a small electric fish with an accessory electric organ at the base of the pectoral fin and a long tail (posterior of anal fin; approximately 1/3 of TL); body colored grey with 17-18 incomplete dark transversal bands; anal fin with <150 rays
Type locality: Marowijne (Maroni) basin, Suriname
Distribution: Amazon and Orinoco River basins and coastal rivers of the Guianas: Brazil, French Guiana, Suriname and Venezuela
Size: 10 cm TL
Habitat: small, shallow forest creeks with sandy bottom, both in the Savanna Belt (black-water) and in the Interior (clear water); hides during the day in submerged root masses
Position in the water column: bentho-pelagic
Diet:
Reproduction:
Remarks: one of the smallest gymnotiform knifefishes.

Hypopygus lepturus (alcohol), scale bar = 1 cm, Para River (Suriname River) (© W. Crampton)

Hypopygus lepturus (live) (© INRA-Le Bail)

FAMILY APTERONOTIDAE (GHOST KNIFEFISHES)

Apteronotid species are readily recognized as the only gymnotiform fishes with a caudal fin and a dorsal organ (a longitudinal strip of fleshy tissue firmly attached to posterodorsal midline). Apteronotids also possess a high frequency wave-type electric organ discharge (more than 750 Hz at maturity). *Apteronotus* species inhabit both floodplain and *terra firme* (non-floodplain) streams and rivers. Like many apteronotids, *Apteronotus* are aggressive predators of small aquatic insect larvae and fishes. Apteronotids are most diverse in the Amazonian floodplain, where many species are specialized to inhabit deep portions of the river channel (Lundberg *et al.*, 1987). Some apteronotids from the main channel are aggressive piscivores (e.g., *Sternarchella*), whereas others are planktivores (*Adontosternarchus*). *Magosternarchus* spp. feed on the tails of other electric fishes. Other species (e.g., *Sternarchorhynchus, Sternarchorhamphus*) have tubular snouts and forage on aquatic insect larvae on the rivers bottom. *Orthosternarchus* is perhaps the most specialized for life at the river bottom, in many ways resembling cave fishes; they are almost entirely blind, with minute, asymmetrically arranged eyes; and they have almost no pigments or scales, appearing bright pink in life due to the underlying blood hemoglobin (Hilton *et al.*, 2007). There is little direct commercial exploitation of Apteronotid species. Many species are ecologically important in Amazonian floodplains, often constituting a significant fraction of the biomass (Crampton, 1996). Two species (*Apteronotus albifrons* and *A. leptorhynchus*) are common in the aquarium trade. Apteronotids are not an important food resource.

1-a Presence of a tubular snout with downward curve; Marowijne River
Sternarchorhynchus galibi
1-b Snout not tubular
2

2-a Body black with two transversal white bands on the posterior part of the body; scales on the flanks all of the same size
Apteronotus albifrons
2-b Body brown-grey with only one, narrow white transversal band on the caudal peduncle; lateral line scales much larger than other scales of the flanks; Marowijne River
Porotergus gymnotus

This page is intentionally left blank.

Apteronotus albifrons (Linnaeus, 1766)
Local name(s): –

Diagnostic characteristics: body black with two transversal white bands on the posterior part of the body; scales on the flanks all of the same size; pectoral fins black, a narrow white line extends from the snout along the back to the vertical through the anterior of the anal fin
Type locality: Suriname
Distribution: northern and central South America: Argentina, Brazil, Ecuador, French Guiana, Guyana, Paraguay, Peru, Suriname and Venezuela
Size: 35 cm TL
Habitat: apparently present in both rapids and creeks with moderate current and sandy bottom substrate
Position in the water column:
Diet: small aquatic insect larvae and fishes
Reproduction:
Remarks: *A. albifrons* is not often collected in Suriname (it is most often present in rotenone collections of forest streams in the Interior). In the aquarium, this species is not aggressive towards other fishes (contrary to most gymnotiforms). Le Bail *et al.* (2000, p. 406) describe a form of *A. albifrons* with the anterior white band on the tail very broad; this form apparently occurs in eastern French Guiana.

Porotergus gymnotus Ellis in Eigenmann, 1912
Local name(s): –

Diagnostic characteristics: body color brown-grey with only one, narrow white transversal band on the caudal peduncle; pectoral fins transparent-grey; lateral line scales much larger than other scales of the flanks (Marowijne River)
Type locality: Amatuk, Guyana
Distribution: Essequibo River drainage, Amazon River basin: Brazil, French Guiana and Guyana; in Suriname, only known from the Marowijne River and (possibly) Suriname River
Size: 85 cm TL
Habitat: apparently present in both rapids and creeks with moderate current and sandy bottoms
Position in the water column:
Diet:
Reproduction:
Remarks: apparently rare and not often collected in Suriname.

Sternarchorhynchus galibi de Santana & Vari, 2010
Local name(s): –

Diagnostic characteristics: the only apteronotid in Suriname with a tubular snout with downward curve, a short gape that terminates posteriorly of the ventral

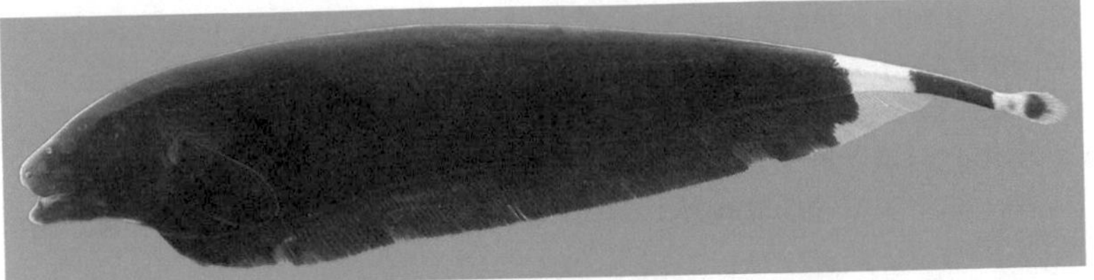

Apteronotus albifrons (alcohol), 109 mm SL, AUM40678 (© M. Sabaj Pérez)

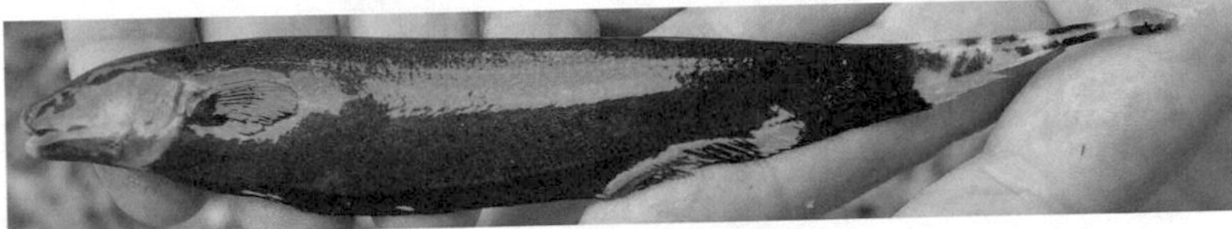

Apteronotus albifrons (live), Mindrineti River (Saramacca River)

Porotergus gymnotus (live), Marowijne River (© INRA-Le Bail)

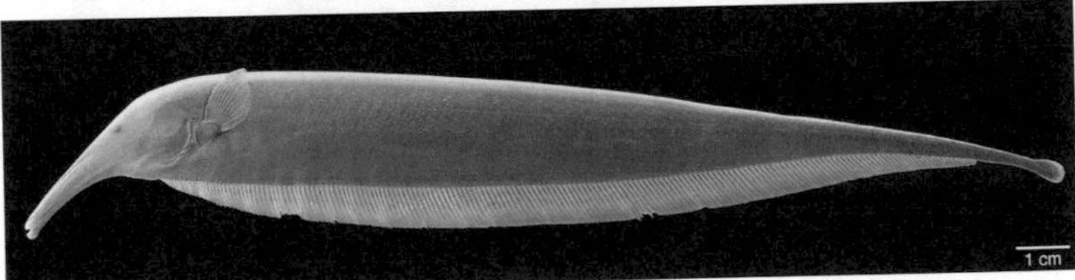

Sternarchorhynchus galibi (live), 194 mm SL, Lawa River (Marowijne River), ANSP187155 (© M. Sabaj Pérez)

through the anterior nares; absence of scales along mid-dorsal region of the bod (Marowijne River)
Type locality: Lawa River in rapids near Anapaikekondre (= Anapaike countr Anapaike at 3°34'N, 109°39'W), Marowijne District, Suriname
Distribution: Lawa River, Marowijne River basin, along border of Suriname an French Guiana
Size: 20 cm TL
Habitat: main river channel
Position in the water column:
Diet: aquatic insect larvae on the rivers bottoms
Reproduction:
Remarks: nothing is known about its biology; rare and not often collected.

This page is intentionally left blank.

FAMILY BATRACHOIDIDAE (TOADFISHES)

Small to medium-sized fishes (to 57 cm) easily recognized by their characteristic body shape: head broad and flattened, often with barbels and/or fleshy flaps; eyes on top of the head, dorsally-directed, and mouth wide. Gill openings are small and restricted to the sides, just in front of pectoral fin base. Two dorsal fins, the first consisting of 2 or 3 strong, sharp spines; the second consisting of a large number of soft rays; pelvic fins jugular, inserted well in advance of pectoral fins, with 1 spine and 2 or 3 soft rays. One to several lateral lines are present on the head and body. The body is naked or covered with small, cycloid scales. Color is mostly drab brown. Toadfishes are marine bottom-dwellers ranging from shallow inshore areas to deep waters; several species enter rivers, and some migrate regularly between shallow and deep waters. They are rather sluggish in their movements and are ambush predators, feeding mainly on mollusks and crustaceans. They may bite when handled. One species, *Batrachoides surinamensis*, is common in estuaries and shallow coastal waters of Suriname.

BATRACHOIDES SURINAMENSIS (Bloch & Schneider, 1801)
Local name(s): lompu (pacuma toadfish)

Diagnostic characteristics: head broad and depressed; two opercular, and two subopercular spines; gill openings small; body covered with small embedded scales; anterior dorsal fin with three spines; posterior dorsal fin with 28-30 soft rays; 25-27 anal fin rays; two lateral lines, 48-66 pores in lower lateral line below the pectoral fins; color brown with several prominent dark transverse bands dorsally on the head and sides of body, belly whitish
Type locality: Suriname
Distribution: brackish and marine waters of the Western Atlantic; in Suriname, known from the estuaries of Corantijn and Suriname rivers, but probably present in all river estuaries and in shallow coastal waters along the Surinamese coast
Size: up to 57 cm TL (3 kg), common to 40 cm TL
Habitat: muddy bottoms of estuaries and shallow inshore coastal waters
Position in the water column: bottom
Diet: sluggish ambush predators that feed on benthic invertebrates (crabs, shrimps, mollusks) and fishes
Reproduction: females mature at 20 cm and males at 25 cm; the females spawn 400-500 eggs with a diameter of 4-5 mm (Keith *et al.*, 2000).
Remarks: at least 4 other lompu species (Batrachoididae) occur in deep water off Suriname (Leopold, 2004); species in the genus *Thalassophryne* have poisonous glands associated with their spines. *B. surinamensis* is a fine food fish and sold at the markets of Paramaribo (but not in large numbers).

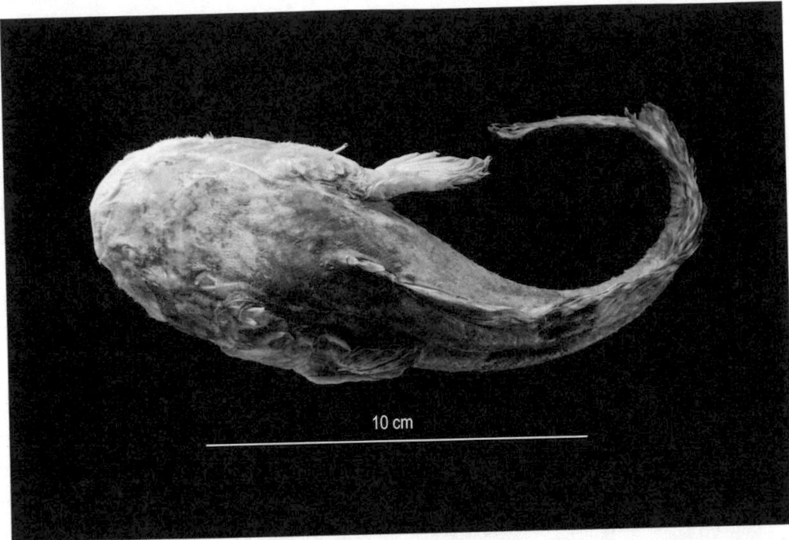

Batrachoides surinamensis (alcohol), Suriname River Estuary, RMNH10727 (© Naturalis)

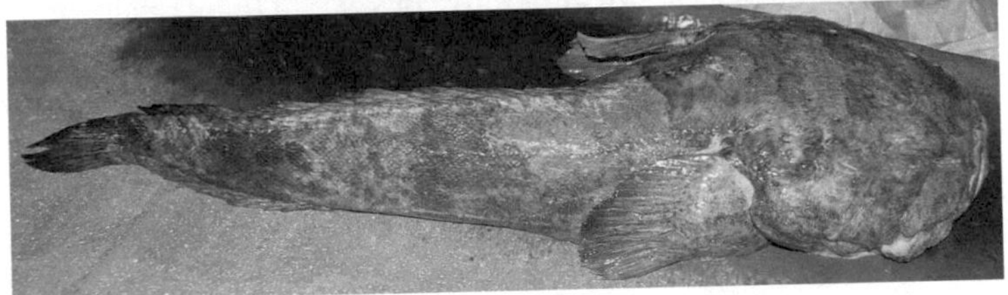

Batrachoides surinamensis (live), Suriname River

Batrachoides surinamensis (live) (© INRA-Le Bail)

FAMILY MUGILIDAE (MULLETS)

The family Mugilidae consists of 62 medium to large-sized species in 14 genera (Thomson, 1997). Mullets are readily distinguished from nearly all other fishes by the presence of two widely separated, short-based, dorsal fins, the absence of a lateral line on the body, an anal fin with three spines, and absence of thread-like filaments associated with the pectoral fin. The head is often broad and flattened dorsally, the eyes are partly covered by adipose 'eyefold' tissue, and the mouth is small, terminal or inferior with small, hidden teeth. The oral and branchial filter-feeding mechanism involves long gill rakers and a specialized 'pharyngobranchial organ' comprising a large, denticulate 'pharyngeal pad' and pharyngeal 'sulcus' on each side of the pharyngeal chamber. Scales are ctenoid and of moderate to large size, with 1 or more longitudinal striae (grooves) on each scale; large modified scales may be present at the insertion of the pectoral and pelvic fins (axillary scales) and the origin of the dorsal fin. Species of the family inhabit brackish water and coastal marine environments of all tropical and temperate seas. Some species stray into river mouths but are not generally considered to be freshwater inhabitants (e.g. several species of the genus *Mugil* occasionally enter into inland waters). Mullet are often found in schools or small groups, feeding in shallow waters either from the surface of the water or just off the bottom. In Suriname, mullets are also caught in estuaries and brackish-water mangrove lagoons (e.g. Bigi Pan Lagoon). Through much of their distributional range, these fishes are highly prized food fishes.

1-a Anal fin with three spines (first spine very short) and 8 soft rays in adults (2 spines and 9 soft rays in specimens <50 cm SL); < 43 scales in longitudinal series, (indistinct) dark longitudinal stripes present

2

1-b Anal fin with three spines (first spine very short) and 9 soft rays in adults (2 spines and 10 soft rays in specimens <50 cm SL); scales small, 42-45 in longitudinal series; absence of (indistinct) longitudinal stripes

Mugil incilis

2-a >35 (usually 38-42) scales in longitudinal series, 14-15 scales in transversal line between first dorsal fin and insertion of pelvic fin; body depth at origin of first dorsal fin 24-28% of SL; pelvic and anal fins and lower lobe of caudal fin grey brown; longitudinal dark stripes indistinct

Mugil cephalus

2-b <35 (usually 29-34) large scales in longitudinal series, 10-12 scales in transversal line between first dorsal fin and insertion of pelvic fin; body depth at origin of first dorsal fin 17-23% of SL; pelvic and pectoral fins pale yellow; regular brown dusky stripes on the flanks

Mugil liza

This page is intentionally left blank.

Mugil cephalus Linnaeus, 1758
Local name(s): aarder, kweriman, prasi (Lijding, 1959) (flathead mullet, striped mullet)

Diagnostic characteristics: anal fin with three spines (first spine very short) and 8 soft rays in adults (2 spines and 9 soft rays in specimens <50 cm SL); >35 (usually 38-42) scales in longitudinal series, 14-15 scales in transversal line between first dorsal fin and insertion of pelvic fin; body depth at origin of first dorsal fin 24-28% of SL; head depth equal to or greater than head width at level of posterior margin operculum; teeth very small, not visible to the naked eye (or appearing as a fine fringe), pelvic and anal fins and lower lobe of caudal fin grey brown; about 6-7 indistinct longitudinal dark stripes along flanks, following rows of scales, less conspicuous ventrally
Type locality: European sea, Europe
Distribution: circumglobal in temperate and tropical seas and estuaries
Size: maximum 120 cm SL, common to 35 cm
Habitat: adults in inshore marine waters, estuaries, lagoons and rivers
Position in the water column: benthopelagic
Diet: juveniles feed on plankton; adults feed on organic detritus and small benthos (crustaceans, worms, mollusks) that are filtered from the mud
Reproduction: spawning is in the sea, approximately 5-7 million eggs per female
Remarks: a cosmopolitan species that lives in schools. Mullets are sold at the markets of Paramaribo and Nieuw Nickerie.

Mugil incilis Hancock, 1830
Local name(s): aarder, kweriman, prasi (Lijding, 1959) (trench mullet)

Diagnostic characteristics: anal fin with three spines (first spine very short) and 9 soft rays in adults (2 spines and 10 soft rays in specimens <50 cm SL); 43-47 scales in longitudinal series, 21-23 circumpeduncular scales; origin of first dorsal fin usually slightly closer to tip of snout than to base of caudal fin; head deeper than wide at posterior margin operculum; teeth very small; color bluish grey or olivaceous dorsally, flanks silvery lacking stripes, and abdomen off-white; pectoral fins with dark spot at origin
Type locality: Guyana
Distribution: Western Atlantic
Size: 40 cm TL
Habitat: adults inhabit inshore marine waters and estuaries, but may enter fresh water
Position in the water column: benthopelagic
Diet:
Reproduction: in the reproductive season (first months of the year in French Guiana; Keith *et al.*, 2000) small groups of *M. incilis* form in the mouth of the rivers

Mugil cephalus (live) (© INRA-Le Bail)

Mugil incilis (live) (© INRA-Le Bail)

and tidal creeks (estuaries); females spawn several millions of eggs; the newly hatched larvae move inshore to lagoons and mangrove forests to feed.
Remarks:

Mugil liza Valenciennes, 1836
Local name(s): aarder, kweriman, prasi (Lijding, 1959) (Lebranche mullet)

Diagnostic characteristics: anal fin with three spines (first spine very short) and 8 soft rays in adults (2 spines and 9 soft rays in specimens <50 cm SL); <35 (usually 29-34) scales in longitudinal series, 10-12 scales in transversal line between first dorsal fin and insertion of pelvic fin; body depth at origin of first dorsal fin 17-23% of SL; color dusky bluish dorsally, flanks silvery, and abdomen off-white; several indistinct longitudinal dark brown stripes along the flanks; pelvic and pectoral fins pale yellow
Type locality: Martinique Island, West Indies
Distribution: widespread in Western Atlantic
Size: up to 100 cm TL, but more common to 40 cm TL
Habitat: adults inhabit marine inshore waters and brackish water lagoons; may occasionally enter fresh water, but never ascends far upriver
Position in the water column: benthopelagic
Diet: probably similar to *M. cephalus*; migrates along the shore in search of food
Reproduction: spawning in offshore waters, several millions of eggs per female
Remarks:

Mugil liza (live) (© INRA-Le Bail)

FAMILY RIVULIDAE (SOUTH AMERICAN ANNUAL FISHES)

Rivulids are small fishes, usually between 50-80 mm of total length (*R. igneus* is an exception with 150 mm TL), usually with a slender, subcylindrical body that can be recognized among other cyprinodontiforms (the 'killifish' of aquarium hobbyists) by the continuous branchiostegal and opercular membranes, the reduced laterosensory system of the head, and a number of synapomorphies related to the bony structures of head and fins. Members of the Rivulidae are typically oviparous fishes with external fertilization. The gaudy and diversified color patterns exhibited by males are the most conspicuous features to identify rivulids among members of the Neotropical ichthyofauna. Although primarily freshwater fishes, some rivulids live in estuarine areas (e.g. the mangrove rivulus *Kryptolebias marmoratus*). The marginal habitats of many rivulid species are unoccupied by other fish except for occasional *Pyrrhulina* species. Most rivulids are good jumpers and will seek new water bodies by moving overland through leaf litter if their habitat becomes unsuitable (Huber, 1992). Many rivulids are annual fishes, uniquely living in seasonal freshwater pools formed during the rainy season. During dry periods, all adults die, but the eggs survive in diapause, hatching in the next rainy season.

1-a Species living in estuarine environment or brackish water (mangroves); coloration generally dull; supracaudal and postopercular ocellus always present
Kryptolebias marmoratus

1-b Freshwater species; generally bright colors in males; supracaudal ocellus sometimes present in females, no postopercular ocellus (sometimes a dark spot)
2

2-a Presence of small (red) spots on the sides, forming longitudinal lines
3

2-b Longitudinal lines, if they are present, restricted to the anterior part of the body; orange and blue chevron marks or irregular blotches on the posterior part of the flanks
5

3-a Large species (>10 cm TL), frontal scalation type D (Hoedeman, 1959), >46 scales in the lateral line row; males with numerous small spots on the caudal fin (colored orange with or without black margins); females without supracaudal ocellus; caudal fin of females with many small spots
Rivulus igneus

3-b Medium-sized species (<10 cm TL), frontal scalation type E; females with supracaudal ocellus
4

4-a <43 scales in the lateral line
*Rivulus stagnatus**

4-b >44 scales in the lateral line
Rivulus aff. *holmiae*

5-a Body coloration with irregular blotches on the flanks, supracaudal ocellus present in females, >45 scales in lateral line (Tafelberg)
Rivulus amphoreus

5-b Body coloration with orange and blue chevron marks on the posterior part of the flanks, <44 scales in lateral line
6

6-a Relatively robust species (>5.5 cm TL), >35 scales in lateral line row, high number of anal rays (14-15), low number of dorsal rays (8-9), supracaudal ocellus absent in both males and females, males and females differ little in coloration
7

6-b Small species (<5.5 cm TL), <35 scales in lateral line row, anal rays <14, supracaudal ocellus present or absent in females
8

7-a Sides yellowish with punctuation forming some longitudinal lines, numerous vertical irregular bars on sides, belly whitish, anal fin with numerous brown spots and subdistal red-colored area, dorsal and caudal pinkish with numerous brown spots, caudal fin with red superior margin, known from Litani River, Upper Marowijne River
Rivulus gaucheri

7-b Strong melanism, body coloration with network of dark markings beginning behind the eye and continuing just posterior to gill cover, a pattern of oblique bars on the flanks, but capacity to change color pattern within seconds; juveniles have a strong orange body coloration, known from Tapanahony River, Upper Marowijne River
Kryptolebias sepia

8-a Caudal ocellus absent in females
9

8-b Caudal ocellus present in females; male with an orange caudal fin with a black lower margin and a yellowish submargin, sides with blue chevron markings, dorsal fin with transparent margin and some transparent blotches
Laimosemion agilae

---

* related species are *R. urophthalmus* Günther, 1866 (Amazon) and *R. lungi* Berkenkamp, 1984 (French Guiana); when preserved, the three species *R. stagnatus*, *R. urophthalmus* and *R. lungi*, cannot be separated, by color pattern, nor by morphometrics (Huber 1992, p. 407). *R.* aff. *lanceolatus* (status uncertain, rows of red spots forming lines, might be close to *R. urophthalmus*) is another ill-defined species that has been reported for Suriname by Huber (1992).

9-a Male with bright orange caudal fin, with transparent posterior margin; 7-8 rays in dorsal fin

Laimosemion frenatus

9-b Male with caudal fin spotted forming about 5 vertical bars

10

10-a Caudal fin of male with black lower margin; 10-12 rays in dorsal fin; sides with brownish dot on each scale, sometimes fused (upper part) forming dark spots arranged in 4-10 irregular vertical bars; frightened males have a wide black band aloing the ventral side of the body from opercle to caudal fin

Laimosemion breviceps

10-b Caudal fin of male lacking black lower margin; males with large orange dots; dots posteriorly arranged in chevron marks against a blueish background; a post-opercular blotch sometimes visible

Laimosemion aff. *geayi*

KRYPTOLEBIAS MARMORATUS (Poey, 1880)
Local name(s): – (mangrove rivulus)

Diagnostic characteristics: both the supracaudal and postopercular ocellus always present in both males and females; 49 scales in the lateral line row; 11-12 rays in the anal fin; unpaired fins with dark irregular speckles; coloration generally dull, dark speckles irregularly distributed on a golden-brown background
Type locality: Rio de Janeiro, Brazil
Distribution: Atlantic drainages of North, Central and South America; estuarine areas of Atlantic coast
Size: 6 cm TL (7.5 cm in females, 4.1 mm in males according to Huber 1992)
Habitat: standing water of brackish-water lagoons and mangrove forests, often in sunny places (the same habitat as that of *Poecilia vivipara*)
Position in the water column:
Diet:
Reproduction: unique reproductive behavior with hermaphroditic 'females' and low sexual dichromatism; primary males (born as males) represent 5% of the population; about 60% of the hermaphroditic individuals change into secondary males after 3-4 years; the proportion of males is dependent on the water temperature, with males in the majority below 20 °C and all hermaphrodites above 25 °C (Harrington, 1971); the hermaphrodite form synchronically produces both sperm and eggs and these fishes are able to fertilize their own eggs internally, producing viable offspring; the eggs (1.6 mm diameter) can survive a diapause
Remarks: the brackish-water habitat explains its wide distribution from Rio de Janeiro along the Atlantic coast up to Guyana (Huber, 1992) or possibly Florida. Keith *et al.* (2000) list this species as *Rivulus ocellatus* for French Guiana, but *Kryptolebias* (*Rivulus*) *ocellatus* has recently been restricted to southeastern Brazil

Kryptolebias marmoratus (live) (© INRA-Le Bail)

(see Costa, 2006); *K. marmoratus* is one of the very few species of fishes capable of self-fertilization (internal, producing clonal populations of homozygous, genetically identical hermaphroditic fish).

KRYPTOLEBIAS SEPIA Vermeulen & Hrbek, 2005
Local name(s): –

Diagnostic characteristics: relatively robust species (females up to 6.5 cm TL), 38-40 scales in lateral line row, high number of 15 anal-fin rays, low number of 7-8 dorsal-fin rays, supracaudal ocellus absent in both males and females, males and females differ little in coloration; strong melanism, body coloration with network of dark markings beginning behind the eye and continuing just posterior to gill cover, a pattern of oblique bars on the flanks, but capacity to change color pattern within seconds; juveniles have a strong orange body coloration, known from Tapanahony River, Upper Marowijne River
Type locality: 5 kilometers downriver from Paloemeu on the right bank, 3°22'43"N, 55°24'42"W, Tapanahony River, Upper Marowijne system, Suriname
Distribution: known only from small forest creeks emptying into the Tapanahony and Paloemeu rivers, southeastern Suriname
Size: 6.5 cm SL
Habitat: extremely shallow parts of small creeks (closed canopy cover) in the Interior, outside the main stream in swampy areas directly adjacent to the creeks themselves, often with a thick layer of leaf litter (see photograph in Vermeulen & Hrbek (2005)); *K. sepia* is found in only a few centimeters of water directly above the leaf litter; the creek proper had permanent running clear water
Position in the water column:
Diet: predator of small fishes and invertebrates; they may also feed on tadpoles (Vermeulen & Hrbek, 2005)
Reproduction: males differ little from females in their body coloration, shape and fins. Eggs (2 mm in diameter, color dark amber) are covered with a sticky substance and placed one by one between roots at the water's edge, or just above the water edge. Very low water levels stimulate mating and water levels decreasing to less than 3 cm instantly result in spawning. Eggs develop in 14-16 days at 24 °C. This species is unlikely to lay eggs capable of undergoing a developmental diapause.
Remarks:

LAIMOSEMION AGILAE (Hoedeman, 1954)
Local name(s): –

Diagnostic characteristics: small species (<5.5 cm TL), <35 scales in lateral line row, <14 rays in anal fin, frontal scalation of F-type, sometimes E-type (Hoedeman, 1959); male with an orange caudal fin with a black lower margin and a yellowish submargin, sides with blue chevron markings, dorsal fin with transparent margin

Laimosemion agilae (© INRA-Le Bail)

and some transparent blotches; females less colorful, but often with supracaudal ocellus present
Type locality: Agila rivulet between Agila and Berlin, tributary to Para River, Suriname River system, Suriname
Distribution: Atlantic coastal river basins: French Guiana, Guyana and Suriname
Size: 5 cm TL (females smaller)
Habitat: prefers small, shallow creeks with slow water current, but also abundant in sunny places with high water temperature (e.g. swamps of the Old Coastal Plain), and even in strong currents below the falls
Position in the water column:
Diet:
Reproduction: males and females mature after 10 months; the female produces about 20 eggs (1.5 mm diameter) that hatch after 4 weeks incubation (at 22 °C); newborn larvae measure 3 mm TL
Remarks: a well-known and popular aquarium fish that breeds easily in captivity and is not much inclined to jump out the aquarium; *Rivulus agilae* is a synonym.

LAIMOSEMION BREVICEPS (Eigenmann, 1909)
Local name(s): –

Diagnostic characteristics: small species (<5.5 cm TL), 35 scales in lateral line row, 12 anal-fin rays, 10-12 dorsal-fin rays, frontal scalation type E; males with caudal fin spotted, with spots forming about 5 vertical bars, black lower margin, sides with brownish dot on each scale, sometimes fused (upper part) forming dark spots arranged in 4-10 irregular vertical bars; frightened males have a wide black band along the ventral side of the body from opercle to caudal fin; females with color pattern similar to that of males, lacking a supracaudal ocellus
Type locality: Shrimp Creek, where the path from Tukeit to the head of the Kaieteur crosses it, Guyana
Distribution: Amazon River basin, Guyana; listed for Suriname by Huber (1992; Table 6), but not in his (Huber 1992) examined material for the species
Size: 3.5 cm TL (males up to 5.5 cm according to Huber, 1992)
Habitat: small, shallow rivulets and clear-water pools close to the stream edge
Position in the water column:
Diet:
Reproduction: eggs 2 mm in diameter; maturity reached in 5 months
Remarks: its occurrence in Suriname is in need of confirmation. *Rivulus breviceps* is a synonym.

LAIMOSEMION FRENATUS (Eigenmann, 1912)
Local name(s): –

Diagnostic characteristics: a small, slender species with a faint to conspicuous black longitudinal band on the sides, depending on mood, and yellow-green dots;

This page is intentionally left blank.

males with bright orange caudal fin, with transparent posterior margin; 32-34 scales in lateral line12-13 anal-fin rays, 7-8 rays in dorsal fin; females subdued, lacking a supracaudal ocellus (sometimes a faint small spot)
Type locality: Gluck Island, Guyana
Distribution: Essequibo River basin, Guyana and Suriname; in Suriname, the only certain collection locality is in the Marowijne River Basin, 37 km to the north of Stoelmanseiland (Huber, 1992)
Size: 5 cm TL
Habitat: shady forest creeks with brown, slow-current waters
Position in the water column:
Diet:
Reproduction: egg diameter is 1.6 mm; maturity is reached in 10 months; males show contrasted colors when sexually excited or aggressively dominating others
Remarks: its occurrence in Suriname is in need of confirmation. *Rivulus frenatus* is a synonym.

LAIMOSEMION aff. *GEAYI* (Vaillant, 1899)
Local name)s): -

Diagnostic characteristics: caudal fin of male spotted forming about 5 vertical bars and lacking a black lower margin; males with the back colored orange and latero-ventrally large orange dots; dots posteriorly arranged in chevron marks; a post-opercular blotch sometimes visible; dorsal and anal fins spotted
Type locality: Carsevenne River, French Guiana, elevation 450 meters
Distribution: Atlantic coastal river drainages and Amazon River basin: Brazil and French Guiana; in Suriname a species close to *L. geayi* occurs in the Paloemeu River (e.g. Vermeulen & Hrbek, 2005)
Size: 5 cm
Habitat: small, shallow rainforest creeks with slow current and leaf litter on the bottom and also in swampy areas along the creek
Position in the water column: found among dead leaves on the bottom of the stream
Diet:
Reproduction: females are even smaller than males; the eggs are 1.6 mm in diameter
Remarks: *Rivulus geayi* is a synonym; although Keith *et al.* (2000) show *Laimosemion geayi* present in eastern French Guiana (and replaced by *L. agilae* to the west, thus in Suriname), *Laimosemion* aff. *geayi* was collected in shallow tributaries of Paloemeu River together with *Kryptolebias sepia* (Vermeulen & Hrbek, 2005; personal observation).

Laimosemion aff. *geayi* (live), Paloemeu River (© K. Wan Tong You)

RIVULUS AMPHOREUS Huber, 1979
Local name(s): –

Diagnostic characteristics: a large (up to 8.5 cm TL for males), deep-bodied species; males with regular (chess-board) to irregular red-brown and yellow-green blotches on the flanks, females subdued, but with supracaudal ocellus present, >45 scales in lateral line11-12 dorsal-fin rays; known principally from Tafelberg Mountain
Type locality: vicinity of Tafelberg Mountain, zone of high hills up to 1000 meters in altitude, 120 km south-southwest of Paramaribo, Suriname
Distribution: Atlantic river basins, Suriname; known to occur on Tafelberg Mountain, and possibly in the Upper Nickerie River (Blanche Marie Falls) and Lely Mountains
Size: 6 cm TL (males up to 8.5 cm; Huber, 1992)
Habitat: occurs in fast currents in addition to the usual *Rivulus* habitat of small, shallow creeks; mainly known from clear-water mountain streams
Position in the water column:
Diet:
Reproduction: eggs of 2 mm diameter; sexual maturity reached after 4 months; dominant males show yellow-green iridescence around the eye
Remarks: possibly restricted to mountain creeks (altitude > 200 m-asl).

RIVULUS GAUCHERI Keith, Nandrin & Le Bail, 2006
Local name(s): –

Diagnostic characteristics: a slender medium-sized species with 14-15 anal-fin rays, 8-9 dorsal-fin rays, 38-40 scales along the lateral line; males with the sides yellowish with punctuation forming some longitudinal lines, numerous vertical irregular bars on sides, belly whitish, anal fin with numerous brown spots and subdistal red-colored area, dorsal and caudal pinkish with numerous brown spots, caudal fin with red superior margin; females with more or less similar coloration as the males, lacking a supracaudal ocellus; known from Litani River, Upper Marowijne River
Type locality: marshes in Alama River, Litani, Marowijne River Basin
Distribution: Upper Marowijne River
Size: 5.7 cm TL
Habitat: found in small groups in a small mountainous creek (2-3 m wide, 60 cm deep) and pools and marshes with clear, shallow (10-20 cm) water along the forest creek; they were generally found under leaves and branches along the edges, at places where current was slow
Position in the water column:
Diet:
Reproduction:
Remarks: a species of highland areas, much like *Kryptolebias sepia* and *Rivulus amphoreus*.

This page is intentionally left blank.

Rivulus aff. *holmiae* Eigenmann, 1909
Local name(s): –

Diagnostic characteristics: a large to medium-sized species (<9 cm TL) with 5-6 rows of discontinuous red lines (only 2-3 remaining near the caudal peduncle), frontal scalation type E, 42-46 scales in the lateral line, 9-10 dorsal-fin rays, 15-17 anal-fin rays; females with prominent supracaudal ocellus
Type locality: creek near Holmia, Guyana
Distribution: *R. holmiae* is only known with certainty from its type locality in the highlands of Guyana at Holmia (>500 m altitude); Amazon River basin, Guyana; Huber (1992, p. 244) remarks that the Suriname material of *R. holmiae* (Suriname and Marowijne rivers) differs in frontal scalation (D-type) and color pattern (more yellow with regular striations and bright yellow submargins on unpaired fins) and may be a separate species
Size: 8 cm TL (9 cm in Huber, 1992)
Habitat: probably opportunistic in small hill streams and lowland creeks
Position in the water column:
Diet: possibly a predator of small and young fishes
Reproduction: eggs 2 mm diameter, maturity in 9 months
Remarks:

Rivulus igneus Huber, 1991
Local name(s): –

Diagnostic characteristics: a very large species (up to 15 cm TL), frontal scalation type D (Hoedeman, 1959), 45-50 scales in the lateral line row, 10-13 dorsal-fin rays, 17-20 anal-fin rays; males with numerous small spots on the caudal fin (colored orange with or without black margins), dorsal and anal fins dotted near base, anal fin with dark distal margin; females same coloration as males, but much subdued, without supracaudal ocellus and a caudal fin with many small spots
Type locality: Montagne des Singes, 5°10'N, 52°70'W [French Guiana]
Distribution: Oyapock River basin and adjacent coastal basins: Brazil and French Guiana; in Suriname, known from high elevation mountain creeks in Lely and Nassau Mountains (Lower Marowijne River Basin)
Size: 15 cm TL
Habitat: prefers stagnant to slow-current waters, shallow with a thick bed of dead leaves and branches
Position in the water column:
Diet: a greedy species, able to swallow cut earthworms
Reproduction: eggs 2.8 mm in diameter, maturity in 13 months; more than 100 fry were derived from a single spawn of a pair, eggs were deposited on the bottom; females always larger than males in the aquarium
Remarks: the largest species in the genus *Rivulus* (looks almost like a stonwalapa or matuli *Erythrinus erythrinus*); strongly territorial in the aquarium.

Rivulus igneus (live) (© INRA-Le Bail)

RIVULUS AFF. *LANCEOLATUS* Eigenmann, 1909
Local name(s): –

Diagnostic characteristics: medium-sized, slender species with rows of red dots against blue background, forming lines; female may be distinct, caudal ocellus present and prominent; front al scalation type E, 6-8 dorsal-fin rays, 11-15 anal-fin rays, 41-47 lateral line scales
Type locality: Rockstone, Guyana
Distribution: Atlantic coastal river basins, Guyana; known only with certainty from its type locality, but *R.* aff. *lanceolatus* is recorded in Suriname from Corantijn and Suriname rivers (Huber, 1992)
Size: 5.5 cm TL
Habitat: probably forest creeks
Position in the water column:
Diet:
Reproduction:
Remarks: *Rivulus lanceolatus* is an ill-defined species with uncertain taxonomic status (Huber, 1992, p. 279-282).

RIVULUS STAGNATUS Eigenmann, 1909
Local name(s): –

Diagnostic characteristics: medium- to large-sized, average shaped species (6 cm TL) with irregular, discontinuous rows of small (red) spots against a blue-ish background on the sides, forming at least 7 longitudinal lines; frontal scalation type E, females subdued and with supracaudal ocellus; 35-44 (mostly <43; 38-42 in material from Suriname) scales in the lateral line, 6-9 dorsal-fin rays, 11-14 anal-fin rays; Surinamese *R. stagnatus* show upper and lower lighter bands in the caudal fin (Huber, 1992)
Type locality: Christianburg (Guyana)
Distribution: Atlantic coastal river basins: Guyana and Suriname; in Suriname, known from Corantijn, Nickerie, Coppename and Suriname rivers
Size: 6 cm TL
Habitat: opportunistic species living in a wide variety of habitats, mostly in stagnant water or slow current; abundant in little pools
Position in the water column:
Diet: feed on insects (ants, flies, beetles) with occasionally a small amount of vegetable matter
Reproduction: eggs 1.5 mm in diameter; maturity is reached in 5 months
Remarks: related species are *R. urophthalmus* Günther, 1866 (Amazon) and *R. lungi* Berkenkamp, 1984 (French Guiana); when preserved, the three species cannot be separated, by color pattern, nor by morphometrics (Huber 1992, p. 407) and possibly they are synonyms.

Rivulus lungi (live) (© INRA-Le Bail)

Rivulus sp. (live), Upper Kabalebo River (© R. Smith)

Rivulus cf. *stagnatus* (live), Upper Nickerie River (© R. Smith)

FAMILY POECILIIDAE (LIVEBEARERS)

Poeciliids are small and laterally compressed cyprinodontiform fishes that can be characterized by: (1) pectoral fins inserted high on the body, (2) pelvic fins that migrate anteriorly during growth, (3) recessed supraorbital pores 2b through 4a, and (4) pleural ribs on the first several haemal arches as well as a series of other internal synapomorphies (see details in Parenti, 1981; Costa, 1998). The body form ranges from extremely elongate (e.g. *Tomeurus*) to deep-bodied; the dorsally-oriented mouth and dorso-ventrally flattened head is adapted for utilization of a well oxygenated surface film at the atmosphere-water interface (i.e. 'aquatic surface respiration'; Kramer & McClure, 1982) to survive in the oxygen-depleted water that often characterizes the habitat of these fishes (Lewis, 1970). Size is extremely variable ranging from the tiny *Fluviphylax palikur* of French Guiana (maximum adult recorded size: 13.9 mm) to the giant of the group, *Belonesox belizanus*, which reaches 200 mm. Poeciliids comprise about 300 valid species inhabiting the fresh and brackish waters of American and African continents (but they are reported to occur in salt waters in coastal areas). The subfamily Poeciliinae is broadly distributed throughout the Americas. Poeciliines are characterized by: (1) the uniquely derived possession of a gonopodium formed by the modified male anal-fin rays 3, 4, and 5 (Parenti, 1981), (2) internal fertilization, (3) viviparity (*Tomeurus gracilis* possess facultative viviparity) (Wourms, 1981; Parenti *et al.*, 2010). The Poeciliinae includes well-known aquarium fishes such as the guppies, mosquito fishes, swordtails, platys and mollies, being very familiar to the non-scientific public. On the other hand, poeciliines are well known from several biological standpoints, being object of study for ecologists, anatomists, embryologists and many others biologist researchers. Notwithstanding, this fish assemblage is disappointingly ill-studied from the perspective of systematics.

1-a Presence of a transparent membrane between the anal fin and the caudal fin; distance between the anal fin and the dorsal fin much larger than the distance between the dorsal fin and the caudal fin; body elongated and transparent

Tomeurus gracilis

1-b Absence of a transparent membrane between the anal fin and the caudal fin; distance between the anal fin and the dorsal fin much smaller than the distance between the dorsal fin and the caudal fin; body not transparent

2

2-a Humeral spot round, positioned high on the body, close to the insertion of the dorsal fin, but always separated from the predorsal line; caudal peduncle deep; adult males with upper and lower margins of caudal fin black; relatively large species (up to 5 cm TL in females); anal fin of the male modified, but not bearing spines; sides with silvery and dark cross bands

Poecilia vivipara

2-b Humeral spot often vertically elongated reaching the predorsal line; depth caudal peduncle not large; absence of black lining of dorsal and ventral part of caudal fin in adults; much smaller species (females <3 cm TL); anal fin of males with short intromittent organ bearing numerous recurved hooks on anterior and posterior margins; males brilliantly colored, much smaller than females

3

3-a Males with vertically-elongated humeral spot; females with dark gestation spot

4

3-b Males without humeral spot of if humeral spot present it is rounded

5

4-a Males with a flame-like horizontally-elongated marking on the caudal fin (sometimes prolonged anteriorly on the caudal peduncle and body), in some forms the males are bright blue, yellow-green or red; females with a very black gestation spot and a rounded humeral spot (if present); species of the Coastal Plain, often in very shallow ditches with brackish water that are in connection with the estuary

Micropoecilia parae

4-b Males with a vertically elongated spot on the caudal peduncle, usually prolonged along the upper and lower margins of the caudal fin; freshwater species of the Savanna Belt and northern part of the Interior

Micropoecilia bifurca

5-a Males with a rounded humeral spot (when present) and a horizontally elongated black spot at base of the upper margin of the caudal fin; females with a small rounded humeral spot (if present) and no dark gestation spot on the belly; body slightly elongated (height 3.8-4.1 times in SL)

Micropoecilia picta

5-b Male with two to four spots of varying size and variously placed along the sides, very variable; females reticulated, without spots

Poecilia reticulata

Micropoecilia bifurca (Eigenmann, 1909)
Local name(s): todobere (however, note that 'todobere' translates to tadpole; both tadpoles and poeciliids are locally known as todobere) (guppy)

Diagnostic characteristics: the smallest poeciliid of Suriname; males yellowish with a vertically elongated black spot on the caudal peduncle, usually prolonged along the upper and lower margins of the caudal fin, a vertically elongated humeral spot, a thin, black line ventrally on the abdomen between the anal and caudal fins, and a long gonopodium (3.8-3.9 times in SL); females with a darkish gestation spot on the belly; a freshwater species of the Savanna Belt and northern part of the Interior
Type locality: Christianburg [Guyana]
Distribution: Orinoco Delta: Venezuela to French Guiana; in Suriname, known from the Saramacca and Suriname rivers, but probably present in other rivers as well
Size: 2.5 cm TL
Habitat: the only poeciliid in Suriname that occurs in freshwater in black-water streams of the Savanna Belt and in clear-water creeks of the (northern part of the) Interior, mostly in shallow water along the shore; in streams with slow current and sandy/muddy bottom and in swamps, often among aquatic vegetation
Position in the water column: surface
Diet:
Reproduction:
Remarks: the smallest *Micropoecilia* species.

Micropoecilia parae (Eigenmann, 1894)
Local name(s): todobere (guppy)

Diagnostic characteristics: males with a flame-like horizontally-elongated marking on the caudal fin (sometimes prolonged anteriorly on the caudal peduncle and body), in some forms (known as melanzonus) the males are bright blue or greenish; females with a very black gestation spot and a rounded humeral spot (if present); species of the Coastal Plain, often in little canals with brackish water that are in connection with the estuary
Type locality: in the ditches of the Rua das Mongubas of Para, Brazil
Distribution: Guyana to Brazil: Brazil, French Guiana, Guyana and Suriname; in Suriname, known from the estuaries of the Suriname, Commewijne and Nickerie rivers, but probably present in the estuary of other rivers as well
Size: 3 cm TL
Habitat: shallow ditches with brackish water that are in connection with the estuary of large rivers, but also in coastal swamps with freshwater, often collected together with *Poecilia vivipara*, *Poecilia reticulata* and *Micropoecilia picta*
Position in the water column: surface
Diet:

Micropoecilia bifurca (live), female (© L. Volkmann)

Micropoecilia bifurca (live), male (© L. Volkmann)

Micropoecilia bifurca (live) (© INRA-Le Bail)

Reproduction: sexual dichromatism, with males (smaller than females) very colorful and females grey-brown with very dark gestation spot on the belly; females give birth to 6-30 young
Remarks: male *M. parae* exhibit one of the most complex polymorphisms known to occur within populations, whereas females are monomorphic; five color morphs and associated size dimorphism have been described (Lindholm *et al.*, 2004): the 'immaculate' morph has the males grey-brown like the female (but lacking the gestation spot on the belly), the 'parae' morph has a colorful tail stripe and vertical bars, while in the 'melanzonus' type morphs the males are bright blue, red or yellow-green with double horizontal black stripes; the frequency of occurrence of the five color morphs may be the result of the combined forces of sexual and natural (predation) selection (Lindholm *et al.*, 2004) as described for *Poecilia reticulata* by Endler (1983).

MICROPOECILIA PICTA (Regan, 1913)
Local name(s): todobere (guppy)

Diagnostic characteristics: males with a rounded humeral spot (when present), a horizontally elongated black/orange spot at base of the upper margin of the caudal fin, a dorsal fin with red and black markings, and a very short gonopodium (4.3-4.5 times in SL); often orange colors dominate to a more or lesser extent in the body coloration of males; females yellow-brown with a small rounded humeral spot (if present), transparent fins and a grey gestation spot on the belly; body slightly elongated (height 3.8-4.1 times in SL)
Type locality: Demerara River, Guyana
Distribution: Trinidad and Venezuela to French Guiana: Brazil, French Guiana, Guyana, Suriname, and Trinidad and Tobago; in Suriname known from estuary of Suriname and Commewijne rivers
Size: 3 cm TL
Habitat: shallow ditches with brackish water in connection with the estuary and coastal swamps
Position in the water column: surface
Diet:
Reproduction: the developing embryos can represent up to 22.5% of the weight of the female; females give birth to 11-25 young
Remarks:

POECILIA RETICULATA Peters, 1859
Local name(s): todobere (guppy)

Diagnostic characteristics: males with two to four spots of varying size and variously placed along the sides, very variable (see Endler, 1983); females grey-reticulated, without spots
Type locality: Caracas, Guayre River, Venezuela
Distribution: Northern South America and Western Atlantic; widely introduced elsewhere

Micropoecilia parae (live), color morphs of males (A-E), female (F) (© F. Breden & A. Lindholm)

Micropoecilia picta (live), red morph (© A. Lindholm)

Poecilia reticulata (live) (© F. Breden)

Size: 3.5 cm TL
Habitat: small, shallow canals (in)directly in connection with river estuaries, often with brackish water; coastal swamps
Position in the water column: surface
Diet:
Reproduction: *P. reticulata* guppies exhibit sexual dimorphism: females are larger and grey in body color, males have splashes, spots, or stripes that can be any of a wide variety of colors. The gestation period of a guppy is 21–30 days, with an average of 28 days, varying according to water temperature. Males possess a modified tubular anal fin called a gonopodium located directly behind the ventral fin which is flexed forward and used as a delivery mechanism for one or more balls of spermatozoa. The male will approach a female and will flex his gonopodium forward before thrusting it into her and ejecting these balls (females may store these for months afterward, being able to give birth long after isolation from any male guppy). After the female guppy is inseminated, a dark area near the anus, known as the gravid/gestation spot, will enlarge and darken. Just before birth, the eyes of fry may be seen through the translucent skin in this area of the female's body. When birth occurs, individual offspring are dropped in sequence over the course of an hour or so. The female guppy has drops of between 2–100 fry, typically ranging between 5 and 30. From the moment of birth, each fry is fully capable of swimming, eating, and avoiding danger. After giving birth, the female is ready for conception again within only a few hours. Guppies have the ability to store sperm, so the females can give birth many times, after only once breeding with a male. Young fry take roughly three or four months to reach maturity.
Remarks: R.J.L. Guppy discovered this tiny fish in Trinidad in 1866, and the fish was named *Girardinus guppii* in his honor by A. Günther later that year. However, the fish had previously been described in America and is now known has *Poecilia reticulata* (with *G. guppii* considered a junior synonym). The common name "guppy" still remains. Guppies have been introduced to many different countries on all continents, except Antarctica. Sometimes this has occurred accidentally, but most often as a means of mosquito control, the hope being that the guppies would eat the mosquito larvae slowing down the spread of malaria. In many cases, these guppies have had a negative impact on native fish faunas. Guppies are very popular aquarium fish that have been bred in a number of ornate strains characterized by color or pattern and by shape and size of the tail and dorsal fins.

Poecilia vivipara Bloch & Schneider, 1801
Local name(s): molly (molly)

Diagnostic characteristics: a relatively large (up to 5 cm TL in females), high-bodied guppy with a small, round humeral spot, positioned high on the body, close to the insertion of the dorsal fin, but always separated from the predorsal line; adult males with upper and lower margins of caudal black; anal fin of the male modified, but not bearing spines; sides often with broad silvery and dark cross bands; females grey-brown; fins transparent, except for the dorsal and caudal fins which may have small dark spots at the base

Poecilia vivipara (alcohol), 42 mm SL, ROM61654 (© E. Holm, ROM)

Poecilia vivipara (live), male (© INRA-Le Bail)

Type locality: Suriname
Distribution: Venezuela to Argentina; also islands of the Lesser Antilles: Argentina, Brazil, French Guiana, Guyana, Puerto Rico and Martinique (introduced), Suriname, Trinidad and Tobago, Uruguay and Venezuela
Size: 5 cm TL
Habitat: in brackish water close to the ocean or estuary; often in large schools
Position in the water column: surface
Diet: predator of mosquito larvae
Reproduction: females give birth to 6-10 fry that follow the mother for some hours after they are born
Remarks: the *Poecilia* species are collectively known as mollies, with the exception of Endler's livebearer (*P. wingei*) and the famous guppy (*P. reticulata*).

TOMEURUS GRACILIS Eigenmann, 1909
Local name(s): –

Diagnostic characteristics: a beautiful, elongated guppy with a transparent body, a short head and a relatively large mouth; the large eye has a diameter > snout length; presence of a transparent membrane between the anal fin and the caudal fin; dorsal fin positioned posteriorly on the body, anal fin positioned anteriorly on the body (distance between the anal fin and the dorsal fin much larger than the distance between the dorsal fin and the caudal fin); gonopodium very long (1/3 of TL)
Type locality: Mud Creek in Aruka River, Guyana
Distribution: Brazil, Guyana, Suriname and Venezuela; in Suriname known to occur in the Corantijn, Saramacca, Suriname and Commewijne rivers, but probably present in lower reaches of other rivers as well
Size: 3.3 cm TL
Habitat: lives in small schools in shallow creeks and canals with muddy or sandy bottom, both in the estuaries and upstream in lower freshwater reaches of the large rivers (e.g. in Corantijn River, *T. gracilis* was collected at a beach 500 m upstream of the village Apoera (i.e. 155 km from the river mouth)
Position in the water column: surface (often floating rather motionless at the surface)
Diet:
Reproduction: internal fertilization is followed by females laying fertilized eggs singly or retaining fertilized eggs until or near hatching (facultative viviparity); females have one or two eggs that are visible through the transparent body wall; sperm are packaged in naked sperm bundles, free sperm are stored in the ovaries (Parenti *et al.*, 2010)
Remarks: interestingly a molecular analysis by Hrbek *et al.* (2007) shows *Xenodexia ctenolepis* sister to all other poeciliids and *T. gracilis* sister to all Poeciliinae but *Xenodexia*, suggesting that internal fertilization and live-bearing may been present in the ancestor of the Poeciliinae, with livebearing being lost in the species *T. gracilis*.

Tomeurus gracilis (alcohol), scale bar = 5 mm, INHS49017 (© M. Sabaj Pérez)

Tomeurus gracilis (live) (© INRA-Le Bail)

FAMILY ANABLEPIDAE
(FOUR-EYED FISHES AND ONE-SIDED LIVEBEARERS)

The family Anablepidae, with the three genera *Anableps, Jenynsia* and *Oxyzygonectes*, is uniquely diagnosed among cyprinodontiform fishes by the possession of laterality in the male urogenital papilla or gonopodium. Both dextral and sinistral individuals occur in all species. Most species exhibit teeth that are to some degree tricuspid. Two of the three species of otherwise distinctive four-eyed fishes have unicuspid teeth. The onesided livebearers and the four-eyed fishes, *Jenynsia* and *Anableps* respectively, are sister taxa (Ghedotti, 1998) and exhibit both internal fertilization and viviparity. Adult males have the anal fin modified as a fleshy tubular gonopodium with the tip angled laterally. The four-eyed fishes, genus *Anableps,* are distributed in Central America and northern South America. *Anableps* are the most distinctive members of the family and grow to the largest sizes of any cyprinodontiform fishes, reaching up to 32 cm in total length. *Anableps* are called four-eyed fishes because they have eyes that are prominently raised above the top of the head (Fig. 12.7) and each eye is divided lengthwise forming two pupils, one dorsal and one ventral, in each eye. *Anableps* individuals often swim with the center of the eye at the water's surface and are capable of simultaneous aerial and aquatic vision. Female *Anableps* exhibit sexual laterality and have a flap of skin covering either the dextral or sinistral surface of the urogenital opening (Ghedotti, 1998). *Anableps* species are known to consume terrestrial insects and *A. anableps* has been seen to occasionally capture insects in the air (Zahl *et al.*, 1977). *Anableps anableps* has been observed consuming tidally exposed silt covered with diatoms (Zahl *et al.*, 1977) and the gut contents of *A. anableps* and *A. microlepis* confirm that these species do consume silt (Ghedotti, 1998). Turner (1938, 1940) reported on viviparity in *Anableps anableps* and noted that, as in poeciliids, *Anableps* has intrafollicular gestation during which nutrients are transferred through a vascular follicular "placenta". Knight *et al.* (1985) determined that *Anableps* embryos had post fertilization weight increases of 298,000% (for *Anableps anableps*) to 843,000% (for *Anableps dowi*) showing the greatest amount of viviparous maternal nutrient transfer in any teleost. Burns & Flores (1981) documented seasonal breeding and superfoetation (embryos in different stages in the ovary) in *Anableps dowi* and noted that the female genital opening closes during gestation.

1-a Interorbital distance equal to or greater than the eye diameter; caudal rounded; <60 scales in lateral line; sides with conspicuous violet stripes
Anableps anableps

1-b Interorbital distance smaller than the diameter of the eye; caudal obliquely rounded; >60 scales in lateral line; sides with obscure stripes
Anableps microlepis

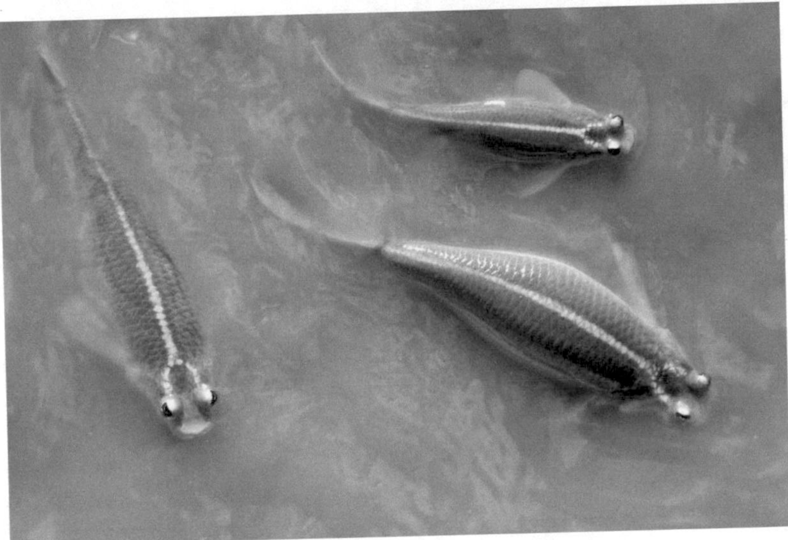

Fig. 12.7. Four-eyed fish (*Anableps* sp.) swimming at the surface in the intertidal habitat of mudflats of the Suriname River Estuary (© W. Kolvoort).

ANABLEPS ANABLEPS (Linnaeus, 1758)
Local name(s): kutai (four-eyes)

Diagnostic characteristics: body elongated and cylindrical in cross section, its depth about 8 times in TL; the eyes are elevated above the top of the head and comprise two pupils separated by a longitudinal dark bar of corneal tissue; interorbital distance equal to or greater than the eye diameter; caudal rounded; 50-55 scales in lateral line; ventral fins large and rounded, dorsal fin very small and positioned posteriorly on the body; sides with 5 conspicuous violet stripes
Type locality: India [in error]
Distribution: Brazil, French Guiana, Guyana, Suriname, Trinidad and Tobago, and Venezuela
Size: 30 cm TL (400 g)
Habitat: estuaries and lower freshwater reaches (under tidal influence) of large rivers, often observed in schools on mudflats at the edge of the water
Position in the water column: surface
Diet: diatoms, other microscopic algae and benthic invertebrates that live in the mud, terrestrial insects and silt (Zahl *et al.*, 1977)
Reproduction: viviparous, females reach sexual maturity at 13 cm, adult males have the anal fin modified as a fleshy tubular gonopodium, reproduction is in the dry season; embryos are retained in modified ovarian follicles during the entire period of gestation; in this prolonged gestation period, the embryos increase in weight 3000-fold and their length at birth is 4.5-5.2 cm (Knight *et al.*, 1985)
Remarks: although easy to spot on the tidally exposed mudflats, *Anableps* are very difficult to catch because, swimming at the surface, they can see simultaneously above and under the water surface and can see a person approaching from a distance of 10 or more meters (they are also most adept at jumping over the float line of a seine net).

ANABLEPS MICROLEPIS Müller & Troschel, 1844
Local name(s): kutai (four-eyes)

Diagnostic characteristics: body form much like *A. anableps*, but differs in the interorbital distance that is much smaller than the diameter of the eye and a more pointed snout; the caudal fin is obliquely rounded; 80-90 scales in lateral line; sides with obscure stripes
Type locality: Guyana
Distribution: Brazil, French Guiana, Guyana, Suriname, Trinidad and Tobago, and Venezuela
Size: 32 cm TL (500 g)
Habitat: strictly a brackish-water / marine species that is not present in the lower freshwater reaches of the rivers; found along sandy beaches in small schools (about 10 individuals) or as one or two, isolated individuals

Anableps anableps (alcohol), INHS49016 (© M. Sabaj Pérez)

Anableps microlepis (live) (© INRA-Le Bail)

Position in the water column: surface
Diet: probably more or less the same as *A. anableps*
Reproduction: see *A. anableps*
Remarks: in French Guiana, *A. microlepis* is less abundant compared to *A. anableps* (Keith *et al.*, 2000).

This page is intentionally left blank.

FAMILY BELONIDAE (NEEDLEFISHES)

Needlefishes are elongate fishes with both upper and lower jaws extended into long beaks filled with sharp teeth and nostrils in a pit anterior to eyes. There are no spines in the fins; dorsal and anal fins posterior in position; pectoral fins short. The lateral line is running down from pectoral fin origin and then along the ventral margin of body. Scales are small, cycloid, and easily detached. These fishes live at the surface (Fig. 4.34) and are protectively colored for this mode of life by being green or blue on the back and silvery white on the lower sides and belly. Usually, there is a dusky or dark blue stripe along the sides, and the fleshy tip of the lower jaw is frequently red or orange. Most species are marine, but some occur in freshwaters and estuaries (e.g. Collette, 1974, 1982). Needlefish are carnivorous, feeding largely on small fishes which they catch sideways in their beaks. Information on the ecology of the three endemic South American genera was presented by Goulding and Carvalho (1984). Needlefishes tend to leap and skitter at the surface and some people have been injured when accidentally struck by them, particularly at night when the fishes are attracted by lights.

1-a Caudal fin rounded, caudal peduncle compressed, with no trace of a vertical keel; 7-8 pectoral-fin rays; presence of elongate nasal barbel; posterior part of body elongate compared to anterior part; dorsal and anal fins very long, with 27-43 and 24-36 rays, respectively; freshwater
Potamorrhaphis guianensis
1-b Caudal fin distinctly divided into two lobes, not rounded, caudal peduncle strongly depressed; 9-10 pectoral-fin rays, nearly twice as long as pelvic fins; nasal papilla spatulate, contained within a deep nasal fossa (more like in marine belonid genera); dorsal fin rays 13-16, anal fin rays 8-11 (usually 9 or 10); lower reaches of rivers, up to the first rapids
Pseudotylosurus microps
1-c Caudal fin emarginate without distinct lobes, caudal peduncle not strongly depressed (round in cross section), no keels on caudal peduncle; large species (up to 70 cm), dorsal-fin rays 12 to 17; body colored light-greenish above, silvery below; occurs in estuaries in brackish or salt water
Strongylura marina

POTAMORRHAPHIS GUIANENSIS (Jardine, 1843)
Local name(s): nanaifisi, naaldvis (freshwater needlefish)

Diagnostic characteristics: elongate fishes with tubular-shaped body, both upper and lower jaws extended into long beaks filled with sharp teeth, and nostrils in a

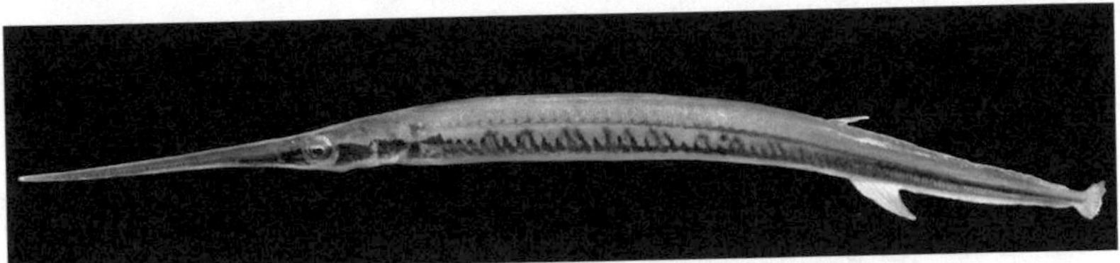

Potamorrhaphis guianensis (live), ANSP179480 (© M. Sabaj Pérez)

Potamorrhaphis guianensis (live), Commewijne River (© R. Covain)

pit anterior to eyes; elongate nasal barbel; no spines in the fins; dorsal and anal fins posterior in position; caudal fin rounded; caudal peduncle compressed, with no trace of a vertical keel; 7-8 pectoral-fin rays; posterior part of body elongate compared to anterior part, dorsal and anal fins very long, with 27-43 and 24-36 rays, respectively; body colored yellowish-brown with a black longitudinal band; freshwater
Type locality: Padauiri River, Guyana
Distribution: South America. Amazon and Orinoco River basins and the Guianas: Brazil, Colombia, Ecuador, French Guiana, Guyana, Peru, Suriname and Venezuela; in Suriname, present in all river systems
Size: 29.1 cm TL
Habitat: in the upstream (upstream of the first rapids) and lower reaches of rivers and medium-sized tributary creeks, in places with slow current, often near the shore under overhanging vegetation; always in fresh water
Position in the water column: surface (Fig. 4.34)
Diet: insectivorous; the arrow-like body-shape is hypothesized to be an adaptation for high velocity strikes at prey from a stationary position (Goulding & Carvalho, 1984)
Reproduction: Keith *et al.* (2000) note that mating *P. guianensis* swim parallel to each other in oblique position (head directed to the bottom); approximately 15-40 eggs are spawned by the female; the eggs attach to submerged (overhanging?) vegetation by 2-3 mm long filaments; they hatch in 9-10 hours in pelagic larvae
Remarks:

Pseudotylosurus microps (Günther, 1866)
Local name(s): nanaifisi

Diagnostic characteristics: caudal fin distinctly divided into two lobes, not rounded; caudal peduncle strongly depressed; nasal papilla spatulate, contained within a deep nasal fossa (more like in marine belonid genera); 9-10 pectoral-fin rays, nearly twice as long as pelvic fins; dorsal fin rays 13-16, anal fin rays 8-11 (usually 9 or 10)
Type locality: Suriname
Distribution: South America: primarily a species of the Orinoco, Guianas, and Lower Amazon (Araguaia, Xingu, Negro, and Branco); one record from the Upper Amazon; in Suriname, known from the Suriname (from Paramaribo up to Brokopondo; Collette, 1974) and Marowijne rivers
Size: 40.7 cm SL
Habitat: lower reaches of rivers, up to the first rapids
Position in the water column: surface
Diet: piscivorous
Reproduction:
Remarks:

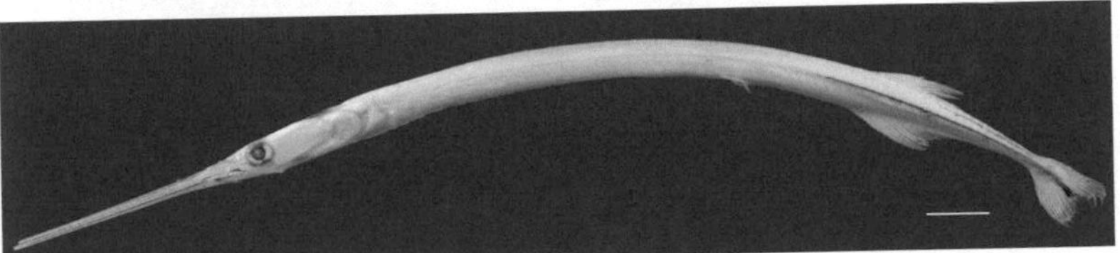

Pseudotylosurus microps (live), scale bar = 1 cm, ANSP179633 (© M. Sabaj Pérez)

Strongylura marina (Walbaum, 1792)
Local name(s): nanaifisi (Atlantic needlefish)

Diagnostic characteristics: caudal fin emarginate without distinct lobes, caudal peduncle not strongly depressed (round in cross section), no keels on caudal peduncle; large species (up to 70 cm); dorsal-fin rays 12 to 17; anal rays 16-20; pectoral fin rays 10-12, usually 11; body colored light-greenish above, silvery below; occurs in estuaries in brackish or salt water
Type locality: no locality stated [New York, U.S.A.]; no types known.
Distribution: Western Atlantic from Massachusetts south to Rio de Janeiro, including Central America; in Suriname, known from the estuary of the Suriname River, but probably present in other estuaries as well
Size: up to 70 cm TL
Habitat: coastal areas and mangrove-lined lagoons, where it is moderately common (especially small individuals) and also enters fresh water
Position in the water column: surface
Diet: small fishes and crustaceans
Reproduction: only right gonad present
Remarks: according to Leopold (2004) *S. marina* is the most common marine/estuarine needlefish in French Guiana.

This page is intentionally left blank.

FAMILY HEMIRAMPHIDAE (HALFBEAKS)

The Hemiramphidae, the halfbeaks, are one of five families of the order Beloniformes. They are the sister-group of the Exocoetidae, the flying fishes, forming the superfamily Exocoetoidea. Most halfbeaks have an elongate lower jaw that distinguishes them from flying fishes which have lost the elongate lower jaw and from needlefishes (Belonidae) which have both jaws elongated. The family is defined by one derived character: the third pair of upper pharyngeal bones ankylosed into a plate. Other diagnostic characters include: pectoral fins short or moderately long; premaxillae pointed anteriorly, forming a triangular upper jaw (except in *Oxyporhamphus*); lower jaw elongate in juveniles of all genera and adults of most genera; parapophyses forked; swimbladder not extending into haemal canal; nostrils in a pit anterior to the eyes; no spines in fins; dorsal and anal fins posterior in position; pelvic fins in abdominal position, with 6 soft rays; lateral line running down from pectoral fin origin and then backward along ventral margin of body (Collette, 2004). Scales are moderately large, cycloid, and easily detached. Halfbeaks live at the surface and are protectively colored for this mode of life being green or blue on the back and silvery white on the sides and ventrally. The tip of the lower jaw is bright red or orange in life in most species. Most species are marine, but some inhabit freshwaters; omnivorous, feeding on floating sea grasses, crustaceans and small fishes. They are prone to leap and skitter at the surface. The flesh is excellent and larger species of halfbeaks are utilized as food in many parts of the world. They are caught with seines or dipnetted under lights at night.

Hyporhamphus roberti (Valenciennes, 1847)
Local name(s): –

Diagnostic characteristics: a slender fish tapering slightly toward head and tail with its lower jaw very long and upper jaw short; dorsal fin (14-16 rays) and anal fin (15 to 17 rays) positioned far back and opposite each other, about equal in length and alike in outline; ventrals stand about midway between a point below the eye and the base of the caudal; the teeth are small and the scales are largest on the upper surface of the head; the beak is much shorter in young fish than it is in adults; color translucent bottle green above with silvery tinge, each side with a narrow but well-defined silvery band running from the pectoral fin to the caudal fin, the sides darkest above and paler below this band; the tip of the lower jaw is crimson in life, with a short filament, and three narrow dark streaks run along the middle of the back; the lining of the belly is black.
Type locality: Cayenne, French Guiana

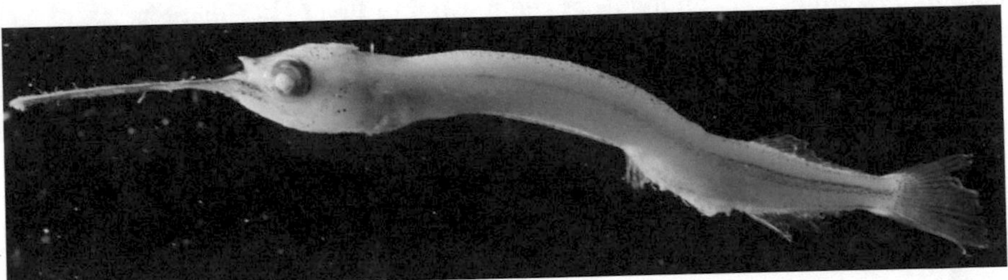

Hyporhamphus roberti (alcohol), early life stage, 39mm TL, Overbridge mid-river sandbank, Lower Suriname River (© A. Gangadin)

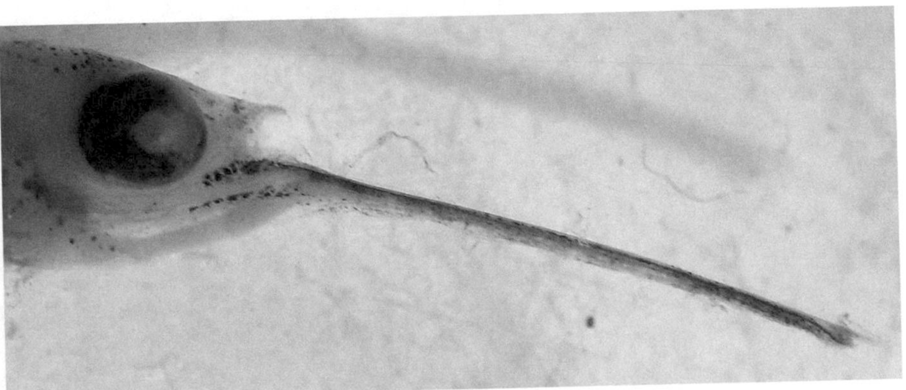

Hyporhamphus roberti (alcohol), detail of the head (© A. Gangadin)

Distribution: Western Atlantic. Two subspecies are recognized (Collette, 2004): *roberti* from Lake Maracaibo, Venezuela, east and south to Iguape, Brazil and *hildebrandi* from the Gulf of Uraba, Colombia, north and west to Belize; in Suriname, known from the Corantijn and Suriname estuaries
Size: 30 cm TL
Habitat: marine, also occurring in estuaries (e.g. Vari, 1982)
Position in the water column: surface
Diet:
Reproduction:
Remarks: early stages of *Hyporhamphus* were collected in *Cabomba/Mayaca* vegetation at a mid-river sandbank in the Suriname River (km 105; Fig. 4.13c); i.e. fresh water.

This page is intentionally left blank.

FAMILY SYNGNATHIDAE (PIPEFISHES AND SEAHORSES)

The family Syngnathidae consists of about 150 species, most of which are near-shore marine inhabitants. Some species are euryhaline and others seem to be restricted to freshwaters. The family includes the elongate pipefishes and the curiously shaped seahorses. All species are encased in bony plates that form rings around the body. Male syngnathids brood eggs in specialized pouches on the ventral surface of either the abdomen or tail (i.e. the males are pregnant; sex role reversal).

PSEUDOPHALLUS AFF. *BRASILIENSIS* Dawson, 1974

Local name(s): –

Diagnostic characteristics: a very small pipefish lacking an anal fin; the superior trunk and tail bony ridges are discontinuous near the rear of the dorsal fin; the inferior trunk and tail ridges are continuous, lateral trunk ridge continuous with lateral tail ridge and confluent with superior tail ridge; mature males with the brood pouch below tail rings 11-22 and a mottled brown color, females with a (faintly) striped pigmentation pattern and an enlarged anal papilla
Type locality: Rio Tocantins, Igarape Ino, faro de Panaquera, Para, Brazil
Distribution: Southwestern Atlantic: Brazil; in Suriname, known to occur in the Corantijn and Suriname rivers (Mol, 2012); taxonomy and distribution of the two related species *P. brasiliensis* and *P. mindii* are unclear (some authors include *P. brasiliensis* in *P. mindii*, but *P. mindii* seems a much larger species)
Size: 7.5 cm TL
Habitat: in shallow near-shore habitats in the main river channel (Corantijn) or close to the mouth of shallow tributaries (Suriname River, 1 specimen); in Corantijn River, most specimens were collected at a sheltered river beach with slow current, clear water, mixed sand/silt bottom and dwarf Amazon sword plant (*Helanthium tenellum*) vegetation (see Mol, 2012; Fig. 4.12)
Position in the water column: benthopelagic, often associated with submerged vegetation
Diet:
Reproduction: mature males (53-72 mm TL) with fertilized eggs in brood pouches (pregnant) were collected in freshwater in Corantijn River (km 155; Fig. 4.12) and these gave birth to about 20 offspring (transparent, 7-9 mm TL, 0.6-0.9 mg wet mass) (Mol, 2012); in Corantijn River, the sex ratio and difference in coloration suggested that female *P.* aff. *brasiliensis* live in a sex-role reversed harem structure with territorial females competing for males that brood their eggs
Remarks: *P.* aff. *brasiliensis* is apparently a sparsely distributed species with low mobility, small home-range, low fecundity and lengthy parental care; syngnathids

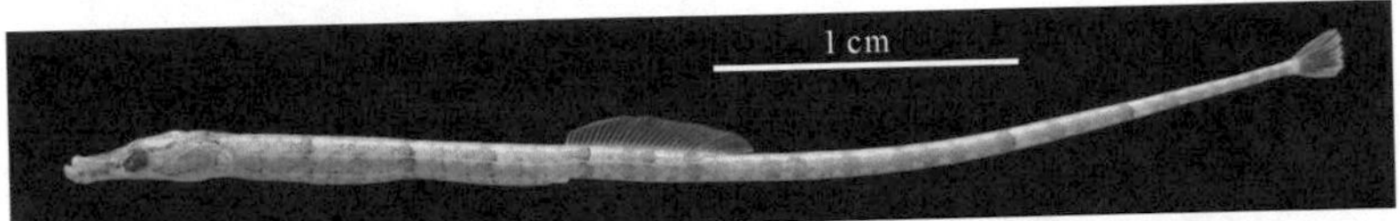

Pseudophallus aff. *brasiliensis* (alcohol), female, Corantijn River, FMNH119605 (© P. Willink)

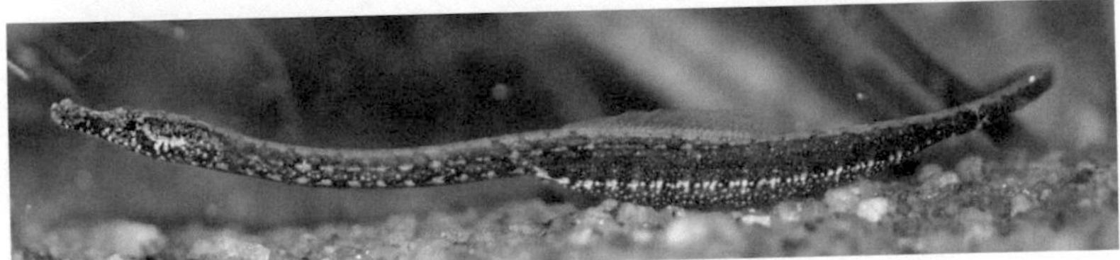

Pseudophallus aff. *brasiliensis* (live), male with ripe brood pouch, Corantijn River

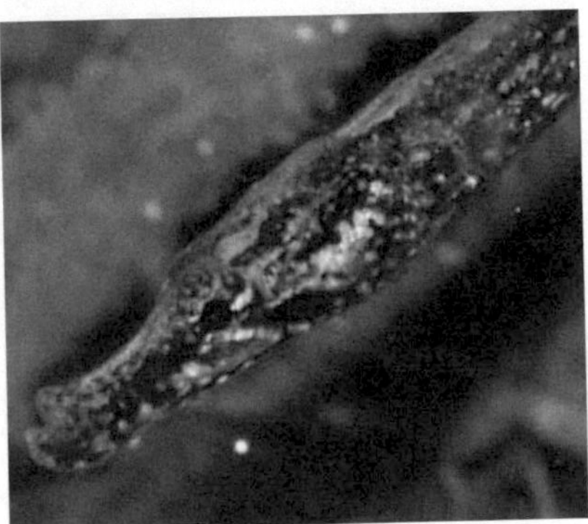

Pseudophallus aff. *brasiliensis* (live), detail of the head, Corantijn River (© P. Ouboter)

are vulnerable to human-induced disturbance and with freshwater habitats among the most threatened in the world, the Corantijn pipefish and its habitat clearly need legal protection.

This page is intentionally left blank.

FAMILY SYNBRANCHIDAE (SWAMP EELS)

Swamp eels are eel-like fishes distributed in fresh water, occasionally in brackish water, throughout tropical and subtropical regions, including southern and eastern mainland Asia, the Indo-Australian Archipelago, West Africa (Liberia), Mexico, Central and South America. The pectoral and pelvic fins are absent, the dorsal and anal fins are rudimentary, and the caudal fin is short, rudimentary or absent. The eyes are minute. The gill membranes fused, leaving a single ventral gill opening appearing as a short transverse slit or pore. There is no swimbladder and ribs are also absent. Vertebral numbers range between 98 and 188. Most species are air breathers and found in swamps or similar conditions with low oxygen levels (Lüling, 1980). Most species are burrowers, some cave dwellers. The largest species is *Synbranchus marmoratus*, reaching 150 cm TL. The family includes 22 species, most of them in Asia. Only four species are recognized from South and Central America: the three species presently known in the genus *Synbranchus* (Favorito *et al.*, 2005) and the blind swamp eel (*Ophisternon infernale*) which is endemic to Mexico where it lives in cave systems. Because of their secluded life style and paucity of strongly expressed external diagnostic characters, species identification and detection is difficult and requires examination of internal characters (e.g. vertebrae number). The number of species will increase with forthcoming revisions (Torres *et al.*, 2005).

SYNBRANCHUS MARMORATUS Bloch, 1795
Local name(s): (zwamp)aal, snekfisi (marbled swamp eel)

Diagnostic characteristics: body eel-like, without scales; pectoral, pelvic and caudal fins absent; dorsal and anal fins reduced to a rayless ridge; a single ventral slit-like opening to the gill chamber; body color grey-brown dorsally and yellow-brown ventrally, spotted
Type locality: Suriname
Distribution: widespread in Central and South America; Mexico to northern Argentina
Size: up to 100 cm TL
Habitat: common in coastal freshwater swamps and rice fields, but synbranchid eels also occur in the Interior; they are never very abundant except in drainage canals of Paramaribo
Position in the water column: bottom
Diet: fish and invertebrates
Reproduction: a protogynous diandric fish; two types of males are present, primary males are born as males, while secondary males are born as females and later in life change into males (Lo Nostro & Guerrero, 1996); this reproductive

Synbranchus marmoratus (live), dorsal, Lawa River (Marowijne River), ANSP187334 (© M. Sabaj Pérez)

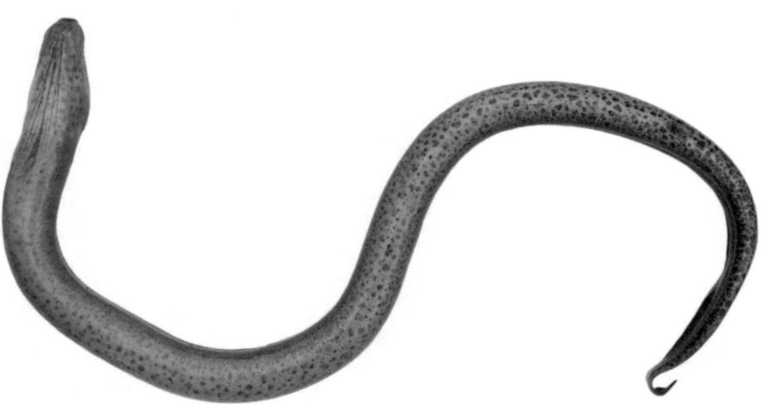

Synbranchus marmoratus (live), ventral, Lawa River (Marowijne River), ANSP187334 (© M. Sabaj Pérez)

strategy including sex reversal enhances reproductive success in a species that often occurs in low population numbers

Remarks: *Synbranchus marmoratus* is a facultative air-breather and active at night, usually at the edge of the water. The swamp eel can travel over land for long distances. They also burrow, especially during the dry season (Lüling, 1980); during the time in the burrow the metabolism slows, but the fish still may flee if disturbed. After the first rains, marbled swamp eels return to larger bodies of water. In Suriname, the swamp eels are often observed when canals in the Coastal Plain are 'cleaned' from vegetation with a Poclain excavator.

This page is intentionally left blank.

FAMILY CENTROPOMIDAE (SNOOK)

The Centropomidae are a single genus family of freshwater and marine fishes in the order Perciformes, including the common snook or róbalo, *Centropomus undecimalis*. Dating from the upper Cretaceous, the centropomids are distinguished by their elongate, compressed body with two-part dorsal fins, a large mouth with unequal jaws (lower jaw protruding beyond the upper, a preoperculum with a serrated posterior and ventral border, a lateral line that extends onto the caudal fin, and, frequently, a triangular, concave shape to the head. First dorsal fin with 8 spines, second dorsal fin with 1 spine and 8-11 soft rays, pelvics with 1 spine and 5 soft rays, anal with 3 strong spines (2nd largest) and 5-8 soft rays; scales ctenoid, moderate to large. All three species that occur in Suriname (fat, swordspine, and common snook) can be easily distinguished from other fishes by their lateral black line. Snook range from 35 to 130 centimeters in length and are found in shallow coastal waters, estuaries and brackish lagoons in the tropics and subtropics. They are good quality food fish, and are a sought after game fish but tricky to catch.

1-a Less than 60 scales in row just above the lateral line (calculated from operculum to end of caudal peduncle; the lateral line extends onto the caudal fin); black pigmentation present between the 2nd and 3rd anal fin spines (2nd anal fin spine very large)

Centropomus ensiferus

1-b More than 66 scales in row above the lateral line; absence of black pigmentation between the 2nd and 3rd anal fin spines

2

2-a More than 78 scales in row above the lateral line; pelvic fins reaching to or past the anus

Centropomus parallelus

2-b 67-77 scales in row above the lateral line; pelvic fins do not reach to the anus

Centropomus undecimalis

CENTROPOMUS ENSIFERUS Poey, 1860
Local name(s): kartasnuku (swordspine snook)

Diagnostic characteristics: the smallest of snook species in the Western Atlantic; moderately deep body (65-72 percent of head length); snout profile nearly straight or slightly concave; mouth large with lower jaw projecting beyond upper; less than 60 (large) scales in the lateral line; second dorsal fin with 1 spine and 10 soft rays; anal fin with 3 spines and 6 soft rays, and black pigmentation present

Centropomus ensiferus (live), Suriname River

Centropomus ensiferus (live) (© INRA-Le Bail)

between the 2nd and 3rd anal fin spines; tips of pelvic fins reaching to or past the anus; color yellow-brown to brown-green above, silvery on sides and below; fins dusky
Type locality: Cuba
Distribution: Western Atlantic: Florida to Central America and to Brazil.
Size: 30 cm SL (0.4 kg)
Habitat: coastal waters, estuaries, brackish lagoons; freshwater, brackish, marine, but usually prefers low salinity or fresh water
Position in the water column: mid-water, over muddy bottoms
Diet: small fish (Engraulidae) and crustaceans (mainly shrimps)
Reproduction: nothing is known where this species spawns, no extensive migrations are known; it may spawn 80,000-2,500,000 small eggs (0.7-0.8 mm diameter), depending on the size of the female; early stages live in shallow water and mangroves
Remarks: the giant second anal spine gives the swordspine snook its common name.

Centropomus parallelus Poey, 1860
Local name(s): snuku (fat snook)

Diagnostic characteristics: a medium-sized snook species, comparatively deep (body depth 67-81 percent of head length), snout profile straight to slightly concave, mouth large, with lower jaw protruding over upper; more than 78 scales (79-89) in row above the lateral line; tips of pelvic fins reaching to or past the anus; color yellow-brown to brown-green above, silvery on sides and below; fins dusky
Type locality: Havana, Cuba
Distribution: Western Atlantic: Florida to Central America and to Brazil
Size: 58 cm SL (3 kg)
Habitat: inhabits coastal waters, estuaries and brackish lagoons, penetrating into freshwater; freshwater, brackish, marine, but usually prefers low salinity or fresh water
Position in the water column:
Diet: small fish (Engraulidae) and crustaceans (shrimps, crabs)
Reproduction: may spawn in river mouths, no extensive migrations known; a female of 1 kg (35 cm) had 900,000 small eggs (0.7-0.9 mm diameter); early stages live in shallow water (mangroves)
Remarks: an inshore species of mangrove habitat that is found commonly in fresh waters; occurs more in Interior waters (as opposed to estuarine waters) than other snook species.

Centropomus parallelus (live) (© R. Covain)

Centropomus undecimalis (Bloch, 1792)
Local name(s): snuku (common snook)

Diagnostic characteristics: the largest and most slender snook species of the Western Atlantic; body depth 59-64 percent of head length; mouth large with protruding lower jaw; 67-77 scales in row above the lateral line; tip of pelvic fins never reaching to anus; color yellow-brown to brown-green above, silvery on sides and below; fins dusky
Type locality: Jamaica
Distribution: Western Atlantic: North Carolina, U.S.A., to Central America and to Brazil
Size: up to 130 cm SL (23 kg)
Habitat: inhabits shallow (<20 m depth) coastal waters, estuaries and brackish lagoons, penetrating into fresh water; freshwater, brackish, marine
Position in the water column: close to the bottom
Diet: fish and crustaceans
Reproduction: *Centropomus undecimalis* is a protandric hermaphrodite fish (Perra-Garcia *et al.*, 2010), with euryhaline and diadromous habits (Taylor *et al.* 2000)
Remarks: a close relationship with rivers and coastal lagoons has been observed for the common snook throughout its distribution range (Peters *et al.* 1998). These systems are used by the species for its periodic migrations when it feeds, grows and reproduces (Stevens *et al.* 2007).

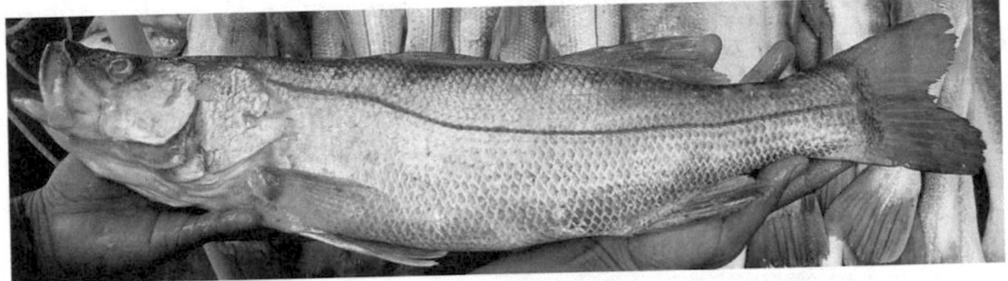

Centropomus undecimalis (live), Suriname River

Centropomus undecimalis (live) (© INRA-Le Bail)

FAMILY SERRANIDAE (SEA BASSES)

Small to enormous fishes (the largest grouper, the jewfish or granmorgu *Epinephelus itajara*, attains a length of 2.5 m and a weight of 400 kg; maximum size for the smallest serranid is about 5 cm total length) with variable body and a deep caudal peduncle. The mouth is moderate to large, terminal, or the lower jaw projecting; teeth on jaws usually small, slender and conical; vomer and palatine bones (on roof of mouth) usually with villiform teeth. The rear edge of opercle with 3 flat spines, the main spine distinct and exposed with one above it and one below it (the latter two often inconspicuous and covered by skin and scales). Preopercle vertical limb generally serrate, the lower (horizontal) limb serrate or undulate, sometimes with strong antrorse (forward-directed) spines. Scales small to moderately large, usually ctenoid, but sometimes nearly smooth; head at least partly scaled, snout and preorbital region usually naked, but cheeks scaly. Dorsal fin usually single with 2 to 11 spines and 10 to 27 soft rays; anal fin with 3 spines and 6 to 17 soft rays; caudal fin forked, lunate, emarginate, truncate, or rounded, with 13 to 16 branched rays; pectoral fins rounded to somewhat pointed, usually longer than pelvic fins; pelvic fin with 1 spine and 5 soft rays, the origin slightly before or behind pectoral-fin base. Lateral line present, but not extended onto caudal fin. Many species are capable of rapid color changes and several species have distinctively colored deep- and shallow-water forms. Color patterns are helpful for identification of species, but one needs to be aware of intraspecific variations.

EPINEPHELUS ITAJARA (Lichtenstein, 1822)
Local name(s): granmorgu (jewfish or goliath grouper)

Diagnostic characteristics: a very large, heavy-bodied fish, nearly round in cross section, with a large head and a large, oblique mouth with completely exposed broad maxilla; 3-5 rows of teeth in the lower jaw; the presence of a number of short weakly developed canine teeth is useful in distinguishing this species from other North Atlantic groupers; operculum with three flat spines, central one largest; dorsal fin with 11 spines and 15-15 soft rays; anal fin with 3 spines and 8 soft rays; pelvic fins much smaller than pectoral fins; ctenoid scales; color brownish yellow, gray, or olive with small dark spots on head, body, and fins; large adults are somber-colored; juveniles tawny with 3 or 4 irregular vertical bands
Type locality: Brazil
Distribution: Atlantic
Size: up to 240 cm (260 kg)
Habitat: inshore marine species found in very shallow marine and brackish water in estuaries and mangrove areas
Position in the water column: in shallow water near the bottom

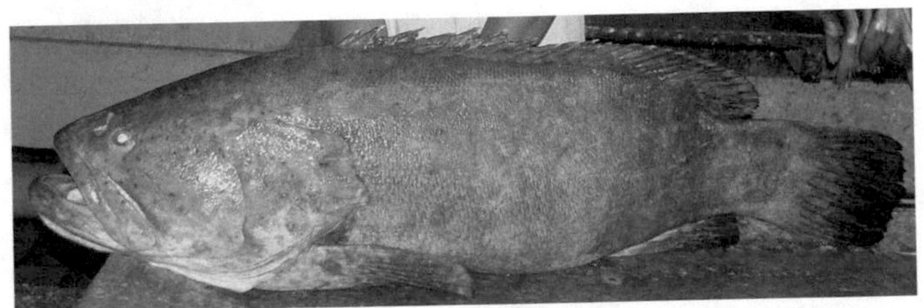

Epinephelus itajara (live) adult, Suriname River Estuary

Epinephelus itajara (live), juvenile (© INRA-Le Bail)

Diet: carnivorous feeding mainly on crustaceans and fish
Reproduction: goliath groupers are believed to be protogynous hermaphrodites, with individuals first maturing as females and only some large adults becoming males. Most groupers follow this pattern, but it has not yet been verified for the goliath. In fact, Bullock *et al.* (1992) found males could be sexually mature at smaller sizes (~1150 mm) and younger ages (4–6 years) than females (~1225 mm and ~6–8 years). They reproduce in river estuaries; the females release eggs while the males release sperm into the open offshore waters; after fertilization, the pelagic eggs are dispersed by water currents; the pelagic larvae are kite-shaped, with the second dorsal-fin spine and pelvic fin spines greatly elongated; the larvae transform into benthic juveniles at lengths of 2.5 cm (25 or 26 days after hatching)
Remarks: a tasty food fish that can be found at the central market of Paramaribo; many grouper species (Epinephelini), including the goliath grouper, are IUCN-redlisted.

This page is intentionally left blank.

FAMILY CARANGIDAE (JACKS AND SCADS)

Small to large (up to 150 cm) fishes with a generally compressed body, but body shape variable ranging from fusiform to strongly compressed (*Selene*). Scales small, sometimes difficult to see, and cycloid (smooth to touch), but ctenoid (rough) in 2 species and needle-like in *Oligoplites*, sometimes extending onto fins; scutes (hard, bony scales in lateral line) present and prominent, or reduced in some species and absent in some genera. Lateral line arched or elevated anteriorly and straight posteriorly, extending onto caudal fin. Up to nine detached finlets sometimes present behind dorsal and anal fins (counts for these rays are included in dorsal and anal fin counts).Two dorsal fins in large juveniles and adults, the first with 4 to 8 spines (short embedded in adults of some species), and the second dorsal fin with 1 spine and 17 to 44 soft rays; anal fin usually with 3 spines and 15-39 soft rays, with the 2 anterior spines detached from the rest of the anal fin; caudal fin widely forked; caudal peduncle slender. Color: darker above (green or blue to blackish) and paler below (silvery to white or yellow-golden), some species almost entirely silvery when alive, others with dark or colored bars or stripes on head, body, or fins, and some can change patterns; young of many species barred or spotted. The family contains many important food fishes. The juveniles of some species extend into estuaries.

1-a Absence of scutes on the straight posterior part of the lateral line on the caudal peduncle

2

1-b Presence of scutes on the straight posterior part of the lateral line on the caudal peduncle

3

2-a Snout pointed, head elongated; last soft rays of dorsal and anal fins present as semi-detached finlets

Oligoplites saliens

2-b Snout rounded; absence of finlets

Trachinotus cayennensis

3-a Body smooth, seemingly without scales and very deep (body depth 1.3-2.3 in Total Length)

Selene vomer

3-b Small scales present on the body; body not very deep (2.4-3.6 in TL)

4

4-a Presence of oval, black spot on the pectoral fins; anal fin yellow

Caranx hippos

4-b Absence of oval black spot on the pectoral fins; anal fin white
Caranx latus

Caranx hippos (Linnaeus, 1766)
Local name(s): zeezalm (jack)

Diagnostic characteristics: body elongate (its depth 2.8-3.2 in fork length) and moderately compressed; snout bluntly pointed; eye moderately large (diameter 3.8-4.2 in head length) with a strong adipose eyelid; mouth large, end of upper jaw extending to below posterior margin of eye or beyond; black spot on posterior part of operculum; dorsal fin with 8 spines followed by 1 spine and 19-21 soft rays; anal fin yellow, with 2 spines followed by 1 spine and 16-17 soft rays; dorsal and anal fin lobes elongate; pectoral fins falcate, longer than head, with oval, black spot; chest scaleless except for small medium patch in front of pelvic fins; lateral line with strong moderately long anterior arch; 23-42 scutes on the posterior part of the lateral line (caudal peduncle); color blue-greenish on the back, silvery-yellow on the belly; juveniles with about 5 dark vertical bars
Type locality: Carolina [South Carolina, U.S.A.].
Distribution: Mediterranean Sea, eastern and Western Atlantic
Size: 115 cm TL (25 kg), common to 60 cm fork length
Habitat: freshwater, brackish, marine; common in brackish water and may ascend rivers
Position in the water column: pelagic, mid-water on shallow flats; occurs in large, fast-moving schools, although larger fish may be solitary in deep offshore water
Diet: primarily fish, but also shrimps and other crustaceans
Reproduction: pelagic eggs, juveniles often occur in shallow water
Remarks: often grunts or croaks when caught.

Caranx latus Agassiz, 1831
Local name(s): –

Diagnostic characteristics: looks much like *C. hippos* in general body shape and coloration, but lacking an oval black spot on the pectoral fins and with the anal fin white; it also has small scales on the chest (*vs* scaleless in *C. hippos*)
Type locality: Atlantic off Brazil
Distribution: Western and eastern Atlantic
Size: 80 cm TL (16 kg), common to 50 cm fork length
Habitat: freshwater, brackish, marine; enters brackish waters and ascends rivers
Position in the water column: usually occurring in small, pelagic schools
Diet: primarily fish, also shrimps and other invertebrates
Reproduction: pelagic eggs
Remarks:

Oligoplites saliens (Bloch, 1793)
Local name(s): – (castin leatherjack)

Diagnostic characteristics: body elongate, slightly deep (depth 3.2-3.5 times in fork length) and greatly compressed; snout pointed, head elongated; lower jaw

Caranx hippos (live) (© INRA-Le Bail)

Oligoplites saliens (live) (© INRA-Le Bail)

expanded with a convex profile; eye small (diameter 4.3-4.4 times in head length); dorsal fin with 4 spines, followed by 1 spine and 20-21 soft rays; anal fin with 2 spines separated from rest of fin, followed by 1 spine and 20-21 soft rays; posterior 11-15 soft rays of dorsal and anal fins present as semi-detached finlets; pectoral fin shorter than head length; scales needle-like and embedded, but visible; lateral line slightly arched over pectoral fin; absence of scutes on the straight posterior part of the lateral line on the caudal peduncle; color fresh, dull bluish-grey above with a dark dorsal midline, sides and belly silvery white; dorsal fin lobe dusky; anal fin mostly clear, caudal fin dark to dusky on scaled portion of base
Type locality: Antilles [Western Atlantic]; no types known
Distribution: Western Atlantic
Size: 43 cm fork length; common to 30 cm FL
Habitat: brackish, marine; probably mainly inshore and in estuaries
Position in the water column: pelagic
Diet: small fishes and crustaceans
Reproduction: pelagic eggs
Remarks: solitary; the specific name '*saliens*' points to spectacular jumps of 2 m out of the water; juveniles may rest immobile at the surface, with head directed downward (Leopold, 2004)

SELENE VOMER (Linnaeus, 1758)
Local name(s): – (Atlantic lookdown)

Diagnostic characteristics: body short, very deep (body depth 1.3-2.3 in Total Length), smooth, seemingly without scales, and extremely compressed, with dorsal and ventral profiles similar and parallel in abdominal area; head very deep, with dorsal profile sharply sloping to a basal terminal mouth with lower jaw protruding; eye small (diameter contained 5.5 to 6.0 times in head length); upper jaw broad at end and ending below and in front of anterior margin of eye; first 4 dorsal-fin spines elongated in small fish (second spine about 2.5 times longer than fork length at about 3.5 cm fork length), these spines becoming shorter and resorbed as the fish grows until the spine length goes about 10 to 25 times into the fork length; second dorsal-fin lobe also elongated at about 2 cm fork length, its length contained about 1.3 times in fork length at 23 cm fork length and 1.5 to 2.0 times at larger sizes; pelvic fins elongated in larvae (longer than pectoral fins to about 5 cm fork length) becoming shorter with growth to about 10 times into pectoral-fin length; lateral-line scutes weak and scarcely differentiated, numbering from 7 to 12 over caudal peduncle; no distinctive color marks, silvery or golden; first prolonged dorsal- and anal-fin ray often blackish; young with pelvic-fin spine and prolonged second and third dorsal-fin spines black, and with dusky, somewhat oblique cross bands; a band over eye continued and tapering below eye; 4 or 5 interrupted bands on body usually very faint (see Leopold, 2004; p.92)
Type locality: America
Distribution: Western Atlantic

Oligoplites saliens (live), Commewijne River

Selene vomer (live), Commewijne River

Size: 40 cm fork length (2 kg); common to 24 cm fork length
Habitat: brackish, marine; this species was collected in a brackish-water aquaculture farm in the district Commewijne (Comfish, Van Alen)
Position in the water column: occurs in small schools often near the bottom in shallow coastal waters over hard or sandy bottoms
Diet: small crustaceans, fish, and worms
Reproduction: pelagic eggs
Remarks: flesh rated from good to excellent.

Trachinotus cayennensis (Cuvier, 1832)
Local name(s): – (Cayenne pompano)

Diagnostic characteristics: body slightly elongate and compressed, with upper and lower profiles similar and head profile sloping to a blunt snout; eye small (diameter contained 3.2 to 4.4 times in head length); upper jaw very narrow at end and extending to below anterior half of eye, lower jaw included; dorsal fin with 5 spines, short and separated from each other in large fish (first spine very small and rudimentary in some fish), followed by 1 spine and 26 to 29 soft rays (usually 27 or 28); anal fin with 2 short spines separated from rest of fin, followed by 1 spine and 23 to 27 soft rays (usually 26 or 27); bases of anal and second dorsal fins about equal in length; pectoral fins short, contained 1.1 to 1.2 times in head length; lateral line slightly arched to below middle of second dorsal fin and then straight; scales small, cycloid (smooth) and partially embedded; absence of scutes on the straight posterior part of the lateral line on the caudal peduncle; color: back dark blue or grey, sides and belly silvery; snout and maxilla dark; large adults with dorsal fin yellowish grey, tip of fin lobe and first fin ray black; anal fin also yellowish grey with the fin lobe darker; pectoral fins dark, inner side and axil almost black; caudal fin yellowish with dark or grey margin
Type locality: Cayenne, French Guiana
Distribution: Western Atlantic
Size: 46 cm fork length; common to 35 cm fork length
Habitat: brackish, marine; adults found in water depths of 16 to 63 m; young found inshore
Position in the water column: mid-water
Diet: crustaceans, small fish and snails (crashed with their pharyngeal jaws)
Reproduction: pelagic eggs
Remarks:

Trachinotus cayennensis (live) (© INRA-Le Bail)

FAMILY LUTJANIDAE (SNAPPERS)

A small family (125 species) of small to medium-sized (to about 160 cm) marine fishes, oblong in shape and moderately compressed. There are two nostrils on each side of the snout and no enlarged pores on the chin. The mouth is terminal and fairly large, with maxilla slipping for most or all of its length under lachrymal when the mouth is closed. Jaws with distinct canines or canine-like teeth; no inciform or molariform teeth. Vomer and palatines with teeth. Cheek and operculum scaly; maxilla with or without scales. Preopercle typically serrate, often finely. Dorsal fin continuous (single) or with shallow notch, with 9-12 spines and 9-18 soft rays. Snappers occur worldwide in warm seas and juveniles of some species enter estuaries and the lower reaches of rivers. They are mostly bottom-associated fishes, occurring from shallow inshore areas to depths of about 550 m, mainly over reefs or rocky outcrops (e.g. the Surinamese red snapper fishing grounds are fossil coral reefs at the edge of the continental shelf). Snappers are active, mostly nocturnal predators feeding on fishes, crustaceans (especially crabs, shrimps, stomatopods, lobsters) and mollusks (gastropods, cephalopods); plankton is particularly important in the diets of those species with reduced dentition and numerous well-developed gill rakers. They are gonochoristic (sexes separate), reaching sexual maturity at about 40 to 50% of maximum length, with big females producing large numbers of eggs. Lutjanids are batch spawners, with individual females usually spawning several times in a reproductive season. Spawning is apparently at night, on some occasions coinciding with spring tides. In those species in which it has been observed, courtship terminates in a spiral swim upward, with gametes released just below the surface. Eggs and larvae are pelagic; the larvae avoid surface waters during the day, but display a more even vertical distribution at night. Snappers are long-lived, slow-growing fishes with relatively low rates of natural mortality and with considerable vulnerability to overfishing; they are important to artisanal fisheries, but seldom the prime interest of major commercial fishing activities. Many species are fine food fishes, frequently found in markets.

1-a Dorsal fin with 10 spines and 12-13 soft rays; presence of a black spot on the lateral line below the anterior part of soft dorsal fin

Lutjanus synagris

1-b Dorsal fin with 10 spines and 14-15 soft rays; absence of black spot on the lateral line below the anterior part of soft dorsal fin; canines at anterior end of upper jaw distinctly larger than anterior teeth in lower jaw

Lutjanus jocu

This page is intentionally left blank.

Lutjanus jocu (Bloch & Schneider, 1801)
Local name(s): – (Dog snapper)

Diagnostic characteristics: body comparatively deep, greatest depth 2.3 to 2.8, usually 2.4 to 2.7, times in standard length; canines at anterior end of upper jaw distinctly larger than anterior teeth in lower jaw ('dog snapper'); membranes of soft dorsal and anal fins with scales; dorsal fin with 10 spines and 14-15 soft rays; tubed lateral-line scales 46 to 49; scales above lateral line 8 to 11; color: back and upper sides olive brown with bronze tinge, sometimes with narrow pale crossbars; lower sides and belly reddish with a coppery cast; no dark lateral spot below anterior part of soft dorsal fin; blue line or series of spots below eye and across opercle; pale triangle below eye (not always apparent); fins mostly yellow-orange, except color of spinous dorsal fin and proximal parts of soft dorsal and caudal fins similar to that of back
Type locality: locality not stated [Havana, Cuba]; no types known
Distribution: Western Atlantic
Size: 90 cm TL, commonly to 60 cm
Habitat: freshwater, brackish, marine; young occur in coastal waters, especially in estuaries, and sometimes rivers, and on occasion enter fresh water
Position in the water column:
Diet: fishes, crustaceans, gastropods, and cephalopods
Reproduction: in the Caribbean, spawning takes place in March-April; a spawning aggregation has been observed off Belize in January
Remarks: a solitary species that appears to have a home range; the young have been collected in a brackish-water aquaculture facility in Commewijne (Van Alen's Comfish)

Lutjanus synagris (Linnaeus, 1758)
Local name(s): – (Lane snapper)

Diagnostic characteristics: dark spot present below anterior part of soft dorsal fin, less than one fourth to none of this spot extending below lateral line in specimens larger than about 6cmstandard length, spot occasionally absent; dorsal fin with 10 spines and 12-13 soft rays; tubed scales in lateral line 47 to 50; color: silvery pink to red with 6 to 8 yellow horizontal stripes and a number of diffuse dark vertical bars; upper part of body with diagonal yellow lines; iris of eye reddish; fins yellowish to reddish, caudal fin with dusky posterior margin.
Type locality: "America septentrionali" [Bahamas]; no types known
Distribution: Western Atlantic
Size: 71 cm TL, commonly to 30 cm
Habitat: mainly marine, found over a variety of bottom types in shallow water (< 30 m)
Position in the water column:
Diet: fishes, crustaceans, worms, gastropods, and cephalopods

Lutjanus jocu (live), juvenile, 30 cm SL, Commewijne River (© N. Martowitono)

Lutjanus synagris (live)

Reproduction: often forms large assemblages, notably during the spawning season. Found in spawning condition from March through August off south Florida; at Trinidad it spawns throughout the year; eggs of 0.6-0.8 mm diameter hatch in about 23 hours at 26 °C
Remarks: one of the most common snappers in the Caribbean fisheries; a related species *L. mahogony* has about 1/4 to 1/2 of the dark lateral spot extending below the lateral line (*vs* spot mostly above the lateral line in *L. synagris*).

This page is intentionally left blank.

FAMILY LOBOTIDAE (TRIPLETAILS)

Only a single species in the Western Atlantic, *Lobotes surinamensis*, that is easy to distinguish from all other species by its typical shape of the body and vertical fins (see below). In some regards it resembles the groupers (Serranidae) but these usually have teeth on the roof of mouth and always an easily visible subocular shelf.

LOBOTES SURINAMENSIS (Bloch, 1790)
Local name(s): pa(pa)uma (Atlantic tripletail)

Diagnostic characteristics: a compressed, deep-bodied perch-like fish with the dorsal and anal fins rounded and symmetrical so that with the tail they appear to be a single three-lobed fin ('tripletail'); head dish-shaped, interorbital space narrow, upper profile concave; eye relatively small; no subocular shelf visible externally; mouth large, slightly oblique, upper jaw protractile; maxilla not slipping under preorbital bone when mouth closed; no teeth on roof of mouth; preopercle with strong dentitions along its margin; dorsal fin single, without a pronounced notch, with 12 spines and 15 or 16 soft rays; anal fin with 3 spines and 11 soft rays; bases of dorsal and anal fins scaled; pectoral fins shorter than pelvic fins; color: varying shades of yellow brown to dark brown with ill defined spots and mottling; the young are often bright yellowish, becoming darker with age
Type locality: Suriname
Distribution: circumglobal in tropical and warm temperate seas, including Mediterranean Sea and Hawaiian Islands, but excluding eastern Pacific
Size: 110 cm (19 kg); common to 50 cm
Habitat: marine, juveniles may drift into estuaries and inshore areas with brackish water
Position in the water column:
Diet:
Reproduction: young tripletails resemble leaves and are often found floating on their side at the surface
Remarks: a sluggish offshore fish that often floats on its side near the surface in the company of floating objects, occasionally drifting into shallow water.

Lobotes surinamensis (live), Suriname River Estuary

FAMILY GERREIDAE (MOJARRAS)

Small to medium-sized fishes (to 41 cm standard length in Western Atlantic) with a compressed body, varying from narrow to deep, a pointed snout, the anterior part of lower head profile concave, a strongly protrusible mouth, pointing downward when protracted, with toothless appearing jaws (with small villiform teeth). Dorsal and anal-fin bases with a high scaly sheath into which the fins can be folded; caudal fin deeply forked; pectoral fin long and pointed; pelvic-fin origin below or somewhat behind pectoral-fin base and bearing a long, scale-like axillary process. Most of head and body covered with conspicuous silvery scales.

DIAPTERUS RHOMBEUS (Cuvier, 1829)
Local name(s): kawfisi (Caitipa or silver mojarra)

Diagnostic characteristics: body rhomboidal, compressed, moderately deep (depth 1.8 to 2.5 in standard length); mouth strongly protrusible; margin of preopercle serrated; second dorsal-fin spine longer than distance between tip of snout and posterior margin of orbit; preorbital bone smooth; sides of body without black longitudinal stripes; second anal-fin spine shorter than anal-fin base, fin spines not greatly thickened; all pharyngeal teeth pointed; 16 to 18, usually 17, gill rakers on lower limb of first gill arch; anal-fin rays typically with 2 spines and 9 soft rays; color: body silvery, somewhat darker above, with bluish reflections; spinous portion of dorsal fin edged with dusky pigment, pectoral fins transparent, pelvic fins and anal fin yellow
Type locality: Martinique Island, West Indies
Distribution: Western Atlantic
Size: 40 cm; common to 30 cm
Habitat: brackish, marine; abundant in mangrove-lined lagoons and also found over shallow mud and sand bottoms in marine areas; may enter fresh water
Position in the water column: mid-water
Diet: small fish feed mainly on plants and microbenthic crustaceans, larger fish eat crustaceans, pelecypods, and polychaete worms in addition to plants
Reproduction:
Remarks: *D. rhombeus* was relatively abundant in a brackish water aquaculture facility in Commewijne; a smaller related species *Diapterus auratus* (28 cm, common to 16 cm) has 12 to 15, usually 12 or 13, gill rakers on lower limb of first gill arch and anal-fin rays typically with 3 spines and 8 soft rays or with 2 spines, 1 unbranched ray, and 8 branched soft rays in small specimens.

Diapterus rhombeus (live), Commewijne River

FAMILY HAEMULIDAE (GRUNTS)

The prior family name, Pomadasyidae, may still be encountered. The systematic status and distribution of several species in South America is unresolved. Grunts are oblong, compressed, perchlike fishes to 60 cm total length with a strongly convex head profile in most species, a small to moderate, non-protrusible mouth with thick lips, and a chin with 2 enlarged pores. Scales are ctenoid (rough to touch), small or moderate, extending onto entire head (except in anterior portion of snout, lips, and chin). Teeth conical, in a narrow band in each jaw, the outer series enlarged but no canines. No teeth on the roof of the mouth. Posterior margin of suborbital not exposed; preopercle with posterior margin slightly concave and serrated; opercle with 1 spine. Dorsal fin single, with 11 to 14 strong spines and generally 11 to 19 soft rays; pectoral fins moderately long; pelvic fins below base of pectoral fins, with 1 spine and 5 soft rays; anal fin with 3 strong spines, the second often very prominent, and 6 to 13 soft rays; caudal fin emarginate to forked. Grunts are fishes of shallow, near-shore waters; nearly all from tropical and subtropical waters. Species of *Pomadasys, Genyatremus,* and *Conodon* are characteristic of mud bottoms and turbid, often brackish water. The name of the family derives from the sound produced by the grinding of pharyngeal teeth. Juveniles typically occur in shallower water than adults and may show several ontogenetic habitat shifts during growth. Most species feed on a variety of benthic invertebrates, particularly crustaceans and polychaetes. Schooling is present in many species, but may become less common in older individuals. The absence of documented spawning events suggests that reproduction typically occurs after sunset. Several grunts are considered good food fish.

GENYATREMUS LUTEUS (Bloch, 1790)
Local name(s): neertje, nerki (Torroto grunt)

Diagnostic characteristics: body ovate, compressed, its depth 41 to 45% of standard length; head small, mouth moderately large, 2 pores, but no median groove on chin; preopercle strongly serrate at angle; dorsal fin high, with 13 spines and about 12 soft rays, the fifth spine the longest; anal fin with 3 spines and 11 soft rays; caudal fin emarginate; vertical fins scaleless; scales small, not parallel with lateral line, arranged obliquely above and horizontally below, largest below the lateral line; pored lateral-line scales 51 to 53; 11 longitudinal rows of scales above and 19 rows below lateral line; color: body silvery with a yellowish cast; preopercular margin yellow; dorsal fin with silvery spines and a black margin; pectoral fins with a yellowish tint; pelvics with a black posterior margin; anal fin yellowish; base of caudal fin yellowish, with a terminal black margin.
Type locality: Antilles [Martinique Island, West Indies]; no types known

Genyatremus luteus (live), Suriname River Estuary

Distribution: Western Atlantic
Size: 37 cm TL (0.8 kg); commonly to 25 cm
Habitat: brackish, marine; found over soft bottom habitats to depths of 40 m, typically, in shallow, brackish waters; ascends rivers in the dry season
Position in the water column:
Diet: crustaceans and small fishes
Reproduction:
Remarks: commonly observed at the markets of Paramaribo.

This page is intentionally left blank.

FAMILY SCIAENIDAE (DRUMS OR CROAKERS)

The croakers or drums, family Sciaenidae, include about 78 genera and 287 species, inhabiting coastal, estuarine and freshwaters in tropical and temperate regions. Dorsal fin long with a deep notch separating the spinous from soft portion, first with 6-13 spines and second with one spine and usually 20-35 soft rays; anal fin with one or two spines (both usually weak, but second may be large) and 6-13 soft rays; lateral line extending to the end of the caudal fin; single barbel or a patch of small barbels on chin of some species; head with large cavernous canals (part of lateral line system) and conspicuous pores on snout and lower jaw; swimbladder usually with many branches; sagitta otolith exceptionally large. Six genera are restricted to freshwaters, but may also occur in estuaries. Three of these genera, *Pachyurus*, *Pachypops*, and *Plagioscion* are widely distributed in South America and also occur in Suriname where they are recognized locally as (ston)kubi. *Plagioscion* species are large piscivores mainly found in the main channel of large rivers and supporting locally important commercial or sport fishing (Goulding, 1980: 179). *Pachyurus* and *Pachypops* are smaller, bentivore species. Many more species of the family are important in the coastal fishery of Suriname and may be landed in coastal ports (see Keith *et al.* (2000) and Leopold (2004) for more information on the coastal species). Coastal species, especially their young (in Suriname known as 'tri'), also occur in the river estuaries and at least one species, blakatere *Cynoscion steindachneri*, occurs in mangrove lagoons. Sciaenids have large earstones (otoliths); in Suriname, these are best known from freshwater *Plagioscion* species ('kubi') and thus known as 'kubi-ston'.

1-a Chin, underside of lower jaw, with small barbel(s)

2

1-b Chin without barbel(s)

6

2-a Pectoral fin long, jet-black reaching beyond anal-fin base; caudal fin long and pointed; eye small, 8 times or more in head length; preopercle margin smooth; soft dorsal-fin rays 31-39 (*Lonchurus*)

3

2-b Pectoral fin short, pale, not reaching beyond anus; caudal fin truncate or rhomboid; eye moderate, less than 5 times in head length; preopercle margin usually serrate; soft dorsal-fin rays 19-30

4

3-a Two slender barbels on tip of lower jaw beside the median mental pore; barbels longer than eye diameter; pectoral fin reaching to caudal peduncle; soft dorsal-fin rays 37-39

Lonchurus lanceolatus

3-b Three pairs of short barbels in tuft on tip of jaw around the median mental pore; barbels in a series of 10-12 pairs along rami of chin; soft dorsal-fin rays 31-34
Lonchurus elegans

4-a Barbels in series of 3-5 pairs along median margins of lower jaw; eye small, 4.5 times or more in head length; caudal fin scales only basal half, never sheath-like; side with series of small spots forming oblique wavy lines along transverse scale rows; estuaries and coastal marine
Micropogonias furnieri

4-b Three miniature barbels on tip of lower jaw; eye large, 3-4 times in head length; small scales cover almost entire caudal fin like a sheath (restricted to fresh water) (*Pachypops*)
5

5-a Absence of longitudinal stripes on sides of body; horizontal eye diameter 2.4-3.1 in head length
Pachypops fourcroi

5-b Three or four dark longitudinal stripes present on sides of body; horizontal eye diameter 3.0-3.7 in head length
Pachypops trifilis

6-a Lateral line with a much thickened appearance, pored lateral-line scales completely concealed by layers of smaller scales; gas bladder with a pair of tubular appendages running from posterior end along lateral wall ending anteriorly in a pair of horns (fresh water) (*Plagioscion*)
7

6-b Lateral line not appearing thickened, pored lateral-line scales with intercalated scales but never concealed by small scales; gas bladder with 1 or 2 chambers, some with variable developed appendages, but never originating from posterior end of gas bladder
8

7-a Distance from anus to origin of anal fin 1.9-3.5 in head length (mean 2.7); second anal-fin spine strong and long, 2 times in head length; black bands on the anal fin
Plagioscion auratus

7-b Distance from anus to origin of anal fin origin 3.6-5.6 in head length (mean 4.4; 10% of individuals of *P. squamosissimus* can show this value ranging from 3.1-3.5; Casatti, 2005); second anal-fin spine 4 times in head length; anal fin hyaline, without black bands
Plagioscion squamosissimus

8-a Preopercle serrate often with 1 or more distinct bony spines at angle or prominent serration on posterior margin
9

8-b Preopercle smooth, never with strong bony spines or serration in adult
11

9-a Head narrower, top cavernous, but usually not translucent under skin, firm to touch; interorbital width 3.5 times or more in head length
Bairdiella ronchus
9-b Head broad, top cavernous, often translucent under skin, hollow or spongy to touch; interorbital width less than 3.5 times in head length (*Stellifer*)
10

10-a Preopercular margin with 2 prominent spines; mouth terminal; gill rakers 36 or more; dorsal fin rays 20-24
Stellifer rastrifer
10-b Preopercular margin with 3 prominent spines; mouth terminal; gill rakers 33-39; dorsal-fin rays 17-20
Stellifer stellifer
10-c Preopercle with 4 or more prominent spines; mouth inferior
Stellifer microps

11-a Mouth small, inferior, snout projecting in front of upper jaw; body elongate; dorsal profile not strongly elevated or arched on nape; body depth more than 4 times in standard length; a freshwater species
Pachyurus schomburgkii
11-b Mouth moderate to large, horizontal to strongly oblique, terminal or lower jaw projecting in front of upper jaw; brackish water or marine species
12

12-a Eyes small, 8-11 times in head length; body rounded in cross-section; mouth large, extremely oblique; top of head cavernous, spongy to touch
Nebris microps
12-b Eyes moderate to large, 3-6 times in head length; body compressed or robust; mouth horizontal to strongly oblique; top of head cavernous, but never spongy to touch
13

13-a Canine-like teeth with arrowhead tips on both jaws, those at tip of upper jaw larger, strongly curved; large canines on lower jaw often exposed externally when mouth closed
Macrodon ancylodon
13-b Canine-like teeth sharp but never arrowheaded, only on upper jaw; teeth on lower jaw conical, usually not exposed externally when mouth closed (*Cynoscion*)
14

14-a Scales on body cycloid, much smaller than pored lateral-line scales; more than 100 in row above lateral line; caudal fin rhomboidal in adults
Cynoscion virescens
14-b Scales on body ctenoid, about the same size or larger than pored lateral-line scales; less than 70 in row above lateral line
15

15-a Body uniformly silvery; caudal fin rhomboidal or double emarginated in adults

Cynoscion jamaicensis

15-b Body with dotted stripes on trunk that run on oblique scale rows; anal fin with 8-10 soft rays; caudal fin lanceolate or pointed in adults

16

16-a Pectoral fin shorter than pelvic fin, 2 times or more in head length; less than 75 scales in lateral-line row; absence of strong canines in upper jaw

Cynoscion steindachneri

16-b Pectoral fin about equal or longer than pelvic fin, less than 2 times in head length; 80-90 scales in lateral-line row; one pair of strong canines in the upper jaw

Cynoscion acoupa

Bairdiella ronchus (Cuvier, 1830)
Local name(s): – (ground croaker)

Diagnostic characteristics: preopercle serrate with few bony spines at angle, lowest spine pointing downward; head narrower, top cavernous, but usually not translucent under skin, firm to touch; interorbital width 3.5 times or more in head length; spinous dorsal fin with 10 (rarely 11) spines, posterior portion with 1 spine and 21-26 (usually 23-24) soft rays; scales on body and top of head ctenoid, 54-59 lateral line scales; color grayish above, silvery below, faint dark oblique to longitudinal streaks on the sides
Type locality: Lake Maracaibo, Venezuela; Dominican Republic; Cuba; Suriname
Distribution: Western Atlantic
Size: 35 cm TL (common to 25 cm)
Habitat: shallow (<40 m) coastal and estuarine waters over muddy and sandy bottoms
Position in the water column: bottom
Diet: crustaceans and fishes
Reproduction:
Remarks: this species was abundant in brackish-water canals of Comfish aquaculture farm, District Commewijne.

Cynoscion acoupa (Lacépède, 1801)
Local name(s): banban (acoupa weakfish)

Diagnostic characteristics: a large, elongated species with a smooth preopercle, never with strong bony spines or serration in adults; mouth large, oblique, lower jaw slightly projecting, one pair of strong canines in the upper jaw; top of head cavernous, but never spongy to touch; scales on body ctenoid, about the same size or larger than pored lateral-line scales; less than 70 transverse rows of scales above lateral line; pectoral fin about equal or longer than pelvic fin, less than 2 times in head length; 80-90 scales in lateral-line row; dorsal fin divided in two parts, first part with 10 spines, posterior part with 1 spine and 17-22 (usually 18-20) soft rays and not scaled; body color uniform grey-silvery, belly yellow-white
Type locality: Cayenne, French Guiana
Distribution: Western Atlantic
Size: 120 cm TL (common to 50 cm)
Habitat: usually found over sandy mud bottoms in shallow (<22 m) coastal waters and estuaries
Position in the water column:
Diet: crustaceans and fishes
Reproduction:
Remarks: an important and popular food fish in Suriname; commonly observed at the markets of Paramaribo.

Bairdiella ronchus (live), Commewijne River

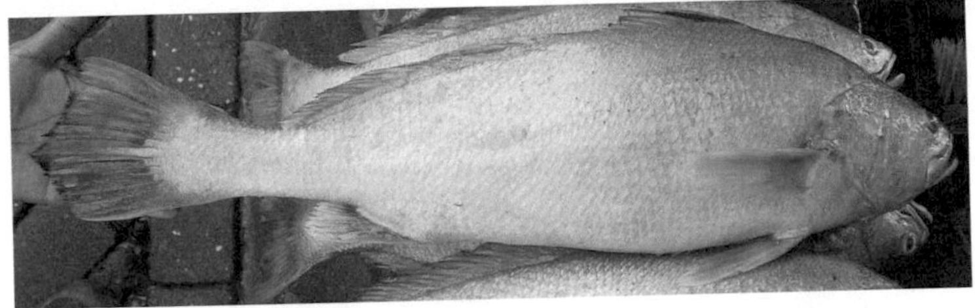

Cynoscion acoupa (live), Paramaribo market

Cynoscion acoupa (live) (© INRA-Le Bail)

Cynoscion steindachneri (Jordan, 1889)
Local name(s): blakatere (smalltooth weakfish)

Diagnostic characteristics: a large species with smooth preopercle, never with strong bony spines or serration in adult; mouth large and oblique, tip of upper jaw without enlarged canines, teeth small, villiform and set in narrow bands; gill rakers long and slender, 11-14 on first arch; top of head cavernous, but never spongy to touch; scales on body large and ctenoid, about the same size or larger than pored lateral-line scales; less than 70 transverse rows of scales above lateral line; less than 75 scales in lateral-line row; dorsal fin divided in two parts, first part with 10 spines, posterior part with 1 spine and 21-24 soft rays; pectoral fin much shorter than pelvic fin, 2 times or more in head length; caudal fin rhomboid in adults; color grayish above, whitish below, dorsal fin dusky, upper margin of pelvic fins orange, inside of mouth orange
Type locality: Ponta Curuca, Para, Brazil
Distribution: Western Atlantic: Guyana to northern Brazil
Size: 110 cm TL (common to 50 cm)
Habitat: in estuaries and brackish-water swamps and lagoons in mangrove forests along the coast; uncommon in typical marine habitats
Position in the water column:
Diet: shrimps, fishes, and sometimes plant material
Reproduction: spawns in sea
Remarks: often observed at the markets of Paramaribo.

Cynoscion virescens (Cuvier, 1830)
Local name(s): kandratiki (green weakfish)

Diagnostic characteristics: a large elongated fish, moderately compressed, with smooth preopercle, never with strong bony spines or serration in adult; mouth large and oblique, lower jaw projecting, teeth sharp, set in narrow bands on both jaws, upper jaw with a pair of large canine-like teeth at the tip and a row of enlarged, sharp outer-row teeth; gill rakers 7-11, moderately long and slender, but shorter than filaments; top of head cavernous, but never spongy to touch; scales on body cycloïd, much smaller than pored lateral-line scales; more than 100 transverse rows above lateral line; caudal fin rhomboidal- in adults; soft dorsal fin unscaled, except 1 or 2 rows of small scales at base; posterior part of dorsal fin with 27-31 soft rays, unscaled except for 2 or 3 rows of scales at base; color greyish to brownish above, silvery below, dorsal fin dusky, its spinous portion black-edged, soft dorsal fin with dark spots on each ray, pectoral, pelvic and anal fins yellowish to orange
Type locality: Suriname
Distribution: Western Atlantic
Size: 95 cm TL (common to 50 cm)

Cynoscion steindachneri (live), Paramaribo market

Cynoscion steindachneri (live) (© INRA-Le Bail)

Cynoscion virescens (live), Paramaribo market

Cynoscion virescens (live) (© INRA-Le Bail)

Habitat: over mud and sandy bottoms in coastal waters near river mouths; juveniles (and in French Guiana adults) inhabit estuaries.
Position in the water column: demersal in daytime, and moves toward the surface at night
Diet: mainly shrimps and occasionally fish
Reproduction:
Remarks: the related species, *Cynoscion microlepidotus*, has the soft dorsal fin almost entirely covered with small scales and a dorsal fin with 22-25 soft rays; *C. virescens* is an important and popular food fish; it is commonly observed at the markets of Paramaribo.

LONCHURUS LANCEOLATUS (Bloch, 1788)
Local name(s): basrabotrofisi (longtail croaker)

Diagnostic characteristics: a small fish, with an elongate and compressed body and a pair of moderately long, slender barbels (longer than eye diameter) on tip of lower jaw beside the median mental pore (the chin) and a smooth preopercle margin; mouth large, but inferior, nearly horizontal, teeth small, but sharp, set in bands on both jaws; pectoral fin very long, jet-black reaching beyond anal-fin base to caudal peduncle; caudal fin long and pointed; eye small, 8 times or more in head length; soft dorsal-fin rays 31-39; soft dorsal-fin rays 37-39; body often brownish to yellowish, slightly darker above, all fins darkish, base of pelvic and anal fins yellowish
Type locality: Suriname
Distribution: Western Atlantic
Size: 30 cm TL
Habitat: found over sandy to muddy bottoms in coastal marine and brackish waters
Position in the water column:
Diet: shrimps and fishes
Reproduction:
Remarks: a related species, manyafisi *Lonchurus elegans*, has three pairs of short barbels in tuft on tip of jaw around the median mental pore; barbels in a series of 10-12 pairs along rami of chin; soft dorsal-fin rays 31-34.

MACRODON ANCYLODON (Bloch & Schneider, 1801)
Local name(s): dagutifi, bangamary (king weakfish)

Diagnostic characteristics: a medium-sized fish with an elongate and moderately compressed body; preopercle smooth, never with strong bony spines or serration in adult; eyes moderate to large, 3-6 times in head length; mouth large and strongly oblique, lower jaw projecting, teeth very sharp with arrowhead, set in narrow ridges on both jaws, upper jaw with pair of large, strongly curved canine-like teeth at tip, lower jaw with several large canine-like teeth at its tip; top of head cavern-

Macrodon ancylodon (live), Paramaribo market

Macrodon ancylodon (live) (© INRA-Le Bail)

ous, but never spongy to touch; large canines on lower jaw often exposed externally when mouth closed; color silvery grayish on back, pale to yellowish below (back punctuated in juveniles)
Type locality: Suriname
Distribution: Western Atlantic
Size: 45 cm TL
Habitat: found over mud or sandy bottoms in coastal waters to about 60 m depth; juveniles enter estuaries and coastal lagoons
Position in the water column:
Diet: shrimps and fish
Reproduction:
Remarks: also an esteemed food fish that is sold at the markets of Paramaribo.

MICROPOGONIAS FURNIERI (Desmarest, 1823)
Local name(s): – (whitemouth croaker)

Diagnostic characteristics: medium to large sized, slightly elongate and moderately compressed; mouth moderately large, subterminal to inferior, teeth villiform in bands; chin with small barbels in series of 3-5 pairs along median margins of lower jaw; pectoral fin short, pale, not reaching beyond anus; caudal fin truncate or rhomboid; preopercle margin strongly serrated, with 2 or 3 sharp spines at angle; soft dorsal-fin rays 19-30; eye 4.5 times or more in head length; caudal fin scales only basal half, never sheath-like; scales ctenoid; side with series of small spots forming oblique wavy lines along transverse scale rows
Type locality: Havana, Cuba
Distribution: Western Atlantic
Size: 90 cm TL (common to 45 cm)
Habitat: estuaries and coastal marine, found over muddy and sandy bottoms, juveniles and young adults may be found year round in estuaries
Position in the water column: bottom
Diet: bottom-dwelling organisms, mainly worms, crustaceans and small fishes
Reproduction:
Remarks: sold at the markets of Paramaribo.

NEBRIS MICROPS Cuvier, 1830
Local name(s): botrofisi, botervis (smalleye croaker)

Diagnostic characteristics: a medium-sized fish with elongate body, rounded in cross-section, tapering to a slender caudal peduncle; smooth preopercle, never with strong bony spines or serration in adult; mouth large, strongly oblique, lower jaw projecting in front of upper jaw; eyes small, 8-11 times in head length; body rounded in cross-section; top of head cavernous, spongy to touch; color more or less uniformly silvery brown to orange, darker above, pectoral, pelvic and anal fins orange with dark tip; juveniles with 5 or 6 saddle-like dark blotches on sides

Micropogonias furnieri (live), Paramaribo market

Nebris microps (live) (© INRA-Le Bail)

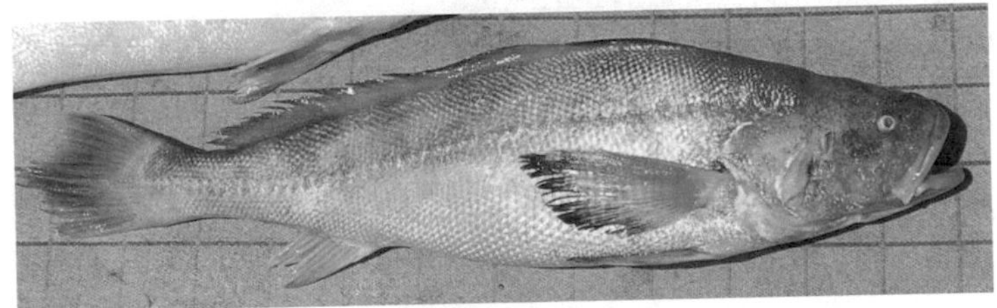

Nebris microps (live), Suriname River Estuary

Type locality: Suriname
Distribution: Western Atlantic
Size: 50 cm TL
Habitat: found over sandy mud bottoms in coastal waters to about 50 m and also entering estuaries, especially the juveniles
Position in the water column: bottom
Diet: shrimps and small crustaceans, but also small fishes
Reproduction:
Remarks: a highly valued food fish that is commonly sold at the markets of Paramaribo.

Pachypops fourcroi (Lacépède, 1802)
Local name(s): stonkubi

Diagnostic characteristics: body moderately elongated, oval in cross-section; mouth small and inferior, villiform teeth; three miniature barbels on tip of lower jaw (sometimes difficult to see); eye large (horizontal eye diameter 2.4-3.1 times in head length); dorsal divided, spinous part with 10 spines, posterior part with 1 spine and 25-27 soft rays; anal with two spines and 6 soft rays; 58 scales in the lateral line row; small scales cover almost entire caudal fin like a sheath (restricted to fresh water)
Type locality: unknown [probably Suriname]
Distribution: Amazon and Orinoco River basins and rivers of the Guianas: Brazil, French Guiana, Guyana, Suriname and Venezuela
Size: 18.7 cm SL
Habitat: both in lower reaches of the river (downstream the first rapids) and in upstream reaches above the rapids, always in the main channel
Position in the water column:
Diet: fish
Reproduction:
Remarks: may live in small groups; the related species *P. trifilis* has three or four dark longitudinal stripes present on sides of body and smaller eyes (diameter 3.0-3.7 in HL).

Pachyurus schomburgkii Günther 1860
Local name(s): –

Diagnostic characteristics: preopercle smooth, never with strong bony spines or serration in adult; mouth small, inferior, snout projecting in front of upper jaw; body elongate; dorsal profile not strongly elevated or arched on nape; body depth more than 4 times in standard length; dorsal fin with 23-28 (27) soft rays; margin of first dorsal fin black; irregular dark spots on the dorsal half of the body
Type locality: Rio Capin [= Capim], Pará State, and Caripe Pará, Brazil

Pachypops fourcroi (live) (© INRA-Le Bail)

Pachyurus schomburgkii (alcohol), scale bar = 1 cm, ANSP162800 (© M. Sabaj Pérez)

Distribution: Amazon and Orinoco River basins: Bolivia, Brazil, Peru and Venezuela; in Suriname present in the upper reaches of the Nickerie, Suriname and Marowijne rivers
Size: 21 cm TL
Habitat: freshwater, upstream of the first rapids
Position in the water column:
Diet: insects and insect larvae
Reproduction:
Remarks:

PLAGIOSCION AURATUS (Castelnau, 1855)
Local name(s): kubi (freshwater croaker)

Diagnostic characteristics: *Plagioscion* species have a lateral line with a much thickened appearance, pored lateral-line scales completely concealed by layers of smaller scales, curved halfway body length, and continued on the caudal fin; head length almost 4 times in total length; mouth moderate and terminal; gas bladder with a pair of tubular appendages running from posterior end along lateral wall ending anteriorly in a pair of horns; pectoral fin short, not reaching the anus; second anal-fin spine strong and long, 2 times in head length; black bands on the anal fin
Type locality: Rio Ucayali, Amazon basin, Peru
Distribution: Amazon, Orinoco and rivers of the Guianas: Brazil, French Guiana, Guyana, Peru, Suriname and Venezuela
Size: 50-70 cm TL (34.6 cm SL)
Habitat: estuaries and lower reaches of rivers (downstream of the first rapids)
Position in the water column: mid-water
Diet: juveniles feed on insect larvae and small crustaceans, adults are essentially piscivorous
Reproduction: in French Guiana reproduction is in June
Remarks: the large sagitta otoliths are locally known as kubi-ston; *P. auratus* is not as common as *P. squamosissimus*. *Plagioscion* spp are important food fishes sold in restaurants and markets of Paramaribo.

PLAGIOSCION SQUAMOSISSIMUS (Heckel, 1840)
Local name(s): kubi (freshwater croaker)

Diagnostic characteristics: lateral line with a much thickened appearance, pored lateral-line scales completely concealed by layers of smaller scales, curved halfway the body and continued on the caudal fin; gas bladder with a pair of tubular appendages running from posterior end along lateral wall ending anteriorly in a pair of horns; pectoral fin short, not reaching the anus, with a black blotch at its base; second anal-fin spine 4 times in head length; anal fin hyaline, without black bands

Plagioscion auratus (live) (© INRA-Le Bail)

Plagioscion squamosissimus (alcohol), 257 mm SL, ANSP177421 (© M. Sabaj Pérez)

Type locality: Rio Negro and Rio Branco rivers, South America
Distribution: Amazon, Orinoco, Paraná, Paraguay and São Francisco basins and rivers of the Guianas: Argentina, Bolivia, Brazil, Colombia, Ecuador, French Guiana, Guyana, Peru, Suriname and Venezuela
Size: 70 cm TL
Habitat: common in estuaries, lower freshwater reaches of rivers and the upper reaches (upstream of the first rapids); also present in Brokopondo Reservoir
Position in the water column: mid-water
Diet: juveniles feed on shrimp (larvae), aquatic insects and copepods; adults are piscivorous
Reproduction:
Remarks: the large otoliths are locally known as kubi-ston; all *Plagioscion* species are popular food fishes in Suriname, but *P. squamosissimus* is the most common kubi species in the landings of river seine ('harititè') fisheries of the lower reaches of Surinamese rivers. *Plagioscion surinamensis* (Bleeker, 1873) is considered a junior synonym of *P. squamosissimus* (Casatti, 2005).

STELLIFER RASTRIFER (Jordan, 1889)
Local name(s): krorokroro (rake stardrum)

Diagnostic characteristics: a small fish with oblong and compressed body; head broad with conspicuous cavernous canals on top, but not spongy to touch; mouth large, oblique and terminal; preopercle with 2 prominent bony spines at angle; interorbital width less than 3.5 times in head length (*Stellifer*); 36 or more gill rakers; posterior portion of dorsal fin with 1 spine and 21-23 soft rays; color yellowish brown, darker above, upper third of spinous dorsal, pectoral and anal fins dusky often with dark tip, inner side of gill cover and roof of mouth black
Type locality: Santos, Sao Paulo State, Brazil
Distribution: Western Atlantic
Size: 25 cm TL (common to 15 cm)
Habitat: found in inshore waters and especially in brackish waters and coastal lagoons over muddy or sandy bottoms.
Position in the water column:
Diet: feeds mainly on small planktonic crustaceans
Reproduction:
Remarks: the related species *Stellifer stellifer* has the preopercular margin with 3 prominent spines, 33-39 gill rakers, and 17-20 dorsal -fin rays, while *Stellifer microps* has the preopercle with 4 or more prominent spines, an inferior mouth and small eyes.

Stellifer rastrifer (live) (© INRA-Le Bail)

Stellifer microps (live), Commewijne River

FAMILY POLYCENTRIDAE (AFRO-AMERICAN LEAF FISHES)

Species of Polycentridae have long been placed in the family Nandidae along with a few African and Asian percoids that share a cryptic, leaf-mimicking color pattern and lurking behavior. Britz (1997) found that the Polycentridae (except the African genus *Afronandus*) share a unique egg surface pattern with narrow ridges running radially from the micropyle, larvae with a multicellular cement gland on top of the head, and a unique adult spawning procedure. There is no evidence of a close relationship between the Polycentridae and the Nandidae + Badidae. South American polycentrids are small fishes, reaching about 6-8 cm SL. The dorsal fin has 16-18 spines and 7-13 soft rays, the anal fin 12-13 spines and 7-14 soft rays. There is no lateral line on the side. The two species have a characteristically large head and a large mouth, with extremely protractile upper jaws in *Monocirrhus*. *Monocirrhus polyacanthus* is strongly compressed laterally and resembles a dead leaf both in color pattern and shape, including a short skin flap projecting from the lower jaw that resembles the stalk of the leaf. It can move slowly towards small prey fish, being mistaken by them for a drifting leaf. In the two Neotropical polycentrid species (*M. polyacanthus* and *Polycentrus schomburgkii*), the pectorals and soft parts of the dorsal and anal fins are highly transparent and are the only fins that move when approaching prey. Both species exhibit male parental care of eggs and larvae. The egg clutch is deposited under leaves of aquatic plants (*Monocirrhus*) or at the roof of small crevices (*Polycentrus*) (Barlow, 1967). The group has never been thoroughly revised, and it is possible that there are more than one species in each genus.

POLYCENTRUS SCHOMBURGKII Müller & Troschel, 1849
Local name(s): kala, bladvis (leaf fish)

Diagnostic characteristics: a small, laterally compressed percoid species with a characteristically large mouth with extremely protractile jaws, large eyes and a large head; a dorsal fin with 16 spines (and 8 branched rays in the transparent posterior part); anal fin with 13 spines (and 7 branched rays in the transparent posterior part); caudal and pectoral fins transparent; absence of a lateral line; body color variable, grey to brown with dark spots (more or less arranged in vertical rows) and often small, whitish spots, three dark eye-markings on the head (a pre-orbital line connects the tip of snout with the eye, another, postorbital line runs from the eye to the insertion of the dorsal fin, and a third suborbital line diagonally runs from the eye down to the rear edge of the gill cover)
Type locality: Suriname
Distribution: South America: Trinidad (W.I.), and Atlantic coastal rivers of Venezuela, Guyana, Suriname, French Guiana and Brazil (state of Amapá); Brazil,

Polycentrus schomburgkii (live), Para River, 22.3 mm SL, FMNHexANSP189014 (© M. Sabaj Pérez)

French Guiana, Guyana, Suriname, Trinidad and Tobago, Venezuela; in Suriname, known from the Corantijn, Coppename, Saramacca, Suriname and Marowijne rivers, but probably present in all rivers
Size: 6 cm SL
Habitat: common in coastal freshwater swamps and black-water creeks of the Savanna Belt, but also present in forest creeks of the Interior (slow current); often associated with submerged aquatic vegetation or leaf litter
Position in the water column: mid-water
Diet: a voracious, piscivorous species that may consume fishes of its own size; prey are sucked in very fast by protraction of its mouth (Liem, 1970)
Reproduction: territorial males are nearly black with prominent white spots, the mature females are yellowish-pink to white, with the abdomen distended by eggs; non-territorial males may show pseudo-female behavior to steal fertilizations ('sneakers'); approximately 300-600 small, yellow eggs are spawned during the night in a cave or at the underside of woody debris or broad leaves; the male guards the brood during the 3-day incubation; the newly hatched larvae remain attached to the leaf or cave ceiling by means of a cement gland on top of their head; around day 6 after hatching the larvae depart from the nest (Barlow, 1967)
Remarks: in an aquarium, the leaf fish *P. schomburgkii* may easily consume tetras and other small species; *P. punctatus* (Linnaeus, 1758) is a synonym.

This page is intentionally left blank.

FAMILY CICHLIDAE (CICHLIDS)

The cichlids are the most species-rich non-Ostariophysan fish family in freshwaters and about 290 valid species are known to occur in South America (Kullander, 1998). Cichlids show much variability in body shape, but can be recognized by several unambiguous anatomical synapomorphies (Kullander, 1998). Among Neotropical fishes they can be recognized externally by a single nostril on each side of the head, the possession of 7-24 (usually 13-16) spines in the dorsal fin, 2-12 (usually 3, rarely more than 5) anal-fin spines, and the lateral line usually divided into one anterior upper portion ending below the end of the dorsal-fin base, and a posterior lower portion running along the middle of the caudal peduncle (Fig. 10.3). Cichlid diversity has been explained both by their advanced brood care and by the versatile design of the pharyngeal jaw complex used for food mastication. There is considerable variation in the shape of the tooth plates and associated dentition, reflecting diet specializations. The oral jaws are generally highly movable and protrusible. Among Neotropical taxa, lengths range from about 25-30 mm adult size in *Apistogramma* to about 1 meter in *Cichla temensis*. Most taxa are in the interval 10-20 cm, however. Most Neotropical cichlids occupy lentic habitats within rivers and streams; but there is also a number of moderately to strongly adapted rheophilic species (e.g. many *Crenicichla* species). The majority of the Neotropical cichlids feed on a variety of invertebrates and some plant matter. *Cichla* and large *Crenicichla* species feed on fishes and large invertebrates. *Chaetobranchus* and *Satanoperca acuticeps* are plankton feeders. Most Neotropical Cichlidae are moderately to strongly sex dimorphic, and breed pairwise. Eggs are typically deposited on a substrate (Fig. 12.8) and both parents guard offspring over several weeks, even for some time after the young are free-swimming (Lowe-McConnell, 1969). Smaller species, particularly in the genus *Apistogramma*, may be strongly sexually dimorphic. Sexes differ in color and the female is smaller than the male and assumes all or most of the care for the eggs and young. Oral incubation, or mouthbrooding, has been recorded for many *Geophagus* and *Satanoperca* species. Mouthbrooding species are usually biparental, and eggs are guarded on a substrate prior to oral incubation which starts with advanced eggs or newly hatched larvae. In South America, cichlids are recorded from virtually all river drainages, but rarely occupy elevations over 500 m-above-sea-level, and generally remain below 200 m-asl. Because of the varied behavior and often attractive colors and moderate size, cichlids are commonly kept as ornamental fish. The traditionally most important aquarium species are *Pterophyllum* and *Symphysodon* species. Sport fishing is concentrated on the *Cichla* species. All the larger species are used as food fish within a traditional artisanal and subsistence fishery, and all local markets in the lowland Amazon, Orinoco, and Guiana Shield drainages offer *Cichla*, *Astronotus*, and other available species of sizes over 10 cm. A scientific general review of the family is provided by Keenleyside (1991). Kullander and Nijssen

Fig. 12.8. Tukunari (*Cichla ocellaris*) guarding its eggs in Brokopondo Reservoir (© W. Kolvoort).

(1989) describe most cichlids of Suriname. In Suriname, many cichlids are known by family or genus-level local names, e.g. krobia (genera *Krobia, Cichlasoma, Aequidens*), tukunari (*Cichla*), datra(fisi) (*Crenicichla*), and kunapari (*Chaetobranchus flavescens*).

1-a Height of dorsal fin larger than depth of body
Pterophyllum scalare
1-b Height of dorsal fin smaller than depth of body
2

2-a Lips of 'African' type, i.e. posterior portion of lower lip fully exposed
3
2-b Lips of 'American' type, i.e. the outline of the posterior portion of the upper lip covered by the snout or lower lip
5

3-a Teeth fine, closely set; posterior dorsal fin spines of approximately equal length; juveniles with black spot on soft dorsal fin; introduced, restricted to more or less brackish water in the Coastal Plain (e.g. mangrove lagoons, canals in Paramaribo and Nieuw Nickerie)
Oreochromis mossambicus
3-b Moderately large conical teeth present in jaws; penultimate dorsal fin spine much shorter than antepenultimate and ultimate, so that the fin margin appears deeply notched just anterior to the soft portion; juveniles without black spot on soft dorsal fin, freshwater species of large black-water streams and rivers in the Interior (*Cichla*)
4

4-a Lateral line divided; space between the two anteriormost transverse bars under the dorsal fin smaller than the width of the transverse bars (specimens > 14 cm); depth of caudal peduncle 8.1-9.7 times in standard length
Cichla monoculus
4-b Lateral line not divided, but continuous; space between the two anteriormost transverse bars under the dorsal fin much larger than the width of the transverse bars (specimens > 14 cm); depth of caudal peduncle 7.6-8.3 times in standard length
Cichla ocellaris

5-a Adpressed pelvic fins not reaching the anus; body elongated, its depth more than 3.7 times in standard length (*Crenicichla*)
23
5-b Adpressed pelvic fins reaching to or over the anus; body moderately deep, its depth less than 3.7 times in standard length
6

6-a Insertion of pelvic fins anterior to insertion of pectoral fins; presence of an oblique black band extending from the tip of the snout to the posterior end of the dorsal fin

Mesonauta guyanae

6-b Insertion of pelvic fins posterior to insertion of pectoral fins; black band absent or if present in horizontal position

7

7-a Numerous (>50) long and slender gill rakers on the first branchial arch

Chaetobranchus flavescens

7-b Gill rakers of the first branchial arch short and less than 20 in number

8

8-a First branchial arch with lobe-like expansion on upper portion

9

8-b Absence of lobe-like expansion on upper portion of first branchial arch

14

9-a Eye diameter less than one time in snout length; distinct midbasal caudal spot; posterior part of the anterior portion of the lateral line separated by ½ scale from the dorsal fin (*Apistogramma*)

10

9-b Eye diameter more than one time in snout length; caudal spot absent or situated on base of dorsal caudal fin lobe; posterior part of the anterior portion of the lateral line separated by at least 1½ scale from the dorsal fin

11

10-a No spot on middle of flank; series of spots along abdominal sides; no prolonged marginal caudal fin rays in adult males; Corantijn River

Apistogramma ortmanni

10-b A large spot on the middle of the flank that may be extended dorsad; no distinct spots in series along sides; adult males with prolonged marginal caudal fin rays

Apistogramma steindachneri

11-a Cheek scaled rostrad to lacrimal; base of anal fin without scales; presence of spot on the dorsal part of the caudal-fin base; no midlateral spot; Nickerie River

Satanoperca leucosticta

11-b Anterior half of cheek scaled; base of anal fin scaled in adults; absence of spot on dorsal part of caudal-fin base; presence of midlateral spot or intense stripe across middle of side

12

12-a Presence of a black dorsal spot at the base of the posterior portion of the dorsal fin; absence of a black midlateral spot below the lateral line; predorsal midline partially naked; Marowijne River

Geophagus harreri

12-b Absence of a black dorsal spot at the base of the posterior portion of the dorsal fin; presence of a black midlateral spot on the flanks; predorsal midline covered by scales

13

13-a Lateral spot on and below 11th-15th lateral line scales; dark stripe over preopercular corner from 5 cm SL; Corantijn and Nickerie rivers

Geophagus brachybranchus

13-b Large spot on and below 11th-13th lateral line scales; no dark maskings on head in adults

Geophagus surinamensis

14-a Principal caudal fin rays 14; 3 preopercular scales; ½ scale between the posterior end of the anterior part of the lateral line and the dorsal fin

Nannacara anomala

14-b Principal caudal-fin rays 16; 0-2 preopercular scales; at least 1½ scale between the posterior end of the anterior part of the lateral line and the dorsal fin

15

15-a Soft dorsal and anal fins extensively scaled

16

15-b Soft dorsal and anal fins with a few basal scales or naked

17

16-a No caudal spot; strong bar from nape to interoperculum through eye; caudal fin immaculate; 3 anal fin spines

Cleithracara maronii

16-b Superior caudal base spot; suborbital stripe in young; caudal fin dotted; 4, rarely 3 anal fin spines

Cichlasoma bimaculatum

17-a Presence of a black spot on the dorsal part of the caudal fin base

18

17-b Absence of a black spot on the dorsal part of the caudal fin base

21

18-a Elongated black spot below the eye; midlateral longitudinal band absent or poorly visible; predorsal squamation with 4 median scales followed by 4 pairs of scales

Aequidens tetramerus

18-b Black band extending from the base of the eye to the preopercle; presence of a dark midlateral longitudinal band; predorsal squamation with single range of median scales only

19

19-a Longitudinal band ends at the base of the caudal fin; Marowijne River

Aequidens paloemeuensis

19-b Posterior end of longitudinal band oriented towards the posterior end of the dorsal-fin base

20

20-a Body depth 41.3-50.8% of standard length; lateral band continued to end of soft dorsal-fin base; two posterior of three facial stripes indistinct

Krobia guianensis

20-b Body depth 40.0-45.3% of standard length; lateral band ending well in advance of end of soft dorsal-fin base; three facial stripes clearly visible; Marowijne River

Krobia itanyi

21-a Large midlateral spot; membrane between spines of anterior portion of dorsal fin ('fin lappets') lacking an extension beyond the spines; Marowijne River

Guianacara oelemariensis

21-b Stripe across side, except in some large specimens in which the stripe is reduced to a dorsal spot; anterior portion of dorsal fin with produced (elongated) fin lappets

22

22-a Flank stripe strongest on and below lateral line; dorsal fin membranes between the anterior dorsal fin spines black

Guianacara owroewefi

22-b Flank stripe wedge shaped, strongest on and above upper lateral line; anterior 3 dorsal fin lappets black in young only; Corantijn River

Guianacara sphenozona

23-a 100-130 scales in horizontal row below the lateral line

24

23-b 52-72 scales in horizontal row below the lateral line

25

24-a 112-130 scales below lateral lines; dark spot just posterior to pectoral fin base; nostrils nearer to postlabial skinfold than to orbit; Corantijn River

Crenicichla lugubris

24-b 100-115 scales below lateral lines; no dark spot posterior to pectoral fin; nostrils almost halfway between postlabial skinfold and orbit; Suriname and Marowijne rivers

Crenicichla multispinosa

25-a Series of dark blotches along lateral lines in males; humeral spot longer than deep in females; Nickerie and Corantijn rivers

Crenicichla nickeriensis

25-b No series of dark blotches along lateral lines in males; humeral spot in females not or only slightly longer than deep

26

26-a Silvery dots present on operculum and cheek in both sexes; Marowijne River
Crenicichla albopunctata
26-b Silvery dots absent
27

27-a Maxilla not reaching orbit or anterior margin of orbit; ground color of body grey; Upper Corantijn River
Crenicichla sipaliwini
27-b Maxilla reaching well beyond orbit; ground color of body tan
28

28-a Humeral spot round with a white ring; males with yellowish dots; Coppename and Saramacca rivers
Crenicichla coppenamensis
28-b Humeral spot with a dorsoanterior notch; males with silvery-white dots
Crenicichla saxatilis

AEQUIDENS PALOEMEUENSIS Kullander & Nijssen, 1989
Local name(s): –

Diagnostic characteristics: elongate, moderately compressed species with a small mouth (posteriorly not reaching the level of the eye orbit); principal caudal-fin rays 16; 0-2 preopercular scales; at least 1½ scale between the posterior end of the anterior part of the lateral line and the dorsal fin; soft dorsal and anal fins with a few basal scales or naked; presence of a black spot on the dorsal part of the caudal fin base; black band extending from the base of the eye to the preopercle, and continuing as an intense dark midlateral longitudinal band that ends at the base of the caudal fin; 5 vertical dark bars; predorsal squamation with single range of median scales only; 23 scales in longitudinal row from operculum to caudal fin; life coloration not known, preserved specimens yellowish brown
Type locality: tributary creek of Paloemeu River, between Trombaka Noord and Trombaka Zuid, Marowijne River, Suriname
Distribution: only known from its type locality, Paloemeu River, Marowijne River drainage, Suriname
Size: 9.5 cm SL
Habitat:
Position in the water column:
Diet:
Reproduction:
Remarks: nothing is known about the biology of this rare species that is known only from Paloemeu River (where it is syntopic with *Krobia itanyi* (Kullander & Nijssen, 1989); *A. paloemeuensis* strongly resembles *A. potaroensis* from the Potaro and Essequibo rivers (Guyana) in color pattern, shape and meristics.

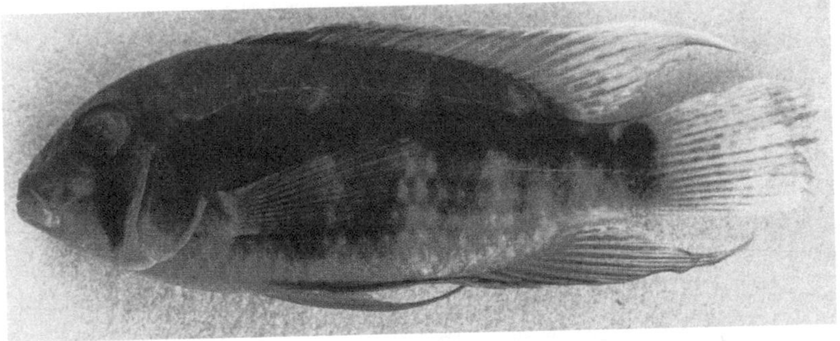

Aequidens paloemeuensis (© INRA-Le Bail)

Aequidens tetramerus (Heckel, 1840)
Local name(s): –

Diagnostic characteristics: principal caudal-fin rays 16; 0-2 preopercular scales; at least 1½ scale between the posterior end of the anterior part of the lateral line and the dorsal fin; soft dorsal and anal fins with a few basal scales or naked; presence of a black spot on the dorsal part of the caudal fin base; elongated black spot below the eye; midlateral longitudinal band absent or poorly visible; predorsal squamation with 4 median scales followed by 4 pairs of scales; 24-25 large scales in a longitudinal row between operculum and caudal fin; body color warmly-brown, turning to yellow on chest and very dark brown along the back; caudal fin and posterior part of anal fin with very fine stripes
Type locality: Rio Branco, Brazil
Distribution: widely distributed in Amazon River basin and Orinoco River basin: Bolivia, Brazil, Colombia, Ecuador, French Guiana, Guyana, Peru, Suriname and Venezuela; in Suriname, known from all rivers (but never common), but conspicuously absent from Saramacca River
Size: 16.2 cm SL
Habitat: small, shallow forest creeks with slow current and clear water
Position in the water column: mid-water
Diet: insects and, to a lesser extent, small fishes and plant matter
Reproduction: very territorial, about 1000 eggs are deposited on stones or wood; the parents guard the newborn
Remarks:

Apistogramma ortmanni (Eigenmann, 1912)
Local name(s): pikin krobia (dwarf cichlid)

Diagnostic characteristics: first branchial arch with lobe-like expansion on upper portion; eye diameter less than one time in snout length; distinct midbasal caudal spot; posterior part of the anterior portion of the lateral line separated by ½ scale from the dorsal fin; no spot on middle of flank; series of spots along abdominal sides; no prolonged marginal caudal fin rays in adult males; Corantijn River
Type locality: Erukin, Guyana
Distribution: Essequibo River drainage: Guyana, Suriname and Venezuela; in Suriname, only known from the Corantijn River
Size: 4.1 cm SL
Habitat: in small to medium-sized forest streams, often among leaf litter in pools with slow current
Position in the water column: benthopelagic
Diet: small invertebrates
Reproduction: all *Apistogramma* species spawn in caves, typically under rocks or in holes in sunken logs or branches; the female is more highly involved with brood

Aequidens tetramerus (live) (© R. Covain)

Aequidens tetramerus (live) (© INRA-Le Bail)

Apistogramma ortmanni (live), Kabalebo River (Corantijn River) (© R. Smith)

care (generally about 100 eggs), whilst the male defends a territory from predators.
Remarks:

APISTOGRAMMA STEINDACHNERI (Regan, 1908)
Local name(s): pikin krobia (dwarf cichlid)

Diagnostic characteristics: body moderately elongate, broadest and deepest just behind the gill cover; first branchial arch with lobe-like expansion on upper portion; eye diameter less than one time in snout length; distinct midbasal caudal spot; posterior part of the anterior portion of the lateral line separated by ½ scale from the dorsal fin; a prominent and characteristic large spot on the middle of the flank that may be extended dorsad; no distinct spots in series along sides; adult males with prolonged marginal caudal fin rays
Type locality: Georgetown, Demerara River, Guyana
Distribution: Essequibo, Demerara and Mahaica River drainages: Guyana and Suriname; in Suriname, widely distributed and known to occur in the Corantijn, Nickerie, Coppename, Saramacca and Suriname rivers
Size: 6.5 cm SL
Habitat: in small to medium-sized forest creeks, often in pools among leaf litter
Position in the water column: benthopelagic
Diet: small invertebrates
Reproduction: males with elongated marginal caudal fin rays or caudal fin long, truncate; females remain smaller than males (largest females measured 4 cm SL)
Remarks: the largest species of the genus *Apistogramma*.

CHAETOBRANCHUS FLAVESCENS Heckel, 1840
Local name(s): kunapari, kumaparu, kamara paru

Diagnostic characteristics: readily distinguished from other Surinamese cichlids by its numerous long and slender gill rakers (>50 on the first branchial arch); tilapia-like in shape, body color grey-green with a spotted effect due to darker areas at the center of each scale, a dark lateral blotch, dark red spots and vertical stripes on the caudal, dorsal, and anal fins, metallic blue-green markings on the ventral part of the head; large, red-brown eye (7.46-9.52 in SL) and large mouth; absence of scales on the base of the unpaired fins, 25-26 scales in longitudinal row between operculum and caudal fin
Type locality: Rio Guaporé, tributary of Rio Negro, Mato Grosso, Brazil
Distribution: widespread in South America: Brazil, French Guiana, Peru, Suriname and Venezuela; in Suriname, known from the Corantijn, Saramacca and Suriname rivers
Size: 21 cm SL

Apistogramma steindachneri (live) (©W. Kolvoort)

Apistogramma steindachneri (© P. Willink)

Chaetobranchus flavescens (live) (© P. Willink)

Habitat: in Suriname, best known from black-water rivers with slow current (Para River, Coesewijne River, Nanni Creek), but occurs also in the Interior, upstream of the first rapids (e.g. Upper Saramacca River; Kullander & Nijssen, 1989)
Position in the water column: mid-water
Diet: probably a zooplanktivorous species (based on the elongated gill rakers), but stomachs so far examined were empty or rotten (Lowe-McConnell, 1969; Knöppel, 1970)
Reproduction: in the wet season (Lowe-McConnell, 1969; Keith *et al.*, 2000)
Remarks: not a common species in Suriname.

Cichla monoculus Spix & Agassiz, 1831
Local name(s): tukunali, tukunari (peacock bass)

Diagnostic characteristics: lips of 'African' type, i.e. posterior portion of lower lip fully exposed; moderately large conical teeth present in jaws, mouth large; penultimate dorsal fin spine much shorter than antepenultimate and ultimate, so that the fin margin appears deeply notched just anterior to the soft portion; juveniles without black spot on soft dorsal fin; lateral line divided; space between the two anteriormost transverse bars under the dorsal fin smaller than the width of the transverse bars (in specimens > 14 cm); depth of caudal peduncle 8.1-9.7 times in standard length
Type locality: mari Brasiliae [Brazilian seas]
Distribution: Amazon River basin and Oyapock River basin: Brazil, Colombia, French Guiana, and Peru; in Suriname, only known to occur in the Coesewijne River (Saramacca River System), Kalebo River (Corantijn River System) and Nickerie River
Size: 33 cm SL (up to 80 cm TL according to Keith *et al.*, 2000)
Habitat: large to medium-sized rivers, both black-water (Coesewijne River) and clear-water (Kabalebo River)
Position in the water column: mid-water
Diet: fish (juveniles feed on shrimps)
Reproduction: large males develop a distinctive adipose hump on the nape just before spawning and become territorial; reproduction is not markedly seasonal
Remarks: this *Cichla* species is not mentioned for Suriname by Kullander and Nijssen (1989); in French Guiana, it is only known from Oyapock River (Keith *et al.*, 2000).

Cichla ocellaris Bloch & Schneider, 1801
Local name(s): tukunali, tukunari (peacock bass)

Diagnostic characteristics: lips of 'African' type, i.e. posterior portion of lower lip fully exposed; moderately large conical teeth present in jaws; penultimate dorsal fin spine much shorter than antepenultimate and ultimate, so that the fin margin appears deeply notched just anterior to the soft portion; juveniles without black

Cichla monoculus (live) (© INRA-Le Bail)

Cichla ocellaris (live), 300 mm SL, Lawa River (Marowijne River), ANSP189088 (© M. Sabaj Pérez)

spot on soft dorsal fin; lateral line not divided, but continuous; space between the two anteriormost transverse bars under the dorsal fin much larger than the width of the transverse bars (specimens > 14 cm); depth of caudal peduncle 7.6-8.3 times in standard length; a very colorful species with brownish green ground color, darker on the back, grading to silvery white on the belly; the longitudinal band on the posterior half of the body of young fish becomes disrupted into isolated blotches as the vertical bars develop, and as the fish grows the vertical bars become broken up into dark blotches which are increasingly ocellated with silver and yellow pigment; the iris is bright orange-red
Type locality: Indian Ocean [actual is South America]
Distribution: Northern South America: Brazil, French Guiana, Guyana and Suriname; in Suriname, *C. ocellaris* is probably present in all rivers
Size: 41 cm SL (up to 50 cm SL according to Keith *et al.*, 2000)
Habitat: large and medium-sized rivers upstream of the first rapids and Brokopondo Reservoir (Fig. 4.46)
Position in the water column: mid-water
Diet: fish and shrimps (Lowe-McConnell, 1969)
Reproduction: large males develop a distinctive adipose hump on the nape just before spawning and become territorial; reproduction is not markedly seasonal; females spawn 6,000-15,000 adhesive eggs (1.4 mm diameter) per kg wet mass in shallow water (20-30 cm deep), preferably on a flat rock surface (Fig. 12.8); very little feeding takes place during spawning and incubation; the eggs hatch in 70-90 hours, depending on the temperature, and the parents transfer the 5-6 mm long newborn (in their mouth) to a circular nest pit (1.5-6 cm depth, 8-20 cm diameter) which they dig in the sandy/muddy bottom with chin and pectoral fins; the male and female guard the spawn, the male generally attacking intruders; the yolk sac is resorbed 5-6 days after spawning and the young start to feed; the shoal then swims out guarded by both parents (the brilliant caudal ocellus of the parents helps the young to orientate themselves to the parents); at 35 mm the young become autonomous; maturation size can be reached in less than one year (Lowe-McConnell, 1969)
Remarks: in Suriname, *Cichla ocellaris* is a very popular sport and food fish, both in Brokopondo Reservoir and in rivers of the Interior of the country.

CICHLASOMA BIMACULATUM (Linnaeus, 1758)
Local name(s): krobia, owruwefi (two-spot cichlid)

Diagnostic characteristics: a round-bodied robust fish; principal caudal-fin rays 16; 0-2 preopercular scales; at least 1½ scale between the posterior end of the anterior part of the lateral line and the dorsal fin; soft dorsal and anal fins extensively scaled; superior caudal base spot; suborbital stripe in young; caudal fin dotted; 4, rarely 3 anal fin spines; color very variable, becoming darker with age; ground color greenish-brown with a large black blotch about the middle of the side, and another black blotch at the base of the caudal fin

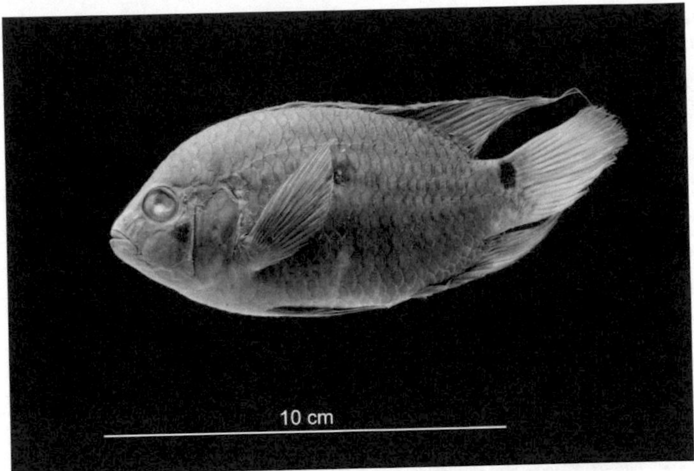

Cichlasoma bimaculatum (alcohol), RMNH37540 (© Naturalis)

Cichlasoma bimaculatum (live), Paramaribo market

Cichlasoma bimaculatum (live) (© INRA-Le Bail)

Type locality: M. Mediterraneo [erroneous, probably from Brazil or Suriname]
Distribution: South America: Brazil, French Guiana, Guyana, Suriname and Venezuela; a relatively common fish in Suriname that occurs in all rivers and in coastal freshwater swamps
Size: 12.3 cm SL
Habitat: common in coastal swamps and canals with standing water, but occurs also in forest creeks of the Interior
Position in the water column: benthopelagic
Diet: omnivorous feeding on shrimps, aquatic and terrestrial insects, vegetable debris, filamentous algae, seeds (Lowe-McConnell, 1969)
Reproduction: reproduction is year-round; males slightly larger than females; females lay up to 700 eggs; both sexes guard the eggs and young (Lowe-McConnell, 1969)
Remarks: *C. bimaculatum* is the 'krobia' cichlid that is most often sold at the markets of Paramaribo.

CLEITHRACARA MARONII (Steindachner, 1881)
Local name(s): krobia, owruwefi (keyhole cichlid)

Diagnostic characteristics: body deep and strongly compressed laterally, with a short caudal peduncle; head short and deep, compressed; snout short, rounded; principal caudal-fin rays 16; 0-2 preopercular scales; at least 1½ scale between the posterior end of the anterior part of the lateral line and the dorsal fin; soft dorsal and anal fins extensively scaled; no caudal spot, but a conspicuous black, round or oblong blotch below the end of the spinous dorsal fin, on, but mostly above the lateral line (this blotch delimited anteriorly and posteriorly by yellowish-light spots); strong bar from nape to interoperculum through eye; caudal fin immaculate; 3 anal fin spines
Type locality: Maroni (Marowijne) River, French Guiana
Distribution: Trinidad Island and northern South America: French Guiana, Guyana, Suriname, Trinidad and Tobago, and Venezuela; in Suriname, known from Corantijn, Nickerie, Coppename, Suriname, Commewijne and Marowijne rivers
Size: 7.1 cm SL
Habitat: mostly in small forest creeks in the lower freshwater reaches of the rivers, but also present upstream of the first rapids in the Upper Marowijne River (where it is less common)
Position in the water column: benthopelagic
Diet: small invertebrates (insect larvae, crustaceans)
Reproduction: approximately 400 eggs are spawned on a flat rock
Remarks: in Suriname, *C. maronii* is nowhere abundant.

Cleithracara maronii (alcohol), 4 cm SL, Commewijne River, RMNH19541 (© Naturalis)

Cleithracara maronii (live) (© INRA-Le Bail)

Crenicichla albopunctata Pellegrin, 1904
Local name(s): datra(fisi) (pike cichlid)

Diagnostic characteristics: body very elongated, with predorsal and preventral contour straight, and snout somewhat rounded in lateral view and from above; adpressed pelvic fins not reaching the anus; 52-72 scales in horizontal row below the lateral line; no series of dark blotches along lateral lines in males; humeral spot in females not or only slightly longer than deep, very ocellated; silvery dots present on operculum and cheek in both sexes; dorsal and anal fins grey with very fine white dots, at least posteriorly (dots smaller than those on operculum and sides of body)
Type locality: Maka Creek, 10 kilometers south of Stoelmanseiland, Lawa River, Marowijne River system, French Guiana
Distribution: Northern South America: French Guiana, Guyana and Suriname; in Suriname, only known from the Marowijne River
Size: 14 cm SL
Habitat: upstream of the first rapids, in the main channel and tributary streams, in reaches with moderate to slow current and sandy bottom
Position in the water column:
Diet:
Reproduction: females have the dorsal fin with dark margin, a red belly, and in some females two black, ocellated dots (sometimes more) are present on the dorsal fin over 11th and 12th spines
Remarks:

Crenicichla coppenamensis Ploeg, 1987
Local name(s): datra(fisi) (pike cichlid)

Diagnostic characteristics: very elongated body 52-72 scales in horizontal row below the lateral line; no series of dark blotches along lateral lines in males; humeral spot in females nor or only slightly longer than deep maxilla reaching well beyond orbit; silvery dots absent; ground color of body tan; humeral spot round with a white ring (less conspicuous in females); males with yellowish dots
Type locality: left bank tributary to Linker Coppename, Coppename River, Suriname
Distribution: Coppename and Saramacca River basins, Suriname
Size: 17.9 cm SL
Habitat: upper courses of Coppename and Saramacca rivers, upstream of the first rapids
Position in the water column:
Diet:
Reproduction: males with a silvery ring around the black humeral spot, a black ocellated spot on the base of the caudal fin (and often a small dark spot at the

Crenicichla albopunctata (live), male (© INRA-Le Bail)

Crenicichla coppenamensis (live), Kabo Creek, Coppename River (© R. Covain)

posterior end of the caudal fin) and yellow dots on the body; females lack a white ring around the caudal fin spot and yellow dots on the body
Remarks:

CRENICICHLA LUGUBRIS Heckel, 1840
Local name(s): datra(fisi) (pike cichlid)

Diagnostic characteristics: a small-scaled species with 112-130 scales in horizontal row below lateral lines; presence of a dark spot just posterior to pectoral fin base; nostrils nearer to postlabial skinfold than to orbit; ground color brown, counter-shaded without preorbital, postorbital or suborbital stripes; five bars between dorsal fin and lateral line, dorsal fin marginated, caudal fin with triangular black blotch above the middle of the base
Type locality: Rio Negro, Brazil
Distribution: Amazon River basin, Essequibo River basin and Corantijn River: Brazil, Guyana and Suriname; in Suriname, only known from the Corantijn River
Size: 24 cm SL
Habitat:
Position in the water column:
Diet:
Reproduction: no sexual dimorphism in color
Remarks:

CRENICICHLA MULTISPINOSA Pellegrin, 1903
Local name(s): datra(fisi) (pike cichlid)

Diagnostic characteristics: a small-scaled species with 100-115 scales in horizontal row below lateral lines; no dark spot posterior to pectoral fin; nostrils almost halfway between postlabial skinfold and orbit; presence of a black ocellus on the upper part of the caudal fin base, less conspicuously ocellated in females
Type locality: Cayenne, French Guiana
Distribution: Maroni (Marowijne) and Mana River basins of Suriname and French Guiana; in Suriname known from the Suriname and Marowijne rivers; also present in Brokopondo Reservoir
Size: 22.5 cm SL
Habitat: rheophilous, in counter-current zones near rapids, but in Brokopondo Reservoir it occurs in standing water
Position in the water column: benthopelagic
Diet: fish
Reproduction: males are beautifully colored with many silvery dots on the posterior part of the body, the dorsal and anal fins whitish-grey with dark margin and scattered with whitish dots, and the caudal fin with many silvery dots
Remarks:

Crenicichla lugubris (live) (© R. Covain)

Crenicichla multispinosa (live), scale bar = 1 cm, Lawa River (Marowijne River), ANSP187101 (© M. Sabaj Pérez)

CRENICICHLA NICKERIENSIS Ploeg, 1987
Local name(s): datra(fisi) (pike cichlid)

Diagnostic characteristics: 52-72 scales in horizontal row below the lateral line; series of dark blotches along lateral lines in males; humeral spot longer than deep in females
Type locality: right bank tributary to Nickerie River, 12 km W.S.W. of Stondansi Fall, Suriname
Distribution: Nickerie and Corantijn River basins, Suriname
Size: 19.1 cm SL
Habitat: both up- and downstream of the first rapids in Nickerie and Corantijn rivers
Position in the water column: benthopelagic
Diet:
Reproduction:
Remarks:

CRENICICHLA SAXATILIS (Linnaeus, 1758)
Local name(s): datra(fisi) (pike cichlid)

Diagnostic characteristics: 52-72 scales in horizontal row below the lateral line; no series of dark blotches along lateral lines in males; humeral spot well marked with an dorso-anterior notch, sometimes ocellated (in breeding fish), and in females not or only slightly longer than deep; supracaudal spot poorly visible in females; maxilla reaching well beyond orbit; silvery dots absent; ground color of body tan; males frequently with silvery-white dots; juveniles with a dark lateral band
Type locality: Carolina Creek, Para River, Suriname River System, Suriname
Distribution: Atlantic coast drainages of northern South America: French Guiana, Guyana, Suriname, Trinidad and Tobago, and Venezuela; in Suriname, present in Nickerie, Saramacca, Suriname, Commewijne and Marowijne rivers
Size: 20 cm SL
Habitat: present in coastal freshwater swamps and canals, in black-water streams of the Savanna Belt and in forest creeks in the Interior upstream of the first rapids
Position in the water column: benthopelagic
Diet: fish, shrimps and aquatic insects
Reproduction: in breeding pairs the male is larger than the female; in aquarium, eggs are deposited in a shallow pit; parents are reputed to move the eggs and young up and down with the water level (Lowe-McConnell, 1969)
Remarks: *C. saxatilis* were found in stomach of *Hoplias malabaricus* and *Cichla ocellaris* (Lowe-McConnell, 1969).

Crenicichla nickeriensis (live), Nickerie River (© R. Smith)

Crenicichla saxatilis (live) (© INRA-Le Bail)

Crenicichla saxatilis (live), male (©W. Kolvoort)

Crenicichla sipaliwini Ploeg, 1987
Local name(s): datra(fisi) (pike cichlid)

Diagnostic characteristics: snout pointed in lateral view, rounded from above; 52-72 scales in horizontal row below the lateral line; no series of dark blotches along lateral lines in males; humeral spot in females not or only slightly longer than deep; silvery dots absent; maxilla not reaching orbit or anterior margin of orbit; ground color of body grey
Type locality: Sipaliwini River near Sipaliwini airstrip, Corantijn River system, Suriname
Distribution: Sipaliwini River basin, Upper Corantijn River basin, Suriname
Size: 17.3 cm SL
Habitat:
Position in the water column: benthopelagic
Diet:
Reproduction: no conspicuous sexual dimorphism
Remarks:

Geophagus brachybranchus Kullander & Nijssen, 1989
Local name(s): songe, agankoi (Lijding, 1959)

Diagnostic characteristics: first branchial arch with lobe-like expansion on upper portion; eye diameter more than one time in snout length; caudal spot absent or situated on base of dorsal caudal fin lobe; posterior part of the anterior portion of the lateral line separated by at least 1½ scale from the dorsal fin; anterior half of cheek scaled; base of anal fin scaled in adults; absence of spot on dorsal part of caudal-fin base; presence of midlateral spot or intense stripe across middle of side; absence of a black dorsal spot at the base of the posterior portion of the dorsal fin; presence of a black midlateral spot on the flanks; predorsal midline covered by scales; lateral spot on and below 11th-15th lateral line scales; stripe over preopercular corner from 5 cm SL; Corantijn and Nickerie rivers
Type locality: Nickerie River, just above Blanch Marie Falls, Suriname
Distribution: Corantijn and Nickerie River drainages, and probably in Essequibo River basin: Guyana and Suriname
Size: 13.8 cm SL
Habitat: streams of the Interior
Position in the water column: bottom
Diet:
Reproduction:
Remarks: a very colorful species.

Crenicichla sipaliwini (live) (© R. Covain)

Geophagus brachybranchus (live), Upper Nickerie River (© R. Smith)

GEOPHAGUS HARRERI Gosse, 1976
Local name(s): songe, agankoi (Lijding, 1959)

Diagnostic characteristics: first branchial arch with lobe-like expansion on upper portion; eye diameter more than one time in snout length; caudal spot absent or situated on base of dorsal caudal fin lobe; posterior part of the anterior portion of the lateral line separated by at least 1½ scale from the dorsal fin; anterior half of cheek scaled; base of anal fin scaled in adults; absence of spot on dorsal part of caudal-fin base; presence of a black dorsal spot at the base of the posterior portion of the dorsal fin; absence of a black midlateral spot below the lateral line; predorsal midline partially naked
Type locality: Ouaqui River at Saut Bali, tributary of Tampok River, Marowijne (Maroni) River system, French Guiana and Suriname
Distribution: endemic to the Marowijne River basin: French Guiana and Suriname
Size: 18.3 cm SL
Habitat: in shallow water with moderate current close to the rapids (Interior)
Position in the water column: bottom
Diet:
Reproduction: males are territorial
Remarks:

GEOPHAGUS SURINAMENSIS (Bloch, 1791)
Local name(s): songe, agankoi (Lijding, 1959)

Diagnostic characteristics: first branchial arch with lobe-like expansion on upper portion; eye diameter more than one time in snout length; caudal spot absent; posterior part of the anterior portion of the lateral line separated by at least 1½ scale from the dorsal fin; anterior half of cheek scaled; base of anal fin scaled in adults; absence of spot on dorsal part of caudal-fin base; absence of a black dorsal spot at the base of the posterior portion of the dorsal fin; presence of a black midlateral spot on the flanks; predorsal midline covered by scales; large spot on and below 11th-13th lateral line scales; no dark maskings on head in adults; one of the most colorful cichlids of Suriname with its characteristic black midlateral spot set against a background of orange-yellow stripes along the body and with peacock-blue stripes on a deep red background on pelvics, caudal, hind dorsal and anal fins
Type locality: Suriname
Distribution: French Guiana and Suriname; in Suriname known to occur in the Corantijn, Coppename, Saramacca, Suriname and Marowijne rivers
Size: 14.8 cm SL
Habitat: rivers of the Interior, upstream of the first rapids
Position in the water column: bottom

Geophagus harreri (live), scale bar = 1 cm, Lawa River (Marowijne River), ANSP187136 (© M. Sabaj Pérez)

Geophagus surinamensis (live) (© INRA-Le Bail)

Diet: omnivorous with preference for plant material, grazing the bottom substrate with its protractible mouth
Reproduction: adult males have blue spots on operculum and cheeks and long reddish filaments on the posterior end of the dorsal, caudal and pelvic fins; the female deposits about 250 eggs on a flat rock surface or in a pit; both parents regularly take the eggs into the mouth until they hatch after three days; parents guard their offspring during several weeks and in this time the larvae return to the mouth cavity of their parents during the night or in moments of danger (Keith *et al.*, 2000)
Remarks: *Geophagus brokopondo* Kullander & Nijssen, 1989 is only known from its type locality Brokopondo Reservoir (dam closed in 1964); Mol *et al.* (2012) point out the possibility that *G. brokopondo* is an ecophenotype of *G. surinamensis*, i.e. a stunted population of *G. surinamensis* (and thus *G. brokopondo* would be a junior synonym of *G. surinamensis*). Stunting was observed in several species of Brokopondo Reservoir (Mol *et al.*, 2007a) and in the cichlids *Krobia guianensis* and *Crenicichla saxatilis* in Petit-Saut Reservoir, French Guiana (Ponton & Mérigoux, 2000). Molecular analysis can resolve the question whether *G. brokopondo* is an ecophenotype of *G. surinamensis* or alternatively a valid species. In the Marowijne River, *G. surinamensis* occurs syntopically with *G. harreri*. *Geophagus surinamensis* can produce a grating noise with its pharyngeal teeth.

Guianacara oelemariensis Kullander & Nijssen, 1989
Local name(s): –

Diagnostic characteristics: soft dorsal and anal fins with a few basal scales or naked; absence of a black spot on the dorsal part of the caudal fin base; presence of a large midlateral spot (not a stripe); membrane between spines of anterior portion of dorsal fin lacking an extension beyond the spines ('fin lappets'); ground color yellowish white, countershaded, dark brown on the back, lighter laterally; flanks with light scale margins, giving a mild patchy appearance, no vertical or horizontal bands, a dark brown stripe of pupil width from eye down across preopercular corner to junction of sub- and interopercula, slightly curved, dorsal fin uniformly brown except for three anterior blackish membranes and posteriormost soft part membranes with colorless dots (these colorless dots are also present on the posteriormost part of the anal fin), caudal fin beyond scaled base clear with about 12 not quite regular vertical series of brownish dots as wide as interspaces
Type locality: small right-bank tributary of Oelemari River, Marowijne River system, Suriname
Distribution: a rare species, endemic to the Marowijne River drainage, Suriname, and only known from its type locality Oelemari River
Size: 8.1 cm SL
Habitat: apparently more or less equal to the habitat of the other *Guianacara* species

Guianacara oelemariensis (© INRA-Le Bail)

Position in the water column: benthopelagic
Diet:
Reproduction:
Remarks: Kullander and Nijssen (1989) note that *Guianacara* bears a strong superficial resemblance to *Geophagus*-like cichlids in the deep preorbital region, the comparatively small scales, emarginate caudal fin and color pattern, but differ from *Geophagus* in lacking an epibranchial lobe.

GUIANACARA OWROEWEFI Kullander & Nijssen, 1989
Local name(s): krobia, owruwefi (Lijding, 1959, for *Aequidens geayi*)

Diagnostic characteristics: soft dorsal and anal fins with a few basal scales or naked; absence of a black spot on the dorsal part of the caudal fin base; stripe across side, except in some large specimens in which the stripe is reduced to a dorsal spot; anterior portion of dorsal fin with produced (elongated) fin lappets; flank stripe strongest on and below lateral line; dorsal fin membranes between the anterior dorsal fin spines black, scales with silvery or pale center
Type locality: Marouini River, below first rapids, Marowijne (Maroni) River system, French Guiana
Distribution: French Guiana and Suriname; *G. owroewefi* appears to be fairly widespread in Suriname, as it is present in the Coppename, Saramacca, Suriname and Marowijne rivers
Size: 10.7 cm SL
Habitat: main channel and large tributaries of the rivers in the Interior (upstream of the first rapids), in shallow, sun-lit water with slow current
Position in the water column:
Diet: feeds on small invertebrates
Reproduction: a pair can reproduce 3-4 times in a year; the spawn (about 300 eggs) is deposited in rock crevices and guarded by the female; the male defends the territory; Keith *et al.* (2000) note that parental care extends to a period of 1-3 months
Remarks: coloration effectively distinguishes *G. owroewefi* from the other two *Guianacara* species of Suriname, especially the vertical black stripe, black membranes of the anteriormost dorsal-fin rays and scales with pale center; in Oelemari River, *G. owroewefi* occurs syntopically with *G. oelemariensis*.

GUIANACARA SPHENOZONA Kullander & Nijssen, 1989
Local name(s): –

Diagnostic characteristics: soft dorsal and anal fins with a few basal scales or naked; absence of a black spot on the dorsal part of the caudal fin base; stripe across side wedge shaped, strongest on and above upper lateral line, except in some large specimens in which the stripe is reduced to a dorsal spot; anterior portion of dorsal fin with produced (elongated) fin lappets; anterior 3 dorsal fin lappets black in young only

Guianacara owroewefi (live) (© P. Willink)

Guianacara sphenozona (live), Sipaliwini River (Corantijn River) (© R. Covain)

Guianacara sphenozona (live), Sipaliwini River (Corantijn River) (© P. Willink)

Type locality: Sipaliwini, Upper Corantijn River drainage, Nickerie District, Suriname
Distribution: Guyana and Suriname; in Suriname, only present in the upper and middle Corantijn River
Size: 8.5 cm SL
Habitat:
Position in the water column:
Diet:
Reproduction:
Remarks: nothing is known about its biology.

Krobia guianensis (Regan, 1905)
Local name(s): krobia, owruwefi (Lijding, 1959; for *Aequidens vittatus*)

Diagnostic characteristics: moderately elongate, moderately compressed laterally, rounded nape, rounded chest; presence of a black spot on the dorsal part of the caudal fin base; black band extending from the base of the eye to the preopercle; presence of a dark midlateral longitudinal band; predorsal squamation with single range of median scales only; posterior end of longitudinal band oriented towards the posterior end of the dorsal-fin base; body depth 41.3-50.8% of standard length; lateral band continued to end of soft dorsal-fin base; two posterior out of three facial stripes indistinct; ground color pale dirty yellowish, cheek light brownish, nape and back brown, sides lighter with light brown vertical bars and feebly lighter interspaces, no evidence of snout stripes as in *K. itanyi*
Type locality: Guyana
Distribution: northern South America: Guyana and Suriname; widely distributed in Suriname, present in the Corantijn, Nickerie, Coppename, Saramacca, Suriname, Commewijne and Lower Marowijne rivers
Size: 12.8 cm SL
Habitat: present in coastal freshwater swamps, black-water streams of the Savanna Belt and small, shallow forest creeks of the Interior (upstream of the first rapids)
Position in the water column: benthopelagic
Diet: omnivorous, but especially eating small crustaceans and insect larvae
Reproduction: in a field study Keenleyside and Bietz (1981) describe its reproductive behavior; the male is slightly larger and more colorful than the female; eggs are deposited on a broad leaf and this leaf with the eggs attached to its surface is displaced (taken into the mouth of the parent) if necessary
Remarks: easily distinguished from *K. itanyi* by having the lateral blotch series extended to the end of the dorsal fin base rather than terminating at the end of the lateral line or shorter, the usually obsolete facial stripes, and more extensively dotted fins; in Marowijne River, *K. guianensis* is replaced by *K. itanyi* above the first rapids; in older literature (e.g. Keenleyside & Bietz, 1981), *K. guianensis* is often identified as *Aequidens vittatus*, a very different species from Paraguay and

Krobia guianensis (live) (© INRA-Le Bail)

Lower Parana drainages, now in the genus *Bujurquina* (see Kullander & Nijssen, 1989; also Lowe-McConnell, 1969, p. 275).

Krobia itanyi (Puyo, 1943)
Local name(s): –

Diagnostic characteristics: presence of a black spot on the dorsal part of the caudal fin base; black band extending from the base of the eye to the preopercle; presence of a dark midlateral longitudinal band; predorsal squamation with single range of median scales only; posterior end of longitudinal band oriented towards the posterior end of the dorsal-fin base; body depth 40.0-45.3% of standard length; lateral band (blotches) ending well in advance of end of soft dorsal-fin base; three facial stripes clearly visible; Marowijne River
Type locality: tributary on left bank of Marowijne River near Manbari Falls, 6 km N of Stoelmanseiland, Suriname
Distribution: endemic to the Marowijne and Mana River drainages: French Guiana and Suriname
Size: 12.5 cm SL
Habitat: small forest creeks of the Interior with slow to moderate current and sandy or rocky bottom
Position in the water column: benthopelagic
Diet: microcrustaceans and insect (larvae)
Reproduction: about 500 eggs are spawned on a flat stone surface; the parents are very territorial during the period that they guard the eggs and newborn
Remarks:

Mesonauta guyanae Schindler, 1998
Local name(s): – (flag cichlid)

Diagnostic characteristics: insertion of pelvic fins anterior to insertion of pectoral fins; presence of a conspicuous oblique black band extending from the tip of the snout to the posterior end of the dorsal fin and presence of a prominent black ocellus on a clear yellow background on the upper half of the caudal peduncle; body color above the diagonal black stripe is yellow-brown, each scale having a darker center; vertical fins are grey spotted with yellow, pectoral fins immaculate, the pelvics with an elongated lemon yellow outer soft ray; this species differs from all other *Mesonauta* species by a characteristic pattern of vertical bars (bar 6 and 7 separated, ventral parts narrow; bar 5 and 6 separated; bar 5 divided).
Type locality: Rockstone, Essequibo River, Guyana
Distribution: Amazon River and Essequibo River basins: Brazil and Guyana; in Suriname, only known from Coropina Creek, Para River, Suriname River System
Size: 10 cm SL

Krobia itanyi (live) (© INRA-Le Bail)

Mesonauta guyanae (live), Para River (Suriname River), FMNHexANSP189590 (© M. Sabaj Pérez)

Habitat: in Suriname, black-water stream in the Savanna Belt; in Guyana it is a common species widely distributed from coastal trenches in Georgetown to Rupununi Savanna pools in the Interior
Position in the water column: benthopelagic
Diet: in Guyana the stomach contents were mainly large algae (*Closterium* spp), zooplankton, bottom debris, and pieces of plant (Lowe-McConnell, 1969)
Reproduction: the breeding male has bright blue on the lower lip and on top of the eyes; females are smaller than males (large males tend to have an elongated snout); in Guyana, *M. guyanae* start spawning before the main rains when the water is low (Lowe-McConnell, 1969) and continue to have broods throughout the rainy season; the pair male-female guards a spawning site (e.g. a stem of an emergent macrophyte) where a mass of about 100-200 small, translucent eggs are fixed; the eggs are fanned by both parents and intruders are chased away; the guarding parents kept within 1 m of the eggs; the eggs hatch after two or three days and at first the 3 mm long larvae remain concentrated in a wriggling mass attached to the stem (about 10 cm below the water surface) by cement glands on their heads. Later both parents were observed to guard a cloud of young (6 mm, black with a conspicuous iridescent eyespot and air bubble in the swim bladder), which kept above them in the water; the young generally kept in a ball about 30 cm in diameter and followed one of the parents, which preceded them with jolting movements of the elongated light-colored, yellow, pelvic fins. The male and female parents took turns at leading the young and chasing away intruding fish. The conspicuous ocellus on the base of the caudal peduncle may play an important part in orientating the young to the parent
Remarks: probably a species introduced to Suriname by Surinamese aquarium hobbyists; it is only known from Coropina Creek, near the international airport of Suriname; *M. guyanae* is not mentioned for Suriname by Kullander and Nijssen (1989); however, this species is present in Guyana (Lowe-McConnell, 1969) and French Guiana (Oyapock River; Keith *et al.*, 2000).

NANNACARA ANOMALA Regan, 1905
Local name(s): – (checkerboard or golden dwarf cichlid)

Diagnostic characteristics: a small stout-bodied (depth 38% of SL) cichlid distinguished from *Apistogramma* dwarf cichlids by the absence of lobe-like expansion on upper portion of first branchial arch; principal caudal fin rays 14; 3 preopercular scales; ½ scale between the posterior end of the anterior part of the lateral line and the dorsal fin; the female is colored light reddish brown to beige with a dark brown horizontal band from the orbit onto caudal fin base and five vertical bars over sides; female is colored similarly to the normal male coloration, except in times of breeding, when she is marked with a coarse checkerboard pattern; male coloration is highly variable, normally with four longitudinal alternating dark and light stripes, the abdomen is nearly white, overlain by a dark stripe along the lateral line, and with another white stripe and a slightly less dark stripe along the

Nannacara anomala (live) (© INRA-Le Bail)

dorsal region; when excited or in nuptial coloration, the adult males are almost entirely translucent metallic blue-green, with a reddish distal margin in the dorsal fin, and the base of the anal and caudal fins with a brown-orange color
Type locality: Essequibo River, Guyana
Distribution: Guianan rivers: Guyana, French Guiana and Suriname; occurs in all rivers of Suriname, mainly in coastal swamps and small tributary creeks in the lower freshwater reaches
Size: 5.6 cm SL
Habitat: often found in leaf litter; common in the coastal freshwater swamps, black-water streams of the Savanna Belt, and tributary forest creeks in the lower reaches of the rivers; apparently rare in upland reaches (only one specimen of Paloemeu River mentioned by Kullander & Nijssen, 1989)
Position in the water column: benthopelagic
Diet: a micropredator, foraging for worms, insects and other invertebrates
Reproduction: male is much bigger than female, generally exhibiting pale blues and reds whilst the smaller female tends to be yellow with a black lateral line on its flank; female will present a checkerboard-esque pattern when in breeding mood; in the aquarium 50-300 eggs are deposited on a substrate previously cleaned by either or both parents, and guarded by the female; the young hatch after two or three days and are gathered by the female in a pit where they remain for about five days on the bottom
Remarks: this peaceful dwarf cichlid is a very popular species in the aquarium hobby.

Oreochromis mossambicus (Peters, 1852)
Local name(s): tilapia (Mozambique tilapia)

Diagnostic characteristics: lips of 'African' type, i.e. posterior portion of lower lip fully exposed; teeth fine, closely set; posterior dorsal fin spines of approximately equal length; males grow slightly larger than females; females and non-breeding males are mainly silver in color with 2-5 blotches along the midline and occasionally the dorsal fin; breeding males are black with white; juveniles with black spot on soft dorsal fin; introduced, restricted to more or less brackish water in the Coastal Plain (e.g. mangrove lagoons, canals in Paramaribo and Nieuw Nickerie)
Type locality: Zambezi River, Mozambique
Distribution: Southern Africa; introduced widely elsewhere
Size: 38 cm SL (in their native range; in Suriname they reach about 22 cm SL)
Habitat: in their native range along the eastern coast of Africa, *Oreochromis mossambicus* occurs in riverine and coastal lagoon habitats; *Oreochromis mossambicus* have a broad salinity tolerance (Trewavas 1983); they can survive from freshwater up to 40 ppt, and are capable of spawning in estuarine waters at salinities as high as 34.5 ppt. In Suriname, *O. mossambicus* is not found in the river proper, but in brackish-water lagoons and drainage canals
Position in the water column:

Oreochromis mossambicus (live), female, Paramaribo market

Oreochromis mossambicus (live), male, Paramaribo market

Diet: *Oreochromis mossambicus* are generalist/opportunistic omnivores that consume detrital material, vegetation ranging from diatoms to macroalgae to rooted plants, invertebrates, and small fish (Bowen, 1981; Trewavas 1983); in Suriname, they feed mainly on filamentous algae and detritus (Mol & Van der Lugt, 1996)
Reproduction: female *Oreochromis mossambicus* mature at approximately 150-160 mm, and males mature at approximately 170-180 mm; males construct nests in sparse to moderately vegetated bottoms where fertilization of the eggs takes place (Bruton & Boltt, 1975); several different females will lay eggs in the nest; females can lay between 50-1,780 eggs, based on individuals' size and environmental conditions (Trewavas, 1983); males are generally aggressive and ritualistic during reproductive season, although male-male confrontations rarely actually become violent (Bruton & Boltt, 1975); once fertilized, the female *Oreochromis mossambicus* takes the eggs into her buccal cavity (mouth) and broods them until hatching; hatching occurs in approximately 3-5 days; once hatched, the females continue to mouth-brood the fry until they are approximately 14-21 days old; male *O. mossambicus* are reported to occasionally mouth-brood eggs and fry as well (Bruton & Boltt, 1975).
Remarks: the Mozambique tilapia, *O. mossambicus*, is native to Africa (Mozambique) but has been introduced to Florida and elsewhere as well; in Suriname, it was introduced in 1954 by the Fisheries Department (Lijding, 1958).

Pterophyllum scalare (Schultze, 1823)
Local name(s): maanvis (freshwater angelfish)

Diagnostic characteristics: an unusually shaped cichlid with greatly laterally compressed, round body and elongated triangular dorsal and anal fins; the height of the dorsal fin is larger than the depth of the body; coloration with vertical black stripes
Type locality: Lower Amazon River, Brazil
Distribution: Brazil, French Guiana, Guyana, Peru and Suriname; in Suriname, probably introduced by aquarium hobbyists in the Para River, Suriname River system (now also present in Blaka-watra Creek, right-bank tributary of Suriname River and in Mindrineti River, a tributary of Saramacca River), but Suriname is within its native distribution range because a not-introduced population of *P. scalare* occurs in northwestern Suriname, in the Kaboeri Creek (Lower Corantijn River)
Size: 7.5 cm SL
Habitat: black-water creeks
Position in the water column:
Diet: angelfish are ambush predators that prey on small fish and macroinvertebrates.
Reproduction: *Pterophyllum scalare* is relatively easy to breed in the aquarium, although it is very difficult to accurately identify the gender of any individual until they are nearly ready to breed; angelfish pairs form long-term relationships where

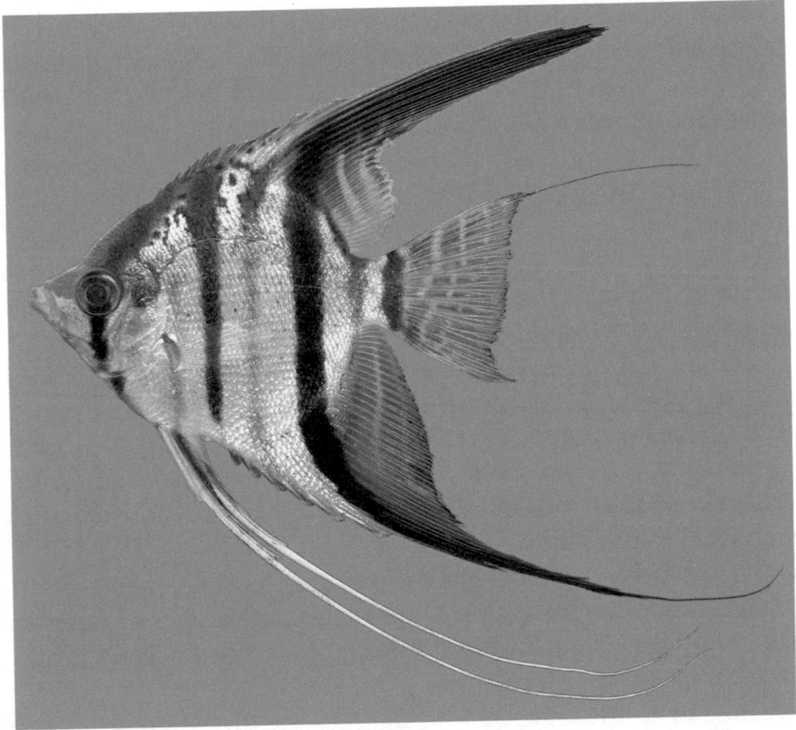

Pterophyllum scalare (live), Para River (Suriname River), ANSP189594 (© M. Sabaj Pérez)

each individual will protect the other from threats and potential suitors; upon the death or removal of one of the mated pair, breeders have experienced both the total refusal of the remaining mate to pair up with any other angelfish and successful breeding with subsequent mates; depending upon aquarium conditions, *P. scalare* reaches sexual maturity at the age of six to twelve months or more; eggs are generally laid on a submerged log or a flattened leaf; brood care is highly developed.

Remarks: a well-known and prized aquarium fish that was probably introduced in Suriname by aquarium hobbyists in Para River, Suriname River system (now also present in Blaka-watra Creek, Suriname River, and Mindrineti River, Saramacca River system); however, *P. scalare* is probably native in Suriname in the Kaboeri Creek, a black-water tributary of the Lower Corantijn River; the body shape allows angelfish to hide among roots and plants, often on a vertical surface and their vertically striped coloration provides additional camouflage; the angelfish is not mentioned for Suriname by Kullander and Nijssen (1989).

Satanoperca leucosticta (Müller & Troschel, 1849)
Local name(s): –

Diagnostic characteristics: first branchial arch with lobe-like expansion on upper portion; eye diameter more than one time in snout length; posterior part of the anterior portion of the lateral line separated by at least 1½ scale from the dorsal fin; cheek scaled rostrad to lacrimal; base of anal fin without scales; presence of spot on the dorsal part of the caudal-fin base; no midlateral spot; general body color pearly grey, sometimes with iridescent scales on the back; vertical fins immaculate or spotted with translucent spots at base, pelvics with elongated outer ray cream colored; Nickerie River

Type locality: probably Rupununi River, Guyana

Distribution: Essequibo River and Nickerie River: Guyana and Suriname; in Suriname, only known from the Nickerie River (Kullander & Nijssen, 1989)

Size: 15 cm SL

Habitat: close or in the first rapids in the Nickerie River (Stondansi Falls); in Guyana they tend to live in pools with soft bottom substrate (Lowe-McConnell, 1969)

Position in the water column: bottom

Diet: in Guyana, they feed on the bottom of muddy pools

Reproduction: in Guyana, *Satanoperca leucosticta* breeds mainly in the rainy season (Lowe-McConnell, 1969); in the aquarium both parents were observed to brood the young in the mouth for up to 37 days (Reid & Atz, 1958); Eigenmann (1912) collected a specimen with young in her mouth. This mouth-brooding habit is unusual among South American cichlids and certain features of mouth-brooding in *S. leucosticta* appear unusual, for instance Reid & Atz (1958) state that *S. leucosticta* waited for practically 24 hours before picking up and incubating the

Satanoperca leucosticta (live) (© W. Kolvoort)

Satanoperca leucosticta (live) (© R. Covain)

spawn (instead of picking them up at once as do African mouth-brooding cichlids). Lowe-McConnell (1969) observed that both sexes cleaned the surface on which the eggs were deposited in batches and fertilized in sequence by the male; the eggs were guarded by both parents for about a day, then picked up and incubated orally by both parents or the female alone. One- or two-day-old young were released into depressions dug into the bottom, and were transferred back and forth between male and female. Free-swimming young returned to the mouth of one or other parent (or were gathered by the parents) when disturbed or at night for as long as 37 days. Unlike other mouth-brooding cichlids, *S. leucosticta* fed while brooding, possibly supplying food to the young in the mouth; Reid & Atz (1958) pointed out that *S. leucosticta* young have relatively little yolk compared with young of the African mouth-brooders.

Remarks: in the related species *Satanoperca* aff. *jurupari* from French Guiana, Loir *et al.* (1989) found ambisexuality with oocytes present in the testes of functional males; it was not determined whether sex inversion occurs, but the fish had not been sampled throughout the year, so that the period of sex inversion could have been missed.

This page is intentionally left blank.

FAMILY ELEOTRIDAE (SLEEPERS)

Eleotridae are small to medium-sized (most do not exceed 20 cm) fishes with a stout body, short, broad head and a blunt snout. Teeth are usually small, conical and in several rows in the jaws. There are two separate dorsal fins, the first with 6 or 7 weak spines, the second with 1 weak spine followed by 6-12 soft rays. The scales are large and there is no lateral line on the body. They are not brightly colored; most are light or dark brown or olive with some metallic glints. The Eleotridae differ from the Gobiidae in the much shorter base of the second dorsal fin (about equal in size to the distance between the posterior end of the second dorsal fin base and the caudal fin base) and in their pelvic fins which are not united to form a ventral disc. Eleotridae are omnivorous bottom-dwelling fishes that typically occur in fresh or brackish waters. Many are relatively inactive, hence the common name of sleeper. The taxonomy of species of *Dormitator* is unresolved and no key to species is available for this genus.

1-a Prominent, ventrally pointed spine on preopercle present (this spine may be difficult to see as it is often covered by skin); 6 spines in the anterior dorsal fin; 40-64 scales in longitudinal scale row (*Eleotris*)
    2

1-b Preopercular spine absent; 7 spines in the anterior dorsal fin; either 25-36 (*Dormitator*) or 90-110 (*Guavina*) scales in longitudinal scale row
    3

2-a 41-46 scales in longitudinal row in Guianas; dark spot present on upper pectoral-fin base (may be covered by opercular membrane), darker than markings on nape at opercular margin (Pezold & Cage, 2002)
    Eleotris amblyopsis

2-b 50-54 scales in longitudinal row in Guianas; no dark spot present on upper pectoral-fin base or if present not strongly-contrasted and not as dark as pigment on nape (Pezold & Cage, 2002)
    Eleotris pisonis

3-a Scales very small, about 110 in longitudinal scale rows
    Guavina guavina

3-b Scales large, about 25 to 35 in longitudinal scale rows
    Dormitator maculatus

This page is intentionally left blank.

Dormitator maculatus (Bloch, 1792)
Local name(s): – (fat sleeper)

Diagnostic characteristics: preopercular spine absent; 7 spines in the anterior dorsal fin; caudal fin rounded; about 25 to 35 relatively large scales in longitudinal scale rows; eye diameter equal to snout length; mouth large and oblique; color dark brown to black above, yellowish below, dorsal, anal and caudal fins hyaline with brownish spots at the base
Type locality: no locality
Distribution: Western Atlantic and Atlantic Slope of U.S.A. to southeastern Brazil
Size: 38 cm TL
Habitat: freshwater, brackish, marine; common in estuaries, canals and mangrove swamps with widely varying salinity (0-21‰); also present in lower freshwater reaches of rivers, but never upstream of the first rapids
Position in the water column: among roots of floating macrophytes (Nordlie, 1981)
Diet: omnivorous, plants, sediments, invertebrates; ontogenetic diet shifts observed (Nordlie, 1981; Teixeira, 1994)
Reproduction: color changes are observed in the reproductive season; the parents guard a nest; eggs (0.3 mm diameter) hatch in 11-16 hours at 27 °C; size-at-maturity was 51 and 46 mm TL for males and females of *D. maculatus*, respectively (Teixeira, 1994)
Remarks: *Dormitator lophocephalus* Hoedeman, 1951 is a small species (to 9 cm) described from Suriname (type locality Paramaribo) that is not well known.

Eleotris amblyopsis (Cope, 1871)
Local name(s): – (Large-scaled spinycheek sleeper)

Diagnostic characteristics: prominent, ventrally pointed spine on preopercle present (this spine may be difficult to see as it is often covered by skin); 6 spines in the anterior dorsal fin; 41-46 scales in longitudinal row in Guianas; large dark spot present on upper pectoral-fin base (may be covered by opercular membrane), darker than markings on nape at opercular; body dark brown laterally, lighter along dorsum or tan with rows of dark brown spots on sides, abdomen and gular region lighter, cheek with two dark streaks radiating posteriorly from eye margin; first dorsal fin with dark band at base and another reticulated band through middle of fin, dark spots generally present on spines above mid-level band, but not at tips, second dorsal fin with 5-6 wavy diagonal bands, sometimes over dusky membrane; caudal fin with 10-14 vertical bars formed by small spots on rays; pectoral and pelvic fins peppered with small spots on rays, membrane may be clear or dusky (Pezold & Cage, 2002)
Type locality: Suriname
Distribution: Western Atlantic; Belize, Costa Rica, Guyana, French Guiana, Panama, Surinam, Trinidad and Tobago, USA, Venezuela
Size: 8 cm SL

Dormitator maculatus (live) (© INRA-Le Bail)

Habitat: mainly in heavily vegetated areas (e.g. floating mats of water hyacinth) in estuaries and brackish lagoons; brackish water, with a few large individuals entering freshwater, upstream of the estuaries
Position in the water column: among roots of floating vegetation
Diet: insect larvae, shrimp post-larvae, but also includes plant material and small fishes (Nordlie, 1981)
Reproduction: breeding probably takes place mainly in lower, more saline parts of the estuaries; larvae pelagic (Nordlie, 1981)
Remarks:

ELEOTRIS PISONIS (Gmelin, 1789)
Local name(s): – (Small-scaled spinycheek sleeper)

Diagnostic characteristics: body elongated, head large and depressed, mouth superior; prominent, ventrally pointed spine on preopercle present (this spine may be difficult to see as it is often covered by skin); 6 spines in the anterior dorsal fin; pelvic fins separated, but positioned very close to each other; 50-54 scales in longitudinal row in Guianas; no dark spot present on upper pectoral-fin base or if present not strongly-contrasted and not as dark as pigment on nape; body tan to dark brown, often with a few scattered brown spots on upper flanks, not forming regular rows or thin stripes, dorsum occasionally lighter than flanks, abdomen and gular region lighter; two streaks radiating across cheek posteriorly from eye, expanding into patch of dark brown pigment on nape along edge of operculum, more pronounced than pigment on upper base of pectoral fin (Pezold & Cage, 2002)
Type locality: South America
Distribution: widespread in Western Atlantic: southeastern U.S.A. to Brazil; Bahamas, Bermuda, Brazil, Dominican Republic, Colombia, Costa Rica, Ecuador, French Guiana, Guyana, Guatemala, Martinique (France), Mexico, Nicaragua, Panama, Puerto Rico, Surinam, Trinidad and Tobago, USA
Size: 11.3 cm SL
Habitat: fresh and brackish water of estuaries (salinity 0-19‰), often sheltering in marginal vegetation (which is also the main feeding area for this species; Nordlie, 1981; Teixeira, 1994)
Position in the water column: both among roots of floating vegetation and benthic in shallow mangrove streams (Nordlie, 1981)
Diet: insect larvae and pupae, crustaceans, snails, polychaetes, and small fishes (Nordlie, 1981)
Reproduction: size at maturity 57 mm TL for males and 43 mm TL for females; reproduction probably linked with rainy season; breeding probably takes place mainly in lower, more saline parts of the estuaries; pelagic larvae (Nordlie, 1981)
Remarks:

Eleotris pisonis (live) (© INRA-Le Bail)

Guavina guavina (Valenciennes 1837)
Local name(s): –

Diagnostic characteristics: preopercular spine absent; 7 spines in the anterior dorsal fin; scales very small, about 110 in longitudinal scale rows
Type locality: Martinique Island, West Indies; Suriname; Cuba
Distribution: Western Atlantic
Size: 23 cm TL
Habitat: freshwater, brackish, marine; apparently not very abundant
Position in the water column:
Diet: invertebrates, mainly crabs (Teixeira, 1994) but also fish (Winemiller & Ponwith, 1998)
Reproduction:
Remarks:

This page is intentionally left blank.

FAMILY GOBIIDAE (GOBIES)

Following the analytical revision of Hoese & Gill (1993), the gobies and related fishes comprising the suborder Gobioidei are distributed over eight families, viz., the Gobiidae with subfamilies Butidinae, Eleotridinae (sleepers), and Gobiinae, the Kraemeriidae, the Microdesmidae with subfamilies Microdesminae and Ptereleotridinae, the Odontobutididae, the Xenisthmidae, and the Rhyacichthyidae. Other authors (e.g. Nelson, 2006) recognize the Eleotridae as a separate family, including the Butinae, and distinguish other subfamilies of the Gobiidae, viz. Oxudercinae (mudskippers), Amplyopinae, Sicydiinae, Gobionellinae, and Gobiinae. Worldwide there are over 2000 species of gobiids, of which about 150 are Eleotridinae or sleepers (Nelson, 2006). The majority of gobies have united pelvic fins forming a ventral disc. Typically, but with many exceptions (e.g. *Gobioides*), the body is stout, the head short and broad, the snout rounded, the teeth usually small, sharp, and conical (in 1 to several rows in the jaws), and gill membranes broadly joined to isthmus. The head typically has a series of sensory canals and pores as well as cutaneous papillae. Two separate dorsal fins, first dorsal fin with 4 to 8 weak spines, second dorsal fin with 1 weak spine followed by 9 to 18 soft rays; caudal fin broad and rounded, comprising 16 or 17 segmented rays; anal fin with 1 weak spine followed by 9 to 18 soft rays; the terminal ray of the second dorsal and anal fins is divided to its base (but only counted as a single element); pelvic fin long with 1 spine and 5 rays, pelvic-fin spines usually joined by fleshy membrane (frenum), and innermost pelvic-fin rays usually joined by membrane, forming a disc; pectoral fin broad with 15 to 22 rays. There is no lateral line on the body. Most of the gobioid species are benthic, shallow water coastal or marine species of relatively small size (most do not exceed 10 cm), but particularly in the tropical regions freshwater species are also common. The smallest known vertebrate is a goby, *Trimmatom nanus*, which matures at 8 mm. Particularly gobiins are sexually dimorphic, with males larger and more colorful, often with the gape much larger than the female. Gobies typically have an elaborate courtship behavior and spawn in pairs. The male guards the eggs, commonly in shelter like in a natural crevice or empty mollusk shell. Among Neotropical gobiins, *Sicydium* and *Awaous* species are found in fast running waters. *Ctenogobius* is recognized as distinct below, but is usually synonymized with *Gobionellus*.

1-a A single continuous dorsal fin; eyes minute, about 10% of head length; body very elongate, eel-like; reaching 50 cm in total length (*Gobioides*)

2

1-b Two dorsal fins; eyes larger, 15% or more of head length; body robust or elongate (but not eel-like); maximum size of adults <40 cm

3

2-a Second dorsal fin with 1 spine and 14 soft rays; anal fin with 1 spine and 13 or 14 soft rays

Gobioides grahamae

2-b Second dorsal fin with 1 spine and 15 soft rays; anal fin with 1 spine and 15 soft rays

Gobioides broussonnetii

3-a Shoulder girdle, under gill cover, with distinct fleshy lobes

Awaous flavus

3-b Shoulder girdle without fleshy lobes

4

4-a Teeth compressed, with bilobed tips; mouth slightly inferior; 2 dusky spots at the base of the caudal fin

Evorthodus lyricus

4-b Teeth conical, pointed-tipped; mouth at end of snout or inferior

5

5-a Long, lateral cephalic canal with 4 pores; numerous elongate gill rakers on both arms of the first gill arch; relatively large species (>20 cm)

Gobionellus oceanicus

5-b Short, lateral cephalic canal with only two pores; no gill rakers or lobes on upper arm of first gill arch, 4 or 5 gill rakers on lower arm, gill rakers short and triangulate; small species (<6 cm) (*Ctenogobius*)

6

6-a Eye greatly reduced, not filling the socket; jaws large, reaching the posterior margin of the eye socket

Ctenogobius thoropsis

6-b Eye normal, not greatly reduced

7

7-a Broad strip of dark pigment crossing lower cheek from lower preopercular angle to just above the corner of the jaw; males often with elongate third spine in first dorsal fin and large, recurved canine tooth midlaterally in lower jaw

Ctenogobius pseudofasciatus

7-b Broad strip not present as described in 7-a; males with or without elongate spine and large, midlateral canine tooth in lower jaw

Ctenogobius phenacus

Awaous flavus (Valenciennes, 1837)
Local name(s): –

Diagnostic characteristics: two dorsal fins; eyes larger, 15% or more of head length; body robust or elongate; maximum size of adults 10 cm; shoulder girdle, under gill cover, with distinct fleshy lobes
Type locality: Suriname
Distribution: South America in coastal streams: Brazil, Colombia, Guyana, Suriname and Venezuela; in Suriname, collected in the lower freshwater reach of Corantijn and Suriname rivers between Orealla and Apoera (km 100-150; Vari, 1982) and at Overbridge (km 105), respectively
Size: 10 cm TL
Habitat: freshwater, brackish; collected in freshwater reaches of the Lower Corantijn and Suriname rivers, but apparently in low numbers
Position in the water column: bottom
Diet:
Reproduction:
Remarks:

Ctenogobius phenacus (Pezold & Lasala, 1987)
Local name(s): –

Diagnostic characteristics: two dorsal fins; eyes larger, 15% or more of head length; body robust or elongate; small species, maximum size of adults <6 cm; teeth conical, pointed-tipped; mouth at end of snout or inferior; broad strip not present as described; males with or without elongate spine and large, midlateral canine tooth in lower jaw
Type locality: Cayenne River estuary, French Guiana; in Suriname, known from the estuaries of the Suriname and Corantijn rivers
Distribution: Northern South America
Size: 5 cm TL
Habitat: brackish, marine
Position in the water column:
Diet:
Reproduction:
Remarks:

Ctenogobius pseudofasciatus (Gilbert & Randall, 1971)
Local name(s): –

Diagnostic characteristics: two dorsal fins; eyes larger, 15% or more of head length; body robust or elongate; small species, maximum size of adults <6 cm; teeth conical, pointed-tipped; mouth at end of snout or inferior; broad strip of dark pigment crossing lower cheek from lower preopercular angle to just above the corner of

Awaous flavus (alcohol), Overbridge mid-river sandbank, Suriname River (© A. Gangadin)

the jaw; males often with elongate third spine in first dorsal fin and large, recurved canine tooth midlaterally on the lower jaw
Type locality: West side of Tortuguero Lagoon, Limón Province, Costa Rica
Distribution: Atlantic slope of Costa Rica, Panama, Suriname and Trinidad; Florida, U.S.A.; in Suriname, known from the Suriname River estuary
Size: 5.3 cm SL
Habitat: freshwater, brackish, marine
Position in the water column:
Diet:
Reproduction:
Remarks:

CTENOGOBIUS THOROPSIS (Pezold & Gilbert, 1987)
Local name(s): –

Diagnostic characteristics: two dorsal fins; eyes larger, 15% or more of head length; body robust or elongate; small species, maximum size of adults <6 cm; teeth conical, pointed-tipped; mouth at end of snout or inferior; eye greatly reduced, not filling the socket; jaws large, reaching posterior margin of socket
Type locality: North of Suriname River, Suriname, depth 10 fathoms
Distribution: Western Atlantic: northern South America; in Suriname, known from the Suriname River estuary
Size: 5.5 cm SL
Habitat: freshwater, brackish, marine
Position in the water column:
Diet:
Reproduction:
Remarks:

EVORTHODUS LYRICUS (Girard, 1858)
Local name(s): –

Diagnostic characteristics: two dorsal fins; eyes larger, 15% or more of head length; body robust or elongate; maximum size of adults <10 cm; teeth compressed, with bilobed tips; mouth slightly inferior; 2 dusky spots at the base of the caudal fin
Type locality: Brazos Santiago, Texas, U.S.A.
Distribution: Western Atlantic
Size: 9 cm TL
Habitat: freshwater, brackish, marine; in Corantijn River, it was collected in fresh water, i.e. far upstream of the estuary (in *H. tenellum* vegetation, 500 m upstream of Apoera, km 155; Fig. 4.12)
Position in the water column: bottom
Diet:
Reproduction:

Evorthodus lyricus (live), Corantijn River

Remarks: with its blunt head, dorsally positioned eyes, and inferior mouth *E. lyricus* superficially resembles a small mudskipper.

Gobioides broussonnetii Lacépède, 1800
Local name(s): –

Diagnostic characteristics: an elongated species with a single continuous dorsal fin; eyes minute, about 10% of head length; body very elongate, eel-like; reaching 50 cm in total length; second dorsal fin with 1 spine and 15 soft rays; anal fin with 1 spine and 15 soft rays
Type locality: no locality stated [? Suriname]
Distribution: North, Central and South America: Belize, Brazil, Colombia, Cuba, French Guiana, Guiana, Mexico, Puerto Rico, Suriname, U.S.A. and Venezuela; in Suriname, known from the Suriname and Corantijn rivers
Size: 46 cm SL
Habitat: freshwater, brackish, marine; collected in the lower freshwater reach of Corantijn River in holes in mud banks near Orealla (km 105) (Fig. 4.15)
Position in the water column: bottom
Diet:
Reproduction:
Remarks:

Gobioides grahamae (Palmer & Wheeler, 1955)
Local name(s): –

Diagnostic characteristics: an elongated species with a single continuous dorsal fin; eyes minute, about 10% of head length; body very elongate, eel-like; reaching 50 cm in total length; second dorsal fin with 1 spine and 14 soft rays; anal fin with 1 spine and 13 or 14 soft rays
Type locality: Marajo Island, Brazil
Distribution: Atlantic coast and river mouths: Costa Rica, Ecuador, Mexico, Panama and Peru; in Suriname, this species was collected in the estuary of Corantijn River (Vari, 1982)
Size: 17 cm SL
Habitat: brackish, marine; collected in the estuary of the Corantijn River by Vari (1982)
Position in the water column: bottom
Diet:
Reproduction:
Remarks: *G. grahamae* seems largely restricted to brackish water in estuaries, as opposed to *G. broussonnetii* which also occurs in freshwater.

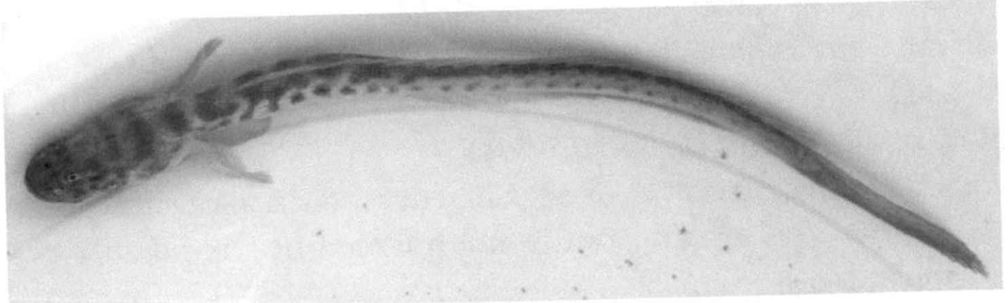

Gobioides broussonnetii (live), Corantijn River

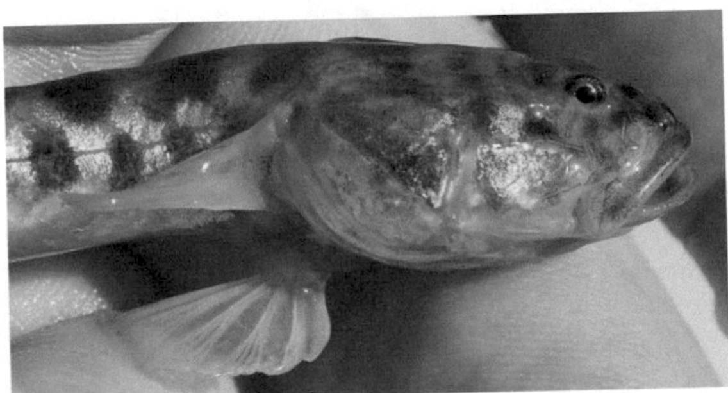

Gobioides broussonnetii (live), detail, Corantijn River

Gobionellus oceanicus (Pallas, 1770)
Local name(s): –

Diagnostic characteristics: two dorsal fins; eyes larger, 15% or more of head length; body robust or elongate; large species, maximum size of adults 30 cm; teeth conical, pointed-tipped; mouth at end of snout or inferior
Type locality: locality unknown [western Atlantic]. Neotype locality: Mindi River, Mindi, Canal Zone, Panama
Distribution: Western Atlantic; in Suriname, it was collected in the estuary of the Corantijn River (Vari, 1982)
Size: up to 30 cm TL
Habitat: freshwater, brackish, marine; it was collected in brackish water in the Corantijn Estuary (Vari, 1982)
Position in the water column: bottom, mainly over muddy substrates in estuaries (Leopold, 2004)
Diet: mainly small shrimps (Leopold, 2004)
Reproduction: females deposit their eggs in a depression excavated in the muddy substrate, but the eggs are apparently not guarded by the parents as in other gobies (Leopold, 2004)
Remarks: lives solitary, apparently not a very active swimmer.

This page is intentionally left blank.

FAMILY EPHIPPIDAE (SPADEFISHES)

Fishes belonging to this family have a very deep, laterally compressed body, a small, protractile mouth, and a series of large blunt gill rakers on the first epibranchial. A single species (*Chaetodipterus faber*) occurs in the Central Western Atlantic; with its notched dorsal fin and the outer jaw teeth larger (and slightly flattened) than inner rows it is not easily confused with other species.

Chaetodipterus faber (Broussonet, 1782)
Local name(s): – (Atlantic spadefish)

Diagnostic characteristics: body deep, included 1.2 to 1.5 times in standard length, orbicular, strongly compressed; mouth small, terminal, jaws provided with bands of brush-like teeth, outer row larger and slightly compressed but pointed at tip; vomer and palatines toothless; preopercular margin finely serrate, opercle ends in blunt point; dorsal fin with 9 spines and 21 to 23 soft rays; spinous portion of dorsal fin low in adults and distinct from the soft-rayed portion; anterior portion of soft dorsal and anal fins prolonged; juveniles with third dorsal fin spine prolonged, becoming proportionately smaller with age; anal fin with 3 spines and 18 or 19 rays; pectoral fins short, about 1.6 in head, with 17 or 18 soft rays; caudal fin emarginated; pelvic fins long, extending to origin of anal fin in adults, beyond that in young; lateral-line scales 45 to 50; head and fins scaled; color: silvery grey with blackish bars (bars may fade in large individuals); juveniles with dark and white mottling
Type locality: Jamaica
Distribution: Western Atlantic; in Suriname, it was collected at a brackish-water aquaculture facility in District Commewijne (Comfish, Van Alen)
Size: maximum to 1 m, commonly to 50 cm
Habitat: freshwater, brackish, marine; inhabits a variety of different habitats along shallow coastal waters, including reefs, mangroves, sandy beaches, harbors, around wrecks and pilings, and under bridges; they are often seen in large schools of more than 500 adult individuals
Position in the water column:
Diet: invertebrates, both benthic and planktonic, as well as algae
Reproduction:
Remarks: juveniles are apt to be encountered around mangroves in their dark coloration with white mottling. This cryptic coloration, when combined with the juveniles' habit of floating tilted on its side, mimics the dead mangrove leaves and possibly other floating objects making the fish difficult to detect.

Chaetodipterus faber (live), Lower Commewijne River

FAMILY TRICHIURIDAE (SCABBARDFISHES OR HAIRTAILS)

Predominantly large fishes (to 1 to 2 m total length) with a remarkably elongate and compressed body (ribbon-like), a single nostril on each side of snout, and a large mouth with very long teeth, usually fang-like at front of upper jaw and sometimes in anterior part of lower jaw. A single extremely long dorsal fin running almost the entire length of the body; its spinous portion either short and continuous with very long soft portion, or moderately long, not shorter than half of soft portion length, and separated from soft portion by a notch. The anal fin is preceded by 2 free spines behind the anus (first inconspicuous and second variously enlarged), with absent or reduced (sometimes restricted to posterior part of fin) soft rays. Pectoral fins with 12 rays, moderately small and situated midlaterally or lower on sides. Pelvic fins absent or reduced to 1 flattened spine and 0 to 1 tiny soft rays. Caudal fin either small and forked, or absent. Lateral line single. Scales absent. No keels on the caudal peduncle. The color of the body is silvery to black with an iridescent tint; fins usually paler.

TRICHIURUS LEPTURUS Linnaeus, 1758
Local name(s): zilverbandvis (Largehead hairtail)

Diagnostic characteristics: body elongate and strongly compressed, ribbon-like, tapering to a point (tip often broken); depth about 15 to 18 in total length; head about 6 to 8 in total length, with upper profile slightly concave, gently rising from snout to dorsal-fin origin; interorbital space and nape convex, with sagittal crest elevated; eye 5 to 7 in head, nearly touching upper profile; dorsal fin moderately high, very long, with 3 spines and 130 to 135 rays, not divided by notch; anal fin reduced to about 100 to 105 minute spinules, usually embedded in skin or slightly breaking through; no caudal and pelvic fins; pectoral fins directed upward, with 1 spine and 11 to 13 rays; color of fresh specimens steel blue with silvery reflection, pectoral fins semitransparent, other fins sometimes tinged with pale yellow; the color becomes uniform silvery grey after death
Type locality: America and China
Distribution: circumglobal in warm seas; in Suriname, known from the estuary of the Suriname River, but probably present in other estuaries as well
Size: 1.2 m TL, common 50 to 100 cm
Habitat: brackish, marine; usually in shallow coastal waters over muddy bottoms, occasionally at surface at night; in Suriname, occasionally observed in estuaries
Position in the water column: benthopelagic
Diet: voracious predators feeding on fishes, squids, and crustaceans; young and immature specimens feed on crustaceans and small fishes; adults more piscivorous
Reproduction: eggs and larvae pelagic
Remarks: this species is eaten by Chinese people in Suriname.

Trichiurus lepturus (live), Suriname River Estuary (© K. Wan Tong You)

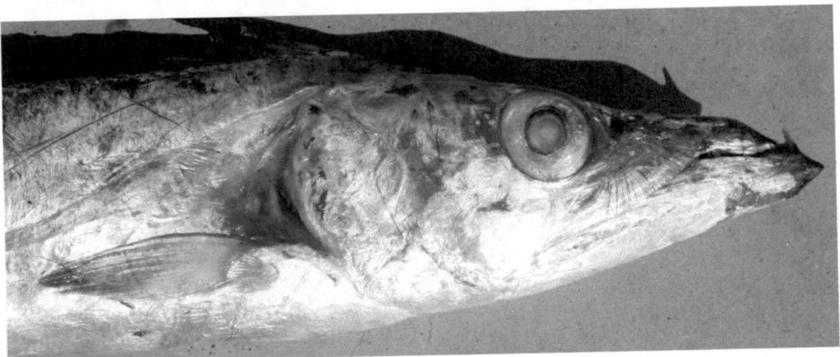

Trichiurus lepturus (live), detail of head, Suriname River Estuary (© K. Wan Tong You)

FAMILY PARALICHTHYIDAE (SAND FLOUNDERS)

Flatfishes with the eyes generally on the left side of the head (sinistral), reversals frequent in some species. No spines present in the fins. Mouth protractile, asymmetrical, lower jaw moderately prominent; teeth in jaws sometimes canine-like (they are also called largetooth flounders); no teeth on vomer. Preopercle exposed, its posterior margin free and visible, not hidden by skin or scales. Dorsal fin long, originating above, lateral to, or anterior to upper eye. Dorsal and anal fins not attached to caudal fin. Both pectoral fins present; rays branched. Both pelvic fins present, with short bases and symmetrical, 5 or 6 rays (6 rays in nearly all species); base of pelvic fin of ocular side on midventral line (*Cyclopsetta* group, including the genus *Syacium*). Caudal fin with 17 or 18 rays, 10 to 13 rays branched (usually 11 or 13, rarely 10 or 12). Lateral line present and obvious on both sides of body. In *Syacium* there is no distinct arch in the lateral line above pectoral fin on ocular side and the lateral line is not prolonged below inferior eye, the urinary papilla is on the blind side, branched caudal-fin rays 11 (rarely 10 or 12), mouth large, maxilla less than 3.5 in head length usually reaching posteriorly to vertical through mid-eye, jaws on blind side not arched, and front teeth in jaws enlarged, larger than lateral teeth. Sand flounders are bottom-dwelling predators, usually burrowing partially or almost entirely in sand or soft mud. They are capable of a rapid change in coloration which allows them to match their background almost perfectly. Most appear to feed on or near the bottom, but some of the larger species will rise off the bottom to capture prey. Most species occur in shallow water.

1-a Body depth usually 48% of standard length or greater (45 to 47% in some specimens from the Caribbean); interorbital width of adults large (dimorphic differences as well as ontogenetic differences, but usually greater than that in *Syacium papillosum* and *Syacium micrurum* of comparable size); length of pectoral fin on blind side >50% of head length; 46 to 55 scales in lateral line; dorsal-fin rays 74 to 85; anal-fin rays 59 to 68

Syacium gunteri

1-b Body depth usually 45% of standard length or less (rarely 47%); length of pectoral fin on blind side <50% of head length; 44 to 69 scales in lateral line; dorsal-fin rays 82 to 94; anal-fin rays 64 to 75

2

2-a 57-69 scales in the lateral line; body depth 39-42% of standard length; in males >150 mm SL: interorbital width <11% of head length; in males 100-150 mm SL: interorbital width <8% of head length

Syacium micrurum

2-b 44-57 scales in the lateral line; body depth 39-47% of standard length; in males >150 mm SL: interorbital width >14% of head length; in males 100-150 mm SL: interorbital width >9% of head length

Syacium papillosum

SYACIUM GUNTERI Ginsburg, 1933
Local name(s): boki (Shoal flounder)

Diagnostic characteristics: body depth usually 48% standard length or greater (45 to 47% on some specimens from the Caribbean); interorbital width of adults large (dimorphic differences as well as ontogenetic differences, but usually greater than that in *Syacium papillosum* and *Syacium micrurum* of comparable size); length of pectoral fin on blind side >50% of head length; 46 to 55 scales in lateral line; dorsal-fin rays 74 to 85; anal-fin rays 59 to 68
Type locality: Off the coast of Louisiana, 12 miles northeast of Barstom Light, Gulf of Mexico, U.S.A.
Distribution: Western Atlantic; in Suriname, known from the estuary of Marowijne River
Size: maximum size 20 cm, commonly to 15 cm total length
Habitat: marine; on shallow, soft bottoms (mostly mud and fine sands with low calcium carbonate and high organic contents), to depths of approximately 95 m (usually less).
Position in the water column: bottom
Diet: diurnal feeding habits; feeds mainly on crustaceans (penaeid shrimps and amphipods), larvae of crustaceans and annelids, and, to a lesser degree, fishes
Reproduction: size at first maturity for females, 6 to 9.6 cm total length; spawning occurs from May to September (Southern Gulf of Mexico); one spawning period per year, perhaps corresponding with rainy season
Remarks: rests at night buried in sand; rare in French Guiana (Leopold, 2004).

SYACIUM MICRURUM Ranzani, 1842
Local name(s): boki (Channel flounder)

Diagnostic characteristics: 57-69 scales in the lateral line; body depth 39-42% of standard length; in males >150 mm SL: interorbital width <11% of head length; in males 100-150 mm SL: interorbital width <8% of head length
Type locality: Brazil.
Distribution: Western Atlantic; in Suriname, known from the estuary of Suriname River
Size: maximum size 30 cm, commonly to 20 cm total length
Habitat: marine; on soft bottom habitats to depths in excess of 400 m, but usually less than 100 m.
Position in the water column: bottom
Diet:
Reproduction:
Remarks: rare in French Guiana (Leopold, 2004).

Syacium micrurum (alcohol) (© INRA-Le Bail)

Syacium papillosum (Linnaeus, 1758)
Local name(s): boki (Dusky flounder)

Diagnostic characteristics: 44-57 scales in the lateral line; body depth 39-47% of standard length; in males >150 mm SL: interorbital width >14% of head length; in males 100-150 mm SL: interorbital width >9% of head length
Type locality: America. No types known
Distribution: Western Atlantic; in Suriname, known from the estuaries of the Corantijn, Suriname and Marowijne rivers
Size: Maximum size 25 cm, commonly to 20 cm total length
Habitat: marine; on shallow soft bottom habitats, usually at depths of 10 to 90 m, but has also been taken in deeper waters (to depths of 140 m)
Position in the water column: bottom
Diet: small fishes and crustaceans
Reproduction:
Remarks: one of the most abundant flatfishes of French Guiana (Leopold, 2004); this is the most important commercial species of the genus because of its acceptable average size and relative abundance.

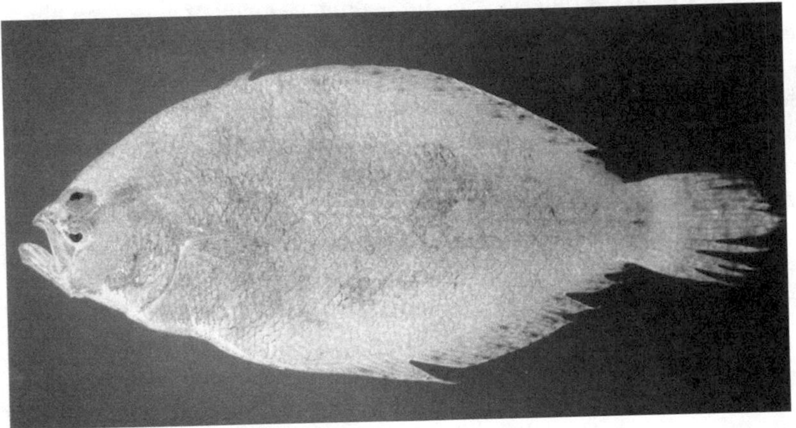

Syacium papillosum (live) (© INRA-Le Bail)

FAMILY ACHIRIDAE (AMERICAN SOLES)

Small flatfishes (usually smaller than 35 cm) with eyes and color pattern on the right side (left-eyed individuals very rare); body round or oval in outline and strongly compressed. Preopercular margin not free, concealed by skin or represented only by a naked superficial groove. Snout rounded, mouth small, oblique and asymmetrical, subterminal; lips fleshy, often fringed with dermal flaps or fleshy convolutions; teeth minute, villiform, difficult to see. Eyes small to minute, without externally prominent bony orbits. Fins without spines, all rays soft; dorsal fin extending forward well in advance of eyes, the anterior rays concealed within a fleshy dermal envelope and difficult to see. Dorsal and anal fins not confluent with the caudal fin. Pectoral fins present or absent, if present that of right side usually longer than left (left pectoral fin usually vestigial or absent on blind side); pelvic fins present bilaterally, either free or joined to anal fin. Lateral line essentially straight, often indistinct, but most readily seen on ocular side, usually crossed at right angles by accessory branches (achirine lines) extending toward dorsal and anal fins; lateral line often ornamented with minute fleshy flaps or cirriform dermal processes. Scales ctenoid (rough to touch). American soles inhabit marine, estuarine, or fluviatile (some species of *Achirus* and *Trinectes*) waters. Most species live close to shore and occur on a variety of soft sandy or muddy sediments. The majority of species feed on benthic invertebrates, with occasional small fishes included in diets of larger species.

1-a Gill openings reduced to narrow slits, separate, not confluent in front of pelvic fins

Apionichthys dumerili

1-b Left and right-side gill openings wide, confluent in front of pelvic fins

2

2-a Interbranchial septum entire, without foramen; ocular-side pectoral fin rudimentary, usually with a single ray (rarely with 2 or 3 fin rays) or absent altogether; blind-side pectoral fin usually absent (or rarely present, with a single ray)

Trinectes paulistanus

2-b Interbranchial septum pierced by a foramen; ocular-side pectoral fin usually with 2 to 8 rays; blind-side pectoral fin either with a single ray or absent (*Achirus*)

3

3-a Dorsal-fin rays 59 to 68; anal-fin rays 43 to 51; ocular-side pectoral fin usually with 3 or 4 rays

Achirus achirus

3-b Dorsal-fin rays 49 to 60; anal-fin rays 38 to 48; ocular-side pectoral fin usually with 5 or 6 rays

Achirus declivis

This page is intentionally left blank.

ACHIRUS ACHIRUS (Linnaeus, 1758)
Local name(s): boki (Drab sole)

Diagnostic characteristics: interbranchial septum pierced by a foramen; ocular-side pectoral fin usually with 3 or 4 rays; blind-side pectoral fin either with a single ray or absent; dorsal-fin rays 59 to 68; anal-fin rays 43 to 51; ocular-side pectoral fin usually with 3 or 4 rays
Type locality: Suriname; no types known
Distribution: off South America; probably Venezuela to northeastern Brazil; in Suriname, known from the Corantijn, Suriname and Marowijne rivers
Size: maximum size 37 cm, commonly to 30 cm
Habitat: freshwater, brackish, marine; occurs on sand-mud bottoms in estuarine waters to fresh water at depths of 20 m or less
Position in the water column: bottom
Diet: small invertebrates (especially crustaceans) and small fishes
Reproduction:
Remarks: growth rate relatively slow; often found completely covered with sediment, presumably for protection from predators and to ambush prey; the related lined sole *Achirus lineatus* has 49 to 60 dorsal-fin rays, 38 to 48 anal-fin rays, 5 or 6 rays in the ocular-side pectoral fin, the caudal fin with numerous dark spots or irregular blotches, and the blind side of body in caudal region darkly shaded. *A. achirus* is commonly sold at the central market of Paramaribo.

ACHIRUS DECLIVIS Chabanaud, 1940
Local name(s): boki

Diagnostic characteristics: dorsal-fin rays 49 to 60; anal-fin rays 38 to 48; ocular-side pectoral fin usually with 5 or 6 rays; caudal fin lacking dark spots or blotches; blind side of body in caudal region not prominently shaded
Type locality: Gulf of Paria, between Trinidad Island, West Indies and Venezuela
Distribution: Western Atlantic; in Suriname, known from the estuaries of the Suriname and Commewijne rivers
Size: 18 cm TL
Habitat: brackish, marine
Position in the water column: bottom
Diet:
Reproduction:
Remarks:

APIONICHTHYS DUMERILI Kaup, 1858
Local name(s): boki (Longtail sole)

Diagnostic characteristics: oval body with minute eyes, barely visible, with diameter much less than interorbital width; body depth about 3 times in standard

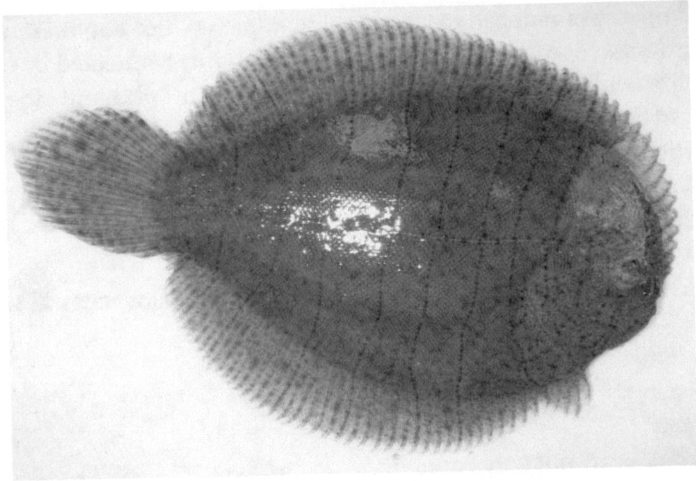

Achirus achirus (live), Corantijn River (km 160) (© R. Smith)

Achirus achirus (live), Paramaribo market

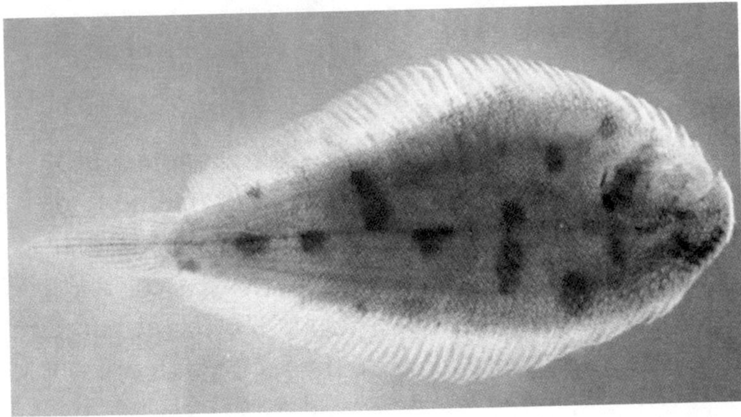

Apionichthys dumerili (live) (© INRA-Le Bail)

length; gill openings reduced to narrow slits, separate, not confluent in front of pelvic fins; dorsal (65-75 rays) and anal (47-58 rays) fins connected by membrane to caudal fin; ocular-side pelvic fin rudimentary, or absent; blind-side pelvic fin distinct from that of ocular-side
Type locality: -
Distribution: widespread in South America; in Suriname, known from the Corantijn and Suriname rivers
Size: maximum size 15 cm, commonly to 11 cm
Habitat: freshwater, brackish; over sandy bottoms in shallow coastal waters and estuaries
Position in the water column: bottom
Diet:
Reproduction:
Remarks: apparently not very common.

Trinectes paulistanus (Mirända-Ribeiro, 1915)
Local name(s): boki (Slipper sole)

Diagnostic characteristics: dorsal-fin rays 54 to 60; anal-fin rays 40 to 45; ocular-side pectoral fin usually with a single ray
Type locality: Santos, Sepetiba, Brazil
Distribution: Western Atlantic; in Suriname, known from the estuaries of the Coppename, Saramacca and Suriname rivers
Size: maximum size 18 cm, commonly to 12 cm
Habitat: brackish, marine; occurs over soft bottoms in estuaries and hypersaline lagoons.
Position in the water column: bottom
Diet:
Reproduction:
Remarks: not very abundant.

This page is intentionally left blank.

FAMILY CYNOGLOSSIDAE (TONGUEFISHES AND TONGUE SOLES)

Lance- or tongue-shaped flatfishes with eyes on left side of body; body highly compressed and tapering to a point posteriorly. Posterior margin of preopercle strongly attached to opercle, without free margin and covered with skin and scales. Dorsal and anal fins confluent with pointed caudal fin; dorsal fin reaching far forward onto head usually in advance of posterior border of upper eye; pectoral fins absent; usually only left pelvic fin (with 4 fin rays) present, located on median line and connected to anal fin by delicate membrane. Mouth small, subterminal, asymmetrical; reaching posteriorly to a point between verticals at anterior and posterior margins of lower eye or slightly posterior to lower eye; jaws moderately curved on ocular side and notably on blind side; teeth minute and usually better developed on blind-side jaws; some species lacking teeth on ocular side jaws; eyes small and usually close together. No spines or spiny fin rays in dorsal, anal, or pelvic fins. Lateral line absent on both ocular and blind sides. Scales ctenoid on both sides of body. Most species are small and not very abundant.

Symphurus plagusia (Bloch & Schneider, 1801)
Local name(s): boki (Duskycheek tonguefish)

Diagnostic characteristics: body relatively deep; greatest depth in anterior 1/3 of body; tapering fairly gradually posterior to midpoint; head wide, head length usually much shorter than head width; snout moderately long, somewhat square; lower eye small, spherical; eyes slightly subequal in position; pupillary operculum absent; maxilla usually reaching posteriorly to point between verticals through posterior margin of lower eye pupil to vertical just slightly posterior to posterior margin of lower eye, ocular-side lower jaw with distinct, fleshy ridge near posterior margin; dorsal-fin rays 89 to 97, dorsal-fin origin far forward, usually at vertical through anterior margin of upper eye, or with first and sometimes second dorsal-fin rays inserting anterior to vertical through anterior margin of upper eye; anal-fin rays 73 to 81; scales absent on blind sides of dorsal- and anal-fin rays; caudal-fin rays usually 12; longitudinal scale rows 79 to 89; color: ocular surface usually uniformly light brown or yellowish, occasionally with 8 to 14, narrow, faint crossbands (crossbands not continued onto dorsal and anal fins), blind side creamy white, without pepper-dots; peritoneum unpigmented; pigmentation of outer surface of ocular-side opercle usually same as that of body, occasionally with dusky blotch on upper opercular lobe due to pigment on inner lining of ocular-side opercle showing through to outer surface; dorsal and anal fins uniformly dusky throughout their lengths, without conspicuous spots or blotches
Type locality: Puerto Yabucoa, .5 miles east of Playa de Guayanes, Puerto Rico

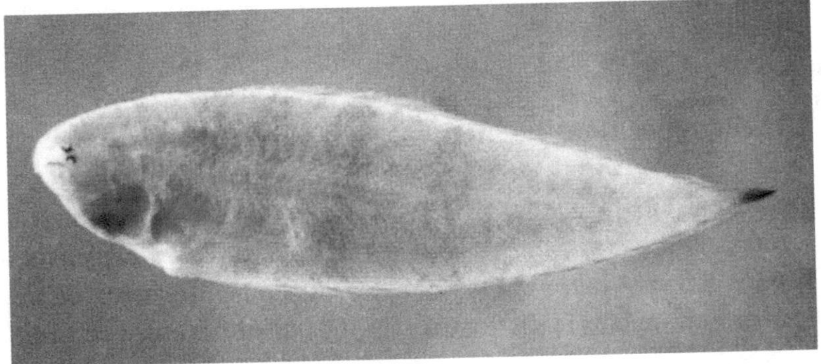

Symphurus plagusia (live) (© INRA-Le Bail)

Distribution: Western Atlantic; in Suriname, known only from the estuary of the Suriname River
Size: maximum about 130 mm standard length
Habitat: brackish, marine; a shallow-water species (1 to 51 m) most commonly inhabiting mud bottoms in estuaries and coastal waters to about 10 m; all life-history stages occur in shallow areas and only occasional individuals taken deeper.
Position in the water column: bottom
Diet: feeds during the night on small benthic invertebrates
Reproduction: males and females attain similar sizes; females mature at sizes larger than 80 mm standard length
Remarks: little is known concerning its ecology.

This page is intentionally left blank.

FAMILY TETRAODONTIDAE (PUFFERFISHES)

Small to moderate-sized fishes (most species < 300 mm), with a heavy blunt body capable of rapid inflation by intake of water (or air) (body inflatable). The body is naked, but commonly features small scales in the shape of prickles. Head large and blunt; jaws modified to form a beak of 4 heavy, powerful teeth, 2 above and 2 below; gill openings without distinct opercular cover, appearing as simple slits anterior to the pectoral fin; eyes located high on the head. Dorsal and anal fins located far posteriorly bearing no spines, but usually with 7 to 15 soft rays; caudal fin usually truncate to slightly rounded; pelvic fins absent. The name derives from their ability, as a defense behavior, to inflate themselves into a nearly globular shape by engulfing either air or water. Most puffers are mildly to strongly toxic, containing an alkaloid poison, tetraodotoxin (tetrodotoxin), which accumulates particularly in the gonads, and which is acutely lethal if eaten (see chapter 8). Several puffer species are known from the coast of South and Central America, but only one species is confined to freshwater (*Colomesus asellus*), and this species is not known from Suriname (it is known from the Essequibo River, Guyana). Two species, although coastal, enter estuaries: *Colomesus psittacus* and *Sphoeroides testudineus*.

1-a Dorsal fin with 10-12 rays; dorsum with 6-7 dark bars; prickles covering most of body (feeling rough when touched)
Colomesus psittacus
1-b Dorsal fin with 7-9 rays; dorsum covered with spots in irregular pattern, absence of distinct dark bands; prickles covering most of body, but usually imbedded and not noticeable to the touch
Sphoeroides testudineus

COLOMESUS PSITTACUS (Bloch & Schneider, 1801)
Local name(s): bosrokoman, tamyaku (Banded puffer)

Diagnostic characteristics: a blunt-headed fish with a stout body, and with heavy jaws forming a beak of 2 teeth in both upper and lower jaws; dorsal and anal fins set far back, near caudal fin, dorsal and anal fins with 10 or 11 soft rays (no spines); pectoral fin with 17 to 19 rays; pelvic fins absent; prickles are present from the snout to posterior margin of the dorsal fin, and chin to near the anus ventrally, and present laterally on the cheeks and to near level of dorsal fin; color: dorsally and laterally, basal pigmentation is a light grey or brown with 6 dark, prominent, uniform, transverse bars; the first extending between the orbits, the sixth across the caudal fin; the lighter areas between bars may sometimes have shading; ventral surface, including the underside of the caudal peduncle, unpigmented

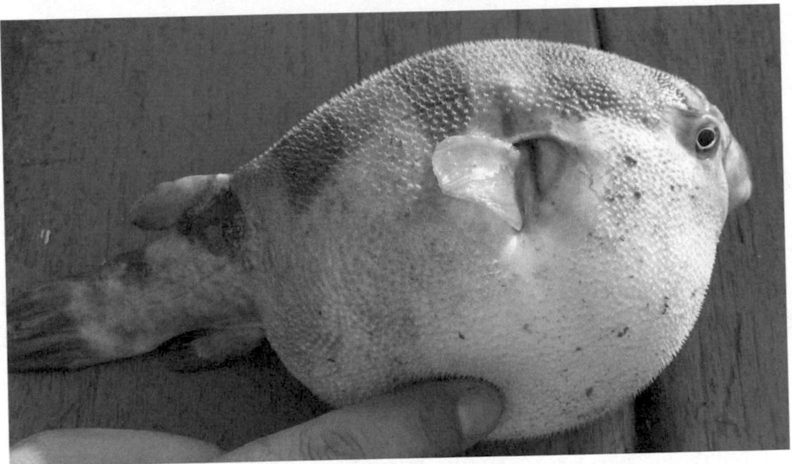

Colomesus psittacus (live), Suriname River Estuary (© K. Wan Tong You)

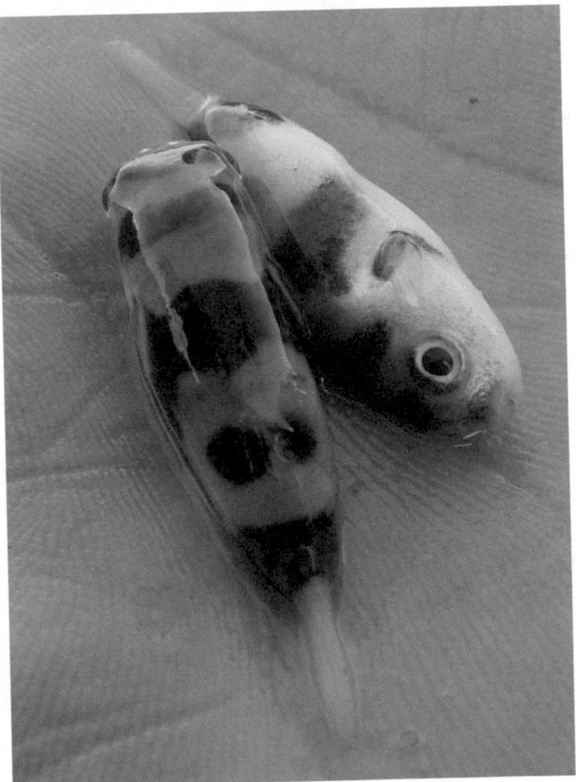

Colomesus psittacus (live), juveniles (© R. Covain)

Type locality: Indian Ocean [in error].
Distribution: Western Atlantic: French Guiana, Guyana, Suriname, Trinidad and Tobago, and Venezuela; in Suriname, known from all estuaries
Size: common to 300 mm, largest known specimens are near 350 mm
Habitat: freshwater, brackish, marine; occasionally entering fresh water
Position in the water column:
Diet:
Reproduction:
Remarks: little else is known of its natural history; toxic (see chapter 8); a freshwater congener from the Amazon (and Guyana), *Colomesus asellus*, is similar but is pigmented on the underside of the caudal peduncle.

SPHOEROIDES TESTUDINEUS (Linnaeus, 1758)
Local name(s): bosrokoman, tamyaku (Checkered puffer)

Diagnostic characteristics: a blunt-headed fish with heavy jaws forming a beak of 2 teeth in both upper and lower jaws; dorsal and anal fins set far back near caudal fin, dorsal fin usually with 8 soft rays (no spines), anal fin with 7 soft rays (no spines), pectoral fins usually with 15 rays, pelvic fins absent; prickles covering most of body, but usually imbedded and not noticeable to the touch; color: upper side chocolate brown to black, with light (yellow or white) bold markings, especially 1 or 2 distinct transverse bars between eyes and a regular geometrical pattern of coarse arches and circular markings on back; belly white to yellow; lower sides heavily spotted
Type locality: India
Distribution: Western Atlantic; in Suriname, known from the estuaries of Corantijn, Coppename, Suriname and Marowijne rivers, but probably present in all estuaries
Size: maximum to 30 cm; common to 20 cm
Habitat: brackish, marine; one of the most common fish species in mangrove areas and estuarine coastlines; confined to very shallow waters over mud or sand bottoms; does not school, but may form huge aggregations
Position in the water column:
Diet: feeds primarily on shellfish through most of its range
Reproduction:
Remarks: avoided because of its toxicity; known to be lethally toxic to humans. Its principle utilization is as poison when fed to pest animals (e.g. feral cats and dogs).

Sphoeroides testudineus (live), Commewijne River

GLOSSARY OF TECHNICAL TERMS

When an entry uses a term that is defined in its own right, this is indicated by an asterisk.

Abdomen – the belly or lower surface of a fish; *ventral area between *pelvic fins and *anus.
Acute – pointed or sharp.
Adipose eyelid – immovable, transparent fleshy tissue that (partially) covers the eyeball in some fishes (e.g. mullets).
Adipose fin – small, fleshy fin, often without spines or rays, on the dorsal midline between the dorsal and caudal fins of some fishes (e.g. catfishes, characoids).
Adnate – joined together congenitally.
Adpressed – pressed down or lying flat.
Air bladder – see swim bladder.
Allochthonous – not indigenous; acquired. Applied to material which did not originate in its present position (e.g. plant material in a deposit, such as stream sediment, which did not grow at that location, but was introduced by some process such as leaves falling in the water). Compare *autochthonous.
Allopatric – occurring in different geographic areas; compare *sympatric, *syntopic.
Anadromous – moving from the ocean into freshwater to spawn, as in some anchovies (Engraulidae).
Anal fin – unpaired median fin supported by rays on the tail, usually just behind the anus.
Anal ring – body segment of a pipefish that contains the anus.
Anterior – pertaining to the front portion often combined with other directional terms such as lateral in a shortened form (anterolateral, referring to front part of sides).
Anus – the posterior external opening of the digestive tract from which wastes are voided; sometimes called the vent.
Apophysis – a protuberance on the surface of the bones.
Attenuate – elongate; extended or drawn out.
Autochthonous – applied to material which originated in its present position (e.g. the plant material in a deposit, such as peat, which actually grew where it is found, rather than being brought in by outside influences). Compare *allochthonous.
Autotroph – an organism that uses carbon dioxide as its main or sole source of carbon (e.g. green plants). Compare *heterotroph.
Axil – the acute, angular area between a fin and the body, usually used in reference to the underside of the pectoral fin toward the base; equivalent to arm-pit.
Axillary scale – an elongate or modified scale that is situated at the insertion of the pelvic or pectoral fins in some fishes.
Band – usually refers to an oblique or irregular marking (compare with *'bar' and *'stripe').
Bar – an elongate nearly straight vertical band of color.
Barbel – elongate, fleshy, tentacle-like sensory projection, found on the head, usually near the mouth (e.g. in catfishes). Barbels are sensitive to touch and taste.
Base of fin – part of the fin that is attached to the body.
Benthic – living on or near the bottom.
Bicuspid – see teeth.
Bifid, bifurcate – separated or divided into 2 branches (forked); see *teeth.

Biodiversity – term used to describe all aspects of biological diversity, especially including *species, *ecosystem complexity, and genetic variation.

Biogeochemical cycle – the movement of chemical elements from organism to physical environment to organism in more or less circular pathways. They are termed 'nutrient cycles' if the elements concerned are essential to life. The form and quantity of an element varies through the cycles, with amounts in the inorganic reservoir pools usually greater than those in the active pools. Exchange between the system components is achieved by physical processes (e.g. weathering) and/or biological processes (e.g. protein synthesis, decomposition). The latter form the vital negative feedback mechanisms that regulate the cycles.

Biogeography – the study of the geographical distribution of plants and animals at different taxonomic levels, the *habitats in which they occur, and the ecological relationships involved.

Biogeographical barrier – a barrier that prevents the migration of species. The various disjunctive geographical groupings of plants and animals are usually delimited by one or more such barriers, which may be climatic, involving temperature and the availability of water, or physical, involving, for example, mountain ranges or expanses of sea water or (for aquatic organisms) land.

Biological productivity – the productivity of organisms and ecosystems, as defined by primary, secondary, and community productivities. See also *primary productivity.

Biomass (*standing crop*) – the total mass of all living organisms or of a particular set (e.g. species), present in an *ecosystem or at a particular *trophic level in a *food-chain, and usually expressed as dry weight or, more accurately, as the carbon, nitrogen, or calorific content per unit area.

Biserial- see *teeth.

Blotch – irregularly shaped color mark. See *speck, *spot.

Body rings – the bony segments of a pipefish that are usually well differentiated by visible sutures.

Bog – a plant community of acidic, wet areas. Decomposition rates in it are slow, favoring peat development. Three types of bog community are commonly distinguished: ombrogenous bogs, raised bogs and valley bogs. These reflect the different physiographic and climatic conditions that may give rise to bog formation. Compare *swamp.

Brackish – applied to water that is *saline, but less so than sea water. According to the Venice system brackish waters are classified by the chlorine they contain as euhaline (1.65-2.2 percentage chlorine), polyhaline (1.0-1.65), mesohaline (0.3-1.0), oligohaline (0.03-0.3) and freshwater (<0.03).

Branchiostegal membranes – membranes on the ventral interior surface of the gill cover supported by branchiostegal rays.

Branchiostegal rays – bony rays supporting the *branchiostegal membranes inside the lower part of the gill cover.

Breast – chest, ventral surface of body in front of the belly, i.e. between the *isthmus and pectoral or pelvic fins.

Buccal – of the mouth or cheek (i.e. the buccal cavity).

Buccal incubation – the incubation of eggs in the mouth of one of the parents. In a number of species of fish, the male or female carries the fertilized eggs in the mouth until some time after hatching (e.g. tilapia and ariid catfishes). During incubation the parent involved takes no food, and so may appear rather lean at the end of the incubation period.

Bucco-pharyngeal papillae – small protuberances on the inner surface of the mouth and beginning of the alimentary canal.

Canine – slender, conical, pointed tooth, usually larger than the surrounding teeth; see *teeth.

Carnivorous – meat eating; feeding on animals.

Cartilage – translucent material that makes up the skeleton of young fishes and which persists in adults of some species (notably sharks and rays) but is largely converted to bone in most fishes.

Catadromous – moving from fresh water to the ocean to spawn, as in tarpon (*Megalops atlanticus*).

Catchment – the area from which a surface watercourse or a groundwater system derives its water. Catchments are separated by divides. A surface catchment area may overlie an aquifer system, but may be unconnected with the aquifer rock itself if there are intervening impermeable aquicludes. In US usage a catchment is often termed a 'watershed'; in UK the term 'drainage basin' is commonly used.

Caudal fin – tail fin, median fin at rear of the body; the term tail alone generally refers to that part of a fish posterior to the anus.

Caudal peduncle – tail base; posterior part of the body between the rear parts of the dorsal and anal fins, and the caudal fin.

Caudal spot – spot at base (origin) of *caudal fin.

Cephalic – pertaining to the head.

Cephalic shield – the large bony dorsal covering on the head of fork-tailed sea catfishes (Ariidae).

Channel – main course of a stream. Also see *thalweg.

Character – any detectable attribute or property of the *phenotype of an organism. Defined heritable differences in the character may exist between individuals within a species.

Cheek – side of head below and slightly behind the eye.

Chlorinity – a measure of the chloride content, by mass, of sea water. It is defined as the amount of chlorine, in grams, in 1 kg of sea water. Chlorinity and *salinity are both measures of the saltness of water. The relationship can be expressed mathematically, as salinity is equivalent to 1.80655 times the chlorinity.

Chromatophore – pigment cell which can be altered in shape to produce color change.

Circumpeduncular scales – the transverse series of scales that completely encircle the tail base.

Cirrus – small, fleshy protuberance.

Clade – in *cladistics, a lineage branch that results from a dichotomous splitting in an earlier lineage. A split produces two distinct new taxa. Each of which is represented as a branch in a phylogenetic diagram. The term is derived from the Greek *klados*, meaning 'twig' or 'branch'.

Cladistics – a special taxonomic system founded in 1966 by W. Hennig, which is applied to the study of evolutionary relationships. It proposes that common origin can be demonstrated by the shared possession of derived characters, characters in any group being either primitive or derived (or synapomorphic). In the cladograms used to portray these relationships, *cladogenesis always creates two equal sister groups: the branching is dichotomous. Thus each pair of sister groups constitutes a *monophyletic group with a common stem *taxon unique to the group, and a parent taxon always gives rise to two daughter taxa which must be given different names from each other and from parent, so the parent species ceases to exist. Monophyletic groups are deduced by identifying synapomorphic character states.

Cladogenesis – in *cladistics, the development of a new *taxon as a result of the splitting of an ancestral lineage to create two equal sister groups, taxonomically separated from the ancestral taxon.

Cleithral – pertaining the cleithrum or area of the cleithrum which is typically the largest bone of a series of bones that support the pectoral fin (pectoral-girdle bones).

Community – a general term applied to any grouping of populations of different species found living together in a particular environment; essentially, the biotic component of an ecosystem. The organisms interact (by competition, *predation, mutualism, etc) and give the community a structure.

Compressed – flattened laterally (i.e. from side to side); a body shape much deeper than wide.

Concave – bowed or curved inward; opposite of *convex.

Conical teeth – see teeth.

Consumer – in the widest sense a *heterotrophic organism that feeds on living or dead organic material. Two main categories are recognized: (a) macroconsumers, mainly animals (*herbivores, *carnivores, and *detritivores), which wholly or partly ingest other living organisms or organic particulate matter; and (b) microconsumers, mainly bacteria and fungi, which feed by breaking down complex organic compounds in dead protoplasm, absorbing some of the decomposition products, and at the same time releasing inorganic and relatively simple organic substances to the environment. Sometimes the term 'consumer' is confined to macroconsumers, microconsumers being known as *'decomposers'. Consumers may be termed 'primary' (herbivores), 'secondary' (herbivore-eating carnivores), and so on, according to their position in the food-chain.

Continental shelf – the gently seaward-sloping surface that extends between the shoreline and the top of the continental slope at about 150 m depth. The average gradient of the shelf is between 1:500 and 1:1000 and, although it varies greatly, the average width is approximately 70 km.

Convex – bowed or curved outward; opposite of *concave.

Coracoid – a (beak-shaped) bone articulating with the scapula and sternum in the pectoral girdle joint

Crenate – having a notched edge.

Crenulate – scalloped or wavy edge.

Cryptic coloration – coloration that makes animals difficult to distinguish against their background, so tending to reduce predation (or enabling a predator to approach its prey unnoticed; e.g. the leaf fish *Polycentrus schomburgkii*). The effect of cryptic coloration may be to cause the appearance of the animal to merge into its background (e.g. the absence of all color in some pelagic fish larvae) or to break up the body outline (e.g. spotted patterns of many bottom-dwelling flat fish). Both effects often occur in the same animal.

Cryptic species – species which are apparently identical phenotypically, often to the point where individuals of such species are themselves unable to make the distinction, but that are incapable of producing hybrid offspring.

Ctenoid scale – scale with tiny tooth-like projections along the posterior margin and on part of the exposed portion. Collectively these little teeth impart a rough texture to the surface of the scales.

Cusps (or *cuspids*) – principal projecting point of a tooth; see *teeth.

Cycloid scale – scale with smooth posterior margin, without spines on posterior margin; therefore, smooth to the touch.

Decomposer – a term that is generally synonymous with 'microconsumer'. In an ecosystem, decomposer organisms (mainly bacteria and fungi) enable nutrient recycling by breaking down the complex organic molecules of dead protoplasm and cell walls into simpler organic and (more importantly) inorganic molecules which may be used again by *primary producers. Recent work suggests that some macroconsumers may also play a role in decomposition (for example, *detritivores, in breaking down litter, speed its bacterial breakdown). In this sense 'decomposer' has a wider meaning than that traditionally implied.

Dentary – the main tooth bearing bone of the lower jaw.

Denticle – small tooth-like structure.

Depressed – flattened from top to bottom; body shape much wider then deep.

Depth – a vertical measurement of the body of a fish; most often employed for the maximum height of the body excluding the fins.

Dermal flap – usually employed in the sense of a bridge or lappet of skin.

Detritivore (*detritus feeder*) – a *heterotrophic animal that feeds on dead material (*detritus). The dead material is most typically of plant origin, but it may include the dead remains of small animals. Since this material may also be digested by decomposer organisms (fungi and bacteria) and forms the *habitat for other organisms (e.g. nematode worms and small insects), these too will form part of the typical detritivore diet. Animals that feed mostly on other dead animals (e.g. hyenas) or that feed mainly on the products (exuviae, e.g. dung) of larger animals are termed scavengers.

Detritus – litter formed from fragments of dead material (e.g. leaf litter, dung, molted feathers, and corpses). In aquatic habitats, detritus from leaf litter or fallen trees (woody debris) provides habitats equivalent to those which occur in soil humus.

Dextral – right-handed, right side; the opposite of *sinistral.

Dichotomous – repeatedly forking or dividing.

Discharge – a measure of the water flow, expressed as volume per unit time, at a particular point (e.g. a river gauging station). Various units of measurement are in common use, depending on the nature of the discharge being measured. River flow may be expressed in cubic meters per second.

Dissolved load – the part of a river's total load that is carried in solution. Five ions normally constitute approximately 90 per cent of the dissolved load: chloride (Cl^-), sulfate (SO_4^{2-}), dissolved bicarbonate (HCO_3^-), sodium (Na^+), and calcium (Ca^{2+}). Generally the load is at its maximum concentration during low-discharge conditions when groundwater is the main source of flow.

Dissolved oxygen level – the concentration of oxygen held in solution in water. Usually it is measured in mg/l or expressed as percentage of the saturation value for a given water temperature. The solubility of oxygen varies inversely with temperature; this is important because the warmer the water the larger the proportion of dissolved oxygen that is used by *poikilotherms. The dissolved oxygen level is an important first indicator of water quality, oxygen levels decline as pollution increases.

Distal – near outer edge; far end from point of attachment or centre of body.

Diurnal – (1) During daytime (as opposed to *nocturnal), as applied to events that occur only during daylight hours or to species that are active only in daylight. (2) At daily intervals, as applied to such daily rhythms as the normal pattern of waking and sleeping, leaf or flower opening and closing, or the characteristic rise and fall of temperatures associated with the hours of light and darkness.

DNA (*deoxyribonucleic acid*) – a nucleic acid, characterized by the presence of the sugar deoxyribose, the pyrimidine bases cytosine and thymine, and the purine bases adenine and guanine. It is the genetic material of organisms, its sequence of paired bases constituting the genetic code. Sequences of both mitochondrial and nuclear DNA are increasingly used in (molecular) systematic studies of fish relationships.

Dorsal – toward the back or upper part of the body; the opposite of *ventral.

Dorsal fin – median fin along the back which is supported by spines and/or rays; in spiny-rayed fishes the dorsal fin is separable into spiny-rayed and soft fins and can be continuous, incised (notched), separate (sometimes the soft-rayed portion has 1 or more spines anteriorly), or with separate spines.

Dorso-ventral – stretching from dorsal to ventral surface.

Drainage basin – see *catchment.

Drought – a period during which rainfall is either totally absent or substantially lower than usual for the area in question, so that there is a resulting shortage of water for human use, agriculture, or natural vegetation and fauna. In Suriname, drought is often associated with the El Niño-Southern Oscillation (ENSO) phenomenon.

Dry season – a period each year during which there is little precipitation. In places in very low latitudes two dry seasons may occur each year, between the northward and southward passage of the equatorial rains.

Ebb tide – falling *tide: the phase of the tide between high water and the succeeding low water (compare *flood tide).

Ecosystem – a term first used by A.G. Tansley (in 1935) to describe a discrete unit that consists of living and non-living parts, interacting to form a stable system. Fundamental concepts include the flow of energy via *food-chains and *food-webs, and the cycling of nutrients *biogeochemically. Ecosystem principles can be applied at all scales. Principles that apply to an ephemeral pond, for example, apply equally to a lake, an ocean, or the whole planet.

Ectopterygoid – one of the series of bones that suspends the jaw; see teeth.

Edentulous – without teeth.

Egg – in egg-laying animals (see *ovipary and *ovovivipary; compare *vivipary), the fertilized ovum and the embryo into which it develops, enclosed within a protective egg membrane. The eggs of animals that are laid out of the water have an impermeable outer covering (e.g. the shells of eggs laid by birds and reptiles) that protects them against drying.

Elongate – extended or drawn out.

Emarginate(d) – margin slightly concave; used to describe the posterior margin of a caudal fin which is inwardly curved.

Embedded – covered by skin (usually refers to scales).

Endemic – unique to a particular locality; restricted to a particular drainage, lake, country, etc.

Epithelial – of a covering or enveloping tissue.

Entire – smooth or straight margin.

Epaxial – referring to the main body muscles (myomeres) of the upper sides.

Estuary – partly enclosed coastal body of water which has a free connection with the open sea and where fresh water, derived from land drainage (river), is mixed with sea water. Estuaries are often subject to tidal action and, where tidal activity is large, ebb and flood tidal currents tend to avoid each other, forming separate channels. In estuaries where tidal activity is small, the invading dense sea water may flow under the lighter fresh water, forming a *salt wedge. A positive estuary is one in which surface salinities are lower within the estuary than in the open sea, owing to freshwater inflow exceeding outflow caused by evaporation. A negative estuary is one in which evaporation exceeds freshwater inflow and therefore hypersaline conditions exist. The action of tidal currents on the large amount of available sediment may give rise to a range of mobile bottom forms, including ebb and flood channels, sandbanks and sand waves.

Eutrophication – the process of nutrient enrichment (usually by nitrates and phosphates) in aquatic ecosystems.

Exotic species – an introduced, non-native species.

Extant – applied to a *taxon some of whose members are living at the present time. Compare *extinct.

Extinct – applied to a *taxon no member of which is living at the present time. Compare *extant.

Extinction – the elimination of a *taxon. This may take place in several ways. In the simplest case the taxon disappears from the record and is not replaced. Alternatively, one taxon may replace another, the earlier group consequently disappearing. Thus there is a process of either subtraction or substitution. Extinction generally takes place at particular times and places, but there are recurring periods when episodes of mass extinction have taken place. Environmental catastrophe, occurring for whatever reason, removes many groups from the environment and ecosystems collapse. Eventually new forms appear and evolution resumes. It would appear that periods of mass extinction control the pattern of evolution.

Extirpation – the bringing of a species to *extinction within part of its range.

Falcate (falciform) – deeply indented or sickle-shaped, e.g. the edge of a fin.

Family – an entity in the classification of animals and plants which consists of a group of related genera. Family names end in 'idea', an example being Gobiidae for the goby family; when used as an adjective, the 'ae' is dropped, hence gobiid fish.

Fauna – the animal life of a region or geological period.

Fecundity – the number of eggs released by a female fish during a spawning bout or breeding season; it varies from one or two in some sharks to tens of millions in tarpon *Megalops atlanticus*; fecundity decreases with increasing egg size and with increasing parental care, but increases with body size in an individual. Mouth-brooders, such as sea catfishes (Ariidae) and some cichlids (e.g. tilapia *Oreochromis mossambicus*) produce only about 100 eggs at a time, and live bearers such as guppies and four-eyes contain only about a dozen embryos. See *fertility.

Fertility – the reproductive capacity of an organism, i.e. the number of eggs that develop in a mated female over a specific period. It is usually calculated at the stage when this number is readily observable (i.e. in *oviparous animals when eggs are laid and in *viviparous animals when young are born), although strictly speaking it applies from the time that fertilization occurs. Sometimes the term fertility is applied only to the production of fertilized eggs (ova), while *fecundity is used for the production of offspring, so excluding those embryos which fail to develop.

Fertilization – the union of two *gametes to produce a zygote, which occurs during sexual reproduction. Fertilization involves the fusion of two haploid nuclei containing genetic material from two distinct individuals (cross-fertilization) or from one individual (self-fertilization). The resulting zygote then develops into a new individual. Most aquatic animals achieve fertilization externally, gametes uniting outside the body of the parents. Terrestrial species, but also livebearers and auchenipterid catfishes among the fishes, have internal fertilization, with the union of gametes inside the female.

Fin base – see *base of fin.

Fin formulae – of the three unpaired fins (dorsal, anal and caudal), only the formula of the first two is usually given (the number of caudal rays shows less variability among species). The first two or three, sometimes more, unbranched (or simple) rays are numbered in small Roman numbers: ii, iii, iv and so on; in certain cases there exists a strong, spinous ray, which is numbered in Roman Capitals: 'I'. The soft, branched rays are numbered in Arabic numbers. The formula of the two paired fins (pectoral and ventral/pelvic) usually shows one simple ray/spinous ray followed by a number of branched rays.

Finlets – small separate dorsal and anal fins.

Fish – broadly speaking, any *poikilothermic, legless, aquatic *vertebrate that possesses a series of gills on each side of the pharynx, a two-chambered heart, no internal nostrils, and at least a median fin as well as a tail fin. If the lampreys and hagfish (Agnatha) are excluded, this definition includes the sharks and rays (Chondrichthyes), in which the skeleton is cartilaginous, as well as the bony fish (Osteichthyes).

Flood plain – the part of a river valley that is made of unconsolidated, river-borne sediment and is periodically flooded. It is built up of relatively coarse debris left behind as a stream channel migrates laterally, and of relatively fine sediment deposited when bankfull flow discharge is exceeded.

Flood tide – the rising tide: the phase of the tide between low water and the next high tide. Compare *ebb tide.

Fontanel – lozenge-shaped or triangular hole in the top of the skull at this point being covered only by skin.

Food chain – the transfer of energy from the *primary producers (green plants) through a series of organisms that eat and are eaten assuming that each organism feeds only one other type of organism (e.g. earthworm – insectivore bird – bird of prey). At each stage much energy is lost as heat, a fact that usually limits the number of steps (*trophic levels) in the chain to four or five. Two basic types of food-chain are recognized: the grazing and detrital pathways. In practice these interact to give a complex *food-web.

Food-web – a diagram that represents the feeding relationships of organisms within an *ecosystem. It consists of a series of interconnecting *food-chains. Only some of the many possible relationships can be shown in such a diagram and it is usual to include only one or two carnivores at the highest levels.

Fork length – length of a fish as measured from the tip of the snout to the fork of the caudal fin (compare *standard length and *total length).

Forked – branched; in reference to the caudal fin, a fin shape with the rear edge distinctly indented, i.e. distinct upper and lower lobes and the posterior margin of each lobe relatively straight or gently curved.

Frontal – a major paired bone of the skull that articulates medially and generally found dorsal to the orbit.

Furcated – forked.

Fusiform – cigar-shaped or spindle-shaped, tapering at both ends (usually refers to body shape).

Gas bladder – see swim bladder.

Genital papilla – small, fleshy projection at the genital pore (immediately behind the anus) in some fishes.

Genotype – the genetic constitution of an organism, as opposed to its physical appearance (*phenotype). Usually this refers to the specific allelic composition of a particular gene or set of genes in each cell of an organism, but it may also refer to the entire *genome.

Genome – the total genetic information carried by a single set of chromosomes (i.e. in a haploid nucleus). A single representative of each of all the chromosome pairs in a nucleus will therefore bear the genome of an individual.

Genus – a taxonomic category including one species or a group of closely related (i.e. of common phylogenetic origin) species; the first part of the scientific name of an animal or plant. The plural is genera.

Gestation period – the length of time from conception to birth in a *viviparous animal; it is usually a relatively fixed period for a particular species.

Gestation spot – a ventral, blackish spot in female poeciliins (at the location of the ovaries with developing embryos) before they give birth to live young.

Gill – breathing organ, organ for exchange of dissolved gasses between water and the blood stream, including the highly vascularized *gill filaments that are used to extract oxygen from water; gill tissues are supported by a *gill arch in fishes.

Gill arch – bony angular skeleton that supports the *gill filaments and *gill rakers.

Gill chamber – cavity where the gills are located.
Gill cover – bony flap covering the outside of the gill chamber; see *operculum.
Gill filaments – principal site of gas exchange in the gill.
Gill membrane – membranes along the posterior and ventral margin of the gill cover (or *operculum).
Gill opening – the opening at the rear end (and often also ventrally) of the head of fishes, from the *gill chamber to the outside, where water of respiration is expelled. Bony fishes have a single such opening on each side; whereas cartilaginous fishes (sharks and rays) have five to seven. The gill openings of sharks and rays are called gill slits.
Gill rakers – bony tooth-like projections along the front edge of the gill arch (on the opposite side from the red gill filaments) that help prevent food from escaping through the gill opening; gill-raker counts are typically taken on the outermost (first) gill arch and are often separated into upper limb and lower limb counts; if a raker straddles the angle or the arch, the count is included in the lower limb; *rudiments are included in counts unless otherwise noted.
Gonads – reproductive organs, ovaries in females, testes in males.
Gonopodium – the front anal-fin rays 3, 4 and 5 of male livebearers (Poeciliidae and Anablepidae), modified to serve as an *intromittent organ for internal fertilization.
Growth – the increase in size of a cell, organ, or organism. This may occur by cell enlargement or by cell division.
Growth band (*growth line*) – a band or line, found in many organisms, which marks growth. Growth lines in *otoliths of fishes can be used to establish the age (in days or years) of individual fish.
Guild – a group of species all members of which exploit similar resources in a similar fashion (e.g. the guild of *piscivorous fishes, *detritivorous fishes etc).
Gular plate – bony plate on the throat, covering the underside of the head as exemplified in elopiform fishes.
Habitat – place where a fish normally lives, characterized by its physical and biotic properties.
Head length – the straight-line measurement of the head taken from the tip of the snout, front of the upper lip or chin – whichever is farthest forward – to the membranous rear edge of the *operculum; sometimes abbreviated as HL.
Herbivorous – vegetarian; feeding on plants or plant parts like flowers, leaves, fruits, seeds etc.
Hermaphrodite – having both male and female sex organs (gonads) in one individual (not necessarily at the same time).
Hermaphroditic fish – fish that, when mature, posses both male (testes) and female (ovary) sex glands at the same time (e.g. the mangrove rivulus *Kryptolebias marmoratus*). In such species cross-fertilization can occur during spawning. In other species the testes develop first, these males turning into females as the fish grow larger and older. Most Serranidae (sea bass) are females first or have both sets of glands equally developed.
Heterocercal – asymmetrical caudal fin in which the vertebral column extends into the upper lobe, which is usually longer (e.g. in sharks).
Heterotroph – an organism that is unable to manufacture its own food from simple chemical compounds and therefore consumes other organisms, living or dead, as its main or sole source of carbon. Often single-celled *autotrophs (e.g. *Euglena*) become heterotrophic in the absence of light.
Holotype – see *type specimen. Compare *lectotype, *neotype, *paratype, and *syntype.
Homocercal – type of caudal fin in which all of the principal rays of the fin attach to the modified last vertebra (the *hypural plate); this type of caudal fin is usually symmetrical with upper and lower lobes of the same size.

Humeral hiatus – see *pseudotympanum.

Humeral spot – a black or dark spot in the humeral region, at the upper edge of the pectoral fin base (on the flank behind the *opercle), occurring in many characids; it is supposed to be a recognition signal.

Humic acid – a mixture of dark-brown organic substances, which can be extracted from soil with dilute alkali and precipitated by acidification to pH 1-2 (in contrast with fulvic acid, which remains soluble in acid solution). Humic and fulvic acids are responsible for the tea color and acidity of black-water streams.

Hyaline – transparent, without marks.

Hydrologic cycle – the flow of water in various states through the terrestrial and atmospheric environments. Storage points (stages) involve groundwater and surface water, ice-caps, oceans, and the atmosphere. Exchanges between stages involve evaporation and transpiration from the Earth's surface, condensation to form clouds, and precipitation followed by run-off.

Hyoid – referring to the series of bones behind the gill cover that suspends the *branchiostegal rays and connects to the gill arches.

Hypomaxilla – see *supramaxilla.

Hypural plate – series of bones that support the caudal-fin rays.

Ichthyology – the study of fishes (ichthyologist: one who studies fishes).

Incisiform teeth – see *teeth.

Incised – notched, cut into; see *dorsal fin.

Incisor – flattened chisel-shaped tooth.

Inferior – below; mouth position on underside of head with snout projecting in front of mouth.

Infraorbital – another term for suborbitals (see *lacrimal); below the eye (as in *infraorbital canal or pores).

Infraorbital canal – the sensorial canal of the *lateral line system that runs around the eye within the small, flat bones, called *infraorbitals, that border the orbit.

Insertion – anterior or posterior point of attachment of a fin to the body.

Integument – referring to the skin.

Interdorsal – space on the back between the bases of the first and second dorsal fin (or the dorsal and adipose fins).

Intermittent stream – a stream which ceases to flow in dry periods. The flow may occur when the water-table is seasonally high, but there will not be flow when the water-table is significantly below the river-channel bed level. One view of the drainage network is that it is made up of *perennial streams, which flow all the time, intermittent streams, which flow when the water-table is seasonally high, and ephemeral streams, which flow only during storm conditions. Such streams tend to have permeable beds and discharge leaking through their beds (transmission losses) is added to the local groundwater.

International Code of Zoological Nomenclature (*International Commission on Zoological Nomenclature, ICZN*) – (1) A set of rules for the formal (scientific) naming of animals, accepted by zoologists, in which the underlying principle is the allocation of a single, unambiguous name to each *taxon. The starting point for naming animals is taken as *Systema naturae* (10th edition) published by *Linnaeus in 1758. (2) The international authority that draws up the regulations and supervises their application.

International Union for Conservation of Nature and Natural Resources (*IUCN*) – an international independent organization, founded in 1948, with headquarters in Switzerland, which promotes and initiates scientifically based conservation measures. Its members include more than 450 government agencies and conservation organizations in more than 100 countries, and it works closely with United Nations agencies. It publishes data books listing endangered species ('red lists'), and in 1980 it published the World Conservation strategy in collaboration with the UN Environment Program, the World Wide Fund for nature, the Food and Agriculture Organization, and UNESCO. Also see *rare species.

Interopercle – lower anterior bone of the gill cover.

Interorbital (distance) – space on top of the head between the eyes.

Intertidal – an area between the highest and lowest tide levels in a coastal region (i.e. the area of the shore covered at high tide and exposed at low tide).

Intertropical convergence zone (ITCZ, equatorial trough) – a low-latitude zone of convergence between air masses coming from either hemisphere at the boundary between north-easterly and south-easterly trade winds. Low-latitude depressions often form along the zone which moves latitudinally with the seasons, their occurrence being mainly in the ocean sectors and sometimes leading to tropical hurricanes when the zone is displaced relatively far from the equator. Over land, continental wind systems (e.g. monsoons) converge at the zone.

Intromittent organ – modified anal fin used for depositing sperm in the female reproductive tract in internal fertilization (e.g. in the families Auchenipteridae and Poeciliidae); see *gonopodium.

Invertebrate – animal lacking a backbone (e.g. insect, shrimp, worm); see *vertebrate.

Isthmus – triangular, frontmost part of the underside of the body, largely separated from the head, in most bony fishes, by the gill openings.

Iteroparity – the condition of an organism that has more than one reproductive cycle during its lifetime (i.e. most fish species). Compare *semelparity.

Jaws – composed of three bones on each side; the *premaxilla in the middle and the maxilla laterally, on the upper jaw, and the mandible (or *dentary bone, that actually is a part of the *mandibulary bone). These bones are usually toothed. However, on the maxilla the teeth are, more often than not, crowded on the upper part of the bone near its joint with the premaxilla.

Jugular – pertaining to the throat region; pelvic fins are jugular when positioned on the underside of the head in front of the pectoral fins.

Juvenile – young of a species; usually a small version of the adult (but lacking ripe *gonads).

Keeled – a shelflike fleshy or bony ridge in the form of a keel, with an acute angle like that of a hull.

Labial – referring to structures which are part of the lips or mouth.

Lacrimal – the most anterior of the series of 6 or fewer bones around the lower margin of the eye that are referred to as suborbital bones, located between the eye and the nostril; the lacrimal is sometimes also referred to as the preorbital.

Lamella – thin plate-like structure.

Lanceolate – spear- or lance-shaped.

Larva – (plural larvae) newborn; the developmental stage of a fish before it becomes a *juvenile.

Lateral – on the side or toward the side.

Lateral band – strip of color along the side of the body.

Lateralis system – sensory system consisting of a series of pores and canals on the head, body, and sometimes caudal fin of a fish, used to detect water movements.

Lateral line – a vibration sensory canal along the midside of the body with a series of pores that communicate to the outside of the body, often through specialized pored lateral-line scales. This canal is the rearward extension of the *lateralis system and contains sense organs which detect pressure changes. The lateral line can be absent, incomplete (only represented by a short anterior part behind the operculum), branched, or, in cichlids, divided into an anterior, dorsally running portion and a lower, posterior portion along the middle of the caudal peduncle; also see *scale formula.

Lateral scales – row of scales along the midside (usually along the lateral line) from rear end of gill cover to base of caudal fin; see *scale formula.

Lateral scale count – number of scales along lateral line, or along midline if lateral line is absent or incomplete; see *scale formula.

Lectotype – one of a collection of *syntypes which, subsequent to publication of the original description, is chosen and designated through published papers to serve as the *type specimen.

Lentic – applied to a freshwater habitat characterized by standing water (e.g. lakes, ponds, swamps, and bogs). Compare *lotic.

Leptocephalus – the elongate, highly compressed, transparent larva of some primitive bony fishes such as tarpon, bonefish and eels.

Life cycle – a series of developmental changes undergone by the individuals comprising a population including *fertilization, reproduction, and the death of those individuals, and their replacement by a new generation. The life cycle in fact is linear in respect of individuals but cyclical in respect of populations.

Lingual – of the tongue.

Linnaeus, Carolus (*Carl von Linné*) (*1707-78*) – a Swedish naturalist remembered for his large contributions to plant classification and his introduction of the binomial system of nomenclature. Trained as a botanist and physician in Uppsala, he went to Holland to continue his studies. While there he published the *Systema Naturae* (1735), a classified list of plants, animals and minerals. The list grew in later editions and the tenth edition (1758) is the starting-point of zoological nomenclature. In the first edition of his *Genera Plantarum* (1737) he gave more details of his artificial classification of plants, based largely on the number of stamens and pistils in a flower and the manner in which they occurred. In 1741 he was appointed to the chair of medicine at Uppsala, but exchanged it within a year for the chair of botany. He then began listing species, grouping them into genera, genera into classes, and classes into orders. In 1749 he introduced binomial classification, a system which used a Latin generic noun followed by a specific adjective. Ironically, it was largely this new system that allowed the development of ideas about the evolution of species, a concept to which Linnaeus was opposed. After his death, his collections and library were sold in 1784 to Sir James Edward Smith, the first president of the Linnean Society of London (founded in 1788) and in 1828 the Society purchased the collection, which it still holds. See also *International Code of Zoological Nomenclature.

Littoral – occurring at or in the immediate vicinity of the shoreline.

Lobule – small lobe or projection of an organ.

Lotic – applied to a freshwater *habitat characterized by flowing ('running') water, e.g. springs, rivers, and streams. Compare *lentic.

Lunate – crescent-shaped; caudal fin shape that is deeply emarginated with narrow lobes.

Mandible – lower jaw.

Marsh – a more or less permanently wet area of mineral soil, as opposed to a peaty area, e.g. around the edges of a lake or on a flood-plain of a river. Colloquially, 'marsh' is often used interchangeably with *swamp and *bog.

Mask – color mark across the eye.

Mating – the union of two individuals of opposite sexual type to accomplish sexual reproduction.

Maxilla – rear bone of the two bones that form the upper jaw, behind/above the *premaxilla. In ancestral fishes the maxilla is the principal bone of the upper jaw that bears teeth; in derived fishes it generally does not bear teeth and serves more to support the premaxilla.

Maxillary barbel – the tentacle-like protuberance attached to each end of the upper lip in catfishes.
Medial – in the middle plane or axis of the body.
Median – middle or toward the midline.
Median fins – unpaired fins that are positioned on the midline (median plane), hence the dorsal, anal and caudal fins.
Melanophore – cell containing melanin, a dark brown or black pigment. When contracted these cells appear as pepperlike dots; when expanded large areas of the fish may become dark.
Membrane – a thin sheet of tissue; often refers to thin sheet of tissue between fin and *branchiostegal rays.
Midlateral scales – refers to the longitudinal series of scales from the upper edge of the operculum or upper pectoral fin base to the base of the caudal fin. Generally used for fishes without a *lateral line such as gobies.
Midwater – in or near the middle water stratum, as opposed to at the surface or on the bottom (see *benthic); applied to fish that swim clear of the bottom and not at or immediately below the surface.
Molar – a low, blunt, rounded tooth for crushing and grinding; see *teeth.
Monophyletic – applied to a group of species that share a common ancestry, being derived from a single interbreeding population, as opposed to a polyphyletic group, which is derived from many such populations. If the members of a given *taxon are descended from a common ancestor they are said to be monophyletic (e.g. the families within a class would be monophyletic if they were all descended from the same family or a lower taxonomic unit). Under the strictest definition they would all have to be descended from a single species.
Monotypic – applied to any *taxon that has only one immediately subordinate taxon. For example, a genus that contains only one species or a family containing only one genus.
Morphology – the external form and structure of individual organisms, as distinct from their anatomy (which involves dissection).
Morphospecies – a group of biological organisms that differs in some morphological respect from all other groups.
Mudflat – an area of a coastline where fine-grained silt or clay is accumulating. Its development is favored by ample sediment, by sheltered conditions, and by the trapping effect of vegetation. It is an early stage in the development of a mangrove forest.
Multicuspid teeth – see *teeth.
Multiserial – arranged in more than one row.
Nape – dorsal part of the body just behind the *occiput or hard dorsal region of the skull, or, in spiny-rayed fishes, the part between the head and the point where the 1st (spiny) dorsal fin begins.
Natural classification – the ordering of organisms into groups on the basis of their evolutionary relationships.
Naturalized – applied to a species that was originally imported from another country but now behaves like a native in that it maintains itself without further human intervention and has invaded native communities.
Natural selection (*'survival of the fittest'*) – a complex process in which the total environment determines which members of a species survive to reproduce and so pass on their genes to the next generation. This need not involve a struggle between organisms. Natural selection is not necessarily the only mechanism for evolution.
Nature reserve – an area of land set aside for nature conservation and associated scientific research, usually with strong legal protection against other uses. Public access may be restricted partially or completely. The precise status and definition of reserves varies

from one country to another. The relative merits of the many criteria that govern the selection of nature reserves have been discussed extensively in recent ecological literature. Often a key objective is to maintain a fully representative range of *habitats and their flora and *fauna.

Neap tide – the tide of small range that occurs every 14 days, near the times of the first and last quarter of the moon, when the Moon, Earth, and Sun are at right angles. The neap tidal range is 10-30 per cent less than the mean tidal range. Compare *spring tide.

Neotype – in taxonomy, the specimen that is chosen to act as the 'type' material subsequent to a published original description. This occurs in cases where the original types have been lost, or where they have been suppressed by the *ICZN.

Niche (*ecological niche*) – the functional position of an organism in its environment, comprising the *habitat in which the organism lives, the periods of time during which it occurs and is active there, and the resources it obtains there. Abstract notion of the milieu on which the animal depends, not in terms of space, but as concerns its food. For example a fish feeding mostly on copepods occupies a different niche from that of another species feeding mostly on other fish. According to most ecologists each species has its own niche and no two species can occupy the same niche.

Nocturnal – active at night.

Nominate – a *taxon (subgenus or subspecies for example) that bears the same name as the equivalent higher taxon (genus or species).

Nostril – nasal opening or nare (fishes usually have 2 on each side).

Nuchal – pertaining to the nape (e.g. a nuchal hump in the subfamily Characinae).

Nuptial – relating to, or occurring during the mating season.

Occipital process – extension of the (supra-)occipital bone at the rear edge of the skull.

Occiput – upper back part of the head or skull.

Ocellus – a round eye-like spot or marking; usually dark, bordered by a ring of light pigment.

Ocular side – in flatfishes (Pleuronectiformes), the side where the eyes are located; the opposite side is called the blind side.

Oligotrophic – applied to waters or soils that are poor in nutrients and with low *primary productivity. The low nutrient content means that plankton blooms are rare and littoral plants are scarce. The low organic content means that dissolved oxygen levels are high. The rivers of Suriname can be considered oligotrophic as they contain few erosion products due to the weathered character of the Precambrian Guiana Shield. Compare *eutrophic.

Omnivorous – feeding on both plants and animals.

Opercle – large posterior upper bone of the gill cover; often bears one to three backward-directed spines in higher (perciform) fishes.

Operculum – gill cover composed of the *preopercle, *opercle, *interopercle, and *subopercle.

Orbital – referring to the eye, particularly the series of bones surrounding the eye including the supraorbital, interorbital, preorbital, and suborbital, which are respectively located above, between, in front of, and below the eye.

Order – a major unit in the classification of organisms; an assemblage of related families. The ordinal word ending in the animal kingdom is 'iformes'.

Origin – anterior point of attachment of fins to the body (anterior insertion); i.e. the point where a fin begins.

Otolith – a small calcareous structure (internal ear bone) in the inner ear of fishes (large in Sciaenidae); concentric growth rings on these structures are often used to determine age.

Ovipary (*oviparous*) – the method of reproduction in which eggs are laid and embryos develop outside the mother's body, nourishment comes from the egg, and each egg eventually hatching into a young animal. Little or no development occurs within the mother's body. Most invertebrates and many vertebrates reproduce in this way.

Ovovivipary (*ovoviviparous*) – the method of reproduction in which young develop from eggs retained within the mother's body but separated from it by egg membranes. The eggs contain considerable yolk, and nourishment for the developing embryo is still derived from the egg. Many insect groups, fish, and reptiles reproduce in this way. Compare *ovipary and *vivipary.

Paired fin – fins found on both sides of the body; collectively, the *pectoral and *pelvic fins.

Palaeospecies (*chronospecies*) – a group of biological organisms, known only from fossils, which differs in some respect from all other groups.

Palate – roof of the mouth.

Palatine – one of a pair of bones on the roof of the mouth; one on each side, between the jaw and the midline; each side of the palate, behind and lateral to the *vomer, often bearing teeth.

Papilla – a small, nipplelike fleshy projection.

Papillose – with papillae.

Paratype – in taxonomy, a specimen, other than a *type specimen, which is used by an author at the time of the original description, and designated as such by the author.

Parental care – all activities that are directed by an animal towards the protection and maintenance of its own offspring or those of a near relative.

Parietal – a bone of the upper posterior part of the skull.

Pectoral fin – paired fins on the sides just behind the gill cover, attached to the shoulder girdle.

Peduncle – a stalk-like process (see *caudal peduncle).

Pelagic – living in open water away from the bottom.

Pelvic fins – paired fins on the lower part of the body in front of the anus, sometimes called the *ventral fins; lower or primitive fishes generally have the pelvic fins in abdominal position while derived (advanced) fishes have the pelvic fins in the *thoracic or *jugular position.

Perennial stream – a stream that normally flows throughout the year, albeit with low dry-weather flows in occasional drought years.

Periphyton (*aufwuchs*) – organisms attached to or clinging to the stems and leaves of plants or other objects (rock, woody debris) projecting above the bottom sediments of freshwater *ecosystems.

pH – a value on a scale 0-14 which gives a measure of the acidity or alkalinity of a medium. A neutral medium has a pH of 7, acidic media have pH values of less than 7, and alkaline media of more than 7. The lower the pH the more acidic, the higher the pH the more alkaline. The pH value is the logarithm of the reciprocal of the hydrogen ion concentration expressed in moles per liter ($pH = \log_{10} 1/H^+$). Most pH values in natural systems lie in the range 4-9. Human blood has a pH of 7.4, ocean water 8.1-8.3, water in saline environments may have a pH around 9.0 or higher, and water in acidic soils and black-water streams may have a pH of 4.0 or less.

Pharyngeal – of, or near, the pharynx.

Pharyngeal teeth – teeth on the elements of the last *gill arch or *pharyngeal arch.

Phenotype – the observable manifestation of a specific *genotype; those observable properties of an organism produced by the genotype in conjunction with the environment. Organisms with the same overall genotype may have different phenotypes because of the effects of the environment and of gene interaction. Conversely, organisms may have the same phenotype but different genotypes, as a result of incomplete dominance, penetrance, or expressivity.

Photosynthesis – the series of metabolic reactions occurring in certain *autotrophs, whereby the energy of sunlight, absorbed by chlorophyll, powers the reduction of carbon dioxide (CO_2) and the synthesis of organic compounds. In green plants, where water (H_2O) acts as both a hydrogen donor and a source of released oxygen, photosynthesis may be summarized by the equation: $6\ CO_2 + 6\ H_2O > C_6H_{12}O_6 + 6\ O_2$; the process in green plants where, under the action of light, carbohydrates are synthesized from carbon dioxide and water (and producing oxygen as a by-product).

Phylogenetics – the taxonomical classification of organisms on the basis of their degree of evolutionary relatedness.

Phylogenetic systematics – the study of biological organisms, and their grouping for purposes of classification, based on their evolutionary descent. See also *cladistics.

Phylogeny – evolutionary relationships within and between taxonomic levels, particularly the patterns of lines of descent, often branching, from one organism to another (i.e. the relationships of groups of organisms as reflected by their evolutionary history).

Piscivorous – fish eating.

Pit lines – very tiny holes in the skin, arranged in parallel, wavy lines and depending on the lateral sensory system; the pit lines are usually situated on the top and sides of the head.

Plankton – small plants (called phytoplankton) and animals (zooplankton) that are mostly free-floating with the currents.

Plicate – with folds of skin.

Poikilotherm (*exotherm*) – an organism whose body temperature varies according to the temperature of its surroundings. Fish are poikilotherms.

Polyandry – the mating of a female with more than one male at one time (usually taken to be during the course of a single breeding season). Simultaneous polyandry occurs where the female mates with more than one male to produce a clutch of eggs or brood of young bearing genes from each male. Successive polyandry occurs where the female mates with one male, lays eggs or gives birth to the young, but plays little part in their parental care (usually cared for by this first male), instead moving away and mating with another male (e.g. in seahorses and pipefishes, family Syngnathidae). The existence of simultaneous polyandry is usually difficult to prove, and polyandry appears to be mainly of the successive type. Polyandry is generally much rarer than *polygyny, presumably because the mother is not usually sexually receptive following successful mating, and since she bears the young she is more involved with their care and less able to seek another mate.

Polygamy – in animals, a pattern of mating in which an individual has more than one sexual partner. See also *polyandry and *polygyny; opposite of monogamy.

Polygyny – in animals, a pattern of mating in which a male has more than one female partner; compare *polyandry.

Polymorphism – phenomenon of genetic origin, causing the individuals of a single species to be not all of the same coloration or, less often, of the same form.

Polytypic – said of a genus with more than one species; the opposite of *monotypic.

Pool – quiet, often relatively deep, segment of a stream; see *riffle, *run.

Pore – tiny opening in the skin, usually involved with sensory perception in fishes.

Pored scale – scale with a pore, e.g. lateral line scales.

Posterior – pertaining to the rear portion.

Postmaxillary process – a broad or finger-like extension of the *premaxilla along the upper edge of the lower arm of this bone.

Post-occipital process – see *occipital process.

Postopercular spot/ocellus – a spot behind the operculum; also see *humeral spot/ocellus.

Postorbital – behind the eye sockets; 1 or more of the suborbital bones, starting with the third suborbital bone and possibly referring also to the fourth, fifth, and sixth suborbital bone (see *lacrimal).

Potamodromous – applied to fish that undertake regular migrations in large freshwater systems (rivers) (e.g. *Prochilodus, Brachyplatystoma*). Compare *anadromous and *catadromous.

Predation – the interaction between species populations in which one organism, the predator, obtains energy (as food) by consuming, usually killing, another, the prey. Most typically, a predator is an animal that catches, kills, and eats its prey; but predation also includes feeding by insectivorous plants and grazing interactions (e.g. the complete consumption of unicellular phytoplankton by zooplankton). Predation is analogous to parasitism in being both a means of feeding and a cause of immediate harm, but differs in that predators are usually larger than prey organisms (but not always, e.g. piranhas) and the prey is usually killed. The distinction is not clear-cut, especially for insect populations. In detail, a continuous gradation of interactions is found from predation to parasitism.

Predorsal scales – the series of scales along the mid-dorsal line between the head and the origin of the dorsal fin.

Premaxilla – anterior bone in upper jaw (see *maxilla); in the higher fishes it extends backward and bears all of the teeth of the jaw. It is this part of the upper jaw which can be protruded by many fishes.

Preopercle – a boomerang-shaped, upper anterior bone of the gill cover, the edges of which form the posterior and lower margins of the cheek. The upper vertical margin is sometimes called the upper limb, and the lower horizontal edge the lower limb; the two limbs meet at the angle of the preopercle.

Preopercular spine – any of the spines along the rear or lower edges of the preopercle.

Preorbital – referring to the region in front of the eye; a suborbital bone in front and below the eye (see *lacrimal).

Prickles – tiny, spiny scales of puffers (Tetraodontidae).

Primary productivity (*primary production*) – the rate at which *biomass is produced by photosynthetic and chemosynthetic autotrophs (mainly green plants) in the form of organic substances some of which are used as food materials. Gross primary productivity (GPP) is the total rate of *photosynthesis and chemosynthesis, including that portion of the organic material produced which is used in respiration during the measurement period. Net primary productivity (NPP) is the rate of production after some has been lost to plant respiration during the measurement period. Net *community productivity is the rate of accumulation of organic material, allowing for both plant respiration and heterotrophy predation during the measurement period. Rates of storage at higher trophic levels are termed secondary productivities. Strictly, the primary production of an *ecosystem (as distinct from its productivity, which is a rate) is the amount of organic material accumulated. It is usual to indicate a period of time, however, since otherwise such data have limited value. Thus in practice the terms 'production' and 'productivity' are often used interchangeably. It is also vital to indicate whether production figures relate to net or gross and to primary or community productivity, or to some portion of these, as with a harvested crop. In the oceans, photosynthesis by *phytoplanktonic algae in the upper 100m (the euphotic zone) accounts for most primary production. Waters of tropical areas are less productive than those of temperate regions, because in tropical areas the water column undergoes no seasonal vertical mixing and so becomes *oligotrophic. Areas of upwelling of nutrient-rich deep waters have high productivity.

Primitive – applied to a *character (as synonym of plesiomorphic) or, occasionally, to a whole organism that preserves the character state of an ancestral stage. See *cladistics.

Principal caudal-fin ray – branched and unbranched caudal-fin rays that reach the rear margin of the fin.

Procurrent caudal-fin ray – small ray (sometimes spinous) at the insertions of the fin that do not reach the rear margin (as opposed to *principal ray).

Producer – in an *ecosystem, an organism that is able to manufacture food from simple inorganic substances (i.e. an autotroph, most typically a green plant). Compare *consumer organism.

Production – (1) The total mass of organic matter that is manufactured in an *ecosystem during a certain period of time. Net production is the yield of the producers and consumers and is the amount of living matter in the ecosystem. (2) In energy-flow studies, that part of the assimilated food or energy which is retained and incorporated in the *biomass of the organism, but excluding the reproductive bodies released by the organism. This may be regarded as growth. In energy-flow measurements, production is expressed as energy per unit time, per unit area. (3) See *primary productivity.

Production/respiration ratio (*P/R ratio*) – the relationship between gross *production and total community respiration. Where P/R = 1 a steady-state community results. If P/R is persistently greater or less than 1 then organic matter either accumulates or is depleted respectively.

Protractile mouth – *protrusible mouth.

Protrusible – capable of projection; in reference to the mouth whose upper lip is not attached to the snout, and which may be extended far forward to catch prey (e.g. in many Perciformes and *Bivibranchia*).

Proximal – part nearest the centre of the body.

Pseudotympanum – translucent, triangular area in the humeral region, on the side behind the *opercle where the muscles are missing and the anterior part of the gas bladder is directly in contact with the skin, which is supposed to improve the hearing (see *Weberian ossicles). Quite often present in very young characoids, the pseudotympanum, also called *humeral hiatus, persists in several species, when adult, chiefly among the Cheirodontinae.

Pterygoid – see *teeth.

Pyloric caecum – (plural caeca) small blind-ending, fingerlike pouches at the junction of the stomach and intestine, opening to the posterior part of the stomach.

Radii (*of scales*) – furrows in the scales, usually radiating from their center; not to be confounded with the circuli, which are thinner, much more numerous and usually concentric, being related to the growth of the fish.

Rarity – the relative abundance of a species and, therefore, its vulnerability to extinction. The *International Union for Conservation of Nature and Natural Resources measures the vulnerability of a species according to five criteria: (a) the rate at which its numbers are observed, inferred, or projected to be declining; (b) in association with (a), whether the species occurs as a single, small population or a few, small, fragmented ones; (c) in association with (a), whether the species occupies a small geographic range or area; (d) the size of the population; (e) a mathematical estimate of the predicted risk of extinction within a specified time. From this assessment, species are allocated a position on a continuum of increasing threat with three categories: as 'critical'; 'endangered'; or 'vulnerable'. Species that are known to be at risk of extinction, but fail to qualify for any of the main categories, are classified as 'susceptible'.

Ray – supporting bony element of fins; flexible, segmented fin ray (often branched); ray is sometimes used as a collective term to designate both soft (branched) rays and spines; it is also sometimes used to designate exclusively, soft rays.

Reproductive isolating mechanism (*RIM*) – the means by which different *species are kept reproductively isolated. These may be: (a) chromosomal (if cross-mating occurs, the incompatibility of the karyotypes makes any hybrid inviable or sterile); (b) mechanical (the two species cannot mate because they are of different sizes, or because the genitalia are shaped differently); (c) ethological (the courtship rituals of the two species diverge at some point so that an incorrect response is given and the sequence is brought to a stop); or (d) ecological (the two species occupy different microhabitats and normally do not meet). Other mechanisms include the breeding seasons being out of phase, or members of one species being unattractive to members of the other. Many RIMs, especially ethological ones, amount merely to mate preference, so that in the absence of a preferred partner (of the same species) a member of a different species will be accepted. In this way hybrids between different species may be bred in captivity and may even be found to be fully fertile.

Reservoir – a surface body of water whose flow is artificially controlled by means of a dam, embankments, or sluice gates in such a way that the water remains static until it is allowed to flow for a specific purpose (e.g. flood control, public water supply or production of hydroelectric energy).

Respiration – (1) Oxidative reactions in cellular metabolism that involve the sequential degradation of food substances and the generation of a high-energy compound, ATP (adenosine triphosphate) in aerobic respiration with the use of molecular oxygen as a final hydrogen acceptor; ATP, carbon dioxide, and water are the products thus formed. (2) The physic-chemical processes involved in the transportation of oxygen to and carbon dioxide from the tissues. (3) (external respiration) The act of breathing.

Reticulate – color markings in a chainlike pattern or network.

Rheophilous – applied to an organism that thrives in running water.

Riffle – fast-flowing, shallow segment of a stream where the surface of the water is broken over rocks or debris. See *pool, *run.

Riparian – pertaining to a river-bank (from the Latin *ripa*, meaning 'bank').

River profile – the slope along the bed of a river, expressed as a graph of distance-from-source against height. In detail it is typically compound, with the profiles of individual segments reflecting the local rock types. It may be broken stepwise by knick points (e.g. at rapid complexes).

River terrace (*stream terrace*) – a fragment of a former valley floor that now stands well above the level of the present *flood-plain and is usually covered by fluvial deposits. It is caused by stream incision, which may be caused by uplift of the land, a fall in sea level, or a change in climate.

Rostral/rostrum – towards the front of the fish / the area of the snout; an extension of the snout (e.g. in sawfish or Pristidae).

Rounded – a caudal-fin shape with the terminal border smoothly convex.

Rudiment – a poorly developed structure, usually small and minimally functional at best; these include small, incompletely developed unbranched fin rays and gill rakers at the ends of a gill arch.

Rugose (*scale*) – in certain species the surface of the scale is not smooth but spread with tiny tubercles or spines; this is usually in relation to a *ctenoid (dented) apical border.

Run – transitional segment of a stream between a *riffle and a *pool, with moderate current and depth.

Saddle – color mark, more or less rectangular, on the back.

Salinity – a measure of the total quantity of dissolved solids in water, in parts per thousand (ptt) by weight, when all organic matter has been completely oxidized, all carbonate has been converted to oxide, and bromide and iodide to chloride. The salinity of ocean water is in the range 33-38 part per thousand, with an average of 35 ppt.

Salt wedge – an intrusion of sea water into a tidal *estuary in the form of a wedge along the bed of the estuary. The lighter fresh water from riverine sources overrides the denser salt water from marine sources unless mixing of the water masses is caused by estuarine topography. Salt wedges are found in *estuaries where a river discharges through a relatively narrow channel.

Savanna – a tropical vegetation dominated by grasses with varying admixtures of tall bushes and/or trees in open formation. Savanna occurs in diverse tropical environments, although most experience a dry season. Much savanna is no doubt climatic climax, although extensive tracts are anthropogenic fire climaxes and others are edaphically (soil) controlled; it is generally difficult to distinguish one type from the other.

Savanna woodland – a savanna in which trees and shrubs form a generally light canopy. The trees and bushes are generally deciduous, yet evergreens are usually also well represented. Some tall trees occur, but most are stunted and gnarled. They frequently have thick, corky, fire-resistant bark.

Scalation – arrangement of scales; squamation; see *scale formula.

Scale formula – most of the species descriptions give the following data concerning the number of scales on the body: (1) longitudinal: range of the number of scales along the lateral line from end of opercle to beginning of caudal fin (example: Sq.L.30-33); when the lateral line is not complete the number of perforated scales is given between parentheses and followed by the total number (example: Sq.L. (7-8)30-33). (2) Transverse: range of the number of scales in a transverse line usually between the dorsal and pelvic fins, the scale of the lateral line not being counted; when there is no lateral line the corresponding row is treated as if there might be a lateral line, or the number of transverse scales is given in full (example Sq.tr.7-8). (3) Predorsal: range of the number of scales along the predorsal line from occipital process to the dorsal fin or, if the line is irregular, along a side line. (4) Circumpeduncular: number of scales around the middle part of the caudal peduncle counting both sides plus one scale above and below. (5) Preventral and post ventral scales: these formulae are rarely used.

Scapula/scapular – a flat bone on the upper part of the pectoral girdle / pertaining to the shoulder region.

School – loosely, an aggregation of fish or marine mammals which are observed swimming together, possibly in response to a threat from a predator. More strictly, a grouping of fish, drawn together by social attraction, whose members are usually of the same species, size, and age, the members of the school moving in unison along parallel paths in the same direction. Sudden changes at the leading edge of the school are followed almost instantaneously by the remainder of the group, with the fish on the flank becoming the new leaders. In general, the overall size and shape of the school, as well as the cruising depth and speed, vary from one species to another.

Scute – a modified scale that can be enlarged, hardened, ridged, keeled, or spiny.

Semelparity (*big-bang reproduction*) – the condition of an organism that has only one reproductive cycle during its life-time. Compare *iteroparity.

Serrae (*abdominal*) – sawlike notches along an edge; also called abdominal spines, the serrae are formed by the scales of the ventral region, which became spinous forming a saw along the trenchant belly of certain groups. The number of serrae in front and behind the pelvic fins is of systematic importance in the Serrasalminae.

Serrate – with saw-like teeth along a margin.

Setae – bristles or hardened hair-like projections.

Sexual dimorphism – external differences in the form (or coloration) between the two sexes of a species; the characters showing this difference are called secondary sexual characters.

Sexual hooklets – tiny hooks along the foremost rays of the pelvic and anal fins of the males of most Characidae (rarely in other groups), sometimes on the whole fin or even on caudal and dorsal fins. These hooks, which may belong to the so-called contact organs, are supposed to act as a coupling device between the two sexes during mating to insure the best fertilization.

Sinistral – left-handed, left side; the opposite of *dextral.

Snout – the region of the head in front of the eye and above the mouth. Snout length is measured from the front of the upper lip to the anterior edge of the eye.

Soft dorsal fin – the portion of the dorsal fin supported by soft rays.

Soft ray – a fin support element that is composed of 2 halves (paired laterally), segmented, and usually flexible and branched. Rarely, soft rays can be pointed and stiff and appear to be a spine.

Spatulate – flattened, sometimes used to describe head or tooth shape.

Species (*singular and plural*) – the fundamental unit in the classification of animals and plants; literally, a group of organisms that resemble one another closely: the term derives from the Latin *speculare*, 'to look'. In *taxonomy it is applied to one or more groups (populations) of individuals that can interbreed within the group but cannot exchange genes with other groups (populations), or, in other words an interbreeding group of biological organisms which is isolated reproductively from all other organisms. A species can be made up of groups in which members do not actually exchange genes with members of other groups (though in principle they could do so), as, for example, at the two extremes of a continuous geographical range. However, if some gene flow occurs along a continuum, the formation of another species is unlikely to occur. Where barriers to gene flow arise (e.g. physical barriers, such as sea, mountains, or areas of unfavorable habitat) this reproductive isolation may lead by either local selection or random genetic drift to the formation of morphologically distinct forms termed races or subspecies. These could interbreed with other races of the same species if they were introduced to one another. Once this potential is lost, through some further evolutionary divergence, the races may be recognized as species, although this concept is not a rigid one. Most species cannot interbreed with others; a few can, but produce infertile offspring; a smaller number may actually produce fertile offspring. The term cannot be applied precisely to organisms whose breeding behavior is unknown. See *morphospecies and *palaeospecies.

Speck – very small blotch; see *blotch, *spot.

Sphenotic – a bone of the skull above and behind the orbit.

Spine – (1) sharp, bony projection, usually on the head; (2) hard, unbranched ray in a fin, i.e. a fin support element that is unpaired laterally, unsegmented, unbranched and usually stiff and pointed.

Spinous dorsal fin – the anterior portion of the dorsal fin that is supported by spines.

Spinule – a small spine.

Spiracle – an opening between the eye and the first gill slit of sharks and rays which leads to the gill chamber in sharks, rays and certain primitive bony fishes. (Not the gill opening.)

Spot – regular circular or elongated color mark. See *blotch, *speck.

Spring tide – a tide of greater than the mean range (i.e. the water level rises markedly above and falls markedly below the mean tide level). Spring tides occur about every two weeks, when the Moon is full or new, and are at their maximum when the Moon and the Sun are in the same plane as the Earth. Compare *neap tide.

Standard length – straight-line distance from the anteriormost point of the fish (tip of the snout, the front of the upper lip, chin) to the posterior end of the vertebral column that is generally equivalent to the end of the *hypural plate (and recognized externally by the crease between the tail and caudal fin when the caudal fin is bent laterally); sometimes abbreviated as SL. Used as a standard measure of the length of the fish (where the caudal fin is often damaged).

Stream order – a measure of the position of a stream (defined as the reach between successive tributaries) within the hierarchy of the drainage network. A commonly used approach allocates order '1' to unbranched tributaries, '2' to the stream after the junction of the first '1' tributaries, and so on. It is the basis for quantitative analysis of the network.

Stripe – generally refers to a horizontal nearly straight band of color; see *band, *bar.

Submandibular pores – pores along the underside of the mandible (lower jaw).

Subopercle – lower rear bone in the gill cover.

Suborbital bones – below the eye sockets; see *lacrimal.

Subterminal – in reference to the position of the mouth: the opening of the mouth positioned posterior and below the tip of the snout; see *inferior, *terminal, *upturned.

Sulcus – a groove or fissure.

Superior – above or on upper surface; a mouth position with the snout behind the anterior opening of the mouth.

Supracaudal spot – a spot on the upper part of the caudal peduncle or caudal-fin base.

Supramaxilla – 1 or 2 bones above the maxilla, between the posterior tip of the premaxilla and the expanded blade of the maxilla, found in primitive bony fishes (e.g. in Clupeiformes); also known as hypomaxilla.

Supra-occipital – a bone on the dorsal side of the great foramen of the skull, usually forming a part of the occipital in the adult, but distinct in the young.

Suspended load – the part of the total load of a stream that is carried in suspension. It is made up of relatively fine particles that settle at a lower rate than the upward velocity of water eddies and its highest concentration is in the zone of greatest turbulence, near the bed. It reaches a maximum in shallow streams of high velocity. Compare *dissolved load.

Swamp – a wet area that is normally covered by water all year and is not subject to drying out during the dry season.

Swim bladder – a gas-filled sac located between the backbone and the abdominal cavity, used in buoyancy (and in some fishes as a secondary respiratory organ); also referred to as air bladder or gas bladder.

Sympatric – occurring in same geographic area; see *allopatric.

Symphysis – the articulation between two bones; often refers to the anterior juncture between the two halves of either jaw.

Synonym – an invalid scientific name of an organism proposed after the accepted name; also referred to as junior synonym.

Syntopic – occurring at exactly the same location, e.g. two fish species are said to occur syntopically if they actually live in the same creek.

Syntype – in taxonomy, all specimens in a type series in which no *type specimen was designated.

Systematics – the part of biology that deals with the classification of living things and their evolution, with the following three categories: Nomenclature, or the naming of entities in accordance with the International Code; Taxonomy, or classifying them in answer to our innate need of order; and Systematics in the strict sense dealing with the relationships of species and higher categories, mostly from an evolutionary point of view.

Tail rings – body segments of a pipefish located between the anus and the base of the caudal fin.

Taxon (*pl. taxa*) – a group of organisms of any taxonomic rank (e.g. order, family, genus, or species).

Taxonomy – the scientific classification of organisms.

Teeth (formula) – teeth play an important role in the classification of many fish groups (e.g. characoids). The teeth are usually present on the *jaws (*premaxillary, *maxillary and *mandibular teeth), in the throat (*pharyngeal teeth), and sometimes on the palate (*palatine teeth) where they are situated mostly on the part called *ectopterygoid. When the fish is large enough, the teeth can be observed under a stereomicroscope, or even a simple magnifying lens. When the fish is very small (and the teeth are of great systematic importance), the dissection of the jaws on one side is necessary; the soft tissues destroyed with a drop of dilute chloride bleach, while watching under the microscope, and simply rinsed before the chemical has attacked the tiny bones. When there is sufficient biological material, it is possible to stain one specimen with alizarin (which stains the bone red) after having cleared the soft tissues, usually with diluted potash (KOH). The teeth are usually counted on the premaxilla first, then on the maxilla and on the mandible, occasionally on the palate, and their number is given from one side only (although both sides are counted for verification). When there is more than one row (teeth said to be biserial, instead of uniserial), which occurs mostly in the Characidae, the two (eventually three) rows are separated by an oblique bar or other system (example of a characoid tooth formula: pmx 2-3/5 (rarely 6); mx 2-3; mdb 5 + several (smaller teeth on side)). The teeth have only one point or *cusp, or several, in which case they are called pluricuspid or multicuspid; with two points, bicuspid or bifid, with three points tricuspid, then quadricuspid, pentacuspid, etc. With more than five cusps one usually gives, for brevity, the range of the number of points, which may be as many as 9-12, or sometimes even more in certain 'hand-like', flat teeth of the Hemiodontidae. Independent of the number of cusps, the tooth may be more or less conical and pointed, in which case it is called caniniform (without cusps), or even canine when very big; it may be compressed with the cutting edge straight or denticulated, like our incisors (incisiform), or broad with the crushing part more or less flat like our molars (Géry, 1977). Loricariid catfishes often have slender, threadlike (filiform) bilobed teeth.

Terminal – pertaining to at the end, or situated at the end; a mouth position with the opening of the mouth even with the tip of the snout, the upper and lower jaws being equally far forward; see *subterminal, *upturned, *superior.

Territorial – defending a particular area.

Thalweg – a line joining the lowest points of successive cross-sections, either along a river channel or, more generally, along a valley that it occupies. The word is German and means 'valley course'.

Thoracic – referring to the breast region; pelvic fins are thoracic in position when directly below the pectoral fins.

Tidal flat – an area of intertidal sand flat, mud flat, and marsh developed in some lagoons in mesotidal areas, and in protected bays and estuarine areas along macrotidal coasts. In tropical areas tidal flats tend to be colonized by mangrove swamps.

Tidal range – the difference in height between consecutive high and low waters. The tidal range varies from a maximum during *spring tides to a minimum during *neap tides. In tide tables daily high- and low-water heights are given for each geographical locality mentioned.

Tide – the periodic rise and fall of the Earth's oceans, caused by the relative gravitational attraction of the Sun, Moon, and Earth. The effect of the Moon is about twice that of the Sun, given rise to the *spring-*neap tide cycle of tides. Variation in tides is caused by: (a) changes in the relative positions of the Sun, Moon, and Earth; (b) uneven distribution of water on the Earth's surface; and (c) variation in the seabed topography. Semidiurnal tides are those with two high and two low waters (period 12 hours and 25 minutes) during a tidal day (24 hours and 50 minutes). Diurnal tides have one high and one low water during a tidal day.

Total length – the length of a fish from the front of whichever jaw is most anterior to the end of the longest caudal fin ray; sometimes abbreviated as TL.

Trait – any detectable *phenotypic property of an organism; a *character.

Transverse scales – row of scales from anal fin origin to dorsal fin (or middle of back); see *scale formula.

Trophic level – a step in the transfer of food or energy within a chain. There may be several trophic levels within a system, for example, *producers (autotrophs), primary *consumers (herbivores), and secondary consumers (carnivores); further carnivores may form fourth and fifth levels. There are rarely more than five levels since usually by this stage the amount of food or energy is greatly reduced. See also *ecosystem.

Tropical rain forest – a term invented in 1898 by the botanist A.F.W. Schimper (tropische Regenwald) to describe the forests of the permanently wet tropics. His definition still stands: 'evergreen, at least 30 m tall, rich in thick-stemmed lianas, and in woody as well as herbaceous epiphytes'. Nearly all independent plants have the form of trees and at maturity most attain 1-30 m tall. By convention, the trees are divided into four strata (but these are mere abstractions) of which the topmost is the emergent stratum of usually single, huge trees (up to 60 m or more) standing head and shoulders above the continuous canopy.

Truncate – terminating abruptly in a square or flattened end; a caudal-fin shape with a vertically straight terminal border and angular or slightly rounded corners.

Trunk myomeres – body segments.

Trunk rings – body segments of a pipefish located between the pectoral fins and the anus.

Turnover time – the measure of the movement of an element in a *biogeochemical cycle; the reciprocal of turnover rate. Turnover time is calculated by dividing the quantity of nutrient present in a particular nutrient pool or reservoir by the flux rate for that nutrient element into or out of the pool. Turnover time thus describes the time it takes to fill or empty that particular nutrient reservoir.

Type locality – the locality where the *type specimen(s) was (were) collected.

Type specimen (*holotype*) – an individual plant or animal chosen by taxonomists to serve as the basis for naming and describing a new species or variety. Compare *lectotype, *neotype, *paratype, and *syntype.

Uniserial – arranged in a single row.

Upturned – in reference to the position of the mouth, used when the mouth opens above the foremost point of the head; see *superior, *subterminal, *inferior and *terminal.

Vent – see anus.

Ventral – below, on the lower half of the head or body, or abdominal part of the body (i.e. that part of the body furthest removed from the backbone); the opposite of *dorsal.

Ventral fins – see *pelvic fins.

Vertebrae – bones of the vertebral column or back bone; vertebral counts are often given as a formula: precaudal vertebrae + caudal vertebrae, where precaudal vertebrae typically have paired ventrolateral extensions that support ribs and caudal vertebrae have a single ventrally directed spine (haemal spine) and does not support ribs.

Vertebrate – animal with a backbone (e.g. fish, frog, bird, pig).

Vertical fins – median fins; the dorsal, caudal, and anal fins.

Vestige – small or underdeveloped structure, as in a rudiment.

Villiform – many small slender outgrows, usually in a close-set patch or carpet; often refers to slender teeth forming velvety bands.

Vivipary – the method of reproduction in which development and growth of the embryo is internally with nourishment from the mother. Compare *ovipary and *ovovivipary.

Vomer(ine) – an unpaired median bone in the front of the roof of the mouth; often used to describe location of teeth on this bone.

Watershed – see *catchment.

Weathering – the breakdown of rocks and minerals at and below the Earth's surface by the action of physical and chemical processes. Essentially it is the response of Earth materials to the low pressures, low temperatures, and presence of air and water that characterize the near-surface environment, but which were not typical of the environment of formation.

Weberian ossicles – chain of small bones, derived from the apophyses of the anterior four or five vertebrae, connecting the swim bladder to the inner ear for sound transmission. The possession of Weberian ossicles characterizes an otophysan fish (orders Cypriniformes, Characiformes, Siluriformes, Gymnotiformes).

Wetlands – a general term applied to open-water *habitats and seasonally or permanently waterlogged land areas, including lakes, rivers, and estuarine and freshwater marshes. Wetland habitats, especially *marsh and *bog areas, are among the most vulnerable to destruction since they can be drained and reclaimed for agriculture or forestry, drained for pest control (e.g. to eliminate breeding grounds for malaria-carrying mosquitoes), or modified for water supply, flood control, hydroelectric power schemes, waste disposal etc.

Zoogeography – the study of the geographical distribution of animals at different taxonomic levels, from the order down to *species level. Emphasis is given to the explanation of distinctive patterns in terms of past and/or present factors.

BIBLIOGRAPHY

Abell, R., M.L. Thieme, C. Revenga, M. Bryer, M. Kottelat, N. Bogutskaya, B. Coad, N. Mandrak, S.C. Balderas, W. Bussing, M.L.J. Stiassny, P. Skelton, G.R. Allen, P. Unmack, A. Naseka, R. Ng, N. Sindorf, J. Robertson, E. Armijo, J.V. Higgins, T.J. Heibel, E. Wikramanayake, D. Olson, H.L. Lopez, R.E. Reis, J.G. Lundberg, M.H. Sabaj Perez & P. Petry, 2008. Freshwater ecoregions of the world: a new map of biogeographic units for freshwater biodiversity conservation. *BioScience* 58, 403-414.

Albert, J.S., 2001. Species diversity and phylogenetic systematics of American knifefishes (Gymnotiformes, Teleostei). *Misc. Publ. Mus. Zool. University of Michigan* 190, 1-129.

Albert, J.S. & R. Campos-da-Paz, 1998. Phylogenetic systematics of Gymnotiformes with diagnoses of 58 clades: a review of available data. In *Phylogeny and Classification of Neotropical Fishes* (L.R. Malabarba, R.E. Reis, R.P. Vari, Z.M.S. Lucena & C.A.S. Lucena, eds), pp. 419-446. EDIPUCRS, Porto Alegre, Brazil.

Albert, J.S. & W.G.R. Crampton, 2003. Seven new species of the Neotropical electric fish *Gymnotus* (Teleostei, Gymnotiformes) with a redescription of *G. carapo* (Linnaeus). *Zootaxa* 287, 1-54.

Albert, J.S. & W.L. Fink, 2007. Phylogenetic relationships of fossil Neotropical electric fishes (Osteichthyes: Gymnotiformes) from the Upper Miocene of Bolivia. *Journal of Vertebrate Paleontology* 27, 17–25.

Albert, J.S., N.R. Lovejoy & W.G.R. Crampton, 2006. Miocene tectonism and the separation of cis- and trans-Andean river basins: evidence from Neotropical fishes. *Journal of South American Earth Sciences* 21, 14-27.

Aleva, G.J.J. & Th.E. Wong, 1998. History of bauxite exploration and mining in Suriname. In *The History of Earth Sciences in Suriname* (Th.E. Wong, D.R. de Vletter, L. Krook, J.I.S. Zonneveld & A.J. van Loon, eds.), pp. 275-310. Royal Netherlands Academy of Arts and Sciences and Netherlands Institute of Applied Geoscience (TNO), Amsterdam.

Alexander, R. McN., 1963. Frontal foramina and tripodes of the characin *Crenuchus*. *Nature* 200, 1225.

Alexander, R. McN., 1965. Structure and function in the catfish. *Journal of Zoology* 148, 88-152.

Alexandrou, M.A., C. Oliveiras, M. Maillard, R.A.R. McGill, J. Newton, S. Creer & M.I. Taylor, 2011. Competition and phylogeny determine community structure in Müllerian co-mimics. *Nature* 469, 84-89.

Almeida, R.G., 1984. Biologia alimentar de três espécies de *Triportheus* (Pisces: Characoidei, Characidae) do Lago do Castanho, Amazonas. *Acta Amazonica* 14, 48-76.

Amatali, M.A., 1993. Climate and surface water hydrology. In *The Freshwater Ecosystems of Suriname* (P.E. Ouboter, ed.), pp. 29-51. Kluwer, Dordrecht, the Netherlands.

Aquino, A.E., 1994. Secondary sexual dimorphism of the dermal skeleton in two species of the Hypoptopomatine genus *Otocinclus* (Siluriformes: Loricariidae). *Ichthyological Explorations of Freshwaters* 5, 217-222.

Araujo-Lima, C.A.R.M., B.R. Forsberg, R. Victoria & L. Martinelli, 1986. Energy sources for detritivorous fishes in the Amazon. *Science* 234, 1256-1258.

Arrington, D.A., K.O. Winemiller, W.F. Loftus & S. Akin, 2002. How often do fish run on empty? *Ecology* 83, 2145-2151.

Artedi, P., 1738. *Ichthyologia, sive opera omnia piscibus scilicet: bibliotheca ichthyologica. Philosophia ichthyologica. Genera piscium. Synonymia specierum. Descriptions specierum. Omnia in hoc genere perfectiora, quam antea ulla. Posthuma vindicavit, recognovit, coaptavit et edidit Carolus Linnaeus, Med. Doct. Et Ac. Imper. N.C.* Leiden.

Artigas, L.F., P. Vendeville, M. Leopold, D. Guiral & J.F. Ternon, 2003. Marine biodiversity in French Guiana: estuarine, coastal, and shelf ecosystems under the influence of Amazonian waters. *Gayana* 67, 302-326.

Assunção, M.I.S. & H.O. Schwassmann, 1995. Reproduction and larval development of *Electrophorus electricus* on Marajó Island (Pará, Brazil). *Ichthyological Explorations Freshwaters* 6, 175-184.

Attrill, M.J., 1998. The benthic macroinvertebrate communities of the Thames estuary. In *A Rehabilitated Estuarine Ecosystem: the Environment and Ecology of the Thames Estuary* (M.J. Attrill, ed.), pp. 85-113. Kluwer Academic Publishers, Dordrecht, the Netherlands.

Augustinus, P.G.E.F., 1978. The changing shoreline of Surinam (South America). PhD thesis Utrecht University. Natuurwetenschappelijke Studiekring voor Suriname en de Nederlandse Antillen, no. 95, 232 pp.

Augustinus, P.G.E.F., 1980. Actual development of the chenier coast of Suriname (South America). *Sedimentary Geology 26*, 91-113.

Balon, E.K., 1975. Reproductive guilds of fishes: a proposal and definition. *Journal of the Fisheries Research Board of Canada 32*, 821-864.

Barletta, M., A. Barletta-Bergan, U. Saint-Paul & G. Hubold, 2005. The role of salinity in structuring the fish assemblages in a tropical estuary. *Journal of Fish Biology 66*, 45-72.

Barletta-Bergan, A., M. Barletta & U. Saint-Paul, 2002. Structure and seasonal dynamics of larval fish in the Caeté River Estuary in North Brazil. *Estuarine Coastal Shelf Science 54*, 193-206.

Barlow, G.W., 1967. Social behavior of a South American leaf fish, *Polycentrus schomburgkii*, with an account of recurring pseudofemale behavior. *American Midland Naturalist 78*, 215-234.

Barthem, R.B. & M. Goulding, 1997. *The Catfish Connection: Ecology, Migration, and Conservation of Amazon Predators*. Columbia University Press, New York.

Barthem, R.B., M.C.L.B. Ribeiro & M. Petrere, 1991. Life strategies of some long-distance migratory catfish in relation to hydroelectric dams in the Amazon basin. *Biological Conservation 55*, 339-345.

Bennett, K.D., 1990. Milankovitch cycles and their effect on species in ecological and evolutionary time. *Paleobiology 16*, 11-21.

Berra, T.M., 2001. *Freshwater Fish Distribution*. The University of Chicago Press, Chicago.

Bleeker, P., 1862. Description de quelques espèces nouvelles de Silures de Suriname. *Verslagen en Mededelingen van de Koninklijke Akademie van Wetenschappen, afdeling Natuurkunde, Amsterdam 14*, 371-389.

Bleeker, P., 1864. Description des espèces de Silures de Suriname, conservées aux Musées de Leide et d'Amsterdam. *Natuurkundige Verhandelingen van de Hollandsche Maatschappij der Wetenschappen te Haarlem, Ser. 2 20*, 1-104.

Bleeker, P., 1873. Description de deux espèces nouvelles de Sciénoïdes de Surinam. *Archives néerlandaises des sciences exactes et naturelles 8*, 456-461.

Boeseman, M., 1952. A preliminary list of Surinam fishes not included in Eigenmann's enumeration of 1912. *Zoologische Mededelingen 31*, 179-200.

Boeseman, M., 1968. The genus *Hypostomus* Lacépède, 1803, and its Surinam representatives (Siluriformes, Loricariidae). *Zoologische Verhandelingen 99*, 1-89.

Boeseman, M., 1971. The 'comb-toothed' Loricariinae of Surinam, with reflections on the phylogenetic tendencies within the family Loricariidae. *Zoologische Verhandelingen 116*, 1-56.

Boeseman, M., 1972. Notes on the South American catfishes, including remarks on Valenciennes and Bleeker types in the Leiden Museum. *Zoologische Mededelingen 47*, 293-320.

Boeseman, M., 1982. The South American mailed catfish genus *Lithoxus* Eigenmann, 1910, with the description of three new species from Surinam and French Guyana and records of the related species (Siluriformes, Loricariidae). *Proc. Kon. Ned. Acad. Wetensch. C85*, 41-58.

Bonatta, S.L. & F.M. Salvano, 1997. A single and early migration for the peopling of the Americas supported by mitochondrial DNA sequence data. *Proceedings of the National Academy of Sciences USA 94*, 1866–1871.

Bone, Q. & R.H. Moore, 2008. *Biology of Fishes* (*3rd ed.*). Taylor and Francis, New York.

Bossy, J., J. Delage & J. Géry, 1965. Interprétation histo-morphologique de l'organe frontal des Crenuchidae. *C. R. l'Ácademie Sciences Paris 261*, 48370-4840.

Boujard, T. & R. Rojas-Beltrán, 1988. Zonation longitudinale du peuplement ichtyque du fleuve Sinnamary (Guyane française). *Revue Hydrobiologie Tropicale 21*, 47-61.

Boujard, T., P. Keith & P. Luquet, 1990. Diel cycle in *Hoplosternum littorale* (Teleostei): evidence for synchronization of locomotor, air breathing and feeding activity by circadian alternation of light and dark. *Journal of Fish Biology 36*, 133-140.

Boujard, T., D. Sabatier, R. Rojas-Beltran, M.F. Prevost & J.F. Renno, 1990. The food habits of three allochthonous feeding characoids in French Guiana. *Revue d'Ecologie: La Terre et la Vie 45*, 247-258.

Bowen, S.H., 1981. Digestion and assimilation of periphytic detrital aggregate by *Tilapia mossambica*. *Transactions of the American Fisheries Society 110*, 239-245.

Bowen, S.H., 1983. Detritivory in neotropical fish communities. *Environmental Biology of Fishes 9*, 137-144.

Brenner, M. & U. Krumme, 2007. Tidal migration and patterns in feeding of the four-eyed fish *Anableps anableps* L. in a north Brazilian mangrove. *Journal of Fish Biology 70*, 406-427.

Britz, R., 1997. Egg surface structure and larval cement glands in nandid and badid fishes with remarks on phylogeny and biogeography. *American Museum Novitates 3195*, 1-17.

Bruton, M.N. & R.E. Boltt, 1975. Aspects of the biology of *Tilapia mossambica* Peters (Pisces: Cichlidae) in a natural freshwater lake (Lake Sibaya, South Africa). *Journal of Fish Biology 7*, 423-445.

Bubberman, F.C., 1973. De bosbranden van 1964 in Suriname. *Nieuwe West-Indische Gids 49*, 163-173. (in Dutch)

Buckup, P.A., 1993. Review of the characidiin fishes (Teleostei: Characiformes), with descriptions of four new genera and ten new species. *Ichthyological Explorations of Freshwaters 4*, 97-154.

Buckup, P.A., 1998. Relationships of the Characidiinae and phylogeny of Characiform fishes (Teleostei: Ostariophysi). In *Phylogeny and Classification of Neotropical Fishes* (L.R. Malabarba, R.E. Reis, R.P. Vari, Z.M.S. Lucena & C.A.S. Lucena, eds.), pp. 123-144. EDIPUCRS, Porto Alegre, Brazil.

Buitrago-Suárez, U.A. & B.M. Burr, 2007. Taxonomy of the catfish genus *Pseudoplatystoma* Bleeker (Siluriformes: Pimelodidae) with recognition of eight species. *Zootaxa 1512*, 1-38.

Bullock, L.H., M.D. Murphy, M.F. Godcharles & M.E. Mitchell, 1992. Age, growth and reproduction of jewfish *Epinephelus itajara* in the Eastern Gulf of Mexico. *U.S. Fishery Bulletin 90*, 243-249.

Burgess, W.E., 1989. *An Atlas of Freshwater and Marine Catfishes*. T.F.H. Publications, Neptune City, NJ.

Burns, J.R. & J.A. Flores, 1981. Reproductive biology of the cuatro ojos, *Anableps dowi* (Pisces: Anablepidae), from El Salvador and its seasonal variations. *Copeia 1981*, 25-32.

Cadée, G.C., 1975. Primary production off the Guyana coast. *Netherlands Journal of Sea Research 9*, 128-143.

Cardoso, Y.P. & J.I. Montoya-Burgos, 2009. Unexpected diversity in the catfish *Pseudancistrus brevispinis* reveals dispersal routes in a Neotropical center of endemism: the Guyanas Region. *Molecular Ecology 189*, 947-964.

Carolsfeld, J., B. Harvey, C. Ross & A. Baer (eds.), 2003. *Migratory Fishes of South America: Biology, Fisheries and Conservation Status*. International Development Research Centre & the World Bank, Ottawa.

Carter, G.S. & L.C. Beadle, 1930. The fauna of the swamps of the Paraguayan Chaco in relation to its environment. I. Physico-chemical nature of the environment. *Zoological Journal of the Linnean Society 37*, 205-258.

Carter, G.S. & L.C. Beadle, 1931. The fauna of the swamps of the Paraguayan Chaco in relation to its environment. II. Respiratory adaptations in the fishes. *Zoological Journal of the Linnean Society 37*, 327-368.

Carvalho, F.M., 1980. Alimentação do mapará (*Hypophthalmus edentatus* Spix, 1829) do lago Castanho, Amazonas (Siluriformes, Hypophthalmidae). *Acta Amazonica 10*, 545-555.

Carvalho, L.N., R. Arruda & J. Zuanon, 2003. Record of cleaning behavior by *Platydoras costatus* (Siluriformes: Doradidae) in the Amazon Basin, Brazil. *Neotropical Ichthyology 1*, 137-139.

Carvalho, M.R., 2001. Review of: "Freshwater Stingrays from South America" by R. Ross, and F. Schäfer. *Copeia 2001*, 1167-1169.

Carvalho, M.R. & N.K. Lovejoy, 2011. Morphology and phylogenetic relationships of a remarkable new genus and two new species of Neotropical freshwater stingrays from the Amazon basin (Chondrichthyes: Potamotrygonidae). *Zootaxa 2776*, 13-48.

Casatti, L., 2001. Taxonomia do gênero Sul-Americano *Pachyurus* Agassiz, 1831 (Teleostei: Perciformes: Sciaenidae) e descrição de duas novas species. *Comun. Mus. Ciênc. Tecnol. PUCRS, Sér. Zoologia, Porto Alegre 14*, 133-178.

Casatti, L., 2002. Revision of the South American genus *Pachypops* Gill 1861 (Teleostei: Perciformes: Sciaenidae), with the description of a new species. *Zootaxa 26*, 1-20.

Casatti, L., 2005. Taxonomy of the South American freshwater genus *Plagioscion* (Teleostei, Perciformes, Sciaenidae). *Zootaxa 1080*, 39-64.

Castro, R.M.C. & R.P. Vari, 2004. Detritivores of the South American fish family Prochilodontidae (Teleostei: Ostariophysi: Characiformes): a phylogenetic and revisionary study. *Smithsonian Contributions to Zoology 622*, 1-189.

Cervigon, F., R. Cipriani, W. Fischer, L. Garibaldi, M. Hendrickx, A.J. Lemus, R. Marquez, J.M. Poutiers, G. Robiana & B. Rodriguez, 1993. *Field Guide to the Commercial Marine and Brackish-Water Resources of the Northern Coast of South America*. Food and Agricultural Organization, Rome, 513 pp.

Charlier, P., 1988. State of exploitation and development strategies for the fishery resources in Suriname (fin-fish). *De Surinaamse Landbouw 36*, 1-18.

Church, J.E. & W.C. Hodgson, 2002. The pharmacological activity of fish venoms. *Toxicon 40*, 1083-1093.

Cockerell, T.D.A., 1925. A fossil fish of the family Callichthyidae. *Science 62*, 397-398.

Cohen, M.A., W.R.C. Beaumont & N.C. Thorp, 1999. Movement and activity patterns of the black piranha. *Environmental Biology of Fishes 54*, 45-52.

Collette, B.B., 1974. South American freshwater needlefishes (Belonidae) of the genus *Pseudotylosurus*. *Zoologische Mededelingen 48*,169-186.

Collette, B.B., 1982. South American freshwater needlefishes of the genus *Potamorrhaphis* (Beloniformes: Belonidae). *Proceedings Biological Society Washington 95*, 714-747.

Collette, B.B., 2004. Family Hemiramphidae Gill 1859 – halfbeaks. *California Academy of Sciences Annotated Checklists 22*, 1-35.

Cordani, U.G. & K. Sato, 1999. Crustal evolution of the South American Platform, based on Nd isotopic systematics on granitoid rocks. *Episodes 22*, 167-173.

Cordier, S., C. Grasmick, M. Pasquier-Passelaigue, L. Mandereau, J.P. Weber & M. Jouan, 1998. Mercury exposure in French Guiana: levels and determinants. *Archives of Environment Health 53*, 299-303.

Costa, W.J.E.M., 1998. Phylogeny and classification of the Cyprinodontiformes (Euteleostei: Atherinomorpha): a reappraisal. In *Phylogeny and Classification of Neotropical Fishes* (L.R. Malabarba, R.E. Reis, R.P. Vari, Z.M.S. Lucena & C.A.S. Lucena, eds.), pp. 537-560. EDIPUCRS, Porto Alegre, Brazil.

Costa, W.J.E.M., 2006. Redescription of *Kryptolebias ocellatus* (Hensel) and *K. caudomarginatus* (Seegers) (Teleostei: Cyprinodontiformes: Rivulidae), two killifishes from mangroves of south-eastern Brazil. *J. Ichthyol. Aquat. Biol. 11*, 5-12.

Covain, R. & S. Fisch-Muller, 2012. Molecular evidence for the paraphyly of *Pseudancistrus sensu lato* (Siluriformes: Loricariidae), with revalidation of several genera. *Cybium 36*, 229-246.

Covain, R., S. Fisch-Muller, J. Montoya-Burgos, J. Mol, P.Y. Le Bail & S. Dray, 2012. The Harttiini (Siluriformes, Loricariidae) from the Guianas: a multitable approach to assess their diversity, evolution, and distribution. *Cybium 36*, 115-161.

Crabtree, R., E.C. Cyr, R.E. Bishop, L.M. Falkenstein & J.M. Dean, 1992. Age and growth of tarpon, *Megalops atlanticus*, larvae in the eastern Gulf of Mexico, with notes on relative abundance and probable spawning areas. *Environmental Biology of Fishes 35*, 361-370.

Crampton, W.G.R., 1996. Gymnotiform fish: an important component of Amazonian floodplain fish communities. *Journal of Fish Biology 48*, 298–301.

Crampton, W.G.R., 1998. Effects of anoxia on the distribution, respiratory strategies and electric signal diversity of gymnotiform fishes. *Journal of Fish Biology 53* (Supplement A), 307-330.

Crampton, W.G.R. & J.S. Albert, 2003. Redescription of *Gymnotus coropinae* (Gymnotiformes, Gymnotidae), an often misidentified species of Neotropical electric fish, with notes on natural history and electric signals. *Zootaxa 348*, 1-20.

Crampton, W.G.R. & C.D. Hopkins, 2005. Nesting and parental care in the weakly electric fish *Gymnotus* (Gymnotiformes: Gymnotidae) with descriptions of larval and adult electric organ discharges of two species. *Copeia 2005*, 48-60.

Crul, R. & L. Reyrink (1980). *Ecologische studie Kabalebo project. Hydrobiologisch onderzoek in het stroomgebied van de Corantijn en Kabalebo rivier in verband met de aanleg van het Devisvallen stuwmeer.* Hydraulic Research Division (Waterloopkundige Afdeling), Ministry of Public Works, Paramaribo, Suriname, 69 pp. (in Dutch)

Da Silva, G.C., J. Sabino, C.J.R. Alho, V.L.B. Nunes & V. Haddad-Jr, 2010. Injuries and envenoming by aquatic animals in fishermen of Coxim and Corumbá municipalities, State of Mato Grosso do Sul, Brazil: identification of the causative agents, clinical aspects and first aid measures. *Revista da Sociedade Brasileira da Medicina Tropical 43*, 1-5.

De Chambrier, S. & J.I. Montoya-Burgos, 2008. *Pseudancistrus corantijniensis*, a new species from the Guyana Shield (Siluriformes: Loricariidae) with a molecular and morphological description of the *Pseudancistrus barbatus* group. *Zootaxa 1918*, 45-58.

De Kom, J.F.M., M.A. Vrede & F.A. De Wolff, 2001. Tetrodotoxin poisoning due to consumption of the brackwater fish 'Bosrokoman'. In *Human Toxicology in Suriname*. Pp. 113-115. PhD thesis J.F.M. de Kom, University of Leiden, the Netherlands.

De Merona, B., J. Mol, R. Vigouroux & P. de Tarso Chaves, 2009. Phenotypic plasticity in fish life-history traits in two neotropical reservoirs: Petit-Saut Reservoir in French Guiana and Brokopondo Reservoir in Suriname. *Neotropical Ichthyology* 7, 683-692.

De Pinna, M.C.C., 1998. Phylogenetic relationships of Neotropical Siluriformes (Teleostei: Ostariophysi): historical overview and synthesis of hypotheses. In *Phylogeny and Classification of Neotropical Fishes* (L.R. Malabarba, R.E. Reis, R.P. Vari, Z.M.S. Lucena & C.A.S. Lucena, eds.), pp. 279-330. EDIPUCRS, Porto Alegre, Brazil.

De Santana C.D. & W.G.R. Crampton, 2011. Phylogenetic interrelationships, taxonomy, and reductive evolution in the Neotropical electric fish genus *Hypopygus* (Teleostei, Ostariophysi, Gymnotiformes). *Zoological Journal of the Linnean Society* 163, 1096-1156.

De Santana, C.D. & R.P. Vari, 2010. Electric fishes of the genus *Sternarchorhynchus* (Teleostei, Ostariophysi, Gymnotiformes); phylogenetic and revisionary studies. *Zoological Journal of the Linnean Society* 159, 223-371.

De Souza, L.S., J.W. Armbruster & D.C. Werneke, 2012. The influence of the Rupununi portal on distribution of freshwater fish in the Rupununi District, Guyana. *Cybium* 36, 31-43.

De Vletter, D.R. & A.L. Hakstege, 1998. The search for gold in Suriname. In *The History of Earth Sciences in Suriname* (Th.E. Wong, D.R. de Vletter, L. Krook, J.I.S. Zonneveld & A.J. van Loon, eds.), pp. 311-350. Royal Netherlands Academy of Arts and Sciences and Netherlands Institute of Applied Geoscience (TNO), Amsterdam.

Deynat, P., 2006. *Potamotrygon marinae* n.sp., a new species of freshwater stingrays from French Guiana (Myliobatiformes, Potamotrygonidae). *Comptes Rendus Biologies* 329, 483-493. (In French)

Diogo, R., M. Chardon & P. Vandewalle, 2004. Osteology and myology of the cephalic region and pectoral girdle of *Batrochoglanis raninus*, with a discussion on the synapomorphies and phylogenetic relationships of the Pseudopimelodinae and Pimelodidae (Teleostei: Siluriformes). *Animal Biology* 54, 261-280.

Dudgeon, D., A.H. Arthington, M.O. Gessner, Z-I. Kawabata, D.J. Knowler, C. Lévêque, R.J. Naiman, A.H. Prieur-Richard, D. Soto, M.L.J. Stiassny & C.A. Sullivan, 2006. Freshwater biodiversity: importance, threats, status, and conservation challenges. *Biological Reviews* 81, 163-182.

Dumas, P., 2006. Tidal migration patterns of juvenile penaeid shrimps in a French Guianese coastal mangrove. *Ann. Limnol. Int. J. Lim.* 42, 157-163.

Dyer, K.R., 1997. *Estuaries: a Physical Introduction*. John Wiley, New York.

Eigenmann, C.H., 1912. The freshwater fishes of British Guiana, including a study of the ecological groupings of species and the relation of the fauna of the plateau to that of the lowlands. *Memoirs of the Carnegie Museum* 5, 1-578.

Eisma, D., P.G.E.F. Augustinus & C. Alexander, 1991. Recent and subrecent changes in the dispersal of Amazon mud. *Netherlands Journal of Sea Research* 28, 181-192.

Emanuels, J.A., 1977. Visserij. In *Encyclopedie van Suriname* (C.F.A. Bruijning & J. Voorhoeve, eds.), pp. 636-640. Elsevier, Amsterdam. (in Dutch)

Endler, J.A., 1983. Natural and sexual selection on color patterns in poeciliid fishes. *Environmental Biology of Fishes* 9, 173-190.

Eschmeyer, W.N. & R. Fricke (eds.), 2011. Catalog of fishes database. On-line version. http://research.calacademy.org/ichthyology/catalog/fishcatsearch.html.

Favorito, S.E., A.M. Zanata & M.I. Assumpção, 2005. A new *Synbranchus* (Teleostei: Synbranchiformes: Synbranchidae) from ilha de Marajó, Pará, Brazil, with notes on its reproductive biology and larval development. *Neotropical Ichthyology* 3, 319-328.

Fernholm, B. & A. Wheeler, 1983. Linnean fish specimens in the Swedish Museum of Natural History, Stockholm. *Zoological Journal of the Linnean Society* 78, 199-286.

Ferraris, C.J. & R.P. Vari, 1999. The South American catfish genus *Auchenipterus* Valenciennes, 1840 (Ostariophysi: Siluriformes: Auchenipteridae): monophyly and relationships, with a revisionary study. *Zoological Journal of the Linnean Society* 126, 387-450.

Figueiredo, J., C. Hoorn, P. van der Ven & E. Soares, 2009. Late Miocene onset of the Amazon River and the Amazon deep-sea fan: evidence from the Foz do Amazonas Basin. *Geology* 37, 619-622.

Fisch-Muller, S., J.I. Montoya-Burgos, P.Y. Le Bail & R. Covain, 2012. Diversity of the Ancistrini (Siluriformes: Loricariidae) from the Guianas: the Panaque group, a molecular appraisal with descriptions of new species. *Cybium 36*, 163-193.

Fittkau, E.J., 1967. On the ecology of Amazonian rainforest streams. *Atas do Simpósio sôbre a Biota Amazônica 3*, 97-108.

Forsberg, B., C.A. Araujo-Lima, L.A. Martinelli, R.L. Victoria & J.A. Bonassi, 1993. Autotrophic carbon sources for fish of the Central Amazon. *Ecology 74*, 643-652.

Garrone-Neto, D. & I. Sazima, 2009. Stirring, charging, and picking: hunting tactics of potamotrygonid rays in the upper Paraná River. *Neotropical Ichthyology 7*, 113-116.

Gautier, J.Y., P. Planquette & Y. Rouger, 1988. Etude éthologique de la relation mâle-femelle au cours du cycle de reproduction chez *Hoplosternum littorale*. *Revue d'Ecologie: La Terre et la Vie 43*, 389-398.

Gayet, M. & F.J. Meunier, 1998. Maastrichtian to early Late Paleocene freshwater Osteichthyes of Bolivia: additions and comments. In *Phylogeny and Classification of Neotropical Fishes* (L.R. Malabarba, R.E. Reis, R.P. Vari & Z.M.S. Lucena, eds.), pp. 85-100. EDIPUCRS, Porto Alegre, Brazil.

Gee, J.H. & J.B. Graham, 1978. Respiratory and hydrostatic functions of the intestine of the catfishes *Hoplosternum thoracatum* and *Brochis splendens* (Callichthyidae). *Journal of Experimental Biology 74*, 1-16.

Geiger, S.P., J.J. Torres & R.E. Crabtree, 2000. Air breathing and gill ventilation frequencies in juvenile tarpon, *Megalops atlanticus*: responses to changes in dissolved oxygen, temperature, hydrogen sulfide, and pH. *Environmental Biology of Fishes 59*, 181-190.

Geijskes, D.C., 1942. Observations on temperature in a tropical river. *Ecology 23*, 106-110.

Geijskes, D.C., 1943. *De visscherij in het district Nickerie*. Landbouwproefstation Report No. 3. Ministry of Agriculture, Paramaribo. (in Dutch)

Géry, J., 1961. Notes on the Ichthyology of Surinam and other Guianas 7. *Hyphessobrycon georgetti* sp. nov., a dwarf species from southern Surinam. *Bulletin of Aquatic Biology 2*, 121-128.

Géry, J., 1962. Notes on the Ichthyology of Surinam and other Guianas 10. The distribution pattern of the genus *Hemibrycon*, with a description of a new species from Surinam and an incursion into ecotaxonomy. *Bulletin of Aquatic Biology 3*, 65-80.

Géry, J., 1966. Notes on characoid fishes collected in Surinam by Mr. H.P. Pijpers, with descriptions of new forms. *Bijdragen tot de Dierkunde 35*, 101-129.

Géry, J., 1972. Poissons characoïdes des Guyanes. I. Généralités. II. Famille des Serrasalmidae. *Zoologische Verhandelingen 122*, 3-266.

Géry, J., 1976. Les genres de Serrasalmidae (Pisces, Characoidei). *Bulletin Zoologisch Museum Universiteit van Amsterdam 5*, 47-54.

Géry, J., 1977. *Characoids of the World.* TFH Publications, Neptune City, USA, 672 pp.

Géry, J. & P. Planquette, 1983. Une nouvelle espèce de *Leporinus* (Poissons characoïdes, Anostomidés) de la Guyane et du Surinam: *Leporinus lebaili* n.sp. *Revue fr. Aquariol. 10*, 65-70.

Géry, J., P. Planquette & P.Y. Le Bail, 1995. Une espèce nouvelle de *Moenkhausia* de la Guyane (Teleostei, Ostariophysi, Characidae), d'écailles nombreuses. *Revue fr. Aquariol. 22*, 67-70.

Géry, J., P. Planquette & P.Y. Le Bail, 1996. Nouvelles espèces guyanaises d'*Astyanax* S.L. (Teleostei, Characiformes, Characidae) a épines pelviennes, avec une introduction concernant le groupe. *Cybium 20*, 3-36.

Géry, J., P.Y. Le Bail & P. Keith, 1999. *Cynodon meionactis* sp.n., un nouveau characidé endémique du bassin du Haut Maroni en Guyane, avec une note sur la validité du genre *Cynodon* (Teleosteri : Ostariophysi : Characiformes) *Revue fr. Aquariol. 25*, 69-77.

Ghedotti, M.J., 1998. Phylogeny and classification of the Anablepidae (Teleostei: Cyprinodontiformes). In *Phylogeny and Classification of Neotropical Fishes* (L.R. Malabarba, R.E. Reis, R.P. Vari, Z.M.S. Lucena & C.A.S. Lucena, eds.), pp. 560-582. EDIPUCRS, Porto Alegre, Brazil.

Gibbs, A.K. & C.N. Barron, 1993. *The Geology of the Guiana Shield.* Oxford University Press, New York, 246 pp.

Gibbs, R.J., 1976. Amazon river sediment transport in the Atlantic Ocean. *Geology 4*, 45-48.

Goulding, M., 1980. *The Fishes and the Forest: Explorations in Amazonian Natural History*. University of California Press, Berkeley and Los Angeles, 280 pp.

Goulding, M., 1981. *Man and Fisheries on an Amazon Frontier*. Junk, The Hague, 137 pp.

Goulding, M., 1989. *Amazon - The Flooded Forest.* BBC Books, London, 208 pp.

Goulding, M. and M.L. Carvalho. 1984. Ecology of Amazonian needlefishes (Belonidae). *Rev. Bras. Zool., São Paulo 2*, 99-111.

Goulding, M., M.L. Carvalho & E.G. Ferreira, 1988. *Rio Negro: Rich Life in Poor Water*. SPB Academic Publishing, The Hague.

Graham, J.B., 1997. *Air-Breathing Fishes: Evolution, Diversity, and Adaptation*. Academic Press, San Diego.

Gregory-Wodzicki, K.M., 2000. Uplift history of the Central and Northern Andes: a review. *Geological Society of America Bulletin 122*, 1091-1105.

Gronovius, L.T., 1754. *Museum Ichthyologicum sistens Piscium indigenorum & quorumdam exoticarum qui in Museo Laurentii Theodorii Gronovii, J.U.D. adservantur, descriptiones ordine systematico*. Leiden, [8 pp +] 70 pp.

Gronovius, L.T., 1756. *Musei Ichthyologici tomus secundus sistens Piscium indigenorum & nonnullarum exoticorum, quorum maxima pars in Museo Laurentii Theodori Gronovii, J.U.D. adservantur, nec non quorumdam in aliis Museis observatorum descriptiones*. Leiden, [6pp] + 88pp [Fish section 46 pp + pp. 86-88].

Haddad-Jr, V., D. Garrone-Neto, J.B. Paula-Neto, F.P.L. Marques & K.C. Barbaro, 2004. Freshwater stingrays: study of epidemiologic, clinic and therapeutic aspects based on 84 envenomings in humans and some enzymatic activities of the venom. *Toxicon 43*, 287-294.

Haddad-Jr, V., R.A. de Souza & P.S. Auerbach, 2008. Marine catfish sting causing fatal heart perforation in a fisherman. *Wilderness and Environmental Medicine 19*, 114 118.

Hammond, D.S., 2005. Biophysical features of the Guiana Shield. In *Tropical Forests of the Guiana Shield* (D.S. Hammond, ed.), pp. 15-194. CABI, Cambridge.

Haripersad-Makhanlal, A. & P.E. Ouboter, 1993. Limnology: physico-chemical parameters and phytoplankton composition. In *The Freshwater Ecosystems of Suriname* (P.E. Ouboter, ed.), pp. 53-75. Kluwer, Dordrecht, the Netherlands.

Harrington, R.W., 1971. How ecological and genetic factors interact to determine when self-fertilizing hermaphrodites of *Rivulus marmoratus* change into functional secondary males, with a reappraisal of the modes of intersexuality among fishes. *Copeia 1971*, 389-431.

Helfman, G.S., 2007. *Fish Conservation: a Guide to Understanding and Restoring Global Aquatic Biodiversity and Fishery Resources*. Island Press, Washington.

Henderson, P.A. & I. Walker, 1990. Spatial organization and population density of the fish community of the litter banks within a central Amazonian blackwater stream. *Journal of Fish Biology 37*, 401-411.

Heyde, H., 1986. *Surinaamse Vissen*. Westfort, Paramaribo, 154 pp.

Hilton, E.J., C. Cox Fernandes, J.P. Sullivan, J.G. Lundberg & R. Campos-da-Paz, 2007. Redescription of *Orthosternarchus tamandua* (Boulenger, 1898) (Gymnotiformes Apteronotidae) with reviews of its ecology, electric organ discharges, external morphology, osteology, and phylogenetic affinities. *Proceedings of the Academy of Natural Sciences Philadelphia 156*, 1-25.

Hoedeman, J.J., 1959. Rivulid fishes of Suriname and other Guyanas, with a preliminary review of the genus *Rivulus*. *Studies of the Fauna of Suriname and Other Guyanas 3*, 44-98.

Hoese, D.F. & A.C. Gill, 1993. Phylogenetic relationships of eleotridid fishes (Perciformes: Gobioidei). *Bulletin of Marine Science 52*, 415-440.

Holmlund, C.M. & M. Hammer, 1999. Ecosystem services generated by fish populations. *Ecological Economics 29*, 253-268.

Holthuis, L.N., 1959. The Crustacea Decapoda of Suriname (Dutch Guiana). *Zoologische Verhandelingen, 44*, 1-296.

Hoogmoed, M.S., 1973. *Notes on the herpetofauna of Surinam. IV. The lizards and amphisbaenians*. Junk, Den Haag.

Hoorn, C. & F. Wesselingh (eds.), 2010. *Amazonia: Landscape and Species Evolution*. Wiley-Blackwell, Chichester, UK.

Hoorn, C., J. Guerrero, G.A. Sarmiento & M.A. Lorente, 1995. Andean tectonics as a cause for changing drainage patterns in Miocene northern South-America. *Geology 23*, 237-240.

Hopkins, C.D. & W. Heiligenberg, 1978. Evolutionary designs for electric signals and electroreceptors in gymnotoid fishes of Surinam. *Behavioral Ecology and Sociobiology 3*, 113-134.

Horeau, V., P. Cerdan, A. Champeau & S. Richard, 1998. Importance of aquatic invertebrates in the diet of rapid-dwelling fish in the River Sinnamary, French Guiana. *Journal of Tropical Ecology 14*, 851-864.

Houttuyn, M., 1764. *Natuurlyke historie of uitvoerige beschryving der dieren, planten en mineraalen, volgens het samenstel van den Heer Linnaeus, met naauwkeurige afbeeldingen. Part I. Vol. 7. De Visschen*. F. Houttuyn, Amsterdam, 446 p.

Hrbek, T., J. Seckinger & A. Meyer, 2007. A phylogenetic and biogeographic perspective on the evolution of poeciliiid fishes. *Molecular Phylogenetics and Evolution 43*, 986-998.

Hu C., E.T. Montgomery, R.W. Schmitt & F.E. Muller-Karger, 2004. The dispersal of the Amazon and Orinoco River water in the tropical Atlantic and Caribbean Sea: observation from space and S-PALACE floats. *Deep-Sea Research II 51*, 1151-1171.

Huber, J.H., 1992. *Review of Rivulus: Ecobiogeographic Relationships*. Cybium. Societé Française d'Ichthyologie, Paris, 526 pp.

Hubert, N. & J.-F. Renno, 2006. Historical biogeography of South American freshwater fishes. *Journal of Biogeography 33*, 1414-1436.

Hurd, P.L., 1997. Cooperative signaling between opponents in fish fights. *Animal Behaviour 54*, 1309-1315.

Ibarra, M. & D.J. Stewart, 1989. Longitudinal zonation of sandy beach fishes in the Napo River Basin, eastern Ecuador. *Copeia 1989*, 364-381.

Isbrücker, I.J.H., 1992a. Uberblick über die güttigen (Unter-) Gattungsnamen der Harnischwelse (Loricariidae) und ihre Synonyme. *Die Aquarien- und Terrarien Zeitschrift (DATZ) 1992*, 71-72.

Isbrücker, I.J.H., 1992b. Der verborgene Fundort von *Hemiancistrus medians* (Kner, 1854). *Die Aquarien- und Terrarien Zeitschrift*, Sonderheft, Eugen Ulmer, Stuttgart, 56-57.

Isbrücker, I.J.H. & H. Nijssen, 1988. Review of the South American Characiform fish genus *Chilodus*, with description of a new species, *C. gracilis* (Pisces, Characiformes, Chilodontidae). *Beaufortia 38*, 47-56.

Isbrücker, I.J.H. & H. Nijssen, 1992. *Corydoras breei*, a new species of callichthyid catfish from the Corantijn river basin in Suriname (Pisces, Siluriformes, Callichthyidae). *Beaufortia 43*, 9-14.

Isbrücker, I.J.H. & H. Nijssen, 1992. Sexualdimorphismus bei Harnischwelsen (Loricariidae). Odontoden, Zähne, Lippen, Tentakel, Genitalpapillen und Flossen. *DATZ-Sonderheft (Harnischwelse)*: 19-33.

Jégu, M. & G.M. Dos Santos, 2002. Révision du statut de *Myleus setiger* Müller & Troschel, 1844 et de *Myleus knerii* (Steindachner, 1881) (Teleostei: Characidae: Serrasalminae) avec une description complémentaire des deux espèces. *Cybium 26*, 33-57.

Jégu, M. & P. Keith, 1999. Le bas Oyapock limite septentrionale ou simple étape dans la progression de la faune des poissons d'Amazonie occidentale. *Comptes Rendus Académie Sciences Paris, Sciences de la Vie 322*: 1133–1143.

Jégu, M. & P. Keith, 2005. Endangered freshwater fish: *Tometes lebaili* Jégu & Keith, 2002. *Environmental Biology of Fishes 72*(4), 37.

Jégu, M., P. Keith & E. Belmont-Jégu, 2002. Une nouvelle espèce de *Tometes* (Teleostei: Characidae: Serrasalminae) du bouclier Guyanais, *Tometes lebaili* n.sp. *Bulletin Français de la Pêche et de la Pisciculture 364*, 23-48.

Jégu, M., P. Keith & P.Y. Le Bail, 2003. *Myleus planquettei* n.sp. (Teleostei, Characidae), une nouvelle espèce de grand Serrasalminae phytophage du bouclier guyanais. *Revue suisse de Zoologie 110*, 833-853.

Jepsen, D.B., 1997. Fish species diversity in sand bank habitats of a neotropical river. *Environmental Biology of Fishes 49*, 449-460.

Junk, W.J., 1973. Investigations on the ecology and production-biology of the floating meadows (Paspalo – Echinochloetum) on the Middle Amazon. Part II. The aquatic fauna in the root zone of floating vegetation. *Amazoniana 4*, 9-102.

Junk, W.J., 1984. Ecology, fisheries and fish culture in Amazonia. In *The Amazon: Limnology and Landscape Ecology of a Mighty Tropical River and its Basin* (H. Sioli, ed.), pp. 443-476. Junk, Dordrecht, the Netherlands.

Junk, W.J., 1985. Temporary fat storage, an adaptation of some fish species to the water level fluctuations and related environmental changes of the Amazon River. *Amazoniana 9*, 315- 351.

Junk, W., P.B. Bayley & R.E. Sparks, 1989. The flood pulse concept in river-floodplain systems. *Canadian Special Publication Fisheries Aquatic Sciences 106*, 110-127.

Junqueira, M.E.P., A.C. Mondin & C.A.M. Lopes, 2006. Microbiota characterization of the catfish (*Cathorops agassizii* and *Genidens genidens*) sting venom. *Braz. J. vet. Res. Anim. Sci. Sao Paulo 43*, 793-796.

Keenleyside, M.H.A., 1955. Some aspects of the schooling behaviour of fish. *Behaviour 8*, 183-248.

Keenleyside, M.H.A. (ed.), 1991. *Cichlid fishes. Behaviour, Ecology and Evolution*. Chapman & Hall, London, 378 pp.

Keenleyside, M.H.A. & B.F. Bietz, 1981. The reproductive behaviour of *Aequidens vittatus* (Pisces, Cichlidae) in Surinam, South America. *Environmental Biology of Fishes 6*, 87-94.

Keith, P. & F.J. Meunier, 2000. *Rhabdolichops jegui*, une nouvelle espèce de Sternopygidae (Gymnotiformes) de Guyane française. *Cybium 24*, 401-410.

Keith, P., P.Y. Le Bail & P. Planquette, 2000. *Atlas des Poissons d'Eau Douce de Guyane. Tome 2 – fascicule I. Batrachoidiformes, Mugiliformes, Beloniformes, Cyprinodontiformes, Synbranchiformes, Perciformes, Pleuronectiformes, Tetraodontiformes*. Museum National d'Histoire Naturelle, Paris, 429 pp.

Keith, P., L. Nandrin & P.Y. Le Bail, 2006. *Rivulus gaucheri*, a new species of rivuline (Cyprinodontiformes: Rivulidae) from French Guiana. *Cybium 30*, 133-137.

King, LC., D.K. Hobday & M. Mellody, 1964. *Cyclic denudation in Surinam*. Internal Report University of Natal on behalf of Geological Mining Service (GMD) of Suriname, 12 p.

Kirschbaum, F. & C. Schugardt, 2002. Reproductive strategies and developmental aspects of mormyrid and gymnotiform fishes. *Journal Physiology Paris 96*, 557-566.

Klinge, H., 1967. Podzol soils: a source of black-water rivers in Amazonia. *Atas do Simpósio sôbre a Biota Amazônica 3*, 117-125.

Knight, F.K., J. Lombardi, J.P. Wourms & J.R. Burns, 1985. Follicular placenta and embryonic growth of the viviparous four-eyed fish (*Anableps*). *Journal of Morphology 185*, 131-142.

Knöppel, H.A., 1970. Food of Central Amazonian fishes. *Amazoniana 2*, 257-352.

Kohda, M., M. Tanimura, M. Kikue-Nakamura & S. Yamagishi, 1995. Sperm drinking by female cat fishes: a novel mode of insemination. *Environmental Biology of Fishes 42*, 1-6.

Kramer, D.L., 1978. Ventilation of the respiratory gas bladder in *Hoplerythrinus unitaeniatus* (Pisces, Characoidei, Erythrinidae). *Canadian Journal of Zoology 56*, 931-938.

Kramer, D.L. & J.B. Graham, 1976. Synchronous air breathing, a social component of respiration in fishes. *Copeia 1976*, 689-697.

Kramer, D.L. & M. McClure, 1980. Aerial respiration in the catfish *Corydoras aeneus* (Callichthyidae). *Canadian Journal of Zoology 58*, 1984-1991.

Kramer, D.L. & M. McClure, 1981. The transit cost of aerial respiration in the catfish *Corydoras aeneus* (Callichthyidae). *Physiological Zoology 54*, 189-194.

Kramer, D.L. & M. McClure, 1982. Aquatic surface respiration, a widespread adaptation to hypoxia in tropical freshwater fishes. *Environmental Biology of Fishes 7*, 47-55.

Kramer, D.L., C.C. Lindsey, G.E.E. Moodie & E.D. Stevens, 1978. The fishes and the aquatic environment of the central Amazon basin, with particular reference to respiratory patterns. *Canadian Journal of Zoology 61*, 1964-1967.

Kullander, S.O., 1998. A phylogeny and classification of the South American Cichlidae (Teleostei: Perciformes). In *Phylogeny and Classification of Neotropical Fishes* (L.R. Malabarba, R.E. Reis, R.P Vari, Z.M. Lucena & C.A.S. Lucena, eds.), pp. 461-498. EDIPUCRS, Porto Alegre, 603 pp.

Kullander, S.O. & E.J.G. Ferreira, 2006. A review of the South American cichlid genus *Cichla*, with descriptions of nine new species (Teleostei: Cichlidae). *Ichthyological Explorations of Freshwaters 17*, 289-399.

Kullander, S.O. & H. Nijssen, 1989. *The Cichlids of Surinam, Teleostei: Labroidei*. E.J. Brill, Leiden, 256 pp.

Le Bail, P.Y., P. Keith & P. Planquette, 2000. *Atlas des Poissons d'Eau Douce de Guyane. Tome 2 – fascicule II. Siluriformes*. Museum National d'Histoire Naturelle, Paris, 307 pp.

Le Bail, P.Y., R. Covain, M. Jégu, S. Fisch-Muller, R. Vigouroux & P. Keith, 2012. Updated checklist of the freshwater and estuarine fishes of French Guiana. *Cybium 36*, 293-319.

Lecomte, F., F.J. Meunier & R. Rojas-Beltran, 1985. Mise en evidence d'un double cycle de croissance annuel chez un silure de Guyane, *Arius couma* (Val., 1839) (Teleostei, Siluriforme, Ariidae) a partir de l'etude squelettochronologique des epines des nageoires. *Comptes Rendus de l'Academie des Sciences Paris Serie III 300*, 181-184.

Lecomte, F., F.J. Meunier & R. Rojas-Beltran, 1986. Donnees preliminaires sur la croissance de deux Teleosteens de Guyane, *Arius proops* (Ariidae, Siluriformes) et *Leporinus friderici* (Anostomidae, Characoidei). *Cybium 10*, 121-134.

Lecomte, F., F.J. Meunier & R. Rojas-Beltran, 1989. Some data on the growth of *Arius proops* (Ariidae, Siluriformes) in the estuaries of French Guiana. *Aquatic Living Resources 2*, 63-68.

Lecomte, F., T. Boujard, F.J. Meunier, J.F. Renno & R. Rojas-Beltran, 1993. The growth of *Myleus rhomboidalis* (Cuvier, 1817) (Characiforme, Serrasalmidae) in two rivers of French Guiana. *Revue d'Ecologie (La Terre et la Vie) 48*, 431-444.

Leopold, L.B., M.G. Wolman & J.P. Miller, 1964. *Fluvial Processes in Geomorphology*. Freeman, San Francisco.

Leopold, M., 2004. *Guide des Poissons de Mer de Guyane*. IFREMER, Plouzane, France, 214 pp.

Lewis, W.M., 1970. Morphological adaptations of cyprinodontoids for inhabiting oxygen deficient waters. *Copeia 1970*, 319-326.

Lewis, W.M., 2008. Physical and chemical features of tropical flowing waters. In *Tropical Stream Ecology* (D. Dudgeon, ed.), pp. 1-24. Elsevier-Academic Press, New York.

Lewis, W.M., S.K. Hamilton & J.F. Saunders, 1995. Rivers of Northern South America. In *Ecosystems of the World: Rivers* (C. Cushing & K. Cummins, eds.), pp. 219-256. Elsevier, New York.

Liem, K.F., 1970. Comparative functional anatomy of the Nandidae (Teleostei). *Fieldiana (Zoology) 56*, 1-166.

Lijding, H.W., 1958. Proeven met tilapia in Suriname. (Experimental breeding of tilapia in Suriname). *De Surinaamse Landbouw 6*, 183-194. (in Dutch with English summary)

Lijding, H.W., 1959. Voorlopige lijst van garnalen, krabben en vissen in Suriname. Preliminary list of Surinam shrimps, crabs and fishes. *De Surinaamse Landbouw 7*, 70-97. (in Dutch)

Lindholm, A.K., R. Brooks & F. Breden, 2004. Extreme polymorphism in a Y-linked sexually selected trait. *Heredity 93*, 156-162.

Linnaeus, C., 1758. *Systema naturae per regna tria naturae, secundum classes, ordines, genera, species, cum characteribus, differentiis, synonymis, locis. Tomus I. ii+824 p. 10th edition*. Stockholm.

Linnaeus, C., 1766. *Systema naturae per regna tria naturae, secundum classes, ordines, genera, species, cum characteribus, differentiis, synonymis, locis. Tomus I. 12th edition*. Stockholm.

Lissmann, H.W., 1958. On the function and evolution of electric organs in fish. *Journal of Experimental Biology 35*, 156-191.

Loneragan, N.R., S.E. Bunn & D.M. Kellaway, 1997. Are mangroves and seagrasses sources of organic carbon for penaeid prawns in a tropical Australian estuary? A multiple stable-isotope study. *Marine Biology 130*, 289-300.

Loir, M., C. Cauty & P.Y. Le Bail, 1989. Ambisexuality in a South American cichlid: *Satanoperca* aff. *leucosticta*. *Aquatic Living Resources 2*, 185-187.

Lo Nostro, F. & G. Guerrero, 1996. Presence of primary and secondary males in a population of *Synbranchus marmoratus*, Bloch 1995, a protogynous fish (Teleostei – Synbranchiformes). *Journal of Fish Biology 49*, 788-800.

Loubens, G. & J. Panfili, 2000. Biologie de *Pseudoplatystoma fasciatum* et *P. tigrinum* (Teleostei: Pimelodidae) dans le bassin du Mamoré (Amazonie Bolivienne). *Ichthyological Explorations of Freshwaters 11*, 13-34.

Lovejoy, N.R. & B.B. Collette, 2001. Phylogenetic relationships of New World needlefishes (Teleostei: Belonidae) and the biogeography of transition between marine and freshwater habitats. *Copeia 2001*, 324-338.

Lovejoy, N.R., E. Bermingham & A.P. Martin, 1998. Marine excursions into South America. *Nature 396*, 421-422.

Lovejoy, N.R., S.C. Willis & J.S. Albert, 2010. Molecular signatures of Neogene biogeographical events in the Amazon fish fauna. In *Amazonia: Landscape and Species Evolution* (Hoorn, C. & F. Wesselingh, eds.), pp. 405-417. Blackwell, Chichester, UK.

Lowe-McConnell, R.H., 1962. The fishes of the British Guiana continental shelf, Atlantic coast of South America, with notes on their natural history. *Zoological Journal of the Linnean Society 44*, 669-700.

Lowe-McConnell, R.H., 1964. The fishes of the Rupununi savanna district of British Guiana, South America. Part 1. Ecological groupings of fish species and effects of the seasonal cycle on the fish. *Zoological Journal of the Linnean Society 45*, 103-144.

Lowe-McConnell, R.H., 1969. The cichlid fishes of Guyana, South America, with notes on their ecology and breeding behaviour. *Zoological Journal of the Linnean Society 48*, 255-302.

Lowe-McConnell, R.H., 1987. *Ecological Studies in Tropical Fish Communities*. Cambridge University Press, Cambridge, 382 pp.

Lowe-McConnell, R.H., 2000. *Land of Waters: Explorations in the Natural History of Guyana, South America*. The Book Guild, Lewes (UK).

Lucanus, O., 2009. *The Amazon Below Water*. Panta Rhei, Hannover (Germany), 345 p.

Lucena, Z.M.S. & Malabarba, 2010. Description of nine new species of *Phenacogaster* (Ostariophysi: Characiformes: Characidae) and notes on the other species of the genus. *Zoologia 27*, 263-304.

Lujan, N.K., 2008. Description of a new *Lithoxus* (Siluriformes: Loricariidae) from the Guayana Highlands with a discussion of Guiana Shield biogeography. *Neotropical Ichthyology 6*, 413-418.

Lujan, N.K. & J.W. Armbruster, 2011. The Guiana Shield. In *Historical Biogeography of Neotropical Freshwater Fishes* (J.S. Albert & R.E. Reis, eds.), pp. 211-224. University of California Press, Berkeley.

Lüling, K.H., 1980. Biotop, Begleitfauna und amphibische Lebensweise von *Synbranchus marmoratus* (Pisces, Synbranchidae) in Seitengewässern des mittleren Paraná (Argentinien). *Bonner Zoologische Beitrage 31*, 111-143.

Lundberg, J.G., 1998. The temporal context for the diversification of Neotropical fishes. In *Phylogeny and Classification of Neotropical Fishes* (L.R. Malabarba, R.E. Reis, R.P Vari, Z.M. Lucena & C.A.S. Lucena, eds.), pp. 49-68. Edipucrs, Porto Alegre, 603 pp.

Lundberg, J.G., W.M. Lewis, J.F. Saunders & F. Mago-Leccia, 1987. A major food component in the Orinoco River: evidence from planktivorous electric fishes. *Science 237*, 81-83.

Lundberg, J.G., L.G. Marshall, J. Guerrero, B. Horton, M.C.S.L. Malabarba & F. Wesselingh, 1998. The stage for Neotropical fish diversification: a history of tropical South American rivers. In *Phylogeny and Classification of Neotropical Fishes* (L.R. Malabarba, R.E. Reis, R.P Vari, Z.M. Lucena & C.A.S. Lucena, eds.), pp. 13-48. EDIPUCRS, Porto Alegre, 603 pp.

Lundberg, J.G., M. Kottelat, G.R. Smith, M. Stiassny & T. Gill, 2000. So many fishes, so little time: an overview of recent ichthyological discoveries in fresh waters. *Annals of the Missouri Botanical Garden 87*, 26-62.

Lundberg, J.G., J.P. Sullivan, R. Rodiles-Hernández & D.A. Hendrickson, 2007. Discovery of African roots for the Mesoamerican Chiapas catfish, *Lacantunia enigmatica*, requires an ancient intercontinental passage. *Proceedings of the Academy of Natural Sciences of Philadelphia 156*, 39-53.

Lundberg, J.G., M.H. Sabaj-Pérez, W.S. Dahdul & O.A. Aguilera, 2010. The Amazonian Neogene fish fauna. In *Amazonia: Landscape and Species Evolution* (Hoorn, C. & F. Wesselingh, eds.), pp. 281-301. Blackwell, Chichester, UK.

Lundberg, J.G., R. Covain, J.P. Sullivan & S. Fisch-Muller, 2012. Phylogenetic position and notes on the natural history of *Pimelabditus moli* Parisi & Lundberg, 2009 (Teleostei: Siluriformes), a recently discovered pimelodid catfish from the Maroni River basin. *Cybium 36*, 105-114.

Mago-Leccia, F., 1994. *Electric Fishes of the Continental Waters of America*. Fundacion para el Desarrollo de las Ciencias Fisicas, Matematicas y Naturales, Caracas. 206 p., 16 unnumb. tables.

Maisey, J.G., 2000. Continental break up and the distribution of fishes of Western Gondwana during the Early Cretaceous. *Cretaceous Research 21*, 281-314.

Malabarba, M.C.S.L., 2004. Revision of the Neotropical genus *Triportheus* Cope, 1872 (Characiformes: Characidae). *Neotropical Ichthyology 2*, 167-204.

Malabarba, M.C., O. Zuleta & C. Del Papa, 2006. *Proterocara argentina*, a new fossil cichlid from the Lumbrera Formation, Eocene of Argentina. *Journal of Vertebrate Paleontology 26*, 267-275.

Marceniuk, A.P. & N.A. Menezes, 2007. Systematics of the family Ariidae (Ostariophysi, Siluriformes), with a redefinition of the genera. *Zootaxa 1416*, 1-126.

Marrenga, M. & C. Ruleman, 2008. *Rapids of the Suriname River*. Ralicon/Buanda, Paramaribo. (http://www.buanda.org)

Mattias, A.T., S.E. Moron & M.N. Fernandes, 1996b. Aquatic respiration during hypoxia of the facultative air-breathing *Hoplerythrinus unitaeniatus*. A comparison with water-breathing *Hoplias malabaricus*. In *Physiology and Biochemistry of Fishes of the Amazon* (A.L. Val, V.M.F. Almeida-Val & D.J. Randall, eds.), pp. 203-211. Instituto Nacional de Pesquisas da Amazonia (INPA), Manaus.

Mattox, G.M.T., M. Toledo-Piza & O.T. Oyakawa, 2006. Taxonomic study of *Hoplias aimara* (Valenciennes, 1846) and *Hoplias macrophthalmus* (Pellegrin, 1907) (Ostariophysi, Characiformes, Erythrinidae). *Copeia 2006*, 516-528.

Mazzoldi, C., V. Lorenzi & M.B. Rasotto, 2007. Variation of male reproductive apparatus in relation to fertilization modalities in the catfish families Auchenipteridae and Callichthyidae (Teleostei: Siluriformes). *Journal of Fish Biology 70*, 243-256.

Mees, G.F., 1967. Freshwater fishes of Suriname: the genus *Heptapterus* (Pimelodidae). *Zoologische Mededelingen 42*, 215-229.

Mees, G.F., 1974. The Auchenipteridae and Pimelodidae of Suriname (Pisces, Nematognathi). *Zoologische Verhandelingen Leiden 132*, 1-256.

Mees, G.F., 1985. Further records of Auchenipteridae and Pimelodidae from Suriname (Pisces: Nematognathi). *Zoologische Mededelingen 59*, 239-249.

Mees, G.F., 1987. The members of the subfamily Aspredininae in Suriname (Pisces, Nematognathi). *Proceedings Koninklijke Nederlandse Akademie Wetenschappen C 90*, 173-192.

Mees, G.F., 1988. The genera of the subfamily Bunocephalinae (Pisces, Nematognathi, Asp dinidae). *Proceedings Koninklijke Nederlandse Akademie Wetenschappen C 91*, 85-101.

Melo, B.F., R.C. Benine, T.C. Mariguela & C. Oliveira, 2011. A new species of *Tetragonopterus* Cuvi 1816 (Characiformes: Characidae: Tetragonopterinae) from the rio Jari, Amapá, northern Bra *Neotropical Ichthyology 9*, 49-56.

Menezes, N.A. & S.H. Weitzman, 2009. Systematics of the Neotropical fish subfamily Glandu caudinae (Teleostei: Characiformes: Characidae). *Neotropical Ichthyology 7*, 295-370.

Meunier, F.J., M. Jégu & P. Keith, 2011. A new genus and species of neotropical electric fish, *Japig kirschbaum* (Gymnotiformes: Sternopygidae), from French Guiana. *Cybium 35*, 47-53.

Meyer, P.K., 1997. Stingray injuries. *Wilderness and Environmental Medicine 8*, 24-28.

Miller, R.R., 1979. Ecology, habits and relationships of the Middle American cuatro ojos, *Anable dowi* (Pisces: Anablepidae). *Copeia 1979*, 82-91.

Mol, J.H., 1993a. Aquatic invertebrates of the Coastal Plain. In *The Freshwater Ecosystems of Surinan* (P.E. Ouboter, ed.), pp. 113-131. Kluwer Academic Publishers, Dordrecht, the Netherlands.

Mol, J.H., 1993b. Structure and function of floating bubble nests of three armoured catfish (Callichthyidae) in relation to the aquatic environment. In *The Freshwater Ecosystems Suriname* (P.E. Ouboter, ed.), pp. 167-197. Kluwer Academic Publishers, Dordrecht, th Netherlands.

Mol, J.H., 1994. Effects of salinity on distribution, growth and survival of three neotropical armoure catfishes (Siluriformes – Callichthyidae). *Journal of Fish Biology 45*, 763-776.

Mol, J.H., 1995. Ontogenetic diet shifts and diet overlap among three closely related neotropic armoured catfishes. *Journal of Fish Biology 47*, 788-807.

Mol, J.H., 1996a. Reproductive seasonality and nest-site differentiation in three closely relate armoured catfishes (Siluriformes: Callichthyidae). *Environmental Biology of Fishes 45*, 363-381.

Mol, J.H., 1996b. Impact of predation on early stages of the armoured catfish *Hoplosternum thoraca tum* (Siluriformes – Callichthyidae) and implications for the syntopic occurrence with othe related catfishes in a neotropical multi-predator swamp. *Oecologia 107*, 395-410.

Mol, J.H., 2006. Attacks on humans by the piranha *Serrasalmus rhombeus* in Suriname. *Studies o Neotropical Fauna and Environment 41*, 189-195.

Mol, J.H., 2012. Occurrence of a freshwater pipefish *Pseudophallus* cf. *brasiliensis* (Syngnathidae) i Corantijn River, Suriname, with notes on its distribution, habitat, and reproduction. *Cybium 36* 45-53.

Mol, J.H. & P.E. Ouboter, 2004. Downstream effects of erosion from small-scale gold mining on th instream habitat and fish community of a small Neotropical rainforest stream. *Conservatio Biology 18*, 201-214.

Mol, J.H. & D. Ponton, 2003. Growth of young armoured catfish *Megalechis thoracata* in neotropica swamps and a rain-forest creek as revealed by daily micro-increments in otoliths. *Journal o Tropical Ecology 19*, 301-313.

Mol, J.H. & F.L. Van der Lugt, 1996. Distribution and feeding ecology of the African Tilapia *Oreochromis mossambicus* (Teleostei, Perciformes, Cichlidae) in Suriname (South America) with comments on the Tilapia – Kwikwi (*Hoplosternum littorale*) (Teleostei, Siluriformes, Callichthyidae) interaction. *Acta Amazonica 25*, 101-116.

Mol, J.H., W. Atsma, G. Flik, H. Bouwmeester & J.W.M. Osse, 1999. Effect of low ambient mineral concentrations on the accumulation of calcium, magnesium and phosphorus by early life stages of the air-breathing armoured catfish *Megalechis personata* (Siluriformes: Callichthyidae). *Journal of Experimental Biology 202*, 2121-2129.

Mol, J.H., D. Resida, J. Ramlal & C.R. Becker, 2000. Effects of El Niño-related drought on freshwater and brackish-water fish in Suriname, South America. *Environmental Biology of Fishes 59*, 429-440.

Mol, J.H., J.S. Ramlal, C. Lietar & M. Verloo, 2001. Mercury contamination in freshwater, estuarine and marine fishes in relation to small-scale gold mining in Suriname, South America. *Environmental Research 86*, 183-197.

Mol, J.H., P. Willink, B. Chernoff & M. Cooperman, 2006. Fishes of the Coppename River, Central Suriname Nature Reserve, Suriname. In *A Rapid Biological Assessment of the Aquatic Ecosystems of the Coppename River Basin, Suriname* (L.E. Alonso & H.J. Berrenstein, eds), pp. 67-79. Conservation International, Washington, DC.

Mol, J.H., B. De Merona, P.E. Ouboter & S. Sahdew, 2007a. The fish fauna of Brokopondo Reservoir, Suriname, during 40 years of impoundment. *Neotropical Ichthyology 5*, 351-368.

Mol, J.H., K. Wan Tong You, I. Vrede, A. Flynn, P. Ouboter & F. Van der Lugt, 2007b. Fishes of Lely and Nassau Mountains, Suriname. In *A Rapid Biological Assessment of the Lely and Nassau Plateaus, Suriname (with additional information on the Brownsberg Plateau)* (L.A. Alonso & J.H. Mol, eds.), pp. 107-118. Conservation International, Arlington, USA.

Mol, J.H., R.P. Vari, R. Covain, P.W. Willink & S. Fisch-Muller, 2012. Annotated checklist of the freshwater fishes of Suriname. *Cybium 36*, 263-292.

Montoya-Burgos, J.I., S. Muller, C. Weber & J. Pawlowski, 1997. Phylogenetic relationships between Hypostominae and Ancistrinae (Siluroidei: Loricariidae): first results from mitochondrial 12S and 16S rRNA gene sequences. *Revue Suisse de Zoologie 104*, 165-198.

Moreira, C.R. & F.C.T. Lima, 2011. On the name of the lepidophagous characid fish *Roeboexodon guyanensis* (Puyo) (Teleostei; Characiformes; Characidae). *Neotropical Ichthyology 9*, 313-316.

Myers, G.S., 1949. Salt tolerance of freshwater fish groups in relation to zoogeographical problems. *Bijdragen tot Dierkunde 28*, 315-322.

Myers, G.S., 1966. Derivation of the freshwater fish fauna of Central America. *Copeia 1966*, 766-773.

Nagelkerken, I., S.J.M. Blaber, S. Bouillon, P. Green, M. Haywood, L.G. Kirton, J.O. Meynecke, J. Pawlik, H.M. Penrose, A. Sasekumar & P.J. Somerfield, 2008. The habitat function of mangroves for terrestrial and marine fauna: a review. *Aquatic Botany 89*, 155-185.

National Resources Conservation Service, 1998. *Stream Visual Assessment Protocol.* NRCS, Portland, USA.

Near, T.J., D.I. Bolnick & P.C. Wainwright, 2005. Fossil calibrations and molecular divergence time estimates in centrarchid fishes (Teleostei: Centrarchidae). *Evolution 59*, 1768-1782.

Nelson, J.A., M.E. Whitmer, E.A. Johnson & D.J. Stewart, 1999. Wood-eating catfishes of the genus *Panaque*: gut microflora and cellulolytic enzyme activities. *Journal of Fish Biology 54*, 1069-1082.

Nelson, J.S., 2006. *Fishes of the world (4th ed)*. Wiley, New York.

Nijssen, H., 1970. Revision of the Surinam catfishes of the genus *Corydoras* Lacépède, 1803 (Pisces, Siluriformes, Callichthyidae). *Beaufortia 18*, 1-75.

Nijssen, H., I.J.H. Isbrücker & J. Géry, 1976. On the species of *Gymnorhamphichthys* Ellis, 1912, translucent sand-dwelling gymnoid fishes from South America (Pisces, Cypriniformes, Gymnotoidei). *Studies Neotropical Fauna Environment 11*, 37-63.

Nordlie, F.G., 1981. Feeding and reproductive biology of eleotrid fishes in a tropical estuary. *Journal of Fish Biology 18*, 97-110.

Odinetz Collart, O., M. Jégu, V. Thatcher & A.S. Tavares, 1996. Les prairies aquatiques de l'Amazonie bresilienne. *ORSTOM Actualites 49*, 8-14.

Oliveira, C., G.S. Avelino, K.T. Abe, T.C. Mariguela, R.C. Benine, G. Orti, R.P. Vari & R.M.C. Castro, 2011. Phylogenetic relationships within the speciose family Characidae (Teleostei: Ostariophysi: Characiformes) based on multilocus analysis and extensive ingroup sampling. *BMC Evolutionary Biology 11*, 275.

Orti, G., A. Sivasundar, K. Dietz & M. Jégu, 2008. Phylogeny of the Serrasalmidae (Characiformes) based on mitochondrial DNA sequences. *Genetics and Molecular Biology 31*, 343-351.

Ouboter, P.E. & B.P.E. De Dijn, 1993. Changes in a polluted swamp. In *The Freshwater Ecosystems of Suriname* (P.E. Ouboter, ed.), p. 239-260. Kluwer Academic Publishers, Dordrecht, the Netherlands.

Ouboter, P.E. & J.H. Mol, 1993. The fish fauna of Suriname. In *The Freshwater Ecosystems of Suriname* (P.E. Ouboter, ed.), p. 133-154. Kluwer Academic Publishers, Dordrecht, the Netherlands.

Ouboter, P.E., B.P.E. De Dijn & U.P.D. Raghoenandan, 1999. *Biodiversity Inventory Ulemari Area: Preliminary Technical Report.* National Zoological Collection of Suriname & National Herbarium of Suriname, Paramaribo, 64 pp.

Oyakawa, O.T. & G.M.T. Mattox, 2009. Revision of the Neotropical trahiras of the *Hoplias lacerdae* species-group (Ostariophysi: Characiformes: Erythrinidae) with descriptions of two new species. *Neotropical Ichthyology 7*, 117-140.

Parenti, L.R., 1981. A phylogenetic and biogeographic analysis of cyprinodontiform fishes (Teleostei, Atherinomorpha). *Bulletin of the American Museum of Natural History 168*, 335-557.

Parenti, L.R., F.L. LoNostro & H.J. Grier, 2010. Reproductive histology of *Tomeurus gracilis* Eigenmann, 1909 (Teleostei: Atherinomorpha: Poeciliidae) with comments on evolution of viviparity in atherinomorph fishes. *Journal of Morphology 271*, 1399-1406.

Parisi, B.M. & J.G. Lundberg, 2009. *Pimelabditus moli*, a new genus and new species of pimelodid catfish (Teleostei: Siluriformes) from the Maroni River basin of northeastern South America. *Notulae Naturae 480*, 1-11.

Pascal, M., G. Hostache, C. Tessier & P. Vallat, 1994. Cycle de reproduction et fecondité de l'Atipa, *Hoplosternum littorale* (Siluriforme) en Guyane française. *Aquatic Living Resources 7*, 25-37.

Pereira, P.R., C.S. Agostinho, R.J. Oliveira & E.E. Marques, 2007. Trophic guilds of fishes in sandbank habitats of a neotropical river. *Neotropical Ichthyology 5*, 399-404.

Perera-García, M.A., M. Mendoza-Carranza, W.M. Contreras-Sánchez, M. Huerta-Ortíz & E. Pérez-Sánchez, 2010. Reproductive biology of common snook *Centropomus undecimalis* (Perciformes: Centropomidae) in two tropical habitats. *Revista de Biologia Tropical* (*Int. J. Trop. Biol.*) *59*, 669-681.

Peters, K.M., R.E. Matheson & R.G. Taylor. 1998. Reproduction and early life history of common snook, *Centropomus undecimalis* (Bloch), in Florida. *Bulletin of Marine Science 62*, 509-529.

Pezold, F. & B. Cage, 2002.A review of the spinycheek sleepers, genus *Eleotris* (Teleostei: Eleotridae), of the western hemisphere, with comparison to the West African species. *Tulane Studies in Zoology and Botany 31*, 19-63.

Planquette, P., P. Keith & P.Y. Le Bail, 1996. *Atlas des Poissons d'Eau Douce de Guyane* (*Tome 1*). Museum National d'Histoire Naturelle, Paris, 429 pp.

Ploeg, A., 1987. Review of the cichlid genus *Crenicichla* Heckel, 1840 from Surinam, with descriptions of three new species (Pisces, Perciformes, Cichlidae). *Beaufortia 37*, 73-98.

Ponton, D. & S. Mérigoux, 2000. Comparative morphology and diet of young cichlids in the dammed River Sinnamary (French Guiana, South America). *Journal of Fish Biology 56*, 87-102.

Ponton, D. & S. Mérigoux, 2001. Description and ecology of some early life stages of fishes in the river Sinnamary (French Guiana, South America). *Folia Zoologica Monograph 50*, 116 pp.

Ponton, D., J.H. Mol & J. Panfili (2001). Use of otolith microincrements for estimating the age and growth rate of young armoured catfish *Hoplosternum littorale*. *Journal of Fish Biology 10*, 1274-1285.

Power, M.E., 1984. Habitat quality and the distribution of algal-eating catfish in a Panamanian stream. *Journal of Animal Ecology 53*, 357-374.

Prado, C.P.A., L.M. Gomiero & O. Froehlich, 2006. Spawning and parental care in *Hoplias malabaricus* (Teleostei, Characiformes, Erythrinidae) in the southern Pantanal, Brazil. *Braz. J. Biol. 66*, 697-702.

Primavera, J.H., 1998. Mangroves as nurseries: shrimp populations in mangrove and non-mangrove habitats. *Estuarine Coastal and Shelf Science 46*, 457-464.

Putzer, H., 1984. The geological evolution of the Amazon basin and its mineral resources. In *The Amazon: Limnology and Landscape Ecology of a Mighty Tropical River* (H. Sioli, ed.), pp. 15-46. Junk Publishers, Dordrecht, the Netherlands.

Reid, M.J. & J.W. Atz, 1958. Oral incubation in the cichlid fish *Geophagus jurupari* Heckel. *Zoologica N.Y. 43*, 77-88.

Reis, R.E., 1989. Systematic revision of the characid subfamily Stethaprioninae (Pisces, Characiformes). *Coun. Mus. Cienc. PUCRS, ser. Zool., Porto Alegre, 2*, 3-86.

Reis, R.E., 1998. Systematics, biogeography, and the fossil record of the Callichthyidae: a review of the available data. In *Phylogeny and Classification of Neotropical Fishes* (L.R. Malabarba, R.E. Reis, R.P Vari, Z.M. Lucena & C.A.S. Lucena, eds.), pp. 351-362. EDIPUCRS, Porto Alegre, 603 pp.

Reis, R.E., C.J. Ferraris & S.O. Kullander (eds.), 2003. *Check List of the Freshwater Fishes of South and Central America*. EDIPUCRS, Porto Alegre, Brazil, 729 pp.

Reis, R.E., P.Y. Le Bail & J.H. Mol, 2005. New arrangement in the synonymy of *Megalechis* Reis, 1997 (Siluriformes: Callichthyidae). *Copeia 2005*, 678-682.

Renno, J.F., P. Berribi, T. Boujard & R. Guyomard, 1990. Intraspecific genetic differentiation of *Leporinus friderici* (Anostomidae, Pisces) in French Guiana and Brazil: a genetic approach to the refuge theory. *Journal of Fish Biology 36*, 85-95.

Renno, J.F., A. Machardom, A. Blanquer & P. Boursot, 1991. Polymorphism of mitochondrial genes in populations of *Leporinus friderici* (Bloch, 1794): intraspecific structure and zoogeography of the Neotropical fish. *Genetica 84*, 137-142.

Revenga, C., I. Campbell, R. Abell, P. de Villiers & M. Bryer, 2005. Prospects for monitoring freshwater ecosystems towards the 2010 targets. *Philosophical Transactions of the Royal Society B 360*, 397-413.

Richter, C.J.J. & H. Nijssen, 1980. Notes on the fishery potential and fish fauna of the Brokopondo Reservoir (Surinam). *Fisheries Management 11*, 119-130.

Roberts, T.R., 1972. Ecology of fishes in the Amazon and Congo basins. *Bulletin Museum Comparative Zoology Harvard University 143*, 117-147.

Rojas-Beltran, R., 1989. Quelques aspects de l'écologie alimentaire de trois mâchoirans (Teleostei, Siluriformes, Ariidae) de la Guyane. *Cybium* 13, 181-187.

Rondeel, A.J., 1965. The Surinam fish protection legislation. *De Surinaamse Landbouw* 4, 63-64.

Rosa, R.S., M.R. de Carvalho & C. Almeida Wanderley, 2008. *Potamotrygon boesemani* (Chondrichthyes: Myliobatiformes: Potamotrygonidae), a new species of Neotropical freshwater stingray from Surinam. *Neotropical Ichthyology* 6, 1-8.

Ross, R. A. & F. Schäfer, 2000. *Freshwater rays*. Aqualog Verlag, 192 p.

Rull, V., 2008. Speciation timing and neotropical biodiversity: the Tertiary-Quaternary debate in the light of molecular phylogenetic evidence. *Molecular Ecology* 17, 2722-2729.

Sabaj-Pérez, M.H. & J.L.O. Birindelli, 2008. Taxonomic revision of extant *Doras* Lacépède, 1803 (Siluriformes: Doradidae) with description of three new species. *Proceedings of the Academy of Natural Sciences of Philadelphia* 157, 189-233.

Sabaj-Pérez, M.H., J.W. Armbruster & L.M. Page, 1999. Spawning in *Ancistrus* (Siluriformes: Loricariidae) with comments on the evolution of snout tentacles as a novel reproductive strategy: larval mimicry. *Ichthyological Explorations of Freshwaters* 10, 217-229.

Santos, G.M. & P.S. Rosa, 1998. Alimentação de *Anostomus ternetzi* e *Synaptolaemus cingulatus*, duas espécies de peixes amazônicos com boca superior. *Rev. Brasil. Biol.* 58, 255-262.

Sarmento-Soares, L.M. & R.F. Martins-Pinheiro, 2008. A systematic revision of *Tatia* (Siluriformes: Auchenipteridae: Centromochlinae). *Neotropical Ichthyology* 6, 495-542.

Sazima, I., 1983. Scale-eating in characoids and other fishes. *Environmental Biology of Fishes* 9, 87-101.

Sazima, I., L.N. Carvalho, F.P. Mendoça & J. Zuanon, 2006. Fallen leaves on the water-bed: diurnal camouflage of three night-active fish species in an Amazonian streamlet. *Neotropical Ichthyology* 4, 119-122.

Schaefer, S.A., 1986. *Historical biology of the loricariid catfishes: phylogenetics and functional morphology*. Unpubl. Ph.D. dissertation University of Chicago, Chicago, USA.

Schaefer, S.A. & F. Provenzano, 1993. The Guyana Shield *Parotocinclus*: systematics, biogeography, and description of a new Venezuelan species (Siluroidei: Loricariidae). *Ichthyological Explorations of Freshwaters* 4, 39-56.

Schaefer, S.A. & D.J. Stewart, 1993. Systematics of the *Panaque dentex* species group (Siluriformes: Loricariidae), wood-eating armored catfishes from tropical South America. *Ichthyological Explorations of Freshwaters* 4, 309-342.

Schiesari, L., J. Zuanon, C. Azevedo-Ramos, M. Garcia, M. Gordo, M. Messias & E.M. Veira, 2003. Macrophyte rafts as dispersal vectors for fishes and amphibians in the Lower Solimões River, Central Amazon. *Journal of Tropical Ecology* 19, 333-336.

Schindler, I., 1998. *Mesonauta guyanae* spec. nov., a new cichlid fish from the Guyana shield, South America (Teleostei: Cichlidae). *Zeitschrift fur Fischkunde* 5, 3-12.

Sheldon, A.L., 1988. Conservation of stream fishes: patterns of diversity, rarity, and risk. *Conservation Biology* 2, 149-156.

Sidlauskas, B.L. & R.P. Vari, 2008. Phylogenetic relationships within the South American fish family Anostomidae (Teleostei, Ostariophysi, Characiformes). *Zoological Journal of the Linnean Society* 154,70-210.

Sidlauskas, B.L. & R.P. Vari, 2012. Diversity and distribution of anostomoid fishes (Teleostei: Characiformes) throughout the Guianas. *Cybium* 36, 71-103.

Sidlauskas, B.L., J. Mol. & R.P. Vari, 2011. Dealing with allometry in linear and geometric morphometrics: a taxonomic case study in the *Leporinus cylindriformis* group (Characiformes: Anostomidae) with description of a new species from Suriname. *Zoological Journal of the Linnean Society* 162, 103-130.

Silva, T.B. & V.S. Uieda, 2007. Preliminary data on the feeding habitats of the freshwater stingrays *Potamotrygon falkneri* and *Potamotrygon motoro* (Potamotrygonidae) from the Upper Paraná River basin, Brazil. *Biota Neotropica* 7, 221-226.

Sinha, N.K.P., 1968. *Geomorphic evolution of the Northern Rupununi Basin, Guyana*. McGill University Savanna Research Project, Savanna Research Series 11. McGill University, Montreal.

Sioli, H., 1950. Das Wasser im Amazonasgebiet. *Forschungen und Fortschritte* 26, 274-280.

Sioli, H., 1975. Tropical river: the Amazon. In *River Ecology* (B.A. Whitton, ed.), pp. 461-488. University of California Press, Berkeley.

Spotte, S., P. Petry & J.A.S. Zuanon, 2001. Experiments on the feeding behavior of the hematophagous candiru, *Vandellia* cf. *plazaii*. *Environmental Biology of Fishes* 60, 459-464.

Stallard, R.F., 1985. River chemistry, geology, geomorphology, and soils in the Amazon and Orinoco basins. In *The Chemistry of Weathering* (J.I. Drever, ed.), pp. 293-316. Reidel, Dordrecht.

Stevens, P.W., D.A. Blewett & G.R. Poulakis. 2007. Variable habitat use by juvenile common snook, *Centropomus undecimalis* (Pisces: Centropomidae): applying a life-history model in a southwest Florida estuary. *Bulletin of Marine Science 80*, 93-108.

Stichting Volkslectuur Suriname, 1995. Woordenlijst Sranan – Nederlands – Engels (Wordlist Sranan – Dutch – English, with a list of plant and animal names). Vaco, Paramaribo.

Stoddard, P.K. & M.R. Markham, 2008. Signal cloaking by electric fish. *BioScience 58*, 415-425.

Swennen, C., P. Duiven & A.L. Spaans, 1982. Numerical density and biomass of macrobenthic animals living in the intertidal zone of Suriname, South America. *Netherlands Journal of Sea Research 15*, 406-418.

Taylor, R.G., J.A. Whittington & H.J. Grier. 2000. Age growth, maturation and protandric sex reversal in the common snook *Centropomus undecimalis,* from the east and west coasts of south Florida. *Fishery Bulletin 98*, 612-624.

Teixeira, R.L., 1994. Abundance, reproductive period, and feeding habits of eleotrid fishes in estuarine habitats of north-east Brazil. *Journal of Fish Biology 45*, 749-761.

Teunissen, P.A., 1976. Vegetation changes in a dammed up freshwater swamp in NW Suriname. *Acta Amazonica 6*, 117-150.

Teunissen, P.A., 1993. Vegetation and vegetation succession of the freshwater wetlands. In *The Freshwater Ecosystems of Suriname* (P.E. Ouboter, ed.), p. 77-98. Kluwer Academic Publishers, Dordrecht, the Netherlands.

Thomson, J.M., 1997. The Mugilidae of the world. *Mem. Queensland Mus. 41*, 457-562.

Thorson, T.B., 1976. Sexual dimorphism in number of rostral teeth of the sawfish, *Pristis perotteti* Müller and Henle, 1841. *Transactions of the American Fisheries Society 102*, 612-614.

Thorson, T.B., D.R. Brooks & M.A. Mayes, 1983a. The evolution of freshwater adaptation in stingrays. *National Geographic Research Reports 15*, 663-694.

Thorson, T.B., J.K. Langhammer & M.I. Oetinger, 1983b. Reproduction and development of the South American freshwater stingrays, *Potamotrygon circularis* and *P. motoro*. *Environmental Biology of Fishes 9*, 3-24.

Tito de Morais, A. & L. Tito de Morais, 1994. The abundance and diversity of larval and juvenile fish in a tropical estuary. *Estuaries 17*, 216-225.

Tito de Morais, L. & J. Raffray, 1999. Movements of *Hoplias aimara* during the filling phase of the Petit-Saut dam, French Guiana. *Journal of Fish Biology 54*, 627-635.

Torres, R.A., J.J. Roper, F. Foresti & C. Oliveira, 2005. Surprising genomic diversity in the Neotropical fish *Synbranchus marmoratus* (Teleostei: Synbranchidae): how many species? *Neotropical Ichthyology 3*, 277-284.

Trewavas, E., 1983. Tilapiine Fishes of the Genera *Sarotherodon, Oreochromis* and *Danakilia*. British Museum of Natural History, Publication Number 878.Comstock Publishing Associates. Ithaca, New York. 583 pp.

Turner, C.L., 1938. Adaptations for viviparity in embryo and ovary of *Anableps anableps*. *Journal of Morphology 62*, 323-349.

Turner, C.L., 1940. Follicular pseudoplacenta and gut modifications in anablepid fishes. *Journal of Morphology 67*, 91-104.

Van der Hammen, T. & H. Hooghiemstra, 2000. Neogene and Quaternary history of vegetation, climate, and plant diversity in Amazonia. *Quaternary Science Reviews 19*, 725-742.

Van der Heide, J., 1982. Lake Brokopondo. Filling phase limnology of a man-made lake in the humid tropics. PhD thesis Vrije Universiteit Amsterdam, Amsterdam.

Vannote, R.L., G.W. Minshall, K.W. Cummins, J.R. Sedell & C.E. Cushing, 1980. The river continuum concept. *Canadian Journal of Fisheries and Aquatic Sciences 37*, 130-137.

Vari, R.P., 1982. Environmental Impact of the Kabalebo Project. Final Report. Inventory, Biology and Ecology of the Fishes in the Corantijn River System, Suriname. Unpublished report Ministry of Development, Paramaribo, Suriname.

Vari, R.P., 1985. A new species of *Bivibranchia* (Pisces: Characiformes) from Surinam, with comments on the genus. *Proceedings of the Biological Society of Washington 98*, 511-522.

Vari, R.P., 1988. The Curimatidae, a lowland tropical fish family (Pisces: Characiformes): distribution, endemism and phylogenetic biogeography. In *Proceedings of a Workshop on Neotropical Distribution Patterns* (P.E. Vanzolini, and W.R. Heyer, eds.), p. 343-377. Academia brasileira de Ciências, Rio de Janeiro.

Vari, R.P., C.J. Ferraris & P. Keith, 2003. A new *Pseudocetopsis* species (Siluriformes: Cetopsidae) from Suriname and French Guiana. *Proceedings of the Biological Society of Washington 116*, 692-698.

Vari, R.P., C.J. Ferraris & M.C.C. de Pinna, 2005. The Neotropical whale catfishes (Siluriformes: Cetopsidae: Cetopsinae), a revisionary study. *Neotropical Ichthyology 3*, 127-238.

Vari, R.P., C.J. Ferraris, A. Radosavljevic & V.A. Funk (eds.), 2009. Checklist of the freshwater fishes of the Guiana Shield. *Bulletin of the Biological Society of Washington 17*, 1-93.

Vari, R.P., B.L. Sidlauskas & P.Y. Le Bail, 2012. New species of *Cyphocharax* (Ostariophysi: Characiformes: Curimatidae) from Suriname and French Guiana and a discussion of curimatid diversity on the Guiana Shield. *Cybium 36*, 63-69.

Vermeer, K., R.W. Risebrough, A.L. Spaans & L.M. Reynolds, 1974. Pesticide effects on fishes and birds in rice fields of Surinam, South America. *Environmental Pollution 7*, 217-236.

Vermeulen, F.B.M. & T. Hrbek, 2005. *Kryptolebias sepia* n.sp. (Actinopterygii: Cyprinodontiformes: Rivulidae), a new killifish from the Tapanahony River drainage in southeast Surinam. *Zootaxa 928*, 1-20.

Walker, I., 1994. The benthic litter-dwelling macrofauna of the Amazonian forest stream Taruma-Mirim: patterns of colonization and their implications for community stability. *Hydrobiologia 291*, 75-92.

Walker, I., 1995. Amazonian streams and small rivers. In *Limnology in Brazil* (J.G. Tundisi, C.E.M. Bicudo & T. Matsamura-Tundisi, eds.), pp. 167-194. Brazilian Academy of Sciences, Sao Paulo.

Walker, I., 2004. The food spectrum of the cardinal tetra (*Paracheirodon axelrodi*, Characidae) in its natural habitat. *Acta Amazonica 34*, 69-73.

Walker, I., P.A. Henderson & P. Sterry, 1991. On the patterns of biomass transfer in the benthic fauna of an Amazonian blackwater river, as evidenced by ^{32}P label experiment. *Hydrobiologia 215*, 153-162.

Waters, M.R., S.L. Forman, T.A. Jennings, L.C. Nordt, S.G. Driese, J.M. Feinberg, J.L. Keene, J. Halligan, A. Lindquist, J. Pierson, C.T. Hallmark, M.B. Collins & J.E. Wiederhold, 2011. The Buttermilk Creek Complex and the origins of Clovis at the Debra L. Friedkin Site, Texas. *Science 311*, 1599-1603.

Weber, C., 1992. Revision du genre *Pterygoplichthys sensu lato* (Pisces, Siluriformes, Loricariidae). *Revue française d'Aquariologie et d'Herpetologie 19*, 1-36.

Weber, C., R. Covain & S. Fisch-Muller, 2012. Identity of *Hypostomus plecostomus* (Linnaeus, 1758), with an overview of *Hypostomus* species from the Guianas (Teleostei: Suliformes: Loricariidae). *Cybium 36*, 195-227.

Weitzman, S.H. & J.S. Cobb, 1975. A revision of the South American fishes of the genus *Nannostomus* (family Lebiasinidae). *Smithsonian Contributions to Zoology 186*, 1-36.

Weitzman, S.H. & L. Palmer, 1996. Do freshwater hatchetfishes really fly? *Tropical Fish Hobbyist 45*, 195-206.

Weitzman, S.H. & L.R. Malabarba, 1998. Perspectives about the phylogeny and classification of the Characidae (Teleostei: Characiformes). In *Phylogeny and Classification of Neotropical Fishes* (L.R. Malabarba, R.E. Reis, R.P. Vari, Z.M.S. Lucena & C.A.S. Lucena, eds.), pp.161-170. EDIPUCRS, Porto Alegre. Brazil.

Weitzman, S.H. & M. Weitzman, 1982. Biogeography and evolutionary diversification in neotropical freshwater fishes, with comments on the refuge theory. In *Biological Diversification in the Tropics* (G.T. Prance, ed.), pp. 403-422. Columbia University Press, New York.

Weitzman, S.H. & R.P. Vari, 1988. Miniaturization in South American freshwater fishes: an overview and discussion. *Proceedings of the Biological Society of Washington 101*, 444-465.

Weitzman, S.H., N.A. Menezes, H.G. Evers & J.R. Burn, 2005. Putative relationships among inseminating and externally fertilizing characids, with a description of a new genus and species of Brazilian inseminating fish bearing an anal-fin gland in males (Characiformes: Characidae). *Neotropical Ichthyology 3*, 329-360, 2005

Welcomme, R.L., 1979. *Fisheries Ecology of Floodplain Rivers*. Longman, London.

Werkhoven, M.C.M. & G.M.T. Peeters, 1993. Aquatic macrophytes. In *The Freshwater Ecosystems of Suriname* (P.E. Ouboter, ed.), pp. 99-112. Kluwer, Dordrecht, the Netherlands.

Westby, G.W.M., 1988. The ecology, discharge diversity and predatory behaviour of gymnotiforme electric fish in the coastal streams of French Guiana. *Behavioral Ecology and Sociobiology 22*, 341-354.

Wheeler, A., 1958. The Gronovius fish collection: a catalogue and historical account. *Bull. Br. Mus. nat. Hist. Zool. Hist. Ser. 1*, 185-249.

Wheeler, A., 1989. Further notes on the fishes from the collection of Laurens Theodore Gronovius (1730-1777). *Zoological Journal of the Linnean Society 95*, 205-218.

Wiest, F.C., 1995. The specialized locomotory apparatus of the freshwater hatchetfish family Gasteropelecidae. *Journal of Zoology 236*, 571-592.

Wijmstra, T.A., 1971. *The Palynology of the Guiana Coastal Basin*. Ph.D. Thesis, University of Amsterdam, Amsterdam, 62 pp.

Willink, P.W. & B.L. Sidlauskas, 2004. Taxonomic notes on select fishes collected during the 2004 AquaRAP expedition to the Coppename River, Central Suriname Nature Reserve, Suriname. In *A Rapid Biological Assessment of the Aquatic Ecosystems of the Coppename River Basin, Suriname* (L.E. Alonso & H.J. Berrenstein, eds.), pp. 101-111. Conservation International, Washington D.C.

Willink, P.W., J.H. Mol & B. Chernoff, 2010. A new species of suckermouth armored catfish, *Pseudancistrus kwinti* (Siluriformes: Loricariidae) from the Coppename River drainage, Central Suriname Nature Reserve, Suriname. *Zootaxa 2332*, 40-48.

Willink, P.W., K. Wan Tong You & M. Piqué, 2011. Fishes of the Sipaliwini and Kutari rivers. In *A Rapid Biological Assessment of the Kwamalasamutu Region, southwestern Suriname* (B.J. O'Shea, L.E. Alonso & T.H. Larsen, eds), pp. 1189-123. RAP Bulletin of Biological Assessment 63. Conservation International, Arlington, USA.

Winemiller, K.O., 1987. Feeding and reproductive biology of the currito, *Hoplosternum littorale*, in the Venezuelan llanos with comments on the possible function of the enlarged male pectoral spines. *Environmental Biology of Fishes 20*, 219-227.

Winemiller, K.O., 1989. Development of dermal lip protuberances for aquatic surface respiration in South American characid fishes. *Copeia 1989*, 382-390.

Winemiller, K.O., 1989. Ontogenetic diet shifts and resource partitioning among piscivorous fishes in the Venezuelan Llanos. *Environmental Biology of Fishes 26*, 177-199.

Winemiller, K.O. & B.J. Ponwith, 1998. Comparative ecology of eleotrid fishes in Central American coastal streams. *Environmental Biology of Fishes 53*, 373-384.

Winemiller, K.O. & H.Y. Yan, 1989. Obligate mucus-feeding in a South American trichomycterid catfish (Pisces: Ostariophysi). *Copeia 1989*, 511-514.

Winemiller, K.O., H. Lopez-Fernandez, D.C. Taphorn, L.G. Nico & A.B. Duque, 2008. Fish assemblages of the Casiquiare river, a corridor and zoogeographical filter for dispersal between the Orinoco and Amazon basins. *Journal of Biogeography 35*, 1551-1563.

Wourms, J.P., 1981. Viviparity: the maternal-fetal relationship in fishes. *American Zoologist 21*, 473-515.

Wright, J.J., 2009. Diversity, phylogenetic distribution, and origins of venomous catfishes. *BMC Evolutionary Biology 9*, 282-294.

Zahl, P.A., J.J.A. McLaughlin & R.J. Gomprecht, 1977, Visual versatility and feeding of the four-eyed fishes, *Anableps*. *Copeia 1977*, 791-793.

Zanata, A.M., 1997. *Jupiaba*, um novo gênero de Tetragonopterinae com osso pélvico em forma de espinho (Characidae, Characiformes). *Iheringia Sér. Zool. Porto Alegre 83*, 99-136.

Zanata, A.M. & M. Toledo-Piza, 2004. Taxonomic revision of the South American fish genus *Chalceus* (Teleostei: Ostariophysi: Characiformes) with the description of three new species. *Zool. J. Linn. Soc. 40*, 103-135.

Zanata, A.M. & R.P. Vari, 2005. The family Alestidae (Ostariophysi, Characiformes): a phylogenetic analysis of a trans-Atlantic clade. *Zoological Journal of the Linnean Society 145*, 1-144.

Zarske, A. & J. Géry, 2008. Revision der neotropischen Gattung *Metynnis* Cope, 1878. II. Beschreibung zweier neuer Arten und zum Status von *Metynnis goeldii* Eigenmann, 1903 (Teleostei: Characiformes: Serrasalmidae). *Vertebrate Zoology 58*, 173-196.

Zarske, A., P.Y. Le Bail & J. Géry, 2006. New and poorly known characiform fishes from French Guiana. 1. Two new tetras of the genera *Hemigrammus* and *Hyphessobrycon* (Teleostei: Characiformes: Characidae). *Zool. Abh. (Dresden) 55*, 17-30.

Zarske, A., P.Y. Le Bail & J. Géry, 2010. New and poorly known Characiform fishes (Teleostei: Characiformes: Characidae) from French Guyana. A new tetra of the genus *Bryconamericus*. *Vertebrate Zoology 60*, 3-10.

Zonneveld, J.I.S., 1972. Sulas and sula complexes. *Göttinger Geographische Abhandlungen 1972*, 93-101.

Zuanon, J., F.A. Bockmann & I. Sazima, 2006. A remarkable sand-dwelling fish assemblage from central Amazonia, with comments on the evolution of psammophily in South American freshwater fishes. *Neotropical Ichthyology 4*, 107-118.

Useful websites

Fishbase www.fishbase.org/: an international consortium runs this large online database. The records (currently over 30,000 species) include taxonomic information, distribution, recent species status on the IUCN Red List of Threatened Species, size and growth parameters, diet composition, trophic levels, and other features of marine and freshwater fishes of the world.

www.calacademy.org/research/ichthyology/catalog/fishcatsearch.html: online, continuously updated version of the *Catalog of Fishes* by W.N. Eschmeyer (ed.).

www.repository.naturalis.nl: gives access to digitized versions (pdf files) of (old) publications in the journals Zoologische Mededelingen and Zoologische Verhandelingen of the Naturalis Museum in Leiden (many publications on Surinamese freshwater fishes by M. Boeseman and G. Mees).

www.ufrgs.br/ni: gives access to articles on neotropical fishes published in the Brazilian journal Neotropical Ichthyology. Systematics and, to a lesser extent, ecology of mainly freshwater fishes.

SUBJECT INDEX

ENGLISH NAMES INDEX

COMMON NAMES INDEX

SYSTEMATIC INDEX

Printed in the United States
By Bookmasters